STUDENT'S SOLUTIONS MANUAL
TO ACCOMPANY BARNETT AND ZIEGLER
COLLEGE ALGEBRA
Fourth Edition

FRED SAFIER
City College of San Francisco

McGraw-Hill Publishing Company
New York St. Louis San Francisco Auckland Bogotá Caracas
Hamburg Lisbon London Madrid Mexico Milan Montreal New Delhi
Oklahoma City Paris San Juan São Paulo Singapore Sydney Tokyo Toronto

Student's Solutions Manual to Accompany Barnett and Ziegler:
COLLEGE ALGEBRA

4 5 6 7 8 9 0 EDW EDW   9 4 3 2 1

ISBN 0-07-003933-X

The editor was Robert Weinstein;
the production supervisor was Leroy A. Young.
Edwards Brothers, Inc., was printer and binder.

Suggestions for Success in Algebra

1. Go to class, participate in the class discussions, and don't hesitate to ask questions.

2. Do all assigned homework, as soon as possible after class. Try each assigned problem on your own. If you need help, first check the text for similar examples. Once you have found an answer, check it with the answer in the back of the text. If you have difficulty, study the solution given here, then try to work the problem again without looking at the solution.

3. Use the answers in the texts and the solutions in this manual as guides. However, do not always try to work toward the answer in the back of the text. You should be able to do problems without looking at the answers. On a test (and in real life) you are seldom provided with the answers, so you cannot work toward them.

4. At the end of each chapter, work the Chapter Review exercises. These exercises provide an excellent review for a test. In areas of weakness, return to an appropriate section in the chapter for a refresher.

5. If you are still having difficulty, be sure to talk to your instructor and to use any tutorial help that is available to you. However, do not let a tutor do the work for you.

6. Take all tests and quizzes that the instructor gives.

This manual contains solutions for all odd-numbered problems in the text exercises plus all problems in the Chapter Review exercises. The Chapter Review exercises are keyed to corresponding text sections.

TABLE OF CONTENTS

CHAPTER 1: PRELIMINARIES

1-1 Algebra and Real Numbers ... 1
1-2 Polynomials: Basic Operations ... 4
1-3 Polynomials: Factoring ... 9
1-4 Rational Expressions: Basic Operations ... 11
1-5 Chapter Review ... 16

CHAPTER 2: EXPONENTS AND RADICALS ... 23

2-1 Integer Exponents ... 23
2-2 Rational Exponents ... 26
2-3 Radicals ... 29
2-4 Chapter Review ... 33

CHAPTER 3: EQUATIONS AND INEQUALITIES ... 37

3-1 Linear Equations ... 37
3-2 Applications Involving Linear Equations ... 42
3-3 Linear Inequalities ... 47
3-4 Absolute Value in Equations and Inequalities ... 51
3-5 Complex Numbers ... 56
3-6 Quadratic Equations ... 59
3-7 Polynomial and Rational Inequalities ... 69
3-8 Equations Reducible to Quadratic Form ... 80
3-9 Chapter Review ... 85

CHAPTER 4: GRAPHS AND FUNCTIONS ... 97

4-1 Basic Tools; Circles ... 97
4-2 Straight Lines ... 104
4-3 Functions ... 113
4-4 Linear and Quadratic Functions ... 120
4-5 Operations on Functions; Composition ... 130
4-6 Inverse Functions ... 138
4-7 Chapter Review ... 153

CHAPTER 5: EXPONENTIAL AND LOGARITHMIC FUNCTIONS ... 165

5-1 Exponential Functions ... 165
5-2 Exponential Functions with Base e ... 172
5-3 Logarithmic Functions ... 179
5-4 Common and Natural Logarithms ... 184
5-5 Exponential and Logarithmic Equations ... 188
5-6 Chapter Review ... 195

CHAPTER 6: POLYNOMIAL FUNCTIONS: GRAPHS AND ZEROS 205

6-1 Synthetic Division 205
6-2 Remainder and Factor Theorems 210
6-3 Fundamental Theorem of Algebra 215
6-4 Isolating Real Zeros 219
6-5 Rational Zero Theorem 226
6-6 Chapter Review 244

CHAPTER 7: SYSTEMS OF EQUATIONS AND INEQUALITIES 253

7-1 Systems of Linear Equations in Two and Three Variables 253
7-2 Systems and Augmented Matrices 264
7-3 Gauss-Jordan Elimination 269
7-4 Systems Involving Second-Degree Equations 285
7-5 Systems of Linear Inequalities 294
7-6 Linear Programming 301
7-7 Chapter Review 311

CHAPTER 8: MATRICES AND DETERMINANTS 327

8-1 Matrix Addition; Multiplication by a Number 327
8-2 Matrix Multiplication 330
8-3 Inverse of a Square Matrix; Matrix Equations 335
8-4 Determinants 349
8-5 Properties of Determinants 355
8-6 Cramer's Rule 359
8-7 Chapter Review 362

CHAPTER 9: SEQUENCES AND SERIES 371

9-1 Sequences and Series 371
9-2 Mathematical Induction 374
9-3 Arithmetic Sequences and Series 384
9-4 Geometric Sequences and Series 389
9-5 Additional Applications 392
9-6 Binomial Formula 394
9-7 Chapter Review 397

CHAPTER 10: ADDITIONAL TOPICS IN ANALYTIC GEOMETRY 405

10-1 Conic Sections; Parabola 405
10-2 Ellipse 409
10-3 Hyperbola 414
10-4 Translation of Axes 419
10-5 Chapter Review 422

CHAPTER 11: AN INTRODUCTION TO PROBABILITY   429
11-1 Multiplication Principle, Permutations, and Combinations   429
11-2 Sample Spaces and Probability   433
11-3 Empirical Probability   438
11-4 Chapter Review   440

APPENDICES   445
B. Variation   445
C. Rational Functions   447
D. Partial Fractions   463

CHAPTER 1

Exercise 1-1

Key Ideas and Formulas

Basic Properties of the Set of Real Numbers

Let R be the set of real numbers and x, y and z arbitrary elements of R.

ADDITION PROPERTIES

CLOSURE: $x + y$ is a unique element in R.

ASSOCIATIVE: $(x + y) + z = x + (y + z)$

COMMUTATIVE: $x + y = y + x$

IDENTITY: 0 is the additive identity; that is, $0 + x = x + 0 = x$ for all x in R, and 0 is the only element in R with this property.

INVERSE: For each x in R, $-x$ is its unique additive inverse; that is, $x + (-x) = (-x) + x = 0$, and $-x$ is the only element in R relative to x with this property.

MULTIPLICATION PROPERTIES

CLOSURE: xy is a unique element in R.

ASSOCIATIVE: $(xy)z = x(yz)$

COMMUTATIVE: $xy = yx$

IDENTITY: 1 is the multiplicative identity; that is, $1x = x1 = x$, and 1 is the only element in R with this property.

INVERSE: For each x in R, $x \neq 0$, $\frac{1}{x}$ is its unique multiplicative inverse; that is, $x(1/x) = (1/x)x = 1$, and $\frac{1}{x}$ is the only element R relative to x with this property.

COMBINED PROPERTY

DISTRIBUTIVE:

$$x(y + z) = xy + xz \qquad\qquad (x + y)z = xz + yz$$

Properties of Negatives:

For all real numbers a and b

1. $-(-a) = a$
2. $(-a)b = -(ab) = a(-b) = -ab$
3. $(-a)(-b) = ab$
4. $(-1)a = -a$
5. $\dfrac{-a}{b} = -\dfrac{a}{b} = \dfrac{a}{-b} \qquad b \neq 0$
6. $\dfrac{-a}{-b} = -\dfrac{-a}{b} = -\dfrac{a}{-b} = \dfrac{a}{b} \qquad b \neq 0$

Zero Properties:

For all real numbers a and b

1. $a \cdot 0 = 0$
2. $ab = 0$ if and only if $a = 0$ or $b = 0$ or both.

1. Since 4 is an element of $\{3,4,5\}$, the statement is true. T

3. Since 3 is an element of $\{3,4,5\}$, the statement is false. F

5. The statement is false, since not every element of $\{1,2\}$ is an element of $\{1,3,5\}$. 2 is an element of $\{1,2\}$, but not an element of $\{1,3,5\}$. F

7. Since each element of the set $\{7,3,5\}$ is also an element of the set $\{3,5,7\}$, the statement is true. T

9. The commutative property $(+)$ states that, in general,
$$x+y=y+x$$
Comparing this with
$$x+7=?$$
we see that 7 takes the place of y.
Hence, $x+7=7+x$.
$7+x$

11. The associative property $(\cdot)$ states that, in general,
$$(xy)z=x(yz)$$
Comparing this with
$$x(yz)=?$$
we see that $x(yz)=(xy)z$.
$(xy)z$

13. The identity property $(+)$ states that, in general,
$$0+x=x+0=x$$
Comparing this with
$$0+9m=?$$
we see that $9m$ takes the place of x.
Hence, $0+9m=9m$.
$9m$

15. Commutative $(\cdot)$ $ym=my$ is a special case of $xy=yx$.

17. Distributive. $7u+9u=(7+9)u$ is a special case of $xz+yz=(x+y)z$.

19. Inverse $(\cdot)$ $(-2)(\frac{1}{-2})=1$ is a special case of $x(\frac{1}{x})=1$.

21. Inverse $(+)$. $w+(-w)=0$ is a special case of $x+(-x)=0$.

23. Identity $(+)$. $3(xy+z)+0=3(xy+z)$ is a special case of $x+0=x$.

25. Negatives. $\dfrac{-x}{-y}=\dfrac{x}{y}$ is a special case of $\dfrac{-a}{-b}=\dfrac{a}{b}$.

27. The even integers between -3 and 5 are $-2,0,2,$ and 4. Hence, the set is written $\{-2,0,2,4\}$.

29. The letters in "status" are: s,t,a,u. Hence the set is written $\{s,t,a,u\}$, or, equivalently, $\{a,s,t,u\}$.

31. Since there are no months starting with B, the set is empty. ϕ

33. Commutative $(+)$.
$$(3x+5)+7 \;=\; 7+(3x+5)$$
is a special case of
$$x+y \;=\; y+x.$$

35. Associative $(+)$.
$$(3x+2)+(x+5) \;=\; 3x+[2+(x+5)]$$
is a special case of
$$(x+y)+z \;=\; x+(y+z)$$

37. Distributive.
$$x(x-y)+y(x-y) \;=\; (x+y)(x-y)$$
is a special case of
$$xz+yz \;=\; (x+y)z$$

39. Zero.
$$(2x-3)(x+5) \;=\; 0$$
if and only if
$$2x-3=0 \text{ or } x+5 \;=\; 0$$
is a special case of
$$ab \;=\; 0 \text{ if and only if}$$
$$a \;=\; 0 \text{ or } b=0.$$

41. Yes. This restates Zero Property (2).

43. A. True.
B. False, $\frac{2}{3}$ is an example of a real number that is not irrational.
C. True.

45. $\frac{3}{5}$ and -1.43 are two examples of infinitely many.

47. A. -3 is an integer. It belongs to Z, therefore also to Q and R.

 B. 3.14 is a rational number $(\frac{314}{100})$. It belongs to Q, therefore also to R.

 C. π is an irrational number. It belongs to R.

 D. $\frac{2}{3}$ is a rational number. It belongs to Q, therefore also to R.

49. A. True; commutative property for addition.

 B. False; for example $3 - 5 \neq 5 - 3$.

 C. True; commutative property for multiplication.

 D. False; for example $9 \div 3 \neq 3 \div 9$.

51. A. List each element of A. Follow these with each element of B that is not yet listed. $\{1, 2, 3, 4, 6\}$

 B. List each element of A that is also an element of B. $\{2, 4\}$.

53.
$$
\begin{aligned}
\text{Let } c &= 0.090909\cdots \\
\text{Then } 100c &= 9.0909\cdots \\
100c - c &= (9.0909\cdots) - (0.090909\cdots) \\
99c &= 9 \\
c &= \frac{9}{99} = \frac{1}{11}
\end{aligned}
$$

55.
$$
\begin{array}{rl}
23 & \\
12 & \\
\hline
46 & 23 \cdot 2 \\
230 & 23 \cdot 10 \\
\hline
276 &
\end{array}
$$

$$
\begin{aligned}
23 \cdot 12 &= 23(2+10) \\
&= 23 \cdot 2 + 23 \cdot 10 \\
&= 46 + 230 \\
&= 276
\end{aligned}
$$

57. A. $0.88888888\cdots$

 B. $0.27272727\cdots$

 C. $2.23606797\cdots$

 D. $1.37500000\cdots$

Exercise 1-2
Key Ideas and Formulas

For n a natural number and a any real number, $a^n = \underbrace{a \cdot a, \cdots, a}_{n \ factors}$

First Property of Exponents: For any natural numbers m and n, and a any real number: $a^m a^n = a^{m+n}$.

A polynomial in x is an algebraic expression of the form

$$a_n x^n + a_{n_1} x^{n-1} + \cdots + a_1 x + a_0$$

where the coefficients $a_0, a_1, \cdots, a_n$ are real numbers and n is a non-negative integer.

Two terms in a polynomial are called like terms if they have exactly the same variable factors to the same powers. Like terms in a polynomial are combined by adding their numerical coefficients.

$$a - b = a + (-b) \qquad a(b + c) = ab + ac \qquad a(b - c) = ab - ac$$

	F: First Product	O: Outer Product	I: Inner Product	L: Last Product
$(a + b)(c + d) =$	ac	$+ \quad ad$	$+ \quad bc$	$+ \quad bd$

$$(a - b)(a + b) = a^2 - b^2$$
$$(a + b)^2 = a^2 + 2ab + b^2$$
$$(a - b)^2 = a^2 - 2ab + b^2$$

---

1. 3

3. $(2x^3 - 3x^2 + x + 5) \quad + \quad (2x^2 + x - 1)$
$$= 2x^3 - 3x^2 + x + 5 + 2x^2 + x - 1$$
$$= 2x^3 - x^2 + 2x + 4$$

5. $(2x^3 - 3x^2 + x + 5) \quad - \quad (2x^2 + x - 1)$
$$= 2x^3 - 3x^2 + x + 5 - 2x^2 - x + 1$$
$$= 2x^3 - 5x^2 + 6$$

Common Error:
Reversing terms:
$(2x^2 + x - 1) - (2x^3 - 3x^2 + x + 5)$
Should write: (first) - (second)
Common Error:
$2x^3 - 3x^2 + x + 5 - 2x^2 + x - 1$
Should change the sign of each term in the second
parentheses: $-(2x^2 + x - 1) = -2x^2 - x + 1$

7.
$$
\begin{array}{rrrrrr}
2x^3 & - & 3x^2 & + & x & + & 5 \\
3x & - & 2 & & & & \\
\hline
6x^4 & - & 9x^3 & + & 3x^2 & + & 15x \\
& & -4x^3 & + & 6x^2 & - & 2x & - & 10 \\
\hline
6x^4 & - & 13x^3 & + & 9x^2 & + & 13x & - & 10
\end{array}
$$

9.
$$
\begin{aligned}
2(x-1) &+ 3(2x-3) - (4x-5) \\
&= 2x - 2 + 6x - 9 - 4x + 5 \\
&= 4x - 6
\end{aligned}
$$

11.
$$
\begin{aligned}
2y &- 3y[4 - 2(y-1)] \\
&= 2y - 3y[4 - 2y + 2] \\
&= 2y - 3y[6 - 2y] \\
&= 2y - 18y + 6y^2 \\
&= -16y + 6y^2 \\
&= 6y^2 - 16y
\end{aligned}
$$

13. $(m-n)(m+n) = m^2 - n^2$

15.

	First Product	Outer Product	Inner Product	Last Product
$(4t-3)(t-2)$ $=$	$4t^2$	$-8t$	$-3t$	$+6$
$=$	$4t^2 - 11t + 6$			

17.

	First Product	Outer Product	Inner Product	Last Product
$(3x+2y)(x-3y)$ $=$	$3x^2$	$-9xy$	$+2xy$	$-6y^2$
$=$	$3x^2 - 7xy - 6y^2$			

19. $(2m-7)(2m+7) = (2m)^2 - (7)^2 = 4m^2 - 49.$

21.

	First Product	Outer Product	Inner Product	Last Product
$(6x-4y)(5x+3y)$ $=$	$30x^2$	$+18xy$	$-20xy$	$-12y^2$
$=$	$30x^2 - 2xy - 12y^2$			

23. $(3x-2y)(3x+2y) = (3x)^2 - (2y)^2 = 9x^2 - 4y^2$

25. $\begin{aligned}(4x-y)^2 &= (4x-y)(4x-y)\\ &= (4x)^2 - 2(4x)(y) + (y)^2\\ &= 16x^2 - 8xy + y^2\end{aligned}$

27.
$$
\begin{array}{rrrrr}
a^2 & - & ab & + & b^2\\
a & + & b & &\\
\hline
a^3 & - & a^2 b & + & ab^2\\
& & a^2 b & - & ab^2 & + & b^3\\
\hline
a^3 & & & & & + & b^3
\end{array}
$$

33.
$$
\begin{array}{rrrrrrrr}
2x^2 & + & x & & & - & 2 &\\
x^2 & - & 3x & & & + & 5 &\\
\hline
2x^4 & + & x^3 & & & - & 2x^2 &\\
& & & - & 6x^3 & - & 3x^2 & + & 6x\\
& & & & & 10x^2 & + & 5x & - & 10\\
\hline
2x^4 & - & 5x^3 & + & 5x^2 & + & 11x & - & 10
\end{array}
$$

29. $\begin{aligned}(3x+2)^2 &= (3x)^2 + 2(3x)(2) + (2)^2\\ &= 9x^2 + 12x + 4\end{aligned}$

31. $\begin{aligned}2x &- 3\{x + 2[x - (x+5)] + 1\}\\ &= 2x - 3\{x + 2[x - x - 5] + 1\}\\ &= 2x - 3\{x + 2[-5] + 1\}\\ &= 2x - 3\{x - 10 + 1\}\\ &= 2x - 3\{x - 9\}\\ &= 2x - 3x + 27\\ &= -x + 27\end{aligned}$

35. $\begin{aligned}(2x-1)^2 &- (3x+2)(3x-2)\\ &= (2x-1)^2 - [(3x+2)(3x-2)]\\ &= (2x)^2 - 2(2x)(1) + 1^2\\ &\quad -[(3x)^2 - (2)^2]\\ &= 4x^2 - 4x + 1 - [9x^2 - 4]\\ &= 4x^2 - 4x + 1 - 9x^2 + 4\\ &= -5x^2 - 4x + 5\end{aligned}$

37. $\begin{aligned}(2m-n)^3 &= (2m-n)(2m-n)(2m-n)\\ &= [(2m-n)(2m-n)](2m-n)\\ &= [(2m)^2 - 2(2m)(n) + n^2](2m-n)\\ &= [4m^2 - 4mn + n^2](2m-n)\end{aligned}$

The last multiplication is best
performed vertically, so we write:

$$
\begin{array}{rrrrrrr}
4m^2 & - & 4mn & + & n^2 & &\\
2m & - & n & & & &\\
\hline
8m^3 & - & 8m^2 n & + & 2mn^2 & &\\
& & - & 4m^2 n & + & 4mn^2 & - & n^3\\
\hline
(2m-n)^3 = \quad 8m^3 & - & 12m^2 n & + & 6mn^2 & - & n^3
\end{array}
$$

39. The sum of the first two polynomials:
$$(3m^2 - 2m + 5) + (4m^2 - m) = 3m^2 - 2m + 5 + 4m^2 - m = 7m^2 - 3m + 5$$
The sum of the last two polynomials:
$$(3m^2 - 3m - 2) + (m^3 + m^2 + 2) = 3m^2 - 3m - 2 + m^3 + m^2 + 2 = m^3 + 4m^2 - 3m$$
Subtract from the sum of the first two polynomials the sum of the last two.
$$\begin{aligned}(m^3 + 4m^2 &- 3m) - (7m^2 - 3m + 5)\\ &= m^3 + 4m^2 - 3m - 7m^2 + 3m - 5\\ &= m^3 - 3m^2 - 5\end{aligned}$$

41. Since $(x-2)^2 = (x)^2 - 2(x)(2) + (2)^2 = x^2 - 4x + 4,$
$(x-2)^3 = (x-2)^2(x-2) = (x^2 - 4x + 4)(x-2)$

$$
\begin{array}{r}
x^2 - 4x + 4 \\
x - 2 \\
\hline
x^3 - 4x^2 + 4x \\
- 2x^2 + 8x - 8 \\
\hline
x^3 - 6x^2 + 12x - 8
\end{array}
$$

Therefore $2(x-2)^3 - (x-2)^2 - 3(x-2) - 4$
$$
\begin{aligned}
&= 2(x^3 - 6x^2 + 12x - 8) - (x^2 - 4x + 4) - 3(x-2) - 4 \\
&= 2x^3 - 12x^2 + 24x - 16 - x^2 + 4x - 4 - 3x + 6 - 4 \\
&= 2x^3 - 13x^2 + 25x - 18
\end{aligned}
$$

43. We will multiply $(x+2)(x^2-3)$ using the FOIL method.

	First Product	Outer Product	Inner Product	Last Product
$(x+2)(x^2-3) =$	x^3	$-3x$	$+2x^2$	-6
	or			
	$x^3 + 2x^2 - 3x - 6$			

Then $-3x\{x[x - x(2-x)] - (x+2)(x^2-3)\}$
$$
\begin{aligned}
&= -3x\{x[x - 2x + x^2] - [x^3 + 2x^2 - 3x - 6]\} \\
&= -3x\{x[-x + x^2] - [x^3 + 2x^2 - 3x - 6]\} \\
&= -3x\{-x^2 + x^3 - x^3 - 2x^2 + 3x + 6\} \\
&= -3x\{-3x^2 + 3x + 6\} \\
&= 9x^3 - 9x^2 - 18x
\end{aligned}
$$

45. The non-zero term with the highest degree in the polynomial of degree m has degree m. This term is not changed during the addition, since there is no term in the polynomial of degree n with the degree $m(m > n)$. It will still be the highest degree term, in the sum, so the sum will be a polynomial of degree m.

47. There are three quantities in this problem, perimeter, length, and width. They are related by the perimeter formula $P = 2\ell + 2w$. Since $x = $ length of the rectangle, and the width is 5 meters less than the length, $x - 5 = $ width of the rectangle. So $P = 2x + 2(x-5)$ represents the perimeter of the rectangle. Simplifying:
$P = 2x + 2x - 10 = 4x - 10$

49. There are several quantities involved in this problem. It is important to keep them distinct by using enough words. We write:

$$x = \text{number of nickels}$$
$$x - 5 = \text{number of dimes}$$
$$(x - 5) + 2 = \text{number of quarters}$$

This follows because there are five less dimes than nickels $(x - 5)$ and 2 more quarters than dimes (2 more than $x - 5$). Each nickel is worth 5 cents, each dime worth 10 cents, and each quarter worth 25 cents. Hence the value of the nickels is 5 times the number of nickels, the value of the dimes is 10 times the number of dimes, and the value of the quarters is 25 times the number of quarters.

$$\text{value of nickels} = 5x$$
$$\text{value of dimes} = 10(x - 5)$$
$$\text{value of quarters} = 25[(x - 5) + 2]$$
$$\text{The value of the pile} = (\text{value of nickels})$$
$$+ (\text{value of dimes})$$
$$+ (\text{value of quarters})$$
$$= 5x + 10(x - 5) + 25[(x - 5) + 2]$$

Simplifying this expression, we get:

$$\text{The value of the pile} = 5x + 10x - 50 + 25[x - 5 + 2]$$
$$= 5x + 10x - 50 + 25[x - 3]$$
$$= 5x + 10x - 50 + 25x - 75$$
$$= 40x - 125$$

Exercise 1-3
Key Ideas and Formulas

$$
\begin{aligned}
ab + ac &= a(b+c) \\
u^2 + 2uv + v^2 &= (u+v)^2 \qquad \text{Perfect square} \\
u^2 - 2uv + v^2 &= (u-v)^2 \qquad \text{Perfect square} \\
u^2 - v^2 &= (u-v)(u+v) \qquad \text{Difference of squares} \\
u^3 - v^3 &= (u-v)(u^2+uv+v^2) \quad \text{Difference of cubes} \\
u^3 + v^3 &= (u+v)(u^2-uv+v^2) \quad \text{Sum of cubes}
\end{aligned}
$$

Common Errors:

Confusing $u^2 + v^2$ (a prime polynomial) with

$$(u+v)^2 = (u+v)(u+v) = u^2 + 2uv + v^2 (\text{ a perfect square}).$$

Confusing $u^2 - v^2 = (u-v)(u+v)(\text{ difference of squares})$ with

$$(u-v)^2 = (u-v)(u-v) = u^2 - 2uv + v^2(\text{ a perfect square}).$$

---

1. $2x^2(3x^2 - 4x - 1)$

3. $5xy(2x^2 + 4xy - 3y^2)$

5. $(x+1)(5x-3)$

7. $(y-2z)(2w-x)$

9.
$$
\begin{aligned}
x^2 - 2x + 3x - 6 &= (x^2 - 2x) + (3x - 6) \\
&= x(x-2) + 3(x-2) \\
&= (x-2)(x+3)
\end{aligned}
$$

11.
$$
\begin{aligned}
6m^2 + 10m - 3m - 5 &= (6m^2 + 10m) - (3m + 5) = 2m(3m+5) - 1(3m+5) \\
&= (3m+5)(2m-1)
\end{aligned}
$$

13.
$$
\begin{aligned}
2x^2 - 4xy - 3xy + 6y^2 &= (2x^2 - 4xy) - (3xy - 6y^2) = 2x(x-2y) - 3y(x-2y) \\
&= (x-2y)(2x-3y)
\end{aligned}
$$

15.
$$
\begin{aligned}
8ac + 3bd - 6bc - 4ad &= 8ac - 4ad - 6bc + 3bd \\
&= (8ac - 4ad) - (6bc - 3bd) \\
&= 4a(2c-d) - 3b(2c-d) \\
&= (2c-d)(4a-3b)
\end{aligned}
$$

17. $(2x-1)(x+3)$

19. $(x-6y)(x+2y)$

21. Prime   23. $(5m + 4n)(5m - 4n)$

25. $x^2 + 10xy + 25y^2 = (x + 5y)(x + 5y) = (x + 5y)^2$

27. Prime

29. $6x^2 + 48x + 72 = 6(x^2 + 8x + 12) = 6(x + 2)(x + 6)$

31. $2y^3 - 22y^2 + 48y = 2y(y^2 - 11y + 24) = 2y(y - 3)(y - 8)$

33. $16x^2y - 8xy + y = y(16x^2 - 8x + 1) = y(4x - 1)^2$

35. $(3s - t)(2s + 3t)$

37. $x^3y - 9xy^3 = xy(x^2 - 9y^2) = xy(x - 3y)(x + 3y)$

39. $3m(m^2 - 2m + 5)$

41. $(m + n)(m^2 - mn + n^2)$ 43. $(c - 1)(c^2 + c + 1)$

45. $(a - b)^2 - 4(c - d)^2 \; = \; (a - b)^2 - [2(c - d)]^2$
$$= \; [(a - b) - 2(c - d)][(a - b) + 2(c - d)]$$

47. $2am - 3an + 2bm - 3bn \; = \; (2am - 3an) + (2bm - 3bn) = a(2m - 3n) + b(2m - 3n)$
$$= \; (2m - 3n)(a + b)$$

49. Prime

51. $x^3 - 3x^2 - 9x + 27 \; = \; (x^3 - 3x^2) - (9x - 27) = x^2(x - 3) - 9(x - 3) = (x - 3)(x^2 - 9)$
$$= \; (x - 3)(x - 3)(x + 3) = (x - 3)^2(x + 3)$$

53. $a^3 - 2a^2 - a + 2 \; = \; (a^3 - 2a^2) - (a - 2) = a^2(a - 2) - 1(a - 2) = (a - 2)(a^2 - 1)$
$$= \; (a - 2)(a + 1)(a - 1)$$

55. $4(A + B)^2 - 5(A + B) - 6 = [4(A + B) + 3][(A + B) - 2]$

57. $m^4 - n^4 = (m^2)^2 - (n^2)^2 = (m^2 - n^2)(m^2 + n^2) = (m - n)(m + n)(m^2 + n^2)$

59. $s^4t^4 - 8st \; = \; st(s^3t^3 - 8) = st[(st)^3 - 2^3] = st(st - 2)[(st)^2 + (st)(2) + (2)^2]$
$$= \; st(st - 2)[s^2t^2 + 2st + 4]$$

61. $m^2 + 2mn + n^2 - m - n \; = \; (m^2 + 2mn + n^2) - (m + n) = (m + n)^2 - 1(m + n)$
$$= \; (m + n)(m + n - 1)$$

63. $18a^3 - 8a(x^2 + 8x + 16) \; = \; 2a[9a^2 - 4(x^2 + 8x + 16)] = 2a\{(3a)^2 - [2(x + 4)]^2\}$
$$= \; 2a[3a - 2(x + 4)][3a + 2(x + 4)]$$

65. $x^4 + 2x^2 + 1 - x^2 \; = \; (x^4 + 2x^2 + 1) - x^2 = (x^2 + 1)^2 - x^2$
$$= \; (x^2 + 1 - x)(x^2 + 1 + x) = (x^2 - x + 1)(x^2 + x + 1)$$

Exercise 1-4

Key Ideas and Formulas

Fundamental Property of Fractions

If a, b, and k are real numbers with $b, k \neq 0$, then

$$\frac{ak}{bk} = \frac{a}{b} \quad \text{Also} \quad \frac{a}{b} = \frac{ak}{bk}$$

Multiplication and Division

For a, b, c, and d, real numbers,

$$\frac{a}{b} \cdot \frac{c}{d} = \frac{ac}{bd} \quad b, d \neq 0$$

$$\frac{a}{b} \div \frac{c}{d} = \frac{a}{b} \cdot \frac{d}{c} \quad b, c, d \neq 0$$

Addition and Subtraction

For a, b, and c real numbers

$$\frac{a}{b} + \frac{c}{b} = \frac{a+c}{b} \quad b \neq 0$$

$$\frac{a}{b} - \frac{c}{b} = \frac{a-c}{b} \quad b \neq 0$$

Common Errors

There are many possible errors in applying the fundamental property of fractions. Always be sure that each quantity removed is a factor of the entire numerator and a factor of the entire denominator. Examples of errors abound, but here are a few classic types to avoid:

$$\left.\begin{array}{c} \dfrac{m+2}{m-3} \neq \dfrac{2}{-3} \\[2mm] \dfrac{x^2 + 2x - 15}{x^2 + 4x - 5} \neq \dfrac{2x - 15}{4x - 5} \end{array}\right\} \text{cannot "cancel" common terms"}$$

$$\left.\dfrac{x+7}{ax - bx} \neq \dfrac{7}{a - bx}\right\} \text{cannot "cancel" a term against a factor of a term}$$

$$\left.\begin{array}{c} \dfrac{x - (x - 4)}{(x - 4)(x + 4)} \neq \dfrac{x}{x + 4} \\[2mm] \dfrac{x^2 + 8}{8x^3} \neq \dfrac{x^2}{x^3} \end{array}\right\} \text{cannot "cancel" a term against a factor}$$

---

1. $\left(\dfrac{d^5}{3a} \div \dfrac{d^2}{6a^2}\right) \cdot \dfrac{a}{4d^3} = \left(\dfrac{d^5}{3a} \cdot \dfrac{6a^2}{d^2}\right) \cdot \dfrac{a}{4d^3} = \left(\dfrac{\cancel{d^5}^{d^3} \cdot \cancel{6a^2}^{2a}}{3a \cdot \cancel{d^2}}\right) \cdot \dfrac{a}{4d^3} = \dfrac{2ad^3}{1} \cdot \dfrac{a}{4d^3} = \dfrac{\cancel{2ad^3}^{1}}{1} \cdot \dfrac{a}{\cancel{4d^3}_{2}} = \dfrac{a^2}{2}$

3. $\dfrac{2y}{18} - \dfrac{-1}{28} - \dfrac{y}{42} = \dfrac{28y}{252} - \dfrac{-9}{252} - \dfrac{6y}{252} = \dfrac{22y + 9}{252}$

   Note:
   $$\left.\begin{array}{rcl} 18 &=& 2 \cdot 3 \cdot 3 \\ 28 &=& 2 \cdot 2 \cdot 7 \\ 42 &=& 2 \cdot 3 \cdot 6 \end{array}\right\} \text{so L.C.D.} = 2 \cdot 2 \cdot 3 \cdot 3 \cdot 7 = 252$$

5. $\dfrac{3x+8}{4x^2} - \dfrac{2x-1}{x^3} - \dfrac{5}{8x} = \dfrac{2x(3x+8)}{8x^3} - \dfrac{8(2x-1)}{8x^3} - \dfrac{x^2(5)}{8x^3}$

$\qquad\qquad\qquad\qquad\qquad = \dfrac{6x^2+16x-16x+8-5x^2}{8x^3} = \dfrac{x^2+8}{8x^3}$

7. $\dfrac{2x^2+7x+3}{4x^2-1} \div (x+3) = \dfrac{2x^2+7x+3}{4x^2-1} \div \dfrac{(x+3)}{1} = \dfrac{(2x+1)(x+3)}{(2x+1)(2x-1)} \cdot \dfrac{1}{(x+3)}$

$\qquad\qquad\qquad\qquad\quad = \dfrac{\overset{1}{\cancel{(2x+1)}}\,\overset{1}{\cancel{(x+3)}}}{\cancel{(2x+1)}(2x-1)} \cdot \dfrac{1}{\cancel{(x+3)}} = \dfrac{1}{2x-1}$

9. $\dfrac{m+n}{m^2-n^2} \div \dfrac{m^2-mn}{m^2-2mn+n^2} = \dfrac{m+n}{m^2-n^2} \cdot \dfrac{m^2-2mn+n^2}{m^2-mn}$

$\qquad\qquad\qquad\qquad\qquad = \dfrac{m+n}{(m+n)(m-n)} \cdot \dfrac{(m-n)(m-n)}{m(m-n)}$

$\qquad\qquad\qquad\qquad\qquad = \dfrac{\overset{1}{\cancel{(m+n)}}}{\cancel{(m+n)}\underset{1}{\cancel{(m-n)}}} \cdot \dfrac{\overset{1}{\cancel{(m-n)}}\,\overset{1}{\cancel{(m-n)}}}{m\underset{1}{\cancel{(m-n)}}} = \dfrac{1}{m}$

11. $\dfrac{1}{a^2-b^2} + \dfrac{1}{a^2+2ab+b^2} = \dfrac{1}{(a+b)(a-b)} + \dfrac{1}{(a+b)(a+b)}$

$\qquad\qquad\qquad\qquad\qquad = \dfrac{a+b}{(a+b)(a+b)(a-b)} + \dfrac{a-b}{(a+b)(a+b)(a-b)}$

$\qquad\qquad\qquad\qquad\qquad = \dfrac{a+b+a-b}{(a+b)(a+b)(a-b)} = \dfrac{2a}{(a+b)^2(a-b)}$

13. $m-3-\dfrac{m-1}{m-2} = \dfrac{m-3}{1} - \dfrac{m-1}{m-2} = \dfrac{(m-3)(m-2)}{m-2} - \dfrac{(m-1)}{m-2}$

$\qquad\qquad\qquad = \dfrac{(m^2-5m+6)-(m-1)}{m-2} = \dfrac{m^2-5m+6-m+1}{m-2} = \dfrac{m^2-6m+7}{m-2}$

15. $\dfrac{5}{x-3} - \dfrac{2}{3-x} = \dfrac{5}{x-3} - \dfrac{-2}{x-3} = \dfrac{5+2}{x-3} = \dfrac{7}{x-3}$

17. $\dfrac{2}{y+3} - \dfrac{1}{y-3} + \dfrac{2y}{y^2-9} = \dfrac{2(y-3)}{(y+3)(y-3)} - \dfrac{(y+3)\cdot 1}{(y+3)(y-3)} + \dfrac{2y}{(y+3)(y-3)}$

$\qquad\qquad\qquad\qquad\quad = \dfrac{2(y-3)-(y+3)+2y}{(y+3)(y-3)} = \dfrac{2y-6-y-3+2y}{(y+3)(y-3)}$

$\qquad\qquad\qquad\qquad\quad = \dfrac{3y-9}{(y+3)(y-3)} = \dfrac{3\overset{1}{\cancel{(y-3)}}}{(y+3)\underset{1}{\cancel{(y-3)}}} = \dfrac{3}{y+3}$

19. $\dfrac{1-\frac{y^2}{x^2}}{1-\frac{y}{x}} = \dfrac{x^2\left(1-\frac{y^2}{x^2}\right)}{x^2\left(1-\frac{y}{x}\right)} = \dfrac{x^2-y^2}{x^2-xy} = \dfrac{(x+y)\overset{1}{\cancel{(x-y)}}}{x\underset{1}{\cancel{(x-y)}}} = \dfrac{x+y}{x}$

21. $\dfrac{\frac{1}{m}+1}{m+1} = \dfrac{m\left(\frac{1}{m}+1\right)}{m(m+1)} = \dfrac{1+m}{m(m+1)} = \dfrac{\overset{1}{\cancel{(m+1)}}}{m\underset{1}{\cancel{(m+1)}}} = \dfrac{1}{m}$

23. $\dfrac{y}{y^2 - y - 2} \quad - \quad \dfrac{1}{y^2 + 5y - 14} - \dfrac{2}{y^2 + 8y + 7}$

$$= \dfrac{y}{(y-2)(y+1)} - \dfrac{1}{(y+7)(y-2)} - \dfrac{2}{(y+1)(y+7)}$$

$$= \dfrac{y(y+7)}{(y+1)(y-2)(y+7)} - \dfrac{1(y+1)}{(y+1)(y-2)(y+7)} - \dfrac{2(y-2)}{(y+1)(y-2)(y+7)}$$

$$= \dfrac{y^2 + 7y - y - 1 - 2y + 4}{(y+1)(y-2)(y+7)} = \dfrac{y^2 + 4y + 3}{(y+1)(y-2)(y+7)} = \dfrac{\overset{1}{\cancel{(y+1)}}(y+3)}{\underset{1}{\cancel{(y+1)}}(y-2)(y+7)}$$

$$= \dfrac{y+3}{(y-2)(y+7)}$$

25. $\dfrac{9 - m^2}{m^2 + 5m + 6} \cdot \dfrac{m+2}{m-3} = \dfrac{\overset{-1}{\cancel{(3-m)}}\overset{1}{\cancel{(3+m)}}}{\underset{1}{\cancel{(m+2)}}\underset{1}{\cancel{(m+3)}}} \cdot \dfrac{\overset{1}{\cancel{m+2}}}{\underset{1}{\cancel{m-3}}} = -1$

27. $\dfrac{x+7}{ax - bx} + \dfrac{y+9}{by - ay} = \dfrac{x+7}{x(a-b)} + \dfrac{y+9}{y(b-a)} = \dfrac{y(x+7)}{xy(a-b)} + \dfrac{-x(y+9)}{xy(a-b)}$

$$= \dfrac{xy + 7y - xy - 9x}{xy(a-b)} = \dfrac{7y - 9x}{xy(a-b)}$$

29. $\dfrac{x^2 - 16}{2x^2 + 10x + 8} \div \dfrac{x^2 - 13x + 36}{x^3 + 1} = \dfrac{x^2 - 16}{2x^2 + 10x + 8} \cdot \dfrac{x^3 + 1}{x^2 - 13x + 36}$

$$= \dfrac{\overset{1}{\cancel{(x-4)}}\overset{1}{\cancel{(x+4)}}}{2\underset{1}{\cancel{(x+1)}}\underset{1}{\cancel{(x+4)}}} \cdot \dfrac{\cancel{(x+1)}(x^2 - x + 1)}{\underset{1}{\cancel{(x-4)}}(x-9)} = \dfrac{x^2 - x + 1}{2(x-9)}$$

31. $\dfrac{x^2 - xy}{xy + y^2} \div \left(\dfrac{x^2 - y^2}{x^2 + 2xy + y^2} \div \dfrac{x^2 - 2xy + y^2}{x^2 y + xy^2} \right) = \dfrac{x^2 - xy}{xy + y^2} \div \left(\dfrac{x^2 - y^2}{x^2 + 2xy + y^2} \cdot \dfrac{x^2 y + xy^2}{x^2 - 2xy + y^2} \right)$

$$= \dfrac{x^2 - xy}{xy + y^2} \div \left(\dfrac{\overset{1}{\cancel{(x-y)}}\overset{1}{\cancel{(x+y)}}}{\underset{1}{\cancel{(x+y)}}\underset{1}{\cancel{(x+y)}}} \cdot \dfrac{\overset{1}{xy\cancel{(x+y)}}}{\underset{1}{\cancel{(x-y)}}(x-y)} \right)$$

$$= \dfrac{x^2 - xy}{xy + y^2} \div \dfrac{xy}{x-y} = \dfrac{x^2 - xy}{xy + y^2} \cdot \dfrac{x - y}{xy}$$

$$= \dfrac{\overset{1}{\cancel{x}}(x-y)}{y(x+y)} \cdot \dfrac{x-y}{\underset{1}{\cancel{xy}}} = \dfrac{(x-y)^2}{y^2(x+y)}$$

33. $\left[\dfrac{x}{x^2-16}-\dfrac{1}{x+4}\right]\div\dfrac{4}{x+4} = \left[\dfrac{x}{(x-4)(x+4)}-\dfrac{1}{x+4}\right]\div\dfrac{4}{x+4}$

$$= \left[\dfrac{x}{(x-4)(x+4)}-\dfrac{(x-4)}{(x-4)(x+4)}\right]\div\dfrac{4}{x+4}$$

$$= \dfrac{x-x+4}{(x-4)(x+4)}\div\dfrac{4}{x+4}=\dfrac{4}{(x-4)(x+4)}\div\dfrac{4}{x+4}$$

$$= \dfrac{\overset{1}{\cancel{4}}}{(x-4)\cancel{(x+4)}}\cdot\dfrac{\cancel{x+4}}{\underset{1}{\cancel{4}}}=\dfrac{1}{x-4}$$

35. $\dfrac{\frac{1}{x}+\frac{1}{y}}{x+y}=\dfrac{xy\left(\frac{1}{x}+\frac{1}{y}\right)}{xy(x+y)}=\dfrac{y+x}{xy(x+y)}=\dfrac{\overset{1}{\cancel{x+y}}}{xy\underset{1}{\cancel{(x+y)}}}=\dfrac{1}{xy}$

37. $\dfrac{1+\frac{2}{x}-\frac{15}{x^2}}{1+\frac{4}{x}-\frac{5}{x^2}}=\dfrac{x^2\left(1+\frac{2}{x}-\frac{15}{x^2}\right)}{x^2\left(1+\frac{4}{x}-\frac{5}{x^2}\right)}=\dfrac{x^2+2x-15}{x^2+4x-5}=\dfrac{\overset{1}{\cancel{(x+5)}}(x-3)}{\underset{1}{\cancel{(x+5)}}(x-1)}=\dfrac{x-3}{x-1}$

39. $\dfrac{y-\frac{y^2}{y-x}}{1+\frac{x^2}{y^2-x^2}}=\dfrac{(y^2-x^2)\left[y-\frac{y^2}{y-x}\right]}{(y^2-x^2)\left[1+\frac{x^2}{y^2-x^2}\right]}=\dfrac{(y^2-x^2)y-(y^2-x^2)\frac{y^2}{y-x}}{y^2-x^2+(y^2-x^2)\frac{x^2}{y^2-x^2}}$

$$= \dfrac{y^3-x^2y-y^3-xy^2}{y^2-x^2+x^2}=\dfrac{-x^2y-xy^2}{y^2}$$

$$= \dfrac{-xy(x+y)}{\underset{y}{\cancel{y^2}}}=\dfrac{-x(x+y)}{y}$$

41. $2-\dfrac{1}{1-\frac{2}{a+2}}=2-\dfrac{(a+2)\cdot1}{(a+2)\left[1-\frac{2}{a+2}\right]}=2-\dfrac{a+2}{a+2-(a+2)\frac{2}{a+2}}=2-\dfrac{a+2}{a+2-2}$

$$= 2-\dfrac{a+2}{a}=\dfrac{2}{1}-\dfrac{a+2}{a}=\dfrac{2a}{a}-\dfrac{(a+2)}{a}=\dfrac{2a-a-2}{a}=\dfrac{a-2}{a}$$

43. We will write each compound fraction as a simple fraction, starting with $\frac{1}{1-\frac{1}{x}}$.

$$1 - \cfrac{1}{1 - \cfrac{1}{1-\frac{1}{1-\frac{1}{x}}}} = 1 - \cfrac{1}{1 - \cfrac{1}{1-\frac{x(1)}{x(1-\frac{1}{x})}}} = 1 - \cfrac{1}{1 - \cfrac{1}{1-\frac{x}{x-1}}}$$

$$= 1 - \cfrac{1}{1 - \cfrac{(x-1)}{(x-1)(1-\frac{x}{x-1})}} = 1 - \cfrac{1}{1 - \cfrac{x-1}{x-1-x}}$$

$$= 1 - \cfrac{1}{1 - \frac{x-1}{-1}} = 1 - \cfrac{(-1)}{(-1)(1-\frac{x-1}{-1})} = 1 - \frac{-1}{-1-(x-1)} = 1 - \frac{-1}{-1-x+1}$$

$$= 1 - \frac{-1}{-x} = 1 - \frac{1}{x} = \frac{x}{x} - \frac{1}{x} = \frac{x-1}{x}$$

45. A. The multiplicative inverse of x is the unique real number $\frac{1}{x}$ such that $x(\frac{1}{x}) = 1$.
Since $\frac{c}{d} \cdot \frac{d}{c} = \frac{cd}{dc} = \frac{cd}{cd} = 1$, $\frac{d}{c}$ is the multiplicative inverse of $\frac{c}{d}$.
B. By Definition 1 of Section 1-1 of the text $a \div b = a(\frac{1}{b}) = a \cdot$ (multiplicative inverse of b). Hence

$$\frac{\frac{a}{b}}{} \div \frac{c}{d} = \frac{a}{b} \cdot \text{(multiplicative inverse of } \frac{c}{d})$$
$$= \frac{a}{b} \cdot \frac{d}{c} \quad \text{by part A.}$$

Exercise 1-5

CHAPTER REVIEW

1.   A. Since 3 is an element of $\{1, 2, 3, 4, 5\}$, the statement is true. T.

B. Since 5 is not an element of $\{4, 1, 2\}$, the statement is true. T.

C. Since B is not an element of $\{1, 2, 3, 4, 5\}$, the statement is false. F.

D. Since each element of the set $\{1, 2, 4\}$ is also an element of the set $\{1, 2, 3, 4, 5\}$,

the statement is true. T.

E. Since sets B and C have exactly the same elements, the sets are equal.

The statement is false. F.

F. The statement is false, since not every element of the set $\{1, 2, 3, 4, 5\}$

is an element of the set $\{1, 2, 4\}$. For example, 3 is an element of the first,

but not the second. F. $(1-1)$

2.   A. The commutative property $(\cdot)$ states that in general,

$xy = yx$

Comparing this with

$x(y + z) =?$

we see that $x(y + z) = (y + z)x$

$(y + z)x$

B. The associative property $(+)$ states that, in general,

$(x + y) + z = x + (y + z)$

Comparing this with

$? = 2 + (x + y)$

we see that

$(2 + x) + y = 2 + (x + y)$

$(2 + x) + y$

C. The distributive property states that, in general,

$(x + y)z = xz + yz$

Comparing this with

$(2 + 3)x =?$

we see that

$(2 + 3)x = 2x + 3x$

$2x + 3x$ $(1-1)$

3. $(3x - 4) + (x + 2) + (3x^2 + x - 8) + (x^3 + 8) \quad = \quad 3x - 4 + x + 2 + 3x^2 + x - 8 + x^3 + 8$

$\qquad\qquad\qquad\qquad\qquad\qquad\qquad\qquad\qquad = \quad x^3 + 3x^2 + 5x - 2 \quad (1-2)$

4.   sum of first and third $= (3x - 4) + (3x^2 + x - 8) = 3x^2 + 4x - 12$

sum of second and fourth $= (x + 2) + (x^3 + 8) = x^3 + x + 10$

subtract the sum of first and third from the sum of second and fourth

$= (x^3 + x + 10) - (3x^2 + 4x - 12) = x^3 + x + 10 - 3x^2 - 4x + 12 = x^3 - 3x^2 - 3x + 22 \quad (1-2)$

5.

$$
\begin{array}{rrr}
3x^2 & +\ x & -8 \\
x^3 & +\ 8 & \\
\hline
3x^5 & +\ x^4 & -8x^3 \\
\end{array}
$$

$$
\begin{array}{rrr}
& 24x^2 & +\ 8x & -64 \\
\hline
3x^5 \quad +\ x^4 \quad -8x^3 \quad + & 24x^2 & +\ 8x & -64 \\
\end{array}
$$

$$(1-2)$$

6. 3 $(1-2)$
7. 1 $(1-2)$

8. $\begin{aligned}
5x^2 - 3x[4 - 3(x-2)] &= 5x^2 - 3x[4 - 3x + 6] \\
&= 5x^2 - 3x[10 - 3x] \\
&= 5x^2 - 30x + 9x^2 \\
&= 14x^2 - 30x
\end{aligned}$

Common Errors:
$$-3(x-2) \neq -3x - 6$$
$$4 - 3(x-2) \neq 1(x-2)$$
The first is incorrect use of the distributive property, the second is incorrect order of operations.

$$(1-2)$$

9. $(3m - 5n)(3m + 5n) = (3m)^2 - (5n)^2 = 9m^2 - 25n^2$ $(1-2)$

10.

	First Product	Outer Product	Inner Product	Last Product
$(2x + y)(3x - 4y) =$	$6x^2$	$-8xy$	$+3xy$	$-4y^2$
$=$	$6x^2 - 5xy - 4y^2$			

$$(1-2)$$

11. $\begin{aligned}
(2a - 3b)^2 &= (2a)^2 - 2(2a)(3b) + (3b)^2 \\
&= 4a^2 - 12ab + 9b^2 \quad (1-2)
\end{aligned}$

12. $(3x - 2)^2$ $(1-3)$
13. Prime. $(1-3)$
14. $\begin{aligned}
6n^3 - 9n^2 - 15n &= 3n(2n^2 - 3n - 5) \\
&= 3n(2n - 5)(n + 1) \quad (1-3)
\end{aligned}$

15. $\dfrac{2}{5b} - \dfrac{4}{3a^3} - \dfrac{1}{6a^2b^2} = \dfrac{12a^3b}{30a^3b^2} - \dfrac{40b^2}{30a^3b^2} - \dfrac{5a}{30a^3b^2} = \dfrac{12a^3b - 40b^2 - 5a}{30a^3b^2}$ $(1-4)$

16. $\begin{aligned}
\dfrac{3x}{3x^2 - 12x} + \dfrac{1}{6x} &= \dfrac{3x}{3x(x-4)} + \dfrac{1}{6x} \\
&= \dfrac{2(3x)}{6x(x-4)} + \dfrac{1(x-4)}{6x(x-4)} = \dfrac{6x + x - 4}{6x(x-4)} = \dfrac{7x - 4}{6x(x-4)} \quad (1-4)
\end{aligned}$

17. $\dfrac{y-2}{y^2-4y+4} \div \dfrac{y^2+2y}{y^2+4y+4} = \dfrac{y-2}{y^2-4y+4} \cdot \dfrac{y^2+4y+4}{y^2+2y} = \dfrac{1}{\cancel{(y-2)}}{(y-2)} \cdot \dfrac{(y+2)(y+2)}{y(y+2)}$

$$= \dfrac{y+2}{y(y-2)} \quad (1-4)$$

18. $\dfrac{u-\frac{1}{u}}{1-\frac{1}{u^2}} = \dfrac{u^2(u-\frac{1}{u})}{u^2(1-\frac{1}{u^2})} = \dfrac{u^3-u}{u^2-1} = \dfrac{u(u^2-1)}{u^2-1} = u \quad (1-4)$

19. The odd integers between -4 and 2 are $-3, -1,$ and 1. Hence the set is written
$\{-3, -1, 1\} \quad (1-1)$

20. Subtraction.
$(-3) - (-2) = (-3) + [-(-2)]$ is a special case of
$a - b = a + (-b) \quad (1-1)$

21. Commutative $(+)$
$3y + (2x+5) = (2x+5) + 3y$ is a special case of
$\qquad x + y = y + x \quad (1-1)$

22. Distributive
$(2x+3)(3x+5) = (2x+3)3x + (2x+3)5$ is a special case of
$\qquad x(y+z) = xy + xz \quad (1-1)$

23. Associative $(\cdot)$
$3 \cdot (5x) = (3 \cdot 5)x$ is a special case of
$\qquad x(yz) = (xy)z \quad (1-1)$

24. Negatives
$\dfrac{a}{-(b-c)} = -\dfrac{a}{b-c}$ is a special case of
$\qquad \dfrac{a}{-b} = -\dfrac{a}{b} \quad (1-1)$

25. Identity $(+)$
$3xy + 0 = 3xy$ is a special case of
$\qquad x + 0 = x \quad (1-1)$

26. A. T   B. F $(1-1)$

27. 0 and -3 are two examples of infinitely many. $(1-1)$

28. A. a and d B. None $(1-2)$

29.
$$\begin{aligned}
(2x-y)(2x+y) - (2x-y)^2 &= (2x)^2 - y^2 - [(2x)^2 - 2(2x)y + y^2] \\
&= 4x^2 - y^2 - [4x^2 - 4xy + y^2] \\
&= 4x^2 - y^2 - 4x^2 + 4xy - y^2 \\
&= 4xy - 2y^2
\end{aligned}$$

Common Errors :
$(2x-y)^2 \ne 4x^2 - y^2$ (Exponents do not distribute over subtraction)
$-(2x-y)^2 \ne (-2x+y)^2$ (square first, then subtract)
$-(2x-y)^2 \ne -4x^2 - 4xy + y^2$ (must change all signs) $(1-2)$

30. A straightforward way to perform this multiplication is vertically:

$$\begin{array}{rrrrr}
m^2 & + 2mn & -n^2 & & \\
m^2 & - 2mn & -n^2 & & \\
\hline
m^4 & + 2m^3n & -m^2n^2 & & \\
& - 2m^3n & -4m^2n^2 & + 2mn^3 & \\
& & -m^2n^2 & - 2mn^3 & + n^4 \\
\hline
m^4 & & -6m^2n^2 & & + n^4
\end{array}$$

Alternatively, we can notice

$$(m^2 + 2mn - n^2)(m^2 - 2mn - n^2) = [(m^2 - n^2) + 2mn][(m^2 - n^2) - 2mn]$$
$$= (m^2 - n^2)^2 - (2mn)^2 \text{ (difference of squares)}$$
$$= (m^2)^2 - 2m^2n^2 + (n^2)^2 - 4m^2n^2$$
$$= m^4 - 2m^2n^2 + n^4 - 4m^2n^2$$
$$= m^4 - 6m^2n^2 + n^4 \quad (1-2)$$

31.
$$-2x\{(x^2 + 2)(x - 3) - x[x - x(3 - x)]\} = -2x\{(x^2 + 2)(x - 3) - x[x - 3x + x^2]\}$$
$$= -2x\{(x^2 + 2)(x - 3) - x[-2x + x^2]\}$$
$$= -2x\{x^3 - 3x^2 + 2x - 6 + 2x^2 - x^3\}$$
$$= -2x\{-x^2 + 2x - 6\}$$
$$= 2x^3 - 4x^2 + 12x \quad (1-2)$$

32.
$$(x - 2y)^3 = (x - 2y)(x - 2y)(x - 2y)$$
$$= [(x - 2y)(x - 2y)](x - 2y)$$
$$= [x^2 - 2(x)(2y) + (2y)^2](x - 2y)$$
$$= [x^2 - 4xy + 4y^2](x - 2y)$$

The last multiplication is best performed vertically, so we write:

x^2	$-$	$4xy$	$+ 4y^2$	
x	$-$	$2y$		
x^3	$-$	$4x^2y$	$+ 4xy^2$	
	$-$	$2x^2y$	$+ 8xy^2$	$-8y^3$
x^3	$-$	$6x^2y$	$+ 12xy^2$	$-8y^3$

$$(1-2)$$

33.
$$(4x - y)^2 - 9x^2 = (4x - y)^2 - (3x)^2 - [(4x - y) - 3x][(4x - y) + 3x]$$
$$= (x - y)(7x - y) \quad (1-3)$$

34. Prime $\quad (1-3)$

35. $3xy(2x^2 + 4xy - 5y^2) \quad (1-3)$

36. $(y - b)^2 - y + b = (y - b)(y - b) - 1(y - b) = (y - b)(y - b - 1) \quad (1-3)$

37.
$$3x^3 + 24y^3 = 3(x^3 + 8y^3) = 3[x^3 + (2y)^3] = 3(x + 2y)[x^2 - x(2y) + (2y)^2]$$
$$= 3(x + 2y)(x^2 - 2xy + 4y^2) \quad (1-3)$$

38.
$$y^3 + 2y^2 - 4y - 8 = y^2(y + 2) - 4(y + 2) = (y + 2)(y^2 - 4) = (y + 2)(y + 2)(y - 2)$$
$$= (y - 2)(y + 2)^2 \quad (1-3)$$

39. $\dfrac{m-1}{m^2-4m+4} + \dfrac{m+3}{m^2-4} + \dfrac{2}{2-m} = \dfrac{m-1}{(m-2)(m-2)} + \dfrac{m+3}{(m-2)(m+2)} + \dfrac{-2}{m-2}$

$= \dfrac{(m-1)(m+2)}{(m-2)(m-2)(m+2)} + \dfrac{(m+3)(m-2)}{(m-2)(m-2)(m+2)} + \dfrac{-2(m-2)(m+2)}{(m-2)(m-2)(m+2)}$

$= \dfrac{(m-1)(m+2)+(m+3)(m-2)-2(m-2)(m+2)}{(m-2)^2(m+2)}$

$= \dfrac{m^2+m-2+m^2+m-6-2(m^2-4)}{(m-2)^2(m+2)} = \dfrac{2m^2+2m-8-2m^2+8}{(m-2)^2(m+2)} = \dfrac{2m}{(m-2)^2(m+2)}$ $(1-4)$

40. $\dfrac{y}{x^2} \div \left(\dfrac{x^2+3x}{2x^2+5x-3} \div \dfrac{x^3y-x^2y}{2x^2-3x+1} \right) = \dfrac{y}{x^2} \div \left(\dfrac{x^2+3x}{2x^2+5x-3} \cdot \dfrac{2x^2-3x+1}{x^3y-x^2y} \right)$

$= \dfrac{y}{x^2} \div \left(\dfrac{\cancel{x}(x+3)}{(2x-1)(x+3)} \cdot \dfrac{(2x-1)(x-1)}{x^2y(x-1)} \right)$

$= \dfrac{y}{x^2} \div \dfrac{1}{xy} = \dfrac{y}{x^2} \cdot \dfrac{xy}{1} = \dfrac{y^2}{x}$ $(1-4)$

41. $\dfrac{1-\frac{1}{1+\frac{x}{y}}}{1-\frac{1}{1-\frac{x}{y}}} = \dfrac{1-\frac{y(1)}{y(1+\frac{x}{y})}}{1-\frac{y(1)}{y(1-\frac{x}{y})}} = \dfrac{1-\frac{y}{y+x}}{1-\frac{y}{y-x}} = \dfrac{\frac{y+x}{y+x}-\frac{y}{y+x}}{\frac{y-x}{y-x}-\frac{y}{y-x}} = \dfrac{\frac{x}{y+x}}{\frac{-x}{y-x}} = \dfrac{x}{y+x} \div \dfrac{-x}{y-x}$

$= \dfrac{1}{y+x} \cdot \dfrac{y-x}{-1} = \dfrac{-1(y-x)}{y+x} = \dfrac{x-y}{x+y}$ $(1-4)$

42. $\dfrac{a^{-1}-b^{-1}}{ab^{-2}-ba^{-2}} = \dfrac{\frac{1}{a}-\frac{1}{b}}{\frac{a}{b^2}-\frac{b}{a^2}} = \dfrac{a^2b^2(\frac{1}{a}-\frac{1}{b})}{a^2b^2(\frac{a}{b^2}-\frac{b}{a^2})} = \dfrac{ab^2-a^2b}{a^3-b^3} = \dfrac{ab(b-a)}{(a-b)(a^2+ab+b^2)}$

$= \dfrac{-ab}{a^2+ab+b^2}$ $(1-4, 1-5)$

43. Let $c = 0.54545454\ldots$ .Then
        $100c = 54.545454\ldots$
  So $(100c - c) = (54.545454\ldots) - (0.54545454\ldots)$
              $= 54$
       $99c = 54$
         $c = \dfrac{54}{99} = \dfrac{6}{11}$

The number can be written as the quotient of two integers, so it is rational. $(1-1)$

44. A. List each element of M. Follow these with each element of N that is not
yet listed. $\{-4, -3, 2, 0\}$, or, in increasing order, $\{-4, -3, 0, 2\}$.

B. List each element of M that is also an element of N. $\{-3, 2\}$. $(1-1)$

45. $x(2x - 1)(x + 3) = x(2x^2 + 6x - x - 3)$
                $= x(2x^2 + 5x - 3)$
                $= 2x^3 + 5x^2 - 3x$
 $(x - 1)^3 = (x - 1)(x - 1)(x - 1)$
           $= (x^2 - 2x + 1)(x - 1)$

This multiplication we perform vertically:

$$
\begin{array}{rrrr}
x^2 & - \ 2x & + & 1 \\
x & - \ 1 & & \\
\hline
x^3 & - \ 2x^2 & + & x \\
 & - \ x^2 & +2x & -1 \\
\hline
x^3 & - \ 3x^2 & +3x & -1 \\
\end{array}
$$

Hence

$$x(2x-1)(x+3) - (x-1)^3 = 2x^3 + 5x^2 - 3x - (x^3 - 3x^2 + 3x - 1)$$

$$= 2x^3 + 5x^2 - 3x - x^3 + 3x^2 - 3x + 1$$

$$= x^3 + 8x^2 - 6x + 1 \quad (1-2)$$

46.

$$4x(a^2 - 4a + 4) - 9x^3 = x[4(a^2 - 4a + 4) - 9x^2]$$

$$= x[2^2(a-2)^2 - (3x)^2]$$

$$= x[2(a-2) + 3x][2(a-2) - 3x]$$

$$= x(2a - 4 + 3x)(2a - 4 - 3x)$$

$$= x(2a + 3x - 4)(2a - 3x - 4) \quad (1-3)$$

47.

$$\left[x - \frac{1}{1-\frac{1}{x}}\right] \div \left[\frac{x}{x+1} - \frac{x}{1-x}\right] = \left[x - \frac{x(1)}{x\left(1-\frac{1}{x}\right)}\right] \div \left[\frac{x(1-x)}{(x+1)(1-x)} - \frac{x(x+1)}{(x+1)(1-x)}\right]$$

$$= \left[x - \frac{x}{x-1}\right] \div \left[\frac{x(1-x) - x(x+1)}{(x+1)(1-x)}\right]$$

$$= \left[\frac{x(x-1)}{x-1} - \frac{x}{x-1}\right] \div \left[\frac{x - x^2 - x^2 - x}{(x+1)(1-x)}\right]$$

$$- \left[\frac{x^2 - x - x}{x-1}\right] \div \left[\frac{-2x^2}{(x+1)(1-x)}\right]$$

$$= \frac{x^2 - 2x}{x-1} \cdot \frac{(x+1)(1-x)}{-2x^2} = \frac{\overset{1}{\cancel{x}(x-2)}}{\cancel{x-1}_{-1}} \cdot \frac{(x+1)(1\overset{1}{\cancel{-x}})}{-2\cancel{x^2}_{x}}$$

$$= \frac{(x-2)(x+1)}{2x} \quad (1-4)$$

CHAPTER 2
Exercise 2-1

Key Ideas and Formulas

a^n, n an integer and a real

1. For n a positive integer:

$$a^n = a \cdot a, \cdots, a$$
$$\text{n factors of } a$$

2. For $n = 0 : a^0 = 1 \quad a \neq 0$

0^0 is not defined

3. For n a negative integer:

$$a^n = \frac{1}{a^{-n}} \quad a \neq 0$$

Properties of Exponents

For m, n and p integers and a

and b real numbers, then:

1. $a^m a^n = a^{m+n}$

2. $(a^n)^m = a^{mn}$

3. $(ab)^m = a^m b^m$

4. $\left(\frac{a}{b}\right)^m = \frac{a^m}{b^m} \quad b \neq 0$

5. $\dfrac{a^m}{a^n} = \begin{cases} a^{m-n} \\ a^{\frac{1}{n-m}} \end{cases} \quad a \neq 0$

6. $a^{-n} = \dfrac{1}{a^n} \quad a \neq 0$

7. $(a^m b^n)^p = a^{pm} b^{pn}$

8. $\left(\dfrac{a^m}{b^n}\right)^p = \dfrac{a^{pm}}{b^{pn}} \quad b \neq 0$

9. $\dfrac{a^{-n}}{b^{-m}} = \dfrac{b^m}{a^n} \quad a, b \neq 0$

10. $\left(\dfrac{a}{b}\right)^{-n} = \left(\dfrac{b}{a}\right)^n \quad a, b \neq 0$

---

1. $y^{-5} y^5 = y^{-5+5} = y^0 = 1$

3. $(2x^2)(3x^3)(x^4) = (2 \cdot 3)(x^2 x^3 x^4)$
$$= 6x^9$$

5. $(3x^3 y^{-2})^2 = 3^2 x^6 y^{-4} = \dfrac{9x^6}{y^4}$

7. $\left(\dfrac{ab^3}{c^2 d}\right)^4 = \dfrac{a^4 (b^3)^4}{(c^2)^4 d^4} = \dfrac{a^4 b^{12}}{c^8 d^4}$

9. $\dfrac{10^{23} \cdot 10^{-11}}{10^{-3} \cdot 10^{-2}} = \dfrac{10^{12}}{10^{-5}} = 10^{12-(-5)} = 10^{17}$

11. $\dfrac{4x^{-2} y^{-3}}{2x^{-3} y^{-1}} = 2x^{-2-(-3)} y^{-3-(-1)}$

$$= 2x^1 y^{-2} = \dfrac{2x}{y^2}$$

Common Error: $\dfrac{x^{-2}}{x^{-3}} \neq x^{-2-3}$

13. $\left(\dfrac{n^{-3}}{n^{-2}}\right)^{-2} = \dfrac{n^6}{n^4} = n^2$

15. $\dfrac{8 \times 10^3}{2 \times 10^{-5}} = 4 \times 10^{3-(-5)} = 4 \times 10^8$

17. $32,250,000 = 3.\underbrace{2250000}\times 10^7$

$\qquad$ 7 places left positive exponent
$$= 3.225 \times 10^7$$

19. $0.085 = 0.\underbrace{08.}5 \times 10^{-2} = 8.5 \times 10^{-2}$

$\qquad$ 2 places right negative exponent

21. $0.\underbrace{000\,000\,07}29 = 7.29 \times 10^{-8}$

$\qquad$ 8 places right

23. $5 \times 10^{-3} = 0.\underbrace{005.} = 0.005$

$\qquad$ 3 places left

25. $2.69 \times 10^7 = 2\underbrace{.6\,900\,000.}$

$\qquad$ 7 places right
$$= 26,900,000$$

27. $5.9 \times 10^{-10} = 0.000\,000\,000\,59$

29. $\dfrac{27x^{-5} x^5}{18y^{-6} y^2} = \dfrac{27x^{-5+5}}{18y^{-6+2}} = \dfrac{3x^0}{2y^{-4}} = \dfrac{3y^4}{2}$

31. $\left(\dfrac{x^4 y^{-1}}{x^{-2} y^3}\right)^2 = \dfrac{x^8 y^{-2}}{x^{-4} y^6} = x^{8-(-4)} y^{-2-6}$

$$= x^{12} y^{-8} = \dfrac{x^{12}}{y^8}$$

33. $\left(\dfrac{2x^{-3} y^2}{4xy^{-1}}\right)^{-2} = \left(\dfrac{x^{-3} y^2}{2xy^{-1}}\right)^{-2} = \dfrac{x^6 y^{-4}}{2^{-2} x^{-2} y^2}$

$$= 2^2 x^{6-(-2)} y^{-4-2}$$

$$= 4x^8 y^{-6} = \dfrac{4x^8}{y^6}$$

35. $\left[\left(\dfrac{u^3 v^{-1} w^{-2}}{u^{-2} v^{-2} w}\right)^{-2}\right]^2 = \left(\dfrac{u^3 v^{-1} w^{-2}}{u^{-2} v^{-2} w^1}\right)^{-4}$

$$= \dfrac{u^{-12} v^4 w^8}{u^8 v^8 w^{-4}}$$

$$= u^{-20} v^{-4} w^{12} = \dfrac{w^{12}}{u^{20} v^4}$$

37. $(x+y)^{-2} = \dfrac{1}{(x+y)^2}$

39. $\dfrac{1+x^{-1}}{1-x^{-2}} = \dfrac{1+\frac{1}{x}}{1+\frac{1}{x^2}} = \dfrac{x^2(1+\frac{1}{x})}{x^2(1-\frac{1}{x^2})} = \dfrac{x^2+x}{x^2-1} = \dfrac{x(x+1)}{(x-1)(x+1)} = \dfrac{x}{x-1}$

41. $\dfrac{x^{-1}-y^{-1}}{x-y} = \dfrac{\frac{1}{x}-\frac{1}{y}}{x-y} = \dfrac{xy(\frac{1}{x}-\frac{1}{y})}{xy(x-y)} = \dfrac{y-x}{xy(x-y)} = \dfrac{-1}{xy}$

43. $\dfrac{4x^2-12}{2x} = \dfrac{4x^2}{2x} - \dfrac{12}{2x} = 2x - \dfrac{6}{x} = 2x - 6x^{-1}$

45. $\dfrac{5x^3-2}{3x^2} = \dfrac{5x^3}{3x^2} - \dfrac{2}{3x^2} = \dfrac{5x}{3} - \dfrac{2}{3x^2} = \dfrac{5}{3}x - \dfrac{2}{3}x^{-2}$

47. $\dfrac{2x^3-3x^2+x}{2x^2} = \dfrac{2x^3}{2x^2} - \dfrac{3x^2}{2x^2} + \dfrac{x}{2x^2} = x - \dfrac{3}{2} + \dfrac{1}{2x} = x - \dfrac{3}{2} + \dfrac{1}{2}x^{-1}$

49. $\dfrac{12(a+2b)^{-3}}{6(a+2b)^{-8}} = \dfrac{12(a+2b)^{-3-(-8)}}{6} = 2(a+2b)^5$

51. $\dfrac{xy^{-2}-yx^{-2}}{y^{-1}-x^{-1}} = \dfrac{\frac{x}{y^2}-\frac{y}{x^2}}{\frac{1}{y}-\frac{1}{x}} = \dfrac{x^2y^2(\frac{x}{y^2}-\frac{y}{x^2})}{x^2y^2(\frac{1}{y}-\frac{1}{x})} = \dfrac{x^3-y^3}{x^2y-xy^2}$

$$= \dfrac{(x-y)(x^2+xy+y^2)}{xy(x-y)} = \dfrac{x^2+xy+y^2}{xy}$$

53. $\left(\dfrac{x^{-1}}{x^{-1}-y^{-1}}\right)^{-1} = \dfrac{x^{-1}-y^{-1}}{x^{-1}} = \dfrac{\frac{1}{x}-\frac{1}{y}}{\frac{1}{x}} = \dfrac{xy(\frac{1}{x}-\frac{1}{y})}{xy(\frac{1}{x})} = \dfrac{y-x}{y}$

55. $\dfrac{(32.7)(0.000\,000\,008\,42)}{(0.0513)(80,700,000,000)} = \dfrac{(32.7)(8.42 \times 10^{-9})}{(0.0513)(8.07 \times 10^{10})} = 6.65 \times 10^{-17}$

57. $\dfrac{(5,760,000,000)}{(527)(0.000\,007\,09)} = \dfrac{5.76 \times 10^9}{(527)(7.09 \times 10^{-6})} = 1.54 \times 10^{12}$

59. 1.0295×10^{11} 61. -4.3647×10^{-18}

63. $(9,820,000,000)^3 = (9.82 \times 10^9)^3 = 9.4697 \times 10^{29}$

65. mass of earth in pounds

$= $ mass of earth in grams $\times$ number of grams/pound

$= 6.1 \times 10^{27} \times 2.2 \times 10^{-3}$

$= 13.42 \times 10^{24}$

$= 1.342 \times 10^{25}$

$= 1.3 \times 10^{25}$ pounds to two significant digits

67. 1 operation in 10^{-8} seconds means $1 \div 10^{-8}$ operations/second, that is 10^8 operations in 1 second, that is, 100,000,000 or 100 million. Similarly, 1 operation in 10^{-10} seconds means $1 \div 10^{-10}$, that is 10,000,000,000 or 10 billion operations in 1 second.

Since 1 minute is 60 seconds, multiply each number by 60 to get $60 \times 10^8 = 6 \times 10^9$ or 6,000,000,000 or 6 billion operations in 1 minute and $60 \times 10^{10} = 6 \times 10^{11}$ or 600,000,000,000 or 600 billion operations in 1 minute.

69. average amount of tax paid per person =

$$= \frac{\text{amount of tax}}{\text{number of persons}}$$

$$= \frac{349{,}000{,}000{,}000}{242{,}000{,}000}$$

$$= \frac{3.49 \times 10^{11}}{2.42 \times 10^8}$$

$$= 1.44 \times 10^3 \text{ dollars per person}$$

to three significant digits

$$= \$1{,}440 \text{ per person}$$

Exercise 2-2

Key Ideas and Formulas

Note: All properties of exponents treated in Section 2-1 are valid for rational number exponents defined in this section.

For n a natural number and b a real number, $b^{1/n}$ is the principal nth root of b.

1. If n is even and b is positive, then $b^{1/n}$ represents the positive nth root of b.

2. If n is even and b is negative, then $b^{1/n}$ does not represent a real number.

3. If n is odd, then $b^{1/n}$ represents the real nth root of b (there is only one).

4. $0^{1/n} = 0$

For m and n natural numbers and b any real number (except b cannot be negative when n is even):

$$b^{m/n} = (b^{1/n})^m \qquad b^{m/n} = (b^m)^{1/n}$$

$$b^{-m/n} = \frac{1}{b^{m/n}} \quad b \neq 0$$

---

1. $16^{1/2} = 4$

3. $16^{3/2} = (16^{1/2})^3 = 4^3 = 64$

5. $-36^{1/2} = -(36^{1/2}) = -6$

7. $(-36)^{1/2}$ is not a real number

9. $\left(\frac{4}{25}\right)^{3/2} = \frac{4^{3/2}}{25^{3/2}} = \frac{8}{125}$

11. $9^{-3/2} = \frac{1}{9^{3/2}} = \frac{1}{27}$

13. $y^{1/5}y^{2/5} = y^{1/5+2/5} = y^{3/5}$

15. $d^{2/3}d^{-1/3} = d^{2/3+(-1/3)} = d^{1/3}$

17. $(y^{-8})^{1/16} = y^{1/16(-8)} = y^{-8/16}$
$$= y^{-1/2} = \frac{1}{y^{1/2}}$$

19. $(8x^3y^{-6})^{1/3} = 8^{1/3}x^{1/3(3)}y^{1/3(-6)}$
$$= 2x^1y^{-2} = \frac{2x}{y^2}$$

21. $\left(\frac{a^{-3}}{b^4}\right)^{1/12} = \frac{a^{-3/12}}{b^{4/12}} = \frac{a^{-1/4}}{b^{1/3}} = \frac{1}{a^{1/4}b^{1/3}}$

23. $\left(\frac{4x^{-2}}{y^4}\right)^{-1/2} = \frac{4^{-1/2}x^1}{y^{-2}} = \frac{xy^2}{4^{1/2}} = \frac{xy^2}{2}$

Common Error: $\frac{4x^1}{y^{-2}}$. This is wrong; the exponent $\left(-\frac{1}{2}\right)$ applies to constant as well as variable factors.

25. $\left(\frac{8a^{-4}b^3}{27a^2b^{-3}}\right)^{1/3} = \frac{8^{1/3}a^{-4/3}b^1}{27^{1/3}a^{2/3}b^{-1}}$
$$= \frac{2b^{1-(-1)}}{3a^{2/3-(-4/3)}} = \frac{2b^2}{3a^2}$$

27. $\frac{8x^{-1/3}}{12x^{1/4}} = \frac{2}{3x^{1/4-(-1/3)}} = \frac{2}{3x^{7/12}}$

29. $\left(\frac{a^{2/3}b^{-1/2}}{a^{1/2}b^{1/2}}\right)^2 = \frac{a^{4/3}b^{-1}}{a^1b^1}$
$$= \frac{a^{4/3-1}}{b^{1-(-1)}} = \frac{a^{1/3}}{b^2}$$

31. $2m^{1/3}(3m^{2/3} - m^6) = 6m^{1/3+2/3}$
$$- 2m^{\frac{1}{3}+6} = 6m - 2m^{19/3}$$

33. $(a^{1/2} + 2b^{1/2})(a^{1/2} - 3b^{1/2}) =$
$$a^{1/2}a^{1/2} - 3a^{1/2}b^{1/2}$$
$$+2a^{1/2}b^{1/2} - 6b^{1/2}b^{1/2} = a - a^{1/2}b^{1/2}$$
$$- 6b$$

35. $(2x^{1/2} - 3y^{1/2})(2x^{1/2} + 3y^{1/2}) =$
$$(2x^{1/2})^2 - (3y^{1/2})^2$$
$$= 4x - 9y$$

37. $(x^{1/2} + 2y^{1/2})^2 = (x^{1/2})^2 + 2(x^{1/2})(2y^{1/2}) + (2y^{1/2})^2 = x + 4x^{1/2}y^{1/2} + 4y$

39. $\dfrac{12x^{1/2} - 3}{4x^{1/2}} = \dfrac{12x^{1/2}}{4x^{1/2}} - \dfrac{3}{4x^{1/2}} = 3 - \dfrac{3}{4}x^{-1/2}$

41. $\dfrac{3x^{2/3} + x^{1/2}}{5x} = \dfrac{3x^{2/3}}{5x^1} + \dfrac{x^{1/2}}{5x^1} = \dfrac{3}{5}x^{-1/3} + \dfrac{1}{5}x^{-1/2}$

43. $\dfrac{x^2 - 4x^{1/2}}{2x^{1/3}} = \dfrac{x^2}{2x^{1/3}} - \dfrac{4x^{1/2}}{2x^{1/3}} = \dfrac{1}{2}x^{5/3} - 2x^{1/6}$

45. $(a^{3/n}b^{3/m})^{1/3} = a^{1/3(3/n)}b^{1/3(3/m)} = a^{1/n}b^{1/m}$

47. $(x^{m/4}y^{n/3})^{-12} = x^{-12(m/4)}y^{-12(n/3)} = x^{-3m}y^{-4n} = \dfrac{1}{x^{3m}y^{4n}}$

49.   A. Since $(x^2)^{1/2}$ represents the positive real square root of x^2, it will not equal x if x is negative. So any negative value of x, for example $x = -2$, will make the left side $\neq$ right side. $[(-2)^2]^{1/2} = 4^{1/2}2 \neq -2$.

      B. Similarly any positive (or zero) value of x, for example $x = 2$, will make the left side $=$ right side. $[2^2]^{1/2} = 4^{1/2} = 2 = 2$.

51. $\dfrac{(2x-1)^{1/2} - (x+2)(\frac{1}{2})(2x-1)^{-1/2}(2)}{(2x-1)}$

$$= \dfrac{(2x-1)^{1/2} - (x+2)(2x-1)^{-1/2}}{(2x-1)^1}$$

$$= \dfrac{(2x-1)^{1/2} - \frac{x+2}{(2x-1)^{1/2}}}{(2x-1)^1}$$

$$= \dfrac{(2x-1)^{1/2}}{(2x-1)^{1/2}}\dfrac{(2x-1)^{1/2} - \frac{x+2}{(2x-1)^{1/2}}}{(2x-1)^1}$$

$$= \dfrac{(2x-1)^1 - (x+2)}{(2x-1)^{3/2}}$$

$$= \dfrac{2x-1-x-2}{(2x-1)^{3/2}} = \dfrac{x-3}{(2x-1)^{3/2}}$$

Common Error: Do not "cancel" the $(2x-1)^{1/2}$. It is not a common factor of numerator and denominator.

53. $\dfrac{2(3x-1)^{1/3} - (2x+1)(\frac{1}{3})(3x-1)^{-2/3}(3)}{(3x-1)^{2/3}}$

$$= \dfrac{2(3x-1)^{1/3} - (2x+1)(3x-1)^{-2/3}}{(3x-1)^{2/3}}$$

$$= \dfrac{2(3x-1)^{1/3} - \frac{(2x+1)}{(3x-1)^{2/3}}}{(3x-1)^{2/3}}$$

$$= \dfrac{(3x-1)^{2/3}}{(3x-1)^{2/3}}\dfrac{2(3x-1)^{1/3} - \frac{(2x+1)}{(3x-1)^{2/3}}}{(3x-1)^{2/3}}$$

$$= \dfrac{2(3x-1)^1 - (2x+1)}{(3x-1)^{4/3}}$$

$$= \dfrac{6x-2-2x-1}{(3x-1)^{4/3}}$$

$$= \dfrac{4x-3}{(3x-1)^{4/3}}$$

55. $15^{5/4} = 15^{1.25} = 29.52$

57. $103^{-3/4} = 103^{-0.75} = 0.03093$

59. $2.876^{8/5} = 2.876^{1.6} = 5.421$

61. $(0.000\,000\,077\,35)^{-2/7}$

    $= (7.735 \times 10^{-8})^{(-2 \div 7)} = 107.6$

63. We are to calculate
    $N = 10x^{3/4}y^{1/4}$

    given $x = 256$ units of labor and

    $y = 81$ units of capital

    $$
    \begin{aligned}
    N &= 10(256)^{3/4}(81)^{1/4} \\
    &= 10(256^{1/4})^3(81)^{1/4} \\
    &= 10(4^3)(3) \\
    &= 1,920 \text{ units of finished product.}
    \end{aligned}
    $$

65. We are to calculate
    $d = 0.00212v^{7/3}$
    given $v = 70$ miles/hour

    $$
    \begin{aligned}
    d &= 0.0212(70)^{7/3} \\
    &= 0.0212(70)^{(7 \div 3)} \\
    &= 428 \text{ feet (to the nearest foot)}
    \end{aligned}
    $$

Exercise 2-3

Key Ideas and Formulas

$\sqrt[n]{b}$ = the principal nth root of
b = $b^{1/n}$ (b is called the radicand, n the index, in $\sqrt[n]{b}$)

$b^{m/n} = \begin{cases} \sqrt[n]{b^m} \\ (\sqrt[n]{b})^m \end{cases}$

For n a natural number > 1, and x and y positive real numbers,

$\sqrt[n]{x^n} = x$

$\sqrt[n]{xy} = \sqrt[n]{x}\,\sqrt[n]{y}$

$\sqrt[n]{\frac{x}{y}} = \frac{\sqrt[n]{x}}{\sqrt[n]{y}}$

Simplified Radical form conditions

1. No radicand (the expression within the radical sign) contains a factor to a power greater than or equal to the index of the radical.

2. No power of the radicand and the index of the radical have a common factor other than 1.

3. No radical appears in a denominator.

4. No fraction appears within a radical.

---

1. $m^{2/3} = \sqrt[3]{m^2}$ or $(\sqrt[3]{m})^2$.
The first is usually preferred, and we will use this form.

3. $6x^{3/5} = 6(x)^{3/5} = 6\sqrt[5]{x^3}$

5. $(4xy^3)^{2/5} = \sqrt[5]{(4xy^3)^2}$

7. $(x+y)^{1/2} = \sqrt{x+y}$

9. $\sqrt[5]{b} = b^{1/5}$

11. $5\sqrt[4]{x^3} = 5x^{3/4}$

13. $\sqrt[5]{(2x^2y)^3} = (2x^2y)^{3/5}$

15. $\sqrt[3]{x} + \sqrt[3]{y} = x^{1/3} + y^{1/3}$
(not $(x+y)^{1/3}$) *Common Error*

17. $\sqrt[3]{-8} = -2$

19. $\sqrt{9x^8y^4} = \sqrt{9}\sqrt{x^8}\sqrt{y^4}$
$= 3x^4y^2$

21. $\sqrt[4]{16m^4n^8} = \sqrt[4]{16}\sqrt[4]{m^4}\sqrt[4]{n^8}$
$= 2mn^2$

23. $\sqrt{8a^3b^5} = \sqrt{4a^2b^4 \cdot 2ab}$
$= \sqrt{4a^2b^4}\sqrt{2ab}$
$= 2ab^2\sqrt{2ab}$

25. $\sqrt[3]{2^4x^4y^7} = \sqrt[3]{2^3x^3y^6 \cdot 2xy}$
$= \sqrt[3]{2^3x^3y^6}\sqrt[3]{2xy}$
$= 2xy^2\sqrt[3]{2xy}$

27. $\sqrt[4]{m^2} = \sqrt[2\cdot2]{m^{2\cdot1}} = \sqrt[2]{m^1} = \sqrt{m}$

29. $\sqrt[5]{\sqrt[3]{xy}} = \sqrt[15]{xy}$

31. $\sqrt[3]{9x^2}\sqrt[3]{9x} = \sqrt[3]{(9x^2)(9x)}$
$= \sqrt[3]{81x^3} = \sqrt[3]{27x^3 \cdot 3}$
$= \sqrt[3]{27x^3}\sqrt[3]{3} = 3x\sqrt[3]{3}$

33. $\frac{1}{\sqrt{5}} = \frac{1}{\sqrt{5}}\frac{\sqrt{5}}{\sqrt{5}} = \frac{\sqrt{5}}{5}$

35. $\frac{6x}{\sqrt{3x}} = \frac{6x}{\sqrt{3x}}\frac{\sqrt{3x}}{\sqrt{3x}} = \frac{6x\sqrt{3x}}{3x} = 2\sqrt{3x}$

37. $\frac{2}{\sqrt{2}-1} = \frac{2}{\sqrt{2}-1}\frac{\sqrt{2}+1}{\sqrt{2}+1}$
$= \frac{2(\sqrt{2}+1)}{2-1} = \frac{2(\sqrt{2}+1)}{1}$
$= 2(\sqrt{2}+1) = 2\sqrt{2}+2$

39. $\frac{\sqrt{2}}{\sqrt{6}+2} = \frac{\sqrt{2}}{\sqrt{6}+2}\frac{\sqrt{6}-2}{\sqrt{6}-2} = \frac{\sqrt{2}(\sqrt{6}-2)}{6-4}$
$= \frac{\sqrt{2}\sqrt{6}-2\sqrt{2}}{2} = \frac{\sqrt{12}-2\sqrt{2}}{2}$
$= \frac{2\sqrt{3}-2\sqrt{2}}{2}$
$= \frac{2(\sqrt{3}-\sqrt{2})}{2} = \sqrt{3}-\sqrt{2}$

41. $\quad x \sqrt[5]{3^6 x^7 y^{11}} \;=\; x\sqrt[5]{(3^5 x^5 y^{10})(3x^2 y)} = x\sqrt[5]{3^5 x^5 y^{10}}\;\sqrt[5]{3x^2 y}$

$$= x(3xy^2)\sqrt[5]{3x^2 y} = 3x^2 y^2 \;\sqrt[5]{3x^2 y}$$

43. $\quad \dfrac{\sqrt[4]{32m^7 n^9}}{2mn} = \dfrac{\sqrt[4]{16m^4 n^8 \cdot 2m^3 n}}{2mn} = \dfrac{2mn^2\sqrt[4]{2m^3 n}}{2mn} = n\sqrt[4]{2m^3 n}$

45. $\quad \sqrt[6]{a^4(b-a)^2} = \sqrt[2\cdot3]{a^{2\cdot2}(b-a)^{2\cdot1}} = \sqrt[3]{a^2(b-a)}$

47. $\quad \sqrt[3]{\sqrt[4]{a^9 b^3}} = \sqrt[3\cdot4]{a^{3\cdot3} b^{3\cdot1}} = \sqrt[4]{a^3 b}$

49. $\quad \sqrt[3]{2x^2 y^4}\;\sqrt[3]{3x^5 y} \;=\; \sqrt[3]{(2x^2 y^4)(3x^5 y)} = \sqrt[3]{6x^7 y^5} = \sqrt[3]{(x^6 y^3)(6xy^2)}$

$$= \sqrt[3]{x^6 y^3}\;\sqrt[3]{6xy^2} = x^2 y\;\sqrt[3]{6xy^2}$$

51. $\quad \sqrt[3]{a^3 + b^3}$ is in simplified form. There is no correct way to simplify this expression further.

53. $\quad \dfrac{\sqrt{2m}\sqrt{5}}{\sqrt{20}} = \dfrac{\sqrt{10m}}{\sqrt{20}} = \sqrt{\dfrac{10m}{20m}} = \sqrt{\dfrac{1}{2}} = \sqrt{\dfrac{1}{2}\cdot\dfrac{2}{2}} = \sqrt{\dfrac{2}{4}} = \dfrac{\sqrt{2}}{2}$ or $\dfrac{1}{2}\sqrt{2}$

55. $\quad \dfrac{4a^3 b^2}{\sqrt[3]{2ab^2}} = \dfrac{4a^3 b^2}{\sqrt[3]{2ab^2}}\cdot\dfrac{\sqrt[3]{2^2 a^2 b}}{\sqrt[3]{2^2 a^2 b}} = \dfrac{4a^3 b^2\sqrt[3]{4a^2 b}}{\sqrt[3]{2^3 a^3 b^3}} = \dfrac{4a^3 b^2\sqrt[3]{4a^2 b}}{2ab} = 2a^2 b\sqrt[3]{4a^2 b}$

Notice that we multiplied numerator and denominator by $\sqrt[3]{2^2 a^2 b}$ instead of $\sqrt[3]{(2ab^2)^2}$, which would have also rationalized the denominator. We choose the *smallest* exponents that will make the radicand a perfect cube.

57. $\quad \sqrt[4]{\dfrac{3y^3}{4x^3}} = \sqrt[4]{\dfrac{3y^3}{4x^3}\dfrac{4x}{4x}} = \sqrt[4]{\dfrac{12xy^3}{16x^4}} = \dfrac{\sqrt[4]{12xy^3}}{\sqrt[4]{16x^4}} = \dfrac{\sqrt[4]{12xy^3}}{2x}$ or $\dfrac{1}{2x}\sqrt[4]{12xy^3}$

Notice again that we could have multiplied numerator and denominator by $(4x^3)^3$, but the simpler quantity $4x$ is preferred.

59. $\quad \dfrac{3\sqrt{y}}{2\sqrt{y}-3} = \dfrac{3\sqrt{y}}{(2\sqrt{y}-3)}\dfrac{(2\sqrt{y}+3)}{(2\sqrt{y}+3)} = \dfrac{3\sqrt{y}(2\sqrt{y}+3)}{(2\sqrt{y})^2-(3)^2} = \dfrac{6y+9\sqrt{y}}{4y-9}$

61. $\quad \dfrac{2\sqrt{5}+3\sqrt{2}}{5\sqrt{5}+2\sqrt{2}} \;=\; \dfrac{(2\sqrt{5}+3\sqrt{2})}{(5\sqrt{5}+2\sqrt{2})}\dfrac{(5\sqrt{5}-2\sqrt{2})}{(5\sqrt{5}-2\sqrt{2})}$

$$= \dfrac{2\sqrt{5}\cdot5\sqrt{5}-2\sqrt{5}\cdot2\sqrt{2}+3\sqrt{2}\cdot5\sqrt{5}-3\sqrt{2}\cdot2\sqrt{2}}{(5\sqrt{5})^2-(2\sqrt{2})^2}$$

$$= \dfrac{(10)(5)-4\sqrt{10}+15\sqrt{10}-(6)(2)}{(25)(5)-(4)(2)} = \dfrac{50+11\sqrt{10}-12}{125-8} = \dfrac{38+11\sqrt{10}}{117}$$

63.
$$\frac{x^2}{\sqrt{x^2+9}-3} = \frac{x^2}{(\sqrt{x^2+9}-3)} \cdot \frac{(\sqrt{x^2+9}+3)}{(\sqrt{x^2+9}+3)} = \frac{x^2(\sqrt{x^2+9}+3)}{(\sqrt{x^2+9})^2-(3)^2} = \frac{x^2(\sqrt{x^2+9}+3)}{x^2+9-9}$$

$$= \frac{\overset{1}{\cancel{x^2}}(\sqrt{x^2+9}+3)}{\underset{1}{\cancel{x^2}}} = \sqrt{x^2+9}+3$$

65. Here we want to remove the radical terms from the *numerator*.

$$\frac{\sqrt{t}-\sqrt{x}}{t-x} = \frac{(\sqrt{t}-\sqrt{x})}{t-x} \cdot \frac{(\sqrt{t}+\sqrt{x})}{\sqrt{t}+\sqrt{x}} = \frac{(\sqrt{t})^2-(\sqrt{x})^2}{(t-x)(\sqrt{t}+\sqrt{x})}$$

$$= \frac{\overset{1}{\cancel{t-x}}}{\underset{1}{\cancel{(t-x)}}(\sqrt{t}+\sqrt{x})} = \frac{1}{\sqrt{t}+\sqrt{x}}$$

67.
$$\frac{\sqrt{x+h}-\sqrt{x}}{h} = \frac{(\sqrt{x+h}-\sqrt{x})}{h} \cdot \frac{(\sqrt{x+h}+\sqrt{x})}{(\sqrt{x+h}+\sqrt{x})} = \frac{(\sqrt{x+h})^2-(\sqrt{x})^2}{h(\sqrt{x+h}+\sqrt{x})} = \frac{x+h-x}{h(\sqrt{x+h}+\sqrt{x})}$$

$$= \frac{\overset{1}{\cancel{h}}}{\underset{1}{\cancel{h}}(\sqrt{x+h}+\sqrt{x})} = \frac{1}{\sqrt{x+h}+\sqrt{x}}$$

69. $\sqrt{x^2}$ is the principal square root of x^2, that is, the positive (or zero) square root of x^2. The two square roots of x^2 are x and $-x$. $\sqrt{x^2}=-x$ if and only if $-x$ is positive or zero. Hence $\sqrt{x^2}=-x$ if and only if x is negative or zero. $x \leq 0$.

71. $\sqrt[3]{x^3}$ is the real third root of x^3. This is always equal to x if x is real. All real numbers.

73.
$$\frac{1}{\sqrt[3]{a}-\sqrt[3]{b}} = \frac{1}{(\sqrt[3]{a}-\sqrt[3]{b})} \frac{[(\sqrt[3]{a})^2+\sqrt[3]{a}\sqrt[3]{b}+(\sqrt[3]{b})^2]}{[(\sqrt[3]{a})^2+\sqrt[3]{a}\sqrt[3]{b}+(\sqrt[3]{b})^2]}$$

using $(x-y)(x^2+xy+y^2)=x^3-y^3$

$$= \frac{1[\sqrt[3]{a^2}+\sqrt[3]{ab}+\sqrt[3]{b^2}]}{(\sqrt[3]{a})^3-(\sqrt[3]{b})^3} = \frac{\sqrt[3]{a^2}+\sqrt[3]{ab}+\sqrt[3]{b^2}}{a-b}$$

75.
$$\frac{1}{\sqrt{x}-\sqrt{y}+\sqrt{z}} = \frac{1}{[(\sqrt{x}-\sqrt{y})+\sqrt{z}]} \frac{[(\sqrt{x}-\sqrt{y})-\sqrt{z}]}{[(\sqrt{x}-\sqrt{y})-\sqrt{z}]} = \frac{\sqrt{x}-\sqrt{y}-\sqrt{z}}{(\sqrt{x}-\sqrt{y})^2-(\sqrt{z})^2}$$

$$= \frac{\sqrt{x}-\sqrt{y}-\sqrt{z}}{(\sqrt{x})^2-2\sqrt{x}\sqrt{y}+(\sqrt{y})^2-(\sqrt{z})^2} = \frac{\sqrt{x}-\sqrt{y}-\sqrt{z}}{x-2\sqrt{xy}+y-z}$$

$$= \frac{\sqrt{x}-\sqrt{y}-\sqrt{z}}{x-2\sqrt{xy}+y-z} = \frac{\sqrt{x}-\sqrt{y}-\sqrt{z}}{x+y-z-2\sqrt{xy}}$$

$$= \frac{(\sqrt{x}-\sqrt{y}-\sqrt{z})}{[(x+y-z)-2\sqrt{xy}]} \frac{[(x+y-z)+2\sqrt{xy}]}{[(x+y-z)+2\sqrt{xy}]}$$

$$= \frac{(\sqrt{x}-\sqrt{y}-\sqrt{z})[(x+y-z)+2\sqrt{xy}]}{(x+y-z)^2-(2\sqrt{xy})^2}$$

$$= \frac{(\sqrt{x}-\sqrt{y}-\sqrt{z})[(x+y-z)+2\sqrt{xy}]}{(x+y-z)^2-4xy}$$

77. $\dfrac{\sqrt[3]{x+h}-\sqrt[3]{x}}{h} = \dfrac{(\sqrt[3]{x+h}-\sqrt[3]{x})}{h}\dfrac{[(\sqrt[3]{x+h})^2+\sqrt[3]{x+h}\sqrt[3]{x}+(\sqrt[3]{x})^2]}{[(\sqrt[3]{x+h})^2+\sqrt[3]{x+h}\sqrt[3]{x}+(\sqrt[3]{x})^2]}$

$$= \dfrac{(\sqrt[3]{x+h})^3-(\sqrt[3]{x})^3}{h[\sqrt[3]{(x+h)^2}+\sqrt[3]{x(x+h)}+\sqrt[3]{x^2}]}$$

using $(a-b)(a^2+ab+b^2)=a^3-b^3$

$$= \dfrac{x+h-x}{h[\sqrt[3]{(x+h)^2}+\sqrt[3]{x(x+h)}+\sqrt[3]{x^2}]}$$

$$= \dfrac{h}{h[\sqrt[3]{(x+h)^2}+\sqrt[3]{x(x+h)}+\sqrt[3]{x^2}]}$$

$$= \dfrac{1}{\sqrt[3]{(x+h)^2}+\sqrt[3]{x(x+h)}+\sqrt[3]{x^2}}$$

79. $\sqrt[kn]{x^{km}} = (x^{km})^{\frac{1}{kn}} = x^{\frac{km}{kn}} = x^{\frac{m}{n}} = \sqrt[n]{x^m}$

81. $\sqrt{0.049\ 375} = 0.2222$

83. $\sqrt[5]{27.0635} = (27.0635)^{1/5} = (27.0635)^{0.2} = 1.934$

85. $\sqrt[7]{0.000\ 000\ 008\ 066} = (8.066\times10^{-9})^{1/7} = (8.066\times10^{-9})^{(1\div7)}$

$$= 6.979\times10^{-2} = 0.06979$$

87. $\sqrt[3]{7+\sqrt[3]{7}} = (7+7^{(1\div3)})^{(1\div3)} = 2.073$

89. $\sqrt[3]{\sqrt[4]{2}} = (2^{1/4})^{1/3} = (2^{0.25})^{(1\div3)} = 1.059$

$\sqrt[12]{2} = 2^{1/12} = 2^{(1\div12)} = 1.059$

91. $\dfrac{1}{\sqrt[3]{4}} = 4^{-1/3} = 4^{(-1\div3)} = 0.6300$

$\dfrac{\sqrt[3]{2}}{2} = \dfrac{2^{1/3}}{2} = 2^{(1\div3)}\div2 = 0.6300$

93. $\dfrac{M_0}{\sqrt{1-\frac{v^2}{c^2}}} = M_0\div\sqrt{1-\dfrac{v^2}{c^2}}$

$$= M_0\div\sqrt{\dfrac{c^2-v^2}{c^2}} = M_0\div\dfrac{\sqrt{c^2-v^2}}{c}$$

$$= M_0\cdot\dfrac{c}{\sqrt{c^2-v^2}}$$

Now rationalize the denominator

$$= M_0\cdot\dfrac{c}{\sqrt{c^2-v^2}}\dfrac{\sqrt{c^2-v^2}}{\sqrt{c^2-v^2}}$$

$$= \dfrac{M_0c\sqrt{c^2-v^2}}{c^2-v^2}$$

Exercise 2-4

CHAPTER REVIEW

1. $6(xy^3)^5 = 6x^5(y^3)^5 = 6x^5y^{15}$ $(2-1)$

2. $\frac{9u^8v^6}{3u^4v^8} = \frac{3u^4}{v^2}$ $(2-1)$

3. $(2 \times 10^5)(3 \times 10^{-3}) = (2 \cdot 3) \times 10^{5+(-3)} = 6 \times 10^2$ $(2-1)$

4. $(x^{-3}y^2)^{-2} = (x^{-3})^{-2}(y^2)^{-2} = x^6y^{-4} = \frac{x^6}{y^4}$ $(2-1)$

5. $u^{5/3}u^{2/3} = u^{5/3+2/3} = u^{7/3}$ $(2-2)$

6. $(9a^4b^{-2})^{1/2} = 9^{1/2}(a^4)^{1/2}(b^{-2})^{1/2}$
   $\qquad = 3a^2b^{-1} = \frac{3a^2}{b}$ $(2-2)$

7. $3\sqrt[5]{x^2}$ $(2-3)$

8. $-3(xy)^{2/3}$ $(2-3)$

9. $3x \sqrt[3]{x^5y^4} = 3x\sqrt[3]{x^3y^3 \cdot x^2y} = 3x \sqrt[3]{x^3y^3} \sqrt[3]{x^2y} = 3x(xy)\sqrt[3]{x^2y} = 3x^2y \sqrt[3]{x^2y}$ $(2-3)$

10. $\sqrt{2x^2y^5}\sqrt{18x^3y^2} = \sqrt{(2x^2y^5)(18x^3y^2)} = \sqrt{36x^5y^7} = \sqrt{36x^4y^6 \cdot xy} = \sqrt{36x^4y^6}\sqrt{xy}$
    $\qquad\qquad = 6x^2y^3\sqrt{xy}$ $(2-3)$

11. $\frac{6ab}{\sqrt{3a}} = \frac{6ab}{\sqrt{3a}}\frac{\sqrt{3a}}{\sqrt{3a}} = \frac{\overset{2}{6ab}\sqrt{3a}}{3a} = 2b\sqrt{3a}$ $(2-3)$

12. $\frac{\sqrt{5}}{3-\sqrt{5}} = \frac{\sqrt{5}}{(3-\sqrt{5})}\frac{(3+\sqrt{5})}{(3+\sqrt{5})} = \frac{\sqrt{5}(3+\sqrt{5})}{3^2-(\sqrt{5})^2} = \frac{\sqrt{5}(3+\sqrt{5})}{9-5} = \frac{\sqrt{5}(3+\sqrt{5})}{4} = \frac{3\sqrt{5}+5}{4}$ $(2-3)$

13. $\sqrt[8]{y^6} = \sqrt[2\cdot4]{y^{2\cdot3}} = \sqrt[4]{y^3}$ $(2-3)$

14. $\left(\frac{8u^{-1}}{2^2u^2v^0}\right)^{-2}\left(\frac{u^{-5}}{u^{-3}}\right)^3 = \left(\frac{2^3u^{-1}}{2^2u^2}\right)^{-2}\left(\frac{u^{-5}}{u^{-3}}\right)^3 = (2u^{-3})^{-2}\left(\frac{u^{-5}}{u^{-3}}\right)^3 = 2^{-2}u^6\frac{u^{-15}}{u^{-9}}$
    $\qquad\qquad = \frac{1}{2^2}u^6u^{-15-(-9)} = \frac{1}{2^2}u^6u^{-6} = \frac{1}{4}$ $(2-1)$

15. $\frac{5^0}{3^2} + \frac{3^{-2}}{2^{-2}} = \frac{1}{3^2} + \frac{2^2}{3^2} = \frac{1}{9} + \frac{4}{9} = \frac{5}{9}$ $(2-1)$

16. $\left(\frac{27x^2y^{-3}}{8x^{-4}y^3}\right)^{1/3} = \left(\frac{3^3x^{2-(-4)}y^{-3-3}}{2^3}\right)^{1/3} = \left(\frac{3^3x^6y^{-6}}{2^3}\right)^{1/3} = \frac{3x^2y^{-2}}{2} = \frac{3x^2}{2y^2}$ $(2-2)$

17. $(a^{-1/3}b^{1/4})(9a^{1/3}b^{-1/2})^{3/2} = a^{-1/3}b^{1/4}9^{3/2}a^{1/2}b^{-3/4} = a^{-1/3+1/2}b^{1/4-3/4}9^{3/2} = 27a^{1/6}b^{-1/2}$
    $\qquad\qquad = \frac{27a^{1/6}}{b^{1/2}}$ $(2-2)$

18. $(x^{1/2} + y^{1/2})^2 = (x^{1/2})^2 + 2(x^{1/2})(y^{1/2}) + y^{1/2})^2 = x + 2x^{1/2}y^{1/2} + y$
    Common Error:
    $(x^{1/2} + y^{1/2})^2 \neq x + y$. (Exponents do not distribute over addition.) $(2-2)$

19. $(3x^{1/2} - y^{1/2})(2x^{1/2} + 3y^{1/2}) = (3x^{1/2})(2x^{1/2}) + (3x^{1/2})(3y^{1/2}) - (2x^{1/2})(y^{1/2}) - (y^{1/2})(3y^{1/2})$
    $\qquad\qquad = 6x + 9x^{1/2}y^{1/2} - 2x^{1/2}y^{1/2} - 3y$
    $\qquad\qquad = 6x + 7x^{1/2}y^{1/2} - 3y$ $(2-2)$

20. $\dfrac{0.000\,000\,000\,52}{(1,300)(0.000\,002)} = \dfrac{5.2\times10^{-10}}{(1.3\times10^3)(2\times10^{-6})} = \dfrac{\overset{2}{\cancel{5.2}}\times10^{-10}}{\underset{1}{\cancel{2.6}}\times10^{-3}} = 2\times10^{-10-(-3)}$

$\qquad\qquad = 2\times10^{-7} \quad (2-1)$

21. $-2x\sqrt[5]{3^6x^7y^{11}} = -2x\sqrt[5]{3^5x^5y^{10}\cdot 3x^2y} = -2x\sqrt[5]{3^5x^5y^{10}}\sqrt[5]{3x^2y} = -2x\cdot 3xy^2\ \sqrt[5]{3x^2y}$

$\qquad\qquad = -6x^2y^2\ \sqrt[5]{3x^2y} \quad (2-3)$

22. $\dfrac{2x^2}{\sqrt[3]{4x}} = \dfrac{2x^2}{\sqrt[3]{4x}}\dfrac{\sqrt[3]{2x^2}}{\sqrt[3]{2x^2}} = \dfrac{2x^2\sqrt[3]{2x^2}}{\sqrt[3]{8x^3}} = \dfrac{\overset{x}{\cancel{2x^2}}\sqrt[3]{2x^2}}{\underset{1}{\cancel{2x}}} = x\sqrt[3]{2x^2} \quad (2-3)$

23. $\sqrt[5]{\dfrac{3y^2}{8x^2}} = \sqrt[5]{\dfrac{3y^2}{8x^2}\dfrac{4x^3}{4x^3}} = \sqrt[5]{\dfrac{12x^3y^2}{32x^5}} = \dfrac{\sqrt[5]{12x^3y^2}}{\sqrt[5]{32x^5}} = \dfrac{\sqrt[5]{12x^2y^2}}{2x} \quad (2-3)$

24. $\sqrt[9]{8x^6y^{12}} = \sqrt[9]{2^3x^6y^{12}} = \sqrt[3\cdot3]{2^{3\cdot1}x^{3\cdot2}y^{3\cdot4}} = \sqrt[3]{2x^2y^4} = \sqrt[3]{y^3\cdot 2x^2y} = \sqrt[3]{y^3}\sqrt[3]{2x^2y} = y\ \sqrt[3]{2x^2y} \quad (2-3)$

25. $\sqrt{\sqrt[3]{4x^4}} = \sqrt[2\cdot3]{2^{2\cdot1}x^{2\cdot2}} = \sqrt[3]{2x^2} \quad (2-3)$

26. $(2\sqrt{x} - 5\sqrt{y})(\sqrt{x} + \sqrt{y}) = 2\sqrt{x}\sqrt{x} + 2\sqrt{x}\sqrt{y} - 5\sqrt{x}\sqrt{y} - 5\sqrt{y}\sqrt{y} = 2x - 3\sqrt{x}\sqrt{y} - 5y$

$\qquad\qquad = 2x - \sqrt{3}xy - 5y \quad (2-3)$

27. $\dfrac{3\sqrt{x}}{2\sqrt{x}-\sqrt{y}} = \dfrac{3\sqrt{x}}{(2\sqrt{x}-\sqrt{y})}\dfrac{(2\sqrt{x}+\sqrt{y})}{(2\sqrt{x}+\sqrt{y})} = \dfrac{3\sqrt{x}(2\sqrt{x}+\sqrt{y})}{(2\sqrt{x})^2-(\sqrt{y})^2} = \dfrac{3\sqrt{x}(2\sqrt{x}+\sqrt{y})}{4x-y} = \dfrac{6x+3\sqrt{xy}}{4x-y} \quad (2-3)$

28. $\dfrac{2\sqrt{u}-3\sqrt{v}}{2\sqrt{u}+3\sqrt{v}} = \dfrac{(2\sqrt{u}-3\sqrt{v})}{(2\sqrt{u}+3\sqrt{v})}\dfrac{(2\sqrt{u}-3\sqrt{v})}{(2\sqrt{u}-3\sqrt{v})}$

$\qquad = \dfrac{(2\sqrt{u})^2-2(2\sqrt{u})(3\sqrt{v})+(3\sqrt{v})^2}{(2\sqrt{u})^2-(3\sqrt{v})^2}$

$\qquad = \dfrac{4u-12\sqrt{uv}+9v}{4u-9v} \quad (2-3)$

29. $\dfrac{y^2}{\sqrt{y^2+4}-2} = \dfrac{y^2}{(\sqrt{y^2+4}-2)}\dfrac{(\sqrt{y^2+4}+2)}{(\sqrt{y^2+4}+2)} = \dfrac{y^2(\sqrt{y^2+4}+2)}{(\sqrt{y^2+4})^2-(2)^2} = \dfrac{y^2(\sqrt{y^2+4}+2)}{y^2+4-4}$

$\qquad = \dfrac{\overset{1}{\cancel{y^2}}(\sqrt{y^2+4}-2)}{\underset{1}{\cancel{y^2}}} = \sqrt{y^2+4}-2 \quad (2-3)$

30. $\dfrac{\sqrt{t}-\sqrt{5}}{t-5} = \dfrac{(\sqrt{t}-\sqrt{5})}{(t-5)}\dfrac{(\sqrt{t}+\sqrt{5})}{(\sqrt{t}+\sqrt{5})}$

$\qquad = \dfrac{(\sqrt{t})^2-(\sqrt{5})^2}{(t-5)(\sqrt{t}+\sqrt{5})}$

$\qquad = \dfrac{\overset{1}{\cancel{t-5}}}{\underset{1}{\cancel{(t-5)}}(\sqrt{t}+\sqrt{5})}$

$\qquad = \dfrac{1}{\sqrt{t}+\sqrt{5}} \quad (2-3)$

31. $\dfrac{4\sqrt{x}-3}{2\sqrt{x}} = \dfrac{4\sqrt{x}}{2\sqrt{x}} - \dfrac{3}{2\sqrt{x}} = 2 - \dfrac{3}{2x^{1/2}} = 2 - \dfrac{3}{2}x^{-1/2} \quad (2-3)$

32. $\dfrac{8(x-2)^{-3}(x+3)^2}{12(x-2)^{-4}(x+3)^{-2}} = \dfrac{2(x-2)^{-3-(-4)}(x+3)^{2-(-2)}}{3} = \dfrac{2(x-2)^1(x+3)^4}{3}$

$\qquad = \dfrac{2(x-2)(x+3)^4}{3} = \dfrac{2}{3}(x-2)(x+3)^4 \quad (2-1)$

33. $\left(\dfrac{a^{-2}}{b^{-1}} + \dfrac{b^{-2}}{a^{-1}}\right)^{-1} = \left(\dfrac{b}{a^2} + \dfrac{a}{b^2}\right)^{-1} = \left(\dfrac{b^3}{a^2b^2} + \dfrac{a^3}{a^2b^2}\right)^{-1} = \left(\dfrac{b^3+a^3}{a^2b^2}\right)^{-1} = \dfrac{a^2b^2}{b^3+a^3}$ or $\dfrac{a^2b^2}{a^3+b^3} \quad (2-1)$

34. $(x^{1/3}-y^{1/3})(x^{2/3}+x^{1/3}y^{1/3}+y^{2/3}) = (x^{1/3})^3 - (y^{1/3})^3 = x-y \quad (2-2)$

35. $\left(\frac{x^{m^2}}{x^{2m-1}}\right)^{\frac{1}{m-1}} = \left(x^{m^2-(2m-1)}\right)^{\frac{1}{m-1}} = \left(x^{m^2-2m+1}\right)^{\frac{1}{m-1}}$

$$= x^{\frac{m^2-2m+1}{m-1}} = x^{\frac{(m-1)^2}{m-1}} = x^{\frac{\cancel{(m-1)}^2}{\cancel{m-1}}_1}$$

$$= x^{m-1} \quad (2-3)$$

36. $\dfrac{1}{1-\sqrt[3]{x}} = \dfrac{1}{(1-\sqrt[3]{x})}\dfrac{[1+\sqrt[3]{x}+(\sqrt[3]{x})^2]}{[1+\sqrt[3]{x}+(\sqrt[3]{x})^2]}$

Common Error:
$(1-\sqrt[3]{x})(1+\sqrt[3]{x}) \neq 1-x$
This would not be a correct
application of $(a-b)(a+b) = a^2 - b^2$

$= \dfrac{1+\sqrt[3]{x}+\sqrt[3]{x^2}}{(1)^3-(\sqrt[3]{x})^3}$

$= \dfrac{1+\sqrt[3]{x}+\sqrt[3]{x^2}}{1-x} \quad (2-3)$

37. $\dfrac{\sqrt[3]{t}-\sqrt[3]{5}}{t-5} = \dfrac{(\sqrt[3]{t}-\sqrt[3]{5})}{(t-5)}\dfrac{[(\sqrt[3]{t})^2+\sqrt[3]{t}\sqrt[3]{5}+(\sqrt[3]{5})^2]}{[(\sqrt[3]{t})^2+\sqrt[3]{t}\sqrt[3]{5}+(\sqrt[3]{5})^2]}$

$= \dfrac{(\sqrt[3]{t})^3-(\sqrt[3]{5})^3}{(t-5)[\sqrt[3]{t^2}+\sqrt[3]{5t}+\sqrt[3]{t^2}]}$

$= \dfrac{\cancel{t-5}_{\;1}}{\cancel{(t-5)}[\sqrt[3]{t^2}+\sqrt[3]{5t}+\sqrt[3]{25}]}$

$= \dfrac{1}{\sqrt[3]{t^2}+\sqrt[3]{5t}+\sqrt[3]{25}}$

38. $\sqrt[n+1]{x^{n^2}x^{2n+1}} = \sqrt[n+1]{x^{n^2+2n+1}} = \left(x^{n^2+2n+1}\right)^{\frac{1}{n+1}} = x^{\frac{n^2+2n+1}{n+1}} = x^{\frac{(n+1)^2}{n+1}}$

$$= x^{\frac{\cancel{(n+1)}^2}{\cancel{n+1}}_1} = x^{n+1} \quad (2-3)$$

39. $\dfrac{(20,410)(0.000\,003\,477)}{0.000\,000\,022\,09} = \dfrac{(2.041\times10^4)(3.477\times10^{-6})}{(2.209\times10^{-8})} = 3.213\times10^6$

Note that the input is given to 4 significant digits, hence we
interpret 4 significant digits of the calculator answer. $(2-2)$

40. $4.434\times10^{-5} \quad (2-2)$
41. $-4.541\times10^{-6} \quad (2-2)$
42. $82.45^{8/3} = (82.45)^{(8\div3)} = 128,800 \quad (2-2)$
43. $(0.000\,000\,419\,9)^{2/7} = (4.199\times10^{-7})^{(2\div7)} = 0.01507 \quad (2-3)$
44. $\sqrt[5]{0.006604} = (6.604\times10^{-3})^{1/5}$

$= (6.604\times10^{-3})^{0.2}$

$= 0.3664 \quad (2-2)$

45. $\sqrt[3]{3+\sqrt{2}} = (3+\sqrt{2})^{(1\div3)} = 1.640 \quad (2-2, 2-3)$
46. $\dfrac{2^{-1/2}-3^{-1/2}}{2^{-1/3}+3^{-1/3}} = [2^{(-0.5)}-3^{(-0.5)}] \div [2^{(-1\div3)}+3^{(-1\div3)}]$

$= 0.08726 \quad (2-2)$

CHAPTER 3

Exercise 3-1

Key Ideas and Formulas

Any equation that can be written in the form $ax + b = 0$ $\quad a \neq 0$ is called a linear equation.

To solve a linear equation, use properties of equality:

If $a = b$, then $a + c = b + c$
If $a = b$, then $a - c = b - c$
If $a = b$, then $ca = cb$ $\quad c \neq 0$
If $a = b$, then $\frac{a}{c} = \frac{b}{c}$ $\quad c \neq 0$
If $a = b$, then $b = a$

If $a = b$, then either may replace the other in any statement without changing the truth or falsity of the statement.

To solve equations involving variables in the denominator, we must exclude any value of the variable that will make a denominator 0. With these values excluded, we may multiply through by the LCD.

1.
$$3(x + 2) = 5(x - 6)$$
$$3x + 6 = 5x - 30$$
$$-2x = -36$$
$$x = 18$$
Solution: 18

3.
$$5 + 4(t - 2) = 2(t + 7) + 1$$
$$5 + 4t - 8 = 2t + 14 + 1$$
$$4t - 3 = 2t + 15$$
$$2t = 18$$
$$t = 9$$
Solution: 9

5.
$$x(x + 2) = x(x + 4) - 12$$
$$x^2 + 2x = x^2 + 4x - 12$$
$$2x = 4x - 12$$
$$-2x = -12$$
$$x = 6$$
Solution: 6

7.
$$3 - \frac{2x - 3}{3} = \frac{5 - x}{2} \quad lcd = 6$$
$$18 - 6\frac{(2x - 3)}{3} = 6\frac{(5 - x)}{2}$$
$$18 - 2(2x - 3) = 3(5 - x)$$
$$18 - 4x + 6 = 15 - 3x$$
$$-4x + 24 = 15 - 3x$$
$$-x = -9$$
$$x = 9$$
Solution: 9

Common Error
After line 2, students often write
$$18 - \overset{2}{\cancel{6}}\frac{2x - 3}{\cancel{3}} = \cdots$$
$$18 - 4x - 3 = \cdots$$
forgetting to distribute the -2. Put compound numerators in parentheses to avoid this.

9.
$$5 - \frac{2x-1}{4} = \frac{x+2}{3} \quad lcd = 12$$
$$60 - 3(2x - 1) = 4(x + 2)$$
$$60 - 6x + 3 = 4x + 8$$
$$-6x + 63 = 4x + 8$$
$$-10x = -55$$
$$x = \frac{-55}{-10}$$

Solution: $\frac{11}{2}$ or 5.5

11.
$$0.1(x - 7) + 0.05x = 0.8$$
$$0.1x - 0.7 + 0.05x = 0.8$$
$$0.15x - 0.7 = 0.8$$
$$0.15x = 1.5$$
$$x = \frac{1.5}{0.15}$$
$$x = 10$$

Solution: 10

Problems of this type can also be solved by elimination of decimals. Thus we could multiply every term on both sides by 100 to get $10(x - 7) + 5x = 80$ and proceed from here.

13.
$$0.3x - 0.04(x + 1) = 2.04$$
$$0.3x - 0.04x - 0.04 = 2.04$$
$$0.26x - 0.04 = 2.04$$
$$0.26x = 2.08$$
$$x = \frac{2.08}{0.26}$$
$$x = 8$$

Solution: 8

15.
$$\frac{3x}{2} - \frac{2-x}{10} = \frac{5+x}{4} + \frac{4}{5}$$
$$lcd = 20$$
$$20\frac{3x}{2} - 20 \cdot \frac{2-x}{10} = 20 \cdot \frac{5+x}{4} + 20 \cdot \frac{4}{5}$$
$$30x - 2(2 - x) = 5(5 + x) + 16$$
$$30x - 4 + 2x = 25 + 5x + 16$$
$$32x - 4 = 5x + 41$$
$$27x = 45$$
$$x = \frac{45}{27} \text{ or } \frac{5}{3}$$

Solution: $\frac{5}{3}$

17.
$$\frac{1}{m} - \frac{1}{9} = \frac{4}{9} - \frac{2}{3m}$$
Excluded value: $m \neq 0 \quad lcd = 9m$
$$9m \cdot \frac{1}{m} - 9m \cdot \frac{1}{9} = 9m \cdot \frac{4}{9} - 9m \cdot \frac{2}{3m}$$
$$9 - m = 4m - 6$$
$$-5m = -15$$
$$m = 3$$

Solution: 3

19.
$$\frac{5x}{x+5} = 2 - \frac{25}{x+5}$$
Excluded value: $x \neq -5 \quad lcd = x + 5$
$$(x+5)\frac{5x}{x+5} = (x+5)2 - (x+5)\frac{25}{x+5}$$
$$5x = 2x + 10 - 25$$
$$5x = 2x - 15$$
$$3x = -15$$
$$x = -5$$

No Solution : -5 is excluded
Common Error: Do not accept the proposed solution (-5) without checking the list of excluded values.

21.
$$\frac{2x}{10} - \frac{3-x}{14} = \frac{2+x}{5} - \frac{1}{2}$$
$$lcd = 70$$
$$70 \cdot \frac{2x}{10} - 70 \cdot \frac{(3-x)}{14} = 70 \cdot \frac{(2+x)}{5} - 70 \cdot \frac{1}{2}$$
$$14x - 5(3 - x) = 14(2 + x) - 35$$
$$14x - 15 + 5x = 28 + 14x - 35$$
$$19x - 15 = 14x - 7$$
$$5x = 8$$
$$5x = 8$$
$$x = \frac{8}{5}$$

Solution: $\frac{8}{5}$

23.
$$\frac{1}{3} - \frac{s-2}{2s+4} = \frac{s+2}{3s+6} \quad \text{Excluded value} : s \neq -2$$

$$\frac{1}{3} - \frac{s-2}{2(s+2)} = \frac{s+2}{3(s+2)} \quad lcd = 6(s+2)$$

$$6(s+2)\cdot\frac{1}{3} - 6(s+2)\frac{(s-2)}{2(s+2)} = 6(s+2)\frac{(s+2)}{3(s+2)}$$

$$2(s+2) - 3(s-2) = 2(s+2)$$

$$2s+4-3s+6 = 2s+4$$

$$-s+10 = 2s+4$$

$$-3s = -6$$

$$s = 2$$

Solution: 2

25.
$$\frac{3x}{2-x} + \frac{6}{x-2} = 3 \quad \text{Excluded value} : x \neq 2 \ lcd = x-2$$

$$(x-2)\frac{3x}{2-x} + (x-2)\frac{6}{x-2} = (x-2)3$$

$$-3x+6 = (x-2)3$$

$$-3x+6 = 3x-6$$

$$-6x = -12$$

$$x = 2$$

No solution: 2 is excluded

27.
$$\frac{5t-22}{t^2-6t+9} - \frac{11}{t^2-3t} - \frac{5}{t} = 0$$

$$\frac{5t-22}{(t-3)^2} - \frac{11}{t(t-3)} - \frac{5}{t} = 0 \quad \text{Excluded values} : t \neq 0,3$$

$$lcd = t(t-3)^2$$

$$t(t-3)^2\frac{(5t-22)}{(t-3)^2} - t(t-3)^2\frac{11}{t(t-3)} - t(t-3)^2\frac{5}{t} = 0$$

$$t(5t-22) - (t-3)11 - (t-3)^2 5 = 0$$

$$5t^2 - 22t - 11t + 33 - 5(t^2-6t+9) = 0$$

$$5t^2 - 33t + 33 - 5t^2 + 30t - 45 = 0$$

$$-3t - 12 = 0$$

$$-3t = 12$$

$$t = -4$$

Solution: -4

29.
$$\frac{1}{x^2 - x - 2} - \frac{3}{x^2 - 2x - 3} = \frac{1}{x^2 - 5x + 6}$$

$$\frac{1}{(x-2)(x+1)} - \frac{3}{(x-3)(x+1)} = \frac{1}{(x-2)(x-3)}$$

Excluded values : $x \neq -1, 2, 3$

$$lcd = (x+1)(x-2)(x-3).$$

$$(x+1)(x-2)(x-3)\frac{1}{(x-2)(x+1)} - (x+1)(x-2)(x-3)\frac{3}{(x-3)(x+1)}$$

$$= (x+1)(x-2)(x-3)\frac{1}{(x-2)(x-3)}$$

$$(x-3)1 - (x-2)3 = (x+1)1$$

$$x - 3 - 3x + 6 = x + 1$$

$$-2x + 3 = x + 1$$

$$-3x = -2$$

$$x = \frac{2}{3}$$

Solution: $\frac{2}{3}$

31.
$$a_n = a_1 + (n-1)d$$
$$a_1 + (n-1)d = a_n$$
$$(n-1)d = a_n - a_1$$
$$d = \frac{a_n - a_1}{n-1}$$

33.
$$\frac{1}{f} = \frac{1}{d_1} + \frac{1}{d_2} \quad lcd = d_1 d_2 f$$

$$d_1 d_2 f \frac{1}{f} = d_1 d_2 f \frac{1}{d_1} + d_1 d_2 f \frac{1}{d_2}$$

$$d_1 d_2 = d_2 f + d_1 f$$

$$d_2 f + d_1 f = d_1 d_2$$
$$(d_2 + d_1)f = d_1 d_2$$

$$f = \frac{d_1 d_2}{d_2 + d_1}$$

35.
$$A = 2ab + 2ac + 2bc$$
$$2ab + 2ac + 2bc = A$$
$$2ab + 2ac = A - 2bc$$
$$a(2b + 2c) = A - 2bc$$
$$a = \frac{A - 2bc}{2b + 2c}$$

37.
$$y = \frac{2x - 3}{3x + 5}$$
$$(3x + 5)y = 2x - 3$$
$$3xy + 5y = 2x - 3$$
$$5y + 3 = 2x - 3xy$$
$$5y + 3 = x(2 - 3y)$$
$$\frac{5y + 3}{2 - 3y} = x$$
$$x = \frac{5y + 3}{2 - 3y}$$

39.
$$\frac{x - \frac{1}{x}}{1 + \frac{1}{x}} = 3 \quad \text{Excluded value} : x \neq 0$$

$$\frac{x(x - \frac{1}{x})}{x(1 + \frac{1}{x})} = 3$$

$$\frac{x^2 - 1}{x + 1} = 3 \quad \text{Excluded value} : x \neq -1$$

$$\frac{(x-1)(x+1)}{x+1} = 3$$

$$x - 1 = 3$$

$$x = 4$$

Solution: 4

41.
$$y = \frac{a}{1 + \frac{b}{x+c}}$$

$$y = \frac{a(x+c)}{(x+c)(1 + \frac{b}{x+c})}$$

$$y = \frac{a(x+c)}{x+c+b}$$

$$y = \frac{ax+ac}{x+c+b}$$

$$y(x+c+b) = ax+ac$$

$$xy + cy + by = ax + ac$$

$$cy + by - ac = ax - xy$$

$$cy + by - ac = x(a-y)$$

$$\frac{cy + by - ac}{a-y} = x$$

$$x = \frac{cy + by - ac}{a-y}$$

43. Let a, b, and c be any three real numbers. We are to prove that if $a = b$, then $a - c = b - c$.

$$a - c = a - c \quad \text{reflexive property}$$

$$a = b \quad \text{given}$$

$$a - c = b - c \quad \text{substitution principle - in the first line}$$

substitute b for a on the right.

45.
$$3.142x - 0.4835(x - 4) = 6.795$$

$$3.142x - 0.4835x + 1.934 = 6.795$$

$$2.6585x + 1.934 = 6.795$$

$$2.6585x = 4.861$$

$$x = \frac{4.861}{2.6585}$$

$$x = 1.83 \text{ to 3 significant digits}$$

Solution : 1.83

47.
$$\frac{2.32x}{x-2} - \frac{3.76}{x} = 2.32 \quad \text{Excluded values :} x \neq 0, 2 \quad lcd \; x(x-2)$$

$$x(x-2)\frac{2.32x}{x-2} - x(x-2)\frac{3.76}{x} = 2.32x(x-2)$$

$$2.32x^2 - 3.76(x-2) = 2.32x(x-2)$$

$$2.32x^2 - 3.76x + 7.52 = 2.32x^2 - 4.64x$$

$$-3.76x + 7.52 = -4.64x$$

$$7.52 = -0.88x$$

$$x = -8.55$$

Solution: -8.55

Exercise 3-2

1. Let $x =$ the number,

    then 10 less than two thirds the number is one fourth the number

    $$\frac{2}{3}x - 10 \qquad\qquad = \qquad\qquad\qquad\qquad \frac{1}{4}x$$

    $$\frac{2}{3}x - 10 \qquad = \frac{1}{4}x$$

    $$12(\tfrac{2}{3}x) - 12(10) = 12(\tfrac{1}{4}x)$$

    $$8x - 120 \quad = \quad 3x$$

    $$-120 \quad = \quad -5x$$

    $$x \quad = \quad 24 \qquad \text{The number is 24.}$$

3. $$\text{Let } x \ = \ \text{ first of the consecutive even numbers}$$

    $$x + 2 \ = \ \text{ second of the numbers}$$

    $$x + 4 \ = \ \text{ third of the numbers}$$

    $$x + 6 \ = \ \text{ fourth of the numbers}$$

    $$\text{first} + \text{second} + \text{third} \ = \ \text{2 more than twice fourth}$$

    $$x + x + 2 + x + 4 \ = \ 2 + 2(x + 6)$$

    $$3x + 6 \ = \ 2 + 2x + 12$$

    $$3x + 6 \ = \ 2x + 14$$

    $$x \ = \ 8$$

    The four consecutive even numbers are $8, 10, 12, 14$.

5. Let $w \ = \ $ width of rectangle

    $2w - 3 \ = \ $ length of rectangle

    We use the perimeter formula

    $$P \ = \ 2a + 2b.$$

    $$54 \ = \ 2w + 2(2w - 3)$$

    $$54 \ = \ 2w + 4w - 6 \qquad 10 = w$$

    $$54 \ = \ 6w - 6 \qquad\qquad 17 = 2w - 3$$

    $$60 \ = \ 6w \qquad\qquad\qquad \text{dimensions} : \text{17 meters by 10 meters}$$

7. Let $P \ = \ $ perimeter of triangle

    $$16 \ = \ \text{length of one side}$$

    $$\frac{2}{7}P \ = \ \text{length of second side}$$

    $$\frac{1}{3}P \ = \ \text{length of third side}$$

    We use the perimeter formula

    $$P \ = \ a + b + c$$

    $$P \ = \ 16 + \frac{2}{7}P + \frac{1}{3}P$$

    $$21P \ = \ 21(16) + 21(\tfrac{2}{7}P) + 21(\tfrac{1}{3}P)$$

    $$21P \ = \ 336 + 6P + 7P$$

    $$21P \ = \ 336 + 13P$$

    $$8P \ = \ 336$$

    $$P \ = \ 42 \text{ feet}$$

9.

$$\text{Let } P = \text{price before discount}$$
$$0.20P = 20 \text{ percent discount on } P$$

Then price before discount − discount = price after discount

$$P - 0.20P = 72$$
$$0.8P = 72$$
$$P = \frac{72}{0.8}$$
$$P = \$90$$

11.   "Break even" means Cost = Revenue

$$\text{Let } x = \text{number of records sold}$$
$$\text{Revenue} = \text{number of records sold} \times \text{price per record}$$
$$= x(8.00)$$
$$\text{Cost} = \text{Fixed Cost} + \text{Variable Cost}$$
$$= 17,680 + \text{number of records} \times \text{cost per record}$$
$$= 17,680 + x(4.60)$$
$$8.00x = 17,680 + 4.60x$$
$$3.40x = 17,680$$
$$x = \frac{17,680}{3.40}$$
$$= 5,200 \text{ records}$$

13.   Let $x =$ amount invested at 10% $0.1x =$ yield on   amount invested at 10%
Then $12,000 - x =$ amount invested at 15% $0.15(12,000 - x) =$ yield on amount invested at 15%

Yield on Amount Invested at 10%	+	Yield on Amount Invested at 15%	=	Total Yield Yield on 12,000 Invested at 12%

$$0.1x + 0.15(12,000 - x) = 0.12(12,000)$$
$$0.1x + 0.15(12,000) - 0.15x = 0.12(12,000)$$
$$0.1x + 1,800 - 0.15x = 1,440$$
$$-0.05x + 1,800 = 1,440$$
$$0.05x = -360$$
$$x = \$7,200 \text{ invested at } 10\%$$
$$12,000 - x = \$4,000 \text{ invested at } 15\%$$

4800

15.   A. We note: The temperature increased 2.5°C for each additional 100 meters of depth. Hence, the temperature increased 25 degrees for each additional kilometer of depth.
Let $x =$ the depth, then $x - 3 =$ the depth beyond 3 kilometers.
$25(x - 3) =$ the temperature increase for $x - 3$ kilometers of depth.
$T =$ temperature at 3 kilometers + temperature increase.
$T = 30 + 25(x - 3)$

B. We are to find T when $x = 15$. We use the above relationship as a formula.
$T = 30 + 25(15 - 3)$
$\ = 330°C$

C. We are to find x when $T = 280$. We use the above relationship as an equation.
$280 = 30 + 25(x - 3)$

$$280 = 30 + 25x - 75$$
$$280 = -45 + 25x$$
$$325 = 25x$$
$$x = 13 \text{ kilometers}$$

17.   Let D = distance from earthquake to station

Then $\dfrac{D}{5}$ = time of primary wave.

$\dfrac{D}{3}$ = time of secondary wave.

Time difference = time of *slower* secondary wave − time of *faster* primary wave

$$12 = \frac{D}{3} - \frac{D}{5}$$
$$15(12) = 15\left(\frac{D}{3}\right) - 15\left(\frac{D}{5}\right)$$
$$180 = 5D - 3D$$
$$180 = 2D$$
$$D = 90 \text{ miles}$$

Common Error:

Time difference is *not* time of fast wave (short) - time of slow wave (long)

19.   We set up the proportion

$$\frac{\text{marked trout in second sample}}{\text{total number in second sample}} = \frac{\text{marked trout in first sample}}{\text{total trout population}}$$

Let x = total population

$$\frac{8}{200} = \frac{200}{x}$$
$$200x\frac{8}{200} = 200x\frac{200}{x}$$
$$8x = 40,000$$
$$x = 5,000 \text{ trout}$$

21.   Let x = amount of distilled water

50 = amount of 30% solution

Then $50 + x$ = amount of 25% solution

acid in 30% solution + acid in distilled water = acid in 25% solution

$$0.3(50) + 0 = 0.25(50 + x)$$
$$0.3(50) = 0.25(50 + x)$$
$$15 = 12.5 + 0.25x$$
$$2.5 = 0.25x$$
$$x = 10 \text{ gallons}$$

23.
$$\text{Let } x = \text{amount of 20\% solution}$$
$$\text{Then } 100 - x = \text{amount of 80\% solution}$$

acid in 20% solution	+	acid in 80% solution
	=	acid in 62% solution

$$0.2x + 0.8(100 - x) = 0.62(100)$$
$$0.2x + 80 - 0.8x = 62$$
$$-0.6x = -18$$
$$x = \frac{-18}{-0.6}$$
$$x = 30 \text{ liters of 20\% solution}$$
$$100 - x = 70 \text{ liters of 80\% solution}$$

25.
$$\text{Let } t = \text{time for both computers to finish the job.}$$
$$\text{Then } t + 1 = \text{time worked by old computer.}$$
$$t = \text{time worked by new computer}$$

Since the old computer can do 1 job in 5 hours, it works at a rate
$(1 \text{ job}) \div (5 \text{ hours}) = \frac{1}{5}$ job per hour
Similarly the new computer works at a rate $\frac{1}{3}$ job per hour.

Part of job completed by old computer in $t + 1$ hours	+	Part of job completed by new computer in t hours	=	1 whole job.
(Rate of old) (time of old)	+	(Rate of new) (Time of new)	=	1
$\frac{1}{5}(t + 1)$	+	$\frac{1}{3}(t)$	=	1
$15(\frac{1}{5})(t + 1)$	+	$15(\frac{1}{3}t)$	=	15
$3(t + 1)$	+	$5t$	=	15
$3t + 3$	+	$5t$	=	15
		$8t + 3$	=	15
		$8t$	=	12
		t	=	1.5 hours

27. Let d = distance flown north

A. Using $t = \dfrac{d}{r}$, we note:

rate flying north	=	$150 - 30$	=	120 miles per hour
rate flying south	=	$150 + 30$	=	180 miles per hour

$$\text{time flying north } + \text{ time flying south} = 3 \text{ hours}$$
$$\frac{d}{120} + \frac{d}{180} = 3$$
$$360\frac{d}{120} + 360\frac{d}{180} = 3(360)$$
$$3d + 2d = 1080$$
$$5d = 1080$$
$$d = 216 \text{ miles}$$

B. We still use the above ideas, except that rate flying north = rate flying south = 150 miles per hour.

$$\frac{d}{150} + \frac{d}{150} = 3$$

$$\frac{2d}{150} = 3$$

$$\frac{d}{75} = 3$$

$$d = 225 \text{ miles}$$

29. Let x = frequency of second note
    y = frequency of third note

$$\frac{264}{4} = \frac{x}{5} = \frac{y}{6}$$

$$\frac{264}{4} = \frac{x}{5} \qquad \frac{264}{4} = \frac{y}{6}$$

$$66 = \frac{x}{5} \qquad 66 = \frac{y}{6}$$

$$x = 330 \text{ hertz} \quad y = 396 \text{ hertz}$$

31. We are to find d when $p = 40$.

$$40 = -\frac{1}{5}d + 70$$

$$5(40) = 5(-\frac{1}{5}d) + 5(70)$$

$$200 = -d + 350$$

$$-150 = -d$$

$$d = 150 \text{ centimeters}$$

33.   total height = height in sand + height in water + height in air
        Let h = total height

$$h = \frac{1}{5}h + 20 + \frac{2}{3}h$$

$$15h = 3h + 300 + 10h$$

$$15h = 13h + 300$$

$$2h = 300$$

$$h = 150 \text{ feet}$$

35. The hands will meet when the minute hand has made exactly 1 more revolution than the hour hand.

Let $\qquad\qquad\qquad t$ = time after 12 o'clock in hours

$$\text{rate of hour hand} = \frac{1}{12}\text{revolution per hour}$$

$$\text{rate of minute hand} = 1 \text{ revolution per hour}$$

$$\text{distance of minute hand} = 1 + \text{distance of hour hand}$$

$$1(t) = 1 + \frac{1}{12}(t)$$

$$t = 1 + \frac{1}{12}t$$

$$12t = 12 + t$$

$$11t = 12$$

$$t = \frac{12}{11} \text{ hours}$$

That clock will read $12 + \frac{12}{11}$ o'clock, that is $1\frac{1}{11}$ hours after 12, or 1 hour $\frac{60}{11}$ minutes after 12. This will be $\frac{60}{11}$ minutes, or $5\frac{5}{11}$ minutes after 1 PM.

Exercise 3-3

Key Ideas and Formulas

a is less than b or b is greater than a if there exists a positive
$(a < b)$ $\qquad$ $(b > a)$
real number p such that $a + p = b$.
For any two real numbers a and b, $a < b, a > b$, or $a = b$.
For a, b, and c any real numbers with $a < b$, then
$a + c < b + c$
$a - c < b - c$

If c is positive, then $ac < bc$ and $\frac{a}{c} < \frac{b}{c}$
If c is negative, then $ac > bc$ and $\frac{a}{c} > \frac{b}{c}$
Similar properties hold if $a > b$, then $a + c > b + c$ and so on. The order of an inequality reverses
if we multiply or divide both sides of an inequality statement by a negative number.

---

1. $\quad -8 \le x \le 7$

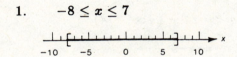

9. $\quad (-7, 8)$

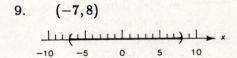

3. $\quad -6 \le x < 6$

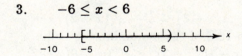

11. $\quad (-\infty, -2]$

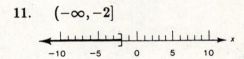

5. $\quad x \ge -6$

13. $\quad [-7, 2); -7 \le x < 2$

15. $\quad (-\infty, 0]; x \le 0$

7. $\quad (-2, 6]$

17. $\quad 7x - 8 \;<\; 4x + 7$
$\qquad\qquad 3x \;<\; 15$
$\qquad\qquad\; x \;<\; 5 \text{ or } (-\infty, 5)$

19.
$$3 - x \geq 5(3 - x)$$
$$3 - x \geq 15 - 5x$$
$$4x \geq 12$$
$$x \geq 3 \text{ or } [3, \infty)$$

21.
$$\frac{N}{-2} > 4$$
$$N < -8 \text{ or } (-\infty, -8)$$

23.
$$-5t < -10$$
$$t > 2 \text{ or } (2, \infty)$$

25.
$$3 - m < 4(m - 3)$$
$$3 - m < 4m - 12$$
$$-5m < -15$$
$$m > 3 \text{ or } (3, \infty)$$

Common Error:
Neglecting to reverse the order after division by -5.

27.
$$-2 - \frac{B}{4} \leq \frac{1+B}{3}$$
$$12\left(-2 - \frac{B}{4}\right) \leq 12\frac{(1+B)}{3}$$
$$-24 - 3B \leq 4(1 + B)$$
$$-24 - 3B \leq 4 + 4B$$
$$-7B \leq 28$$
$$B \geq -4 \text{ or } [-4, \infty)$$

29.
$$-4 < 5t + 6 \leq 21$$
$$-10 < 5t \leq 15$$
$$-2 < t \leq 3 \text{ or } (-2, 3]$$

31. The set includes all integers starting from -4 and proceeding through 2.
$$\{-4, -3, -2, -1, 0, 1, 2\}$$

33. This set includes the integers starting to the right of -4, that is, with -3, and proceeding to 1, just before 2.
$$\{-3, -2, -1, 0, 1\}$$

35.
$$\frac{q}{7} - 3 > \frac{q-4}{3} + 1$$
$$21\left(\frac{q}{7} - 3\right) > 21\left(\frac{(q-4)}{3} + 1\right)$$
$$3q - 63 < 7(q - 4) + 21$$
$$3q - 63 > 7q - 28 + 21$$
$$3q - 63 > 7q - 7$$
$$-4q > 56$$
$$q < -14 \text{ or } (-\infty, -14)$$

37.
$$\frac{2x}{5} - \frac{1}{2}(x - 3) \leq \frac{2x}{3} - \frac{3}{10}(x + 2)$$
The 1cm of the denominators is 30.
$$12x - 15(x - 3) \leq 20x - 9(x + 2)$$
$$12x - 15x + 45 \leq 20x - 9x - 18$$
$$-3x + 45 \leq 11x - 18$$
$$-14x \leq -63$$
$$x \geq 4.5 \text{ or } [4.5, \infty)$$

39.
$$-4 \leq \tfrac{9}{5}x + 32 \leq 68$$
$$-36 \leq \tfrac{9}{5}x \leq 36$$
$$\tfrac{5}{9}(-36) \leq x \leq \tfrac{5}{9}(36)$$
$$-20 \leq x \leq 20 \text{ or } [-20, 20]$$

41.
$$-12 < \tfrac{3}{4}(2 - x) \leq 24$$
$$\tfrac{4}{3}(-12) < 2 - x \leq \tfrac{4}{3}(24)$$
$$-16 < 2 - x \leq 32$$
$$-18 < -x \leq 30$$
$$18 > x \geq -30$$
$$-30 \leq x < 18 \text{ or } [-30, 18)$$

43.
$$16 < 7 - 3x \leq 31$$
$$9 < -3x \leq 24$$
$$-3 > x \geq -8$$
$$-8 \leq x < -3 \text{ or } [-8, -3)$$

45.
$$-6 < -\tfrac{2}{5}(1 - x) \leq 4$$
$$-30 < -2(1 - x) \leq 20$$
$$-30 < -2 + 2x \leq 20$$
$$-28 < 2x \leq 22$$
$$-14 < x \leq 11 \text{ or } (-14, 11]$$

47. $\sqrt{1 - x}$ represents a real number exactly when $1 - x$ is positive or zero. We can write this as an inequality statement and solve for x.

$$1 - x \geq 0$$
$$-x \geq -1$$
$$x \leq 1$$

49. $\sqrt{3x + 5}$ represents a real number exactly when $3x + 5$ is positive or zero. We can write this as an inequality statement and solve for x.

$$3x + 5 \geq 0$$
$$3x \geq -5$$
$$x \geq -\tfrac{5}{3}$$

51. $\dfrac{1}{\sqrt[4]{2x+3}}$ represents a real number exactly when $2x + 3$ is positive. (*not* zero). We can write this as an inequality statement and solve for x.

$$2x + 3 > 0$$
$$2x > -3$$
$$x > -\tfrac{3}{2}$$

53. A. For $ab > 0$, ab must be positive, hence a and b must have the same sign. Either
1. $a > 0$ and $b > 0$
   or
2. $a < 0$ and $b < 0$

B. For $ab < 0$, ab must be negative, hence a and b must have opposite signs. Either
1. $a > 0$ and $b > 0$
   or
2. $a < 0$ and $b > 0$

C. For $\frac{a}{b} > 0$, $\frac{a}{b}$
must be positive, hence
a and b must have the same
sign. Answer as in A.

D. For $\frac{a}{b} < 0$, $\frac{a}{b}$
must be negative, hence
a and b must have
opposite signs. Answer as in B.

55.   A. If $a - b = 1$, then $a = b + 1$.
Therefore, a is greater
than b. >

B. If $u - v = -2$, then $v = u + 2$.
Therefore, u is less than
v. <

57.   If $\frac{b}{a}$ is greater than 1
$\frac{b}{a} > 1$
$a \cdot \frac{b}{a} < a \cdot 1$.
(since a is negative.)
$b < a$
$0 < a - b$
$a - b$ is positive

59.   A. F   B. T   C. T

61.   If $a < b$, then by definition
of $<$, there exists a
positive number p
such that $a + p = b$.
If we divide both sides of
$a + p = b$ by a negative number
c, we obtain
$\frac{a}{c} + \frac{p}{c} = \frac{b}{c}$,
where $\frac{p}{c}$ is negative, or
$\frac{a}{c} = \frac{b}{c} + -\frac{p}{c}$
where $-\frac{p}{c}$ is positive. Hence, by
definition of $<$, we have
$\frac{a}{c} > \frac{b}{c}$

63.
$$
\begin{aligned}
5.23(x - 0.172) &\leq 6.02x - 0.427 \\
5.23x - 0.89956 &\leq 6.02x - 0.427 \\
5.23x - 6.02x - 0.89956 &\leq -0.427 \\
-0.79x - 0.89956 &\leq -0.427 \\
-0.79x &\leq -0.427 + 0.89956 \\
-0.79x &\leq 0.47256 \\
x &\geq -0.60
\end{aligned}
$$

65.
$$
\begin{aligned}
-0.703 &< 0.112 - 2.28x < 0.703 \\
-0.703 - 0.112 &< -2.28x < 0.703 - 0.112 \\
-0.815 &< -2.28x < 0.591 \\
0.357 &> x > -0.259 \\
-0.259 &< x < 0.357
\end{aligned}
$$

67.   We want $200 \leq T \leq 300$.
We are given $T = 30 + 25(x - 3)$.
Substituting, we must solve
$$
\begin{aligned}
200 &\leq 30 + 25(x - 3) \leq 300 \\
200 &\leq 30 + 25x - 75 \leq 300 \\
200 &\leq -45 + 25x \leq 300 \\
245 &\leq 25x \leq 345 \\
9.8 &\leq x \leq 13.8
\end{aligned}
$$
(from 9.8 km to 13.8 km)

69.   We want $R > C$. We are given $R = 2x$
and $C = 300 + 1.5x$. Substituting,
we must solve
$$
\begin{aligned}
2x &> 300 + 1.5x \\
0.5x &> 300 \\
x &> \frac{300}{0.5} \\
x &> 600
\end{aligned}
$$

71.   We want $220 \leq W \leq 2,750$.
We are given $W = 110I$.
Substituting, we must solve
$220 \leq 110I \leq 2,750$
$2 \leq I \leq 25$ or $[2,25]$.

Exercise 3-4

Key Ideas and Formulas

$$|x| = \begin{cases} x \text{ if } x \text{ is positive} \\ 0 \text{ if } x \text{ is } 0 \\ -x \text{ (a positive quantity) if } x \text{ is negative} \end{cases}$$

The distance between two points A and B on a number line with coordinates a and b respectively is given by

$$d(A, B) = |b - a| = |a - b|$$

For $p > 0$

$	x	= p$	is equivalent to	$x = p$ or $x = -p$
$	x	< p$	is equivalent to	$-p < x < p$
$	x	> p$	is equivalent to	$x < -p$ or $x > p$

Common Error: Writing:

$$p < x < -p$$

This is true for no real value of x, since $p < -p$ is false.

To solve:

$	ax + b	= p$,	solve	$ax + b = \pm p$
$	ax + b	< p$,	solve	$-p < ax + b < p$
$	ax + b	> p$,	solve	$ax + b < -p$ or $ax + b > p$

---

1. $\sqrt{5}$

3. $|(-6) - (-2)| = |-4| = 4$

5. $|5 - \sqrt{5}| = 5 - \sqrt{5}$
    since $5 - \sqrt{5}$ is positive.

7. $|\sqrt{5} - 5| = -(\sqrt{5} - 5) = 5 - \sqrt{5}$ since $\sqrt{5} - 5$ is negative.

9. $\begin{aligned} d(A, B) &= |b - a| \\ &= |5 - (-7)| \\ &= |12| \\ &= 12 \end{aligned}$

11. $\begin{aligned} d(A, B) &= |b - a| \\ &= |-7 - 5| \\ &= |-12| \\ &= 12 \end{aligned}$

13. $\begin{aligned} d(A, B) &= |b - a| \\ &= |-25 - (-16)| \\ &= |-9| \\ &= 9 \end{aligned}$

15. $\begin{aligned} d(B, 0) &= |0 - (-4)| \\ &= |4| \\ &= 4 \end{aligned}$

17. $\begin{aligned} d(0, B) &= |-4 - 0| \\ &= |-4| \\ &= 4 \end{aligned}$

19. $\begin{aligned} d(B, C) &= |5 - (-4)| \\ &= |9| \\ &= 9 \end{aligned}$

21. $x = \pm 7$

23. $-7 \le x \le 7$ $[-7, 7]$

25. $x \le -7$ or $x \ge 7$

     $(-\infty, -7] \cup [7, \infty)$

     Common Error: $7 \le x \le -7$

     This is meaningless.

27. $\begin{aligned} |y - 5| &= 3 \\ y - 5 &= \pm 3 \\ y &= 5 \pm 3 \\ y &= 2, 8 \end{aligned}$

29. $\begin{aligned} |y - 5| &< 3 \\ -3 < y - 5 &< 3 \\ 2 < y &< 8 \end{aligned}$ $(2, 8)$

31. $\begin{aligned} |y - 5| &> 3 \\ y - 5 < -3 \text{ or } y - 5 &> 3 \\ y < 2 \text{ or } y &> 8 \end{aligned}$

     $(-\infty, 2) \cup (8, \infty)$

33. $\begin{aligned} |u + 8| &= 3 \\ u + 8 &= \pm 3 \\ u &= -8 \pm 3 \\ u &= -11 \text{ or } -5 \end{aligned}$

35. $\begin{aligned} |u + 8| &\le 3 \\ -3 \le u + 8 &\le 3 \\ -11 \le u &\le -5 \end{aligned}$

     $[-11, -5]$

37. $\begin{aligned} |u + 8| &\ge 3 \\ u + 8 \le -3 \text{ or } u + 8 &\ge 3 \\ u \le -11 \text{ or } u &\ge -5 \end{aligned}$

     $(-\infty, 11] \cup [-5, \infty)$

39. $\begin{aligned} |3x + 4| &= 8 \\ 3x + 4 &= \pm 8 \\ 3x &= -4 \pm 8 \\ x &= \frac{-4 \pm 8}{3} \\ x &= -4, \frac{4}{3} \end{aligned}$

41. $\begin{aligned} |5x - 3| &\le 12 \\ -12 \le 5x - 3 &\le 12 \\ -9 \le 5x &\le 15 \\ -\frac{9}{5} \le x &\le 3 \end{aligned}$

     $[-\frac{9}{5}, 3]$

43. $|2y - 8| > 2$

$2y - 8 < -2 \text{ or } 2y - 8 > 2$

$2y < 6 \text{ or } 2y > 10$

$y < 3 \text{ or } y > 5$

$(-\infty, 3) \cup (5, \infty)$

45. $|5t - 7| = 11$

$5t - 7 = \pm 11$

$5t = 7 \pm 11$

$t = \frac{7 \pm 11}{5}$

$t = -\frac{4}{5}, \frac{18}{5}$

47. $|9 - 7u| < 14$

$-14 < 9 - 7u < 14$

$-23 < -7u < 5$

$\frac{23}{7} > u > -\frac{5}{7}$

$-\frac{5}{7} < u < \frac{23}{7}$

$\left(-\frac{5}{7}, \frac{23}{7}\right)$

49. $\left|1 - \frac{2}{3}x\right| \geq 5$

$1 - \frac{2}{3}x \leq -5 \text{ or}$

$1 - \frac{2}{3}x \geq 5$

$-\frac{2}{3}x \leq -6 \text{ or}$

$-\frac{2}{3}x \geq 4$

$x \geq 9 \text{ or } x \leq -6$

$(-\infty, -6] \cup [9, \infty)$

51. $\left|\frac{9}{5}C + 32\right| < 31$

$-31 < \frac{9}{5}C + 32 < 31$

$-63 < \frac{9}{5}C < -1$

$-35 < C < -\frac{5}{9}$

$\left(-35, -\frac{5}{9}\right)$

53. $\sqrt{x^2} < 2$

$|x| < 2$

$-2 < x < 2 \qquad (-2, 2)$

55. $\sqrt{t^2} \geq 4$

$|t| \geq 4$

$t \leq -4 \text{ or } t \geq 4$

$(-\infty, -4] \cup [4, \infty)$

57. $\sqrt{(1 - 3t)^2} \leq 2$

$|1 - 3t| \leq 2$

$-2 \leq 1 - 3t \leq 2$

$-3 \leq -3t \leq 1$

$1 \geq t \geq -\frac{1}{3}$

$-\frac{1}{3} \leq t \leq 1$

$\left[-\frac{1}{3}, 1\right]$

59. $\sqrt{(2t - 3)^2} > 3$

$|2t - 3| > 3$

$2t - 3 < -3 \text{ or } 2t - 3 > 3$

$2t < 0 \qquad\qquad 2t > 6$

$t < 0 \qquad\qquad t > 3$

$(-\infty, 0) \cup (3, \infty)$

61. $|m - 2| = 4$

The distance between m and 2 is equal to 4. m is 4 units from 2.

63. $|x + 1| = 7$

$|x - (-1)| = 7$

The distance between x and -1 is equal to 7.

x is 7 units from -1.

65. $|u - 4| < 5$

The distance between u and 4 is less than 5.

u is less than 5 units from 4

67. $|z - 2| \leq 3$

The distance between z and 2 is no more than 3.

z is no more than 3 units from 2.

69. The distance between x and 3 is equal to 4.

$|x - 3| = 4$

71. The distance between m and -2 is equal to 5.

$|m - (-2)| = 5$

$|m + 2| = 5$

73. The distance between x and 3 is less than 5.

$|x - 3| < 5$

75. The distance between p and -2 is more than 6.

$|p - (-2)| > 6$

$|p + 2| > 6$

77. The distance between q and 1 is no less than 2.
$$|q-1| \geq 2$$

79. $0 < |x-3| < 0.1$
The distance between x and 3 is less than 0.1 but $x \neq 3$.

$$\begin{aligned} -0.1 &< x-3 < 0.1 \text{ except } x \neq 3 \\ 2.9 &< x < 3.1 \text{ but } x \neq 3 \\ 2.9 &< x < 3 \text{ or } 3 < x < 3.1 \end{aligned}$$
$$(2.9, 3) \cup (3, 3.1)$$

81. $0 < |x-c| < d$
The distance between x and c is less than d but $x \neq c$.

$$\begin{aligned} -d &< x-c < d \text{ except } x \neq c \\ c-d &< x < c+d \text{ but } x \neq c \\ c-d &< x < c \text{ or } c < x < c+d \end{aligned}$$
$$(c-d, c) \cup (c, c+d)$$

83. $|x| = x$ if x is positive or zero. So $|x-5| = x-5$ if $x-5$ is positive or zero. We must solve $x-5 \geq 0$. The solution is $x \geq 5$.

85. $|x| = -x$ if x is negative or zero. So $|x+8| = -(x+8)$ if $x+8$ is negative or zero. We must solve $x+8 \leq 0$. The solution is $x \leq -8$.

87. As in 83, we must solve $4x+3 \geq 0$.
$$\begin{aligned} 4x+3 &\geq 0 \\ 4x &\geq -3 \\ x &\geq -\tfrac{3}{4} \end{aligned}$$

89. As in 85, we must solve $5x-2 \leq 0$.
$$\begin{aligned} 5x-2 &\leq 0 \\ 5x &\leq 2 \\ x &\leq \tfrac{2}{5} \end{aligned}$$

91. We consider four cases for $|3-x| + 3 = |2-3x|$

Case 1: $3-x > 0$ and $2-3x > 0$, that is $x < 3$ and $x < \frac{2}{3}$ (or simply $x < \frac{2}{3}$)
$$|3-x| = 3-x \quad |2-3x| = 2-3x$$

Hence:
$$\begin{aligned} 3-x+3 &= 2-3x \\ -x+6 &= 2-3x \\ 2x &= -4 \\ x &= -2 \end{aligned}$$
which is a possible value for x in this case.

Case 2: $3-x > 0$ and $2-3x < 0$, that is $x < 3$ and $x > \frac{2}{3}$ (or simply $\frac{2}{3} < x < 3$)
$$|3-x| = 3-x \quad |2-3x| = -(2-3x) = 3x-2$$

Hence:
$$\begin{aligned} 3-x+3 &= 3x-2 \\ -x+6 &= 3x-2 \\ -4x &= -8 \\ x &= 2 \end{aligned}$$
which is a possible value for x in this case.

Case: $3-x < 0$ and $2-3x < 0$, that is $x > 3$ and $x > \frac{2}{3}$ (or simply $x > 3$)
$$|3-x| = -(3-x) = x-3 \quad |2-3x| = -(2-3x) = 3x-2$$

Hence:
$$\begin{aligned} x - 3 + 3 &= 3x - 2 \\ x &= 3x - 2 \\ -2x &= -2 \\ x &= 1 \end{aligned}$$

which is not a possible value for x in this case. ($x > 3$).

Case 4: $3 - x < 0$ and $2 - 3x > 0$, that is $x > 3$ and $x < \frac{2}{3}$. These are mutually contradictory, so no solution is possible in this case.

Solution: -2,2

93. There are three possible relations between real numbers a and b; either $a = b, a > b$, or $a < b$. We examine each case separately.

Case 1: $a = b$
$\mid b - a \mid = \mid 0 \mid = 0; \mid a - b \mid = \mid 0 \mid = 0$

Case 2: $a > b$
$\mid b - a \mid = -(b - a) = a - b$
$\mid a - b \mid = a - b$

Case 3: $b > a$
$\mid b - a \mid = b - a$
$\mid a - b \mid = -(a - b) = b - a$
Thus in all three cases $\mid b - a \mid = \mid a - b \mid$.

95. If $m < n$, then $m + m < m + n$ (adding m to both sides)
Also, $m + n < n + n$ (adding n to both sides) Hence,
$$\begin{aligned} m + \ m \ &< m + n < n + n \\ 2m \ &< \ m + n < 2n \\ m \ &< \ \tfrac{m+n}{2} < n \end{aligned}$$

97. *Case 1.* $m > 0$. Then $\mid m \mid = m \quad \mid -m \mid = -(-m) = m$. Hence, $\mid m \mid = \mid -m \mid$
Case 2. $m < 0$. Then $\mid m \mid = -m \quad \mid -m \mid = -m$. Hence $\mid m \mid = \mid -m \mid$
Case 3. $m = 0$. Then $0 = m = -m$, hence $\mid m \mid = \mid -m \mid = 0$.

99. If $n \neq 0, n > 0$ or $n < 0$.
Case 1. $n > 0$ If $m \geq 0 \quad \mid m \mid = m, \frac{m}{n} \geq 0 \quad \mid \frac{m}{n} \mid = \frac{m}{n} \quad \mid n \mid = n$.
Hence: $\mid \frac{m}{n} \mid = \frac{m}{n} = \frac{\mid m \mid}{\mid n \mid}$

If $m < 0 \quad \mid m \mid = -m, \frac{m}{n} < 0 \quad \mid \frac{m}{n} \mid = -\frac{m}{n} \quad \mid n \mid = n$.
Hence: $\mid \frac{m}{n} \mid = -\frac{m}{n} = \frac{-m}{n} = \frac{\mid m \mid}{\mid n \mid}$

Case 2. $n < 0$. If $m > 0 \quad \mid m \mid = m, \frac{m}{n} < 0 \quad \mid \frac{m}{n} \mid = -\frac{m}{n} \quad \mid n \mid = -n$
Hence: $\mid \frac{m}{n} \mid = -\frac{m}{n} = \frac{m}{-n} = \frac{\mid m \mid}{\mid n \mid}$

If $m \leq 0 \quad \mid m \mid = -m \quad \frac{m}{n} > 0 \quad \mid \frac{m}{n} \mid = \frac{m}{n} \quad \mid n \mid = -n$
Hence: $\mid \frac{m}{n} \mid = \frac{m}{n} = \frac{-m}{-n} = \frac{\mid m \mid}{\mid n \mid}$

101. The difference between P and 500 has an absolute value of no more than 20. $\mid P - 500 \mid \leq 20$

103. The difference between N and 2.37 has an absolute value of no more than 0.005
$\mid N - 2.37 \mid \leq 0.005$.

Exercise 3-5

Key Ideas and Formulas

For a, b real numbers.

$$i = \text{the imaginary unit} = \sqrt{-1}$$
$$a + bi = \text{complex number (or imaginary number)}$$
$$0 + bi = \text{pure imaginary number}$$
$$a = a + 0i = \text{real number}$$
$$0 + 0i = \text{zero}$$
$$a - bi = \text{conjugate of } a + bi$$
$$a + bi = c + di \text{ if and only if } a = c \text{ and } b = d$$
$$(a + bi) + (c + di) = (a + c) + (b + d)i$$
$$(a + bi)(c + di) = (ac - bd) + (ad + bc)i$$
$$\frac{a+bi}{c} = \frac{a}{c} + \frac{b}{c}i$$
$$\frac{a+bi}{c+di} = \frac{(a+bi)}{(c-di)}\frac{(c-di)}{(c-di)} \text{which is simplified to } \frac{ac+bd}{c^2+d^2} + \frac{bc-ad}{c^2+d^2}i$$

The principal square root of a negative real number, denoted by $\sqrt{-a}$ where a is positive, is defined by $\sqrt{-a} = i\sqrt{a}$

---

1. $(2 + 4i) + (5 + i) = 2 + 4i + 5 + i = 7 + 5i$

3. $(-2 + 6i) + (7 - 3i) = -2 + 6i + 7 - 3i = 5 + 3i$

5. $(6 + 7i) - (4 + 3i) = 6 + 7i - 4 - 3i = 2 + 4i$

7. $(3 + 5i) - (-2 - 4i) = 3 + 5i + 2 + 4i = 5 + 9i$

9. $(4 - 5i) + 2i = 4 - 5i + 2i = 4 - 3i$

11. $(4i)(6i) = 24i^2 = 24(-1) = -24 \text{ or } -24 + 0i$

13. $-3i(2 - 4i) = -6i + 12i^2 = -6i + 12(-1) = -12 - 6i$

15. $(3 + 3i)(2 - 3i) = 6 - 9i + 6i - 9i^2 = 6 - 3i - 9(-1) = 6 - 3i + 9 = 15 - 3i$

17. $(2 - 3i)(7 - 6i) = 14 - 12i - 21i + 18i^2 = 14 - 33i - 18 = -4 - 33i$

19. $(7 + 4i)(7 - 4i) = 49 - 16i^2 = 49 + 16 = 65$

21. $\dfrac{1}{2 + i} = \dfrac{1}{2 + i}\dfrac{2 - i}{2 - i} = \dfrac{2 - i}{4 - i^2} = \dfrac{2 - i}{4 + 1} = \dfrac{2 - i}{5} = \dfrac{2}{5} - \dfrac{1}{5}i$

23. $\dfrac{3 + i}{2 - 3i} = \dfrac{3 + i}{2 - 3i}\dfrac{2 + 3i}{2 + 3i} = \dfrac{6 + 11i + 3i^2}{4 - 9i^2} = \dfrac{6 + 11i - 3}{4 + 9} = \dfrac{3 + 11i}{13} = \dfrac{3}{13} + \dfrac{11}{13}i$

25. $\dfrac{13 + i}{2 - i} = \dfrac{13 + i}{2 - i}\dfrac{2 + i}{2 + i} = \dfrac{26 + 15i + i^2}{4 - i^2} = \dfrac{26 + 15i - 1}{4 + 1} = \dfrac{25 + 15i}{5} = 5 + 3i$

27. $(2 - \sqrt{-4}) + (5 - \sqrt{-9}) = (2 - i\sqrt{4}) + (5 - i\sqrt{9}) = 2 - 2i + 5 - 3i = 7 - 5i$

29. $(9 - \sqrt{-9}) - (12 - \sqrt{-25}) = (9 - i\sqrt{9}) - (12 - i\sqrt{25})$

                     $= (9 - 3i) - (12 - 5i) = 9 - 3i - 12 + 5i = -3 + 2i$

31. $(3 - \sqrt{-4})(-2 + \sqrt{-49}) = (3 - i\sqrt{4})(-2 + i\sqrt{49})$

                     $= (3 - 2i)(-2 + 7i) = -6 + 25i - 14i^2 = -6 + 25i + 14 = 8 + 25i$

33. $\dfrac{5 - \sqrt{-4}}{7} = \dfrac{5 - i\sqrt{4}}{7} = \dfrac{5 - 2i}{7} = \dfrac{5}{7} - \dfrac{2}{7}i$

35. $\dfrac{1}{2-\sqrt{-9}} = \dfrac{1}{2-i\sqrt{9}} = \dfrac{1}{2-3i} = \dfrac{1}{(2-3i)}\dfrac{(2+3i)}{(2+3i)} = \dfrac{2+3i}{4-9i^2} = \dfrac{2+3i}{4+9} = \dfrac{2+3i}{13} = \dfrac{2}{13} + \dfrac{3}{13}i$

37. $\dfrac{2}{5i} = \dfrac{2}{5i}\cdot\dfrac{i}{i} = \dfrac{2i}{5i^2} = \dfrac{2i}{-5} = -\dfrac{2}{5}i \text{ or } 0 - \dfrac{2}{5}i$

39. $\dfrac{1+3i}{2i} = \dfrac{(1+3i)}{(2i)}\cdot\dfrac{i}{i} = \dfrac{i+3i^2}{2i^2} = \dfrac{i-3}{-2} = \dfrac{3}{2} - \dfrac{1}{2}i$

41. $(2-3i)^2 - 2(2-3i) + 9 \;=\; 4 - 12i + 9i^2 - 4 + 6i + 9$
$$= \; 4 - 12i - 9 - 4 + 6i + 9 = -6i \text{ or } 0 - 6i$$

43. $x^2 - 2x + 2 \;=\; (1-i)^2 - 2(1-i) + 2 = 1 - 2i + i^2 - 2 + 2i + 2$
$$= \left(1 - 2i - 1\right) - 2 + 2i + 2 = 0 \text{ or } 0 + 0i$$

45. $i^{18} \;=\; \underbrace{i^{16}}\cdot i^2 = (i^4)^4 \cdot i^2 = 1^4(-1) = -1$

    largest integer in 18 exactly divisible by 4

    $i^{32} \;=\; (i^4)^8 = 1^8 = 1$

    $i^{67} \;=\; \underbrace{i^{64}}\cdot i^3 = (i^4)^{16} \cdot i^2 \cdot i = 1^{16}(-1)i = -i$

    largest integer in 67 exactly divisible by 4

47. According to the definition of equality for complex numbers

    $(2x - 1) + (3y + 2)i = 5 - 4i = 5 + (-4)i$

    if and only if

    $2x - 1 = 5$ and $3y + 2 = -4$

    So $2x = 6$ and $3y = -6$

    $x = 3 \qquad y = -2$

49. $\sqrt{3-x}$ represents an imaginary number when $3 - x$ is negative. We solve:

    $3 - x \;<\; 0$

    $\qquad 3 \;<\; x \text{ or } x > 3$

51. $\sqrt{2-3x}$ represents an imaginary number when $2 - 3x$ is negative. We solve:

    $2 - 3x \;<\; 0$

    $\quad -3x \;<\; -2$

    $\qquad x \;>\; \frac{2}{3}$

53. $(a + bi) + (c + di) = a + bi + c + di = a + c + bi + di = (a + c) + (b + d)i$

55. $(a + bi)(a - bi) = a^2 - b^2i^2 = a^2 + b^2 \text{ or } (a^2 + b^2) + 0i$

57. $(a + bi)(c + di) \;=\; ac + adi + bci + bdi^2 = ac + (ad + bc)i - bd$
$$= \; (ac - bd) + (ad + bc)i$$

59. $i^{4k} = (i^4)^k = (i^2 \cdot i^2)^k = [(-1)(-1)]^k = 1^k = 1$

61. 1. Definition of addition

    2. Commutative property for addition of real numbers.

    3. Definition of addition (read from right to left).

63. $$\begin{aligned} z\bar{z} &= (x+yi)(x-yi) \\ &= x^2 - (yi)^2 \\ &= x^2 - y^2 i^2 \\ &= x^2 + y^2 \text{ or } (x^2+y^2) + 0i. \end{aligned}$$

This is a real number.

65. To prove a theorem containing the phrase "if and only if", it is often helpful to prove two parts separately. Thus: $\bar{z} = z$ if z is real; $\bar{z} = z$ only if z is real

Hypothesis	z is real	Hypothesis	$\bar{z} = z$
Conclusion	$\bar{z} = z$	Conclusion	z is real
Proof	Assume z is real, then	Proof	Assume $\bar{z} = z$,
	$z = x + 0i = x$		that is, $x - yi = x + yi$
	$\bar{z} = x - 0i = x$		Then by the definition of equality
			$\quad x = x \quad -y = y$
	Hence $z = \bar{z}$.		$\qquad\qquad -2y = 0$
			$\qquad\qquad\quad y = 0$
			Hence $z = x + 0i$, that is, z is real.

67. $$\begin{aligned} \overline{z+w} &= \overline{(x+yi)+(u+vi)} \\ &= \overline{x+yi+u+vi} \\ &= \overline{x+u+(y+v)i} \\ &= (x+u)-(y+v)i \\ &= x+u-yi-vi \\ &= (x-yi)+(u-vi) \\ &= \bar{z}+\bar{w} \end{aligned}$$

69. $$\begin{aligned} \overline{zw} &= \overline{(x+yi)(u+vi)} \\ &= \overline{xu+xvi+yui+yvi^2} \\ &= \overline{xu+(xv+yu)i-yv} \\ &= xu-yv-(xv+yu)i \\ &= xu-xvi-yv-yui \\ &= x(u-vi)-yui+yv(-1) \\ &= x(u-vi)-yui+yvi^2 \\ &= x(u-vi)-yi(u-vi) \\ &= (x-yi)(u-vi) \\ &= \bar{z}\ \bar{w} \end{aligned}$$

71. $$\begin{aligned} (3.17-4.08i)(7.14+2.76i) &= (3.17)(7.14)+(3.17)(2.76i)-(4.08i)(7.14)-(4.08i)(2.76i) \\ &= 22.6338 + 8.7492i - 29.1312i - 11.2608i^2 \\ &= 22.6338 - 20.382i + 11.2608 \\ &= 33.89 - 20.38i \end{aligned}$$

73. $$\begin{aligned} \frac{8.14+2.63i}{3.04+6.27i} &= \frac{(8.14+2.63i)}{(3.04+6.27i)} \cdot \frac{(3.04-6.27i)}{(3.04-6.27i)} \\[2mm] &= \frac{(8.14)(3.04)-(8.14)(6.27i)+(2.63i)(3.04)-(2.63i)(6.27i)}{(3.04)^2+(6.27)^2} \\[2mm] &= \frac{24.7456 - 51.0378i + 7.9952i - 16.4901i^2}{9.2416 + 39.3129} \\[2mm] &= \frac{24.7456 - 43.0426i + 16.4901}{48.55455} = \frac{41.2357 - 43.0426i}{48.5545} = 0.85 - 0.89i \end{aligned}$$

Exercise 3-6

Key Ideas and Formulas

A quadratic equation in one variable is any equation that can be written in the form

$$ax^2 + bx + c = 0 \qquad a \neq 0 \qquad \text{(standard form)}$$

where x is a variable and a, b, c are constants.

To solve a quadratic equation in standard form, check if it can be solved using factoring and the zero property:

$$m \cdot n = 0 \text{ if and only if } m = 0 \text{ or } n = 0 \text{ or both}$$

If not, generally the quadratic formula is applied:

$$x = \frac{-b \pm \sqrt{b^2 - 4ac}}{2a} \quad a \neq 0$$

Occasionally the square root property

$$\text{if } A^2 = C, \text{ then } A = \pm\sqrt{C}$$

can be useful, especially if the equation is in the form

$$ax^2 = b \text{ or } (ax + b)^2 = d$$

The quantity $b^2 - 4ac$, appearing in the quadratic formula, is called the discriminant. If the discriminant is positive, the equation has two distinct real roots. If the discriminant is 0, the equation has one real root (a double root). If the discriminant is negative, the equation has two imaginary roots, one the conjugate of the other.

1.
$$\begin{aligned}
4u^2 &= 8u \\
4u^2 - 8u &= 0 \\
4u(u - 2) &= 0 \\
4u &= 0 \text{ or } u - 2 = 0 \\
u &= 0 \qquad u = 2
\end{aligned}$$

3.
$$\begin{aligned}
9y^2 &= 12y - 4 \\
9y^2 - 12y + 4 &= 0 \\
(3y - 2)^2 &= 0 \\
3y - 2 &= 0 \\
3y &= 2 \\
y &= \tfrac{2}{3} \quad \text{(double root)}
\end{aligned}$$

5.
$$\begin{aligned}
11x &= 2x^2 + 12 \\
0 &= 2x^2 - 11x + 12 \\
2x^2 - 11x + 12 &= 0 \\
(2x - 3)(x - 4) &= 0 \\
2x - 3 &= 0 \quad \text{or } x - 4 = 0 \\
2x &= 3 \quad \text{or } x = 4 \\
x &= \tfrac{3}{2}
\end{aligned}$$

7.
$$\begin{aligned}
m^2 - 12 &= 0 \\
m^2 &= 12 \\
m &= \pm\sqrt{12} \\
m &= \pm 2\sqrt{3}
\end{aligned}$$

9.
$$\begin{aligned}
x^2 + 25 &= 0 \\
x^2 &= -25 \\
x &= \pm 5i
\end{aligned}$$

11.
$$\begin{aligned}
9y^2 - 16 &= 0 \\
9y^2 &= 16 \\
y^2 &= \tfrac{16}{9} \\
y &= \pm\sqrt{\tfrac{16}{9}} \\
y &= \pm\tfrac{4}{3}
\end{aligned}$$

13.
$$4x^2 + 25 = 0$$
$$4x^2 = -25$$
$$x^2 = -\frac{25}{4}$$
$$x = \pm\sqrt{-\frac{25}{4}}$$
$$x = \pm\frac{\sqrt{-25}}{\sqrt{4}}$$
$$x = \pm\frac{5i}{2} \text{ or } \pm\frac{5}{2}i$$

15.
$$(n+5)^2 = 9$$
$$n+5 = \pm\sqrt{9}$$
$$n+5 = \pm 3$$
$$n = -5 \pm 3$$
$$n = -5+3 \text{ or } -5-3$$
$$n = -2, -8$$

17.
$$(d-3)^2 = -4$$
$$d-3 = \pm\sqrt{-4}$$
$$d-3 = \pm 2i$$
$$d = 3 \pm 2i$$

19.
$$x^2 - 10x - 3 = 0$$
$$x = \frac{-b \pm \sqrt{b^2 - 4ac}}{2a}$$
$$a = 1$$
$$b = -10$$
$$c = -3$$
$$x = \frac{-(-10) \pm \sqrt{(-10)^2 - 4(1)(-3)}}{2(1)}$$
$$x = \frac{10 \pm \sqrt{112}}{2} = \frac{10 \pm 4\sqrt{7}}{2}$$
$$= 5 \pm 2\sqrt{7}$$

Common Error:
It is incorrect to
"cancel" this way:
$$\frac{\cancel{10} \pm \sqrt{112}}{\cancel{2}} \neq 5 \pm \sqrt{112}$$

21.
$$x^2 + 8 = 4x$$
$$x^2 - 4x + 8 = 0$$
$$x = \frac{-b \pm \sqrt{b^2 - 4ac}}{2a}$$
$$a = 1$$
$$b = -4$$
$$c = 8$$
$$x = \frac{-(-4) \pm \sqrt{(-4)^2 - 4(1)(8)}}{2(1)}$$
$$x = \frac{4 \pm \sqrt{-16}}{2}$$
$$x = \frac{4 \pm i\sqrt{16}}{2}$$
$$x = \frac{4 \pm 4i}{2}$$
$$x = 2 \pm 2i$$

23.
$$2x^2 + 1 = 4x$$
$$2x^2 - 4x + 1 = 0$$
$$x = \frac{-b \pm \sqrt{b^2 - 4ac}}{2a}$$
$$a = 2$$
$$b = -4$$
$$c = 1$$
$$= \frac{-(-4) \pm \sqrt{(-4)^2 - 4(2)(1)}}{2(2)}$$
$$x = \frac{4 \pm \sqrt{8}}{4}$$
$$x = \frac{4 \pm 2\sqrt{2}}{4}$$
$$x = \frac{2 \pm \sqrt{2}}{2}$$

Common Error:
$$\frac{2 \pm \sqrt{2}}{2} \neq \pm\sqrt{2}$$
$$\neq 1 \pm \sqrt{2}$$
These involve incorrect "cancelling".

25.
$$5x^2 + 2 = 2x$$
$$5x^2 - 2x + 2 = 0$$
$$x = \frac{-b \pm \sqrt{b^2 - 4ac}}{2a}$$
$$a = 5$$
$$b = -2$$
$$c = 2$$
$$x = \frac{-(-2) \pm \sqrt{(-2)^2 - 4(5)(2)}}{2(5)}$$
$$x = \frac{2 \pm \sqrt{-36}}{10}$$
$$x = \frac{2 \pm 6i}{10}$$
$$x = \frac{1}{5} \pm \frac{3}{5}i$$

27.
$$x^2 - 6x - 3 = 0$$
$$x^2 - 6x = 3$$
$$x^2 - 6x + 9 = 12$$
$$(x-3)^2 = 12$$
$$x - 3 = \pm\sqrt{12}$$
$$x = 3 \pm \sqrt{12}$$
$$x = 3 \pm 2\sqrt{3}$$

29.
$$2y^2 - 6y + 3 = 0$$
$$y^2 - 3y + \tfrac{3}{2} = 0$$
$$y^2 - 3y = -\tfrac{3}{2}$$
$$y^2 - 3y + \tfrac{9}{4} = -\tfrac{3}{2} + \tfrac{9}{4}$$
$$\left(y - \tfrac{3}{2}\right)^2 = \tfrac{3}{4}$$
$$y - \tfrac{3}{2} = \pm\sqrt{\tfrac{3}{4}}$$
$$y = \tfrac{3}{2} \pm \tfrac{\sqrt{3}}{\sqrt{4}}$$
$$y = \tfrac{3}{2} \pm \tfrac{\sqrt{3}}{2}$$
$$y = \tfrac{3 \pm \sqrt{3}}{2}$$

31.
$$3x^2 - 2x - 2 = 0$$
$$x^2 - \tfrac{2}{3}x - \tfrac{2}{3} = 0$$
$$x^2 - \tfrac{2}{3}x = \tfrac{2}{3}$$
$$x^2 - \tfrac{2}{3}x + \tfrac{1}{9} = \tfrac{2}{3} + \tfrac{1}{9}$$
$$\left(x - \tfrac{1}{3}\right)^2 = \tfrac{7}{9}$$
$$x - \tfrac{1}{3} = \pm\sqrt{\tfrac{7}{9}}$$
$$x - \tfrac{1}{3} = \pm\tfrac{\sqrt{7}}{3}$$
$$x = \tfrac{1}{3} \pm \tfrac{\sqrt{7}}{3}$$
$$x = \tfrac{1 \pm \sqrt{7}}{3}$$

33.
$$x^2 + mx + n = 0$$
$$x^2 + mx = -n$$
$$x^2 + mx + \tfrac{m^2}{4} = \tfrac{m^2}{4} - n$$
$$\left(x + \tfrac{m}{2}\right)^2 = \tfrac{m^2 - 4n}{4}$$
$$x + \tfrac{m}{2} = \pm\sqrt{\tfrac{m^2 - 4n}{4}}$$
$$x = -\tfrac{m}{2} \pm \tfrac{\sqrt{m^2 - 4n}}{2}$$
$$x = \tfrac{-m \pm \sqrt{m^2 - 4n}}{2}$$

35.
$$12x^2 + 7x = 10$$
$$12x^2 + 7x - 10 = 0$$
$$(4x + 5)(3x - 2) = 0 \quad \text{Polynomial is factorable.}$$
$$4x + 5 = 0 \text{ or } \quad 3x - 2 = 0$$
$$4x = -5 \qquad 3x = 2$$
$$x = -\tfrac{5}{4} \qquad x = \tfrac{2}{3}$$

37.
$$(2y - 3)^2 = 5 \quad \text{Format for the square root method.}$$
$$2y - 3 = \pm\sqrt{5}$$
$$2y = 3 \pm \sqrt{5}$$
$$y = \tfrac{3 \pm \sqrt{5}}{2}$$

39.
$$x^2 = 3x + 1$$
$$x^2 - 3x - 1 = 0 \quad \text{Polynomial is not}$$
factorable, use quadratic
formula
$$x = \tfrac{-b \pm \sqrt{b^2 - 4ac}}{2a}$$
$$a = 1$$
$$b = -3$$
$$c = -1$$
$$x = \tfrac{(-3) \pm \sqrt{(-3)^2 - 4(1)(-1)}}{2(1)}$$
$$x = \tfrac{3 \pm \sqrt{13}}{2}$$

41.
$$7n^2 = -4n$$
$$7n^2 + 4n = 0$$
$$n(7n + 4) = 0 \quad \text{Polynomial is}$$
factorable.
$$n = 0 \text{ or} \qquad 7n + 4 = 0$$
$$7n = -4$$
$$n = -\tfrac{4}{7}$$

43.
$$1 + \tfrac{8}{x^2} = \tfrac{4}{x} \quad \text{Excluded}$$
value : $x \neq 0$
$$x^2 + 8 = 4x$$
$$x^2 - 4x + 8 = 0 \qquad \text{Polynomial is}$$
$$x^2 - 4x = -8 \qquad \text{not factorable,}$$
$$x^2 - 4x = -8 \qquad \text{use quadratic}$$
$$x^2 - 4x + 4 = -4 \qquad \text{formula, or}$$
$$(x - 2)^2 = -4 \qquad \text{complete the}$$
$$x - 2 = \pm\sqrt{-4} \qquad \text{square}$$
$$x - 2 = \pm i\sqrt{4}$$
$$x - 2 = \pm 2i$$
$$x = 2 \pm 2i$$

45. $\frac{24}{10+m} + 1 = \frac{24}{10-m}$

Excluded value: $m \neq -10, 10$; lcm of the denominators is $(10+m)(10-m)$

$$(10+m)(10-m)\frac{24}{10+m} + (10+m)(10-m) = (10+m)(10-m)\frac{24}{10-m}$$

$$24(10-m) + 100 - m^2 = 24(10+m)$$
$$240 - 24m + 100 - m^2 = 240 + 24m$$
$$340 - 24m - m^2 = 240 + 24m$$
$$0 = m^2 + 48m - 100$$
$$m^2 + 48m - 100 = 0 \quad \text{Polynomial is factorable.}$$
$$(m+50)(m-2) = 0$$
$$m + 50 = 0 \quad \text{or} \qquad\qquad m - 2 = 0$$
$$m = -50 \qquad\qquad\qquad m = 2$$

47. $\frac{2}{x-2} = \frac{4}{x-3} - \frac{1}{x+1}$

Excluded values: $x \neq 2, 3, -1$

$$(x-2)(x-3)(x+1)\frac{2}{x-2} = (x-2)(x-3)(x+1)\frac{4}{x-3} - (x-2)(x-3)(x+1)\frac{1}{x+1}$$

$$2(x-3)(x+1) = 4(x-2)(x+1) - (x-2)(x-3)$$
$$2(x^2 - 2x - 3) = 4(x^2 - x - 2) - (x^2 - 5x + 6)$$
$$2x^2 - 4x - 6 = 4x^2 - 4x - 8 - x^2 + 5x - 6$$
$$2x^2 - 4x - 6 = 3x^2 + x - 14$$
$$0 = x^2 + 5x - 8$$
$$x^2 + 5x - 8 = 0 \quad \text{Polynomial is not factorable,}$$
$$\text{use quadratic formula.}$$

$$x = \frac{-b \pm \sqrt{b^2 - 4ac}}{2a}$$
$$a = 1$$
$$b = 5$$
$$c = -8$$
$$x = \frac{-5 \pm \sqrt{(5)^2 - 4(1)(-8)}}{2(1)} = \frac{-5 \pm \sqrt{57}}{2}$$

49. $\frac{x+2}{x+3} - \frac{x^2}{x^2-9} = 1 - \frac{x-1}{3-x}$ Excluded values: $x \neq 3, 3$

$$(x-3)(x+3)\frac{(x+2)}{x+3} - (x-3)(x+3)\frac{x^2}{x^2-9} = (x-3)(x+3) - (x-3)(x+3)\frac{x-1}{3-x}$$

$$(x-3)(x+2) - x^2 = x^2 - 9 + (x-1)(x+3)$$
$$x^2 - x - 6 - x^2 = x^2 - 9 + x^2 + 2x - 3$$
$$-x - 6 = 2x^2 + 2x - 12$$
$$0 = 2x^2 + 3x - 6$$
$$2x^2 + 3x - 6 = 0 \quad \text{Polynomial is not factorable,}$$
$$\text{use quadratic formula}$$

$$x = \frac{-b \pm \sqrt{b^2 - 4ac}}{2a}$$
$$a = 2$$
$$b = 3$$
$$c = -6$$
$$x = \frac{-3 \pm \sqrt{(3)^2 - 4(2)(-6)}}{2(2)}$$
$$x = \frac{-3 \pm \sqrt{57}}{4}$$

51. According to Theorem 6,
    Section 3-4, $|\,3u - 2\,| = u^2$
    is equivalent to $3u - 2 = \pm u^2$,
    Hence we must solve

$3u - 2 = u^2$	$3u - 2 = -u^2$
$0 = u^2 - 3u + 2$	$u^2 + 3u - 2 = 0$
$0 = (u - 1)(u - 2)$	$u = \dfrac{-b \pm \sqrt{b^2 - 4ac}}{2a}$
	$a = 1$
	$b = 3$
	$c = -2$
$u - 1 = 0$ or $u - 2 = 0$	
$u = 1 \qquad u = 2$	$u = -\dfrac{-3 \pm \sqrt{(3)^2 - 4(1)(-2)}}{2(1)}$
	$u = \dfrac{-3 \pm \sqrt{17}}{2}$

53.
$$s = \tfrac{1}{2}gt^2$$
$$\tfrac{1}{2}gt^2 = 2$$
$$gt^2 = 2s$$
$$t^2 = \frac{2s}{g}$$
$$t = \sqrt{\frac{2s}{g}}$$

55.
$$P = EI - RI^2$$
$$RI^2 - EI + P = 0$$
$$I = \frac{-b \pm \sqrt{b^2 - 4ac}}{2a}$$
$$a = R$$
$$b = -E$$
$$c = P$$
$$I = \frac{-(-E) \pm \sqrt{(-E)^2 - 4(R)(P)}}{2(R)}$$
$$I = \frac{E + \sqrt{E^2 - 4RP}}{2R}$$

(positive square root)

57.
$$\sqrt{3}x^2 = 8\sqrt{2}x - 4\sqrt{3}$$
$$\sqrt{3}x^2 - 8\sqrt{2}x + 4\sqrt{3} = 0$$
$$x = \frac{-b \pm \sqrt{b^2 - 4ac}}{2a}$$
$$a = \sqrt{3}$$
$$b = -8\sqrt{2}$$
$$c = 4\sqrt{3}$$
$$x = \frac{-(-8\sqrt{2}) \pm \sqrt{(-8\sqrt{2})^2 - 4(\sqrt{3})(4\sqrt{3})}}{2(\sqrt{3})}$$
$$x = \frac{8\sqrt{2} \pm \sqrt{(64)(2) - (16)(3)}}{2\sqrt{3}}$$
$$x = \frac{8\sqrt{2} \pm \sqrt{80}}{2\sqrt{3}}$$
$$x = \frac{8\sqrt{2} \pm 4\sqrt{5}}{2\sqrt{3}}$$
$$x = \frac{2(4\sqrt{2} \pm 2\sqrt{5})}{2\sqrt{3}} \cdot \frac{\sqrt{3}}{\sqrt{3}}$$
$$x = \frac{4\sqrt{6} \pm 2\sqrt{15}}{3} \text{ or } \tfrac{4}{3}\sqrt{6} \pm \tfrac{2}{3}\sqrt{15}$$

59. $x^2 + 2ix = 3$

$x^2 + 2ix - 3 = 0$

$x = \frac{-b \pm \sqrt{b^2 - 4ac}}{2a}$

$a = 1$

$b = 2i$

$c = -3i$

$x = \frac{-2i \pm \sqrt{(2i)^2 - 4(1)(-3)}}{2(1)}$

$x = \frac{-2i \pm \sqrt{-4 + 12}}{2}$

$x = \frac{-2i \pm \sqrt{8}}{2}$

$x = \frac{-2i \pm 2\sqrt{2}}{2}$

$x = \frac{2(-i \pm \sqrt{2})}{2}$

$x = -i \pm \sqrt{2}$

$x = \sqrt{2} - i, -\sqrt{2} - i$

61.   If a quadratic equation has two roots, they are $\frac{-b + \sqrt{b^2 - 4ac}}{2a}$ and $\frac{-b - \sqrt{b^2 - 4ac}}{2a}$.

If a, b, c are rational, then so are $-b, 2a$, and $b^2 - 4ac$. Then, *either* $\sqrt{b^2 - 4ac}$ is rational,

hence $\frac{-b + \sqrt{b^2 - 4ac}}{2a}$ and $\frac{-b - \sqrt{b^2 - 4ac}}{2a}$ are both rational, *or*, $\sqrt{b^2 - 4ac}$ is irrational,

hence $\frac{-b + \sqrt{b^2 - 4ac}}{2a}$ and $\frac{-b - \sqrt{b^2 - 4ac}}{2a}$ are both irrational, *or*, $\sqrt{b^2 - 4ac}$

is imaginary, hence $\frac{-b + \sqrt{b^2 - 4ac}}{2a}$ and $\frac{-b + \sqrt{b^2 - 4ac}}{2a}$ are both imaginary.

There is no other possibility; one root cannot be rational while the other is irrational. False: F

63. $r_1 = \frac{-b + \sqrt{b^2 - 4ac}}{2a}$ $r_2 = \frac{-b - \sqrt{b^2 - 4ac}}{2a}$

$r_1 r_2 = \frac{(-b + \sqrt{b^2 - 4ac})}{2a} \frac{(-b - \sqrt{b^2 - 4ac})}{2a}$

$= \frac{(-b)^2 - (\sqrt{b^2 - 4ac})^2}{4a^2} = \frac{b^2 - (b^2 - 4ac)}{4a^2} = \frac{b^2 - b^2 + 4ac}{4a^2} = \frac{4ac}{4a^2} = \frac{c}{a}$

65.   The $\pm$ in front still yields the same two numbers even if a is negative.

67. $2.07x^2 - 3.79x + 1.34 = 0$

$x = \frac{-b \pm \sqrt{b^2 - 4ac}}{2a}$

$a = 2.07$

$b = -3.79$

$c = 1.34$

$x = \frac{-(-3.79) \pm \sqrt{(-3.79)^2 - 4(2.07)(1.34)}}{2(2.07)}$

$= \frac{3.79 \pm 1.81}{4.14}$

$x = 1.35, 0.48$

69. $4.83x^2 + 2.04x - 3.18 = 0$

$$x = \frac{-b \pm \sqrt{b^2 - 4ac}}{2a}$$

$a = 4.83$
$b = 2.04$
$c = -3.18$

$$x = \frac{-2.04 \pm \sqrt{(2.04)^2 - 4(4.83)(-3.18)}}{2(4.83)}$$

$$= \frac{-2.04 + 8.10}{9.66}$$

$$x = -1.05, 0.63$$

71. $0.0134x^2 + 0.0414x + 0.0304 = 0$

The discriminant is $b^2 - 4ac$, where $a = 0.0134, b = 0.0414, c = 0.0304$.

$b^2 - 4ac = (0.0414)^2 - 4(0.0134)(0.0304) = 8.45 \times 10^{-5} > 0$

Since the discriminant is positive, the equation has real solutions.

73. $0.0134x^2 + 0.0214x + 0.0304 = 0$

Here $a = 0.0134, \quad b = 0.0214, \quad c = 0.0304$.

$b^2 - 4ac = (0.0214)^2 - 4(0.0134)(0.0304) = -1.17 \times 10^{-3} < 0$

Since $b^2 - 4ac$, the discriminant, is negative, the equation has no real solutions.

75.   Let $\quad x = $ one number. Since their sum is 21,

$\quad 21 - x = $ other number

Then, since their product is 104,

$$x(21 - x) = 104$$
$$21x - x^2 = 104$$
$$0 = x^2 - 21x + 104$$
$$x^2 - 21x + 104 = 0$$
$$(x - 13)(x - 8) = 0$$

$x - 13 = 0 \quad or \quad x - 8 = 0$

$\quad\quad x = 13 \quad\quad\quad\quad x = 8$

The numbers are 8 and 13.

77.   Let $\quad x = $ first of the two consecutive even integers.

Then $x + 2 = $ second of these integers

Since their product is 168

$$x(x+2) = 168$$
$$x^2 + 2x = 168$$
$$x^2 + 2x - 168 = 0$$
$$(x-12)(x+14) = 0$$
$$x - 12 = 0 \quad or \quad x + 14 = 0$$
$$x = 12 \qquad x = -14$$

If $x = 12$, the two consecutive positive even integers must be 12 and 14. We discard the other solution, since the numbers must be positive.

79.
$$\text{Let } x = \text{amount of increase}$$
$$\text{Then } x + 4 = \text{length of new rectangle}$$
$$x + 2 = \text{width of new rectangle}$$

Using $A = ab$, we have

Area of new rectangle $= 2 \times$ Area of old rectangle.

$$(x+4)(x+2) = 2(4)(2)$$
$$x^2 + 6x + 8 = 16$$
$$x^2 + 6x - 8 = 0$$
$$x^2 + 6x = 8$$
$$x^2 + 6x + 9 = 17$$
$$(x+3)^2 = 17$$
$$x + 3 = \pm\sqrt{17}$$
$$x = -3 \pm \sqrt{17}$$

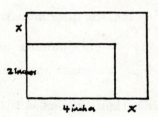

We discard the negative solution and take $x = -3 + \sqrt{17} = 1.12$. Then dimensions of new rectangle are $x + 4$ by $x + 2$ or 5.12 by 3.12 inches.

81. Let $r = $ interest rate. Applying the given formula, we have
$$1440 = 1000(1 + r)^2$$
$$1.44 = (1 + r)^2$$
$$(1 + r)^2 = 1.44$$
$$1 + r = \pm\sqrt{1.44}$$
$$1 + r = \pm 1.2$$
$$r = -1 \pm 1.2$$
Discarding the negative root, we have
$$r = .2 \text{ or } 20\%.$$

83. Let $r = $ rate of slow plane.
Then $r + 140 = $ rate of fast plane.
After 1 hour $r(1) = r = $ distance travelled by slow plane.
$(r + 140)(1) = r + 140 = $ distance travelled by fast plane.

Applying the Pythagorean theorem, we have

$$r^2 + (r + 140)^2 = 260^2$$
$$r^2 + r^2 + 280r + 19,600 = 67,600$$
$$2r^2 + 280r - 48,000 = 0$$
$$r^2 + 140r - 24,000 = 0$$
$$(r + 240)(r - 100) = 0$$
$$r + 240 = 0 \quad \text{or} \quad r - 100 = 0.$$
$$r = -240 \quad r = 100$$

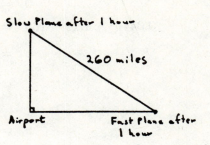

Discarding the negative solution, we have

$$r = 100 \text{ miles per hour} = \text{rate of slow plane}$$
$$r + 140 = 240 \text{ miles per hour} = \text{rate of fast plane.}$$

85.   Let t = time for smaller pipe to fill tank alone

$t - 5$ = time for larger pipe to fill tank alone

5 = time for both pipes to fill tank together

Then $\frac{1}{t}$ = rate for smaller pipe

$\frac{1}{t-5}$ = rate for larger pipe

$\frac{1}{5}$ = rate together

(rate of smaller pipe) + (rate of larger pipe) = (rate together)

$$\frac{1}{t} + \frac{1}{t-5} = \frac{1}{5}$$
$$5t(t-5)\frac{1}{t} + 5t(t-5)\frac{1}{t-5} = 5t(t-5)\frac{1}{5}$$
$$5(t-5) + 5t = t(t-5)$$
$$5t - 25 + 5t = t^2 - 5t$$
$$10t - 25 = t^2 - 5t$$
$$0 = t^2 - 15t + 25$$
$$t^2 - 15t + 25 = 0$$
$$t = \frac{-b \pm \sqrt{b^2 - 4ac}}{2a}$$
$$a = 1$$
$$b = -15$$
$$c = 25$$
$$t = \frac{-(15) \pm \sqrt{(-15)^2 - 4(1)(25)}}{2(1)} = \frac{15 \pm \sqrt{125}}{2}$$
$$t = 13.09, 1.91$$
$$t - 5 = 8.09, -3.09$$

Discarding the answer for t which results in a negative answer for $t - 5$, we have

13.09 hours for smaller pipe alone,

  8.09 hours for larger pipe alone.

87.   Let v = speed of car. Applying the given formula, we have

$$165 = 0.044v^2 + 1.1v$$

$$0 = 0.044v^2 + 1.1v - 165$$

$$0.044v^2 + 1.1v - 165 = 0$$

$$v = \frac{-b \pm \sqrt{b^2 - 4ac}}{2a}$$

$$a = 0.044$$
$$b = 1.1$$
$$c = -165$$

$$v = \frac{-1.1 \pm \sqrt{(1.1)^2 - 4(0.044)(-165)}}{2(0.044)} = \frac{-1.1 \pm 5.5}{0.088}$$

$$v = -75 \text{ or } 50$$

Discarding the negative answer, we have

$v = 50$ miles per hour.

Exercise 3-7
Key Ideas and Formulas

A non-zero polynomial will have a constant sign (either always positive or always negative) within each interval determined by its real zeros plotted on a number line. If a polynomial has no real zeros, then the polynomial is either positive over the whole line or negative over the whole line.

The rational expression P/Q, where P and Q are nonzero polynomials, will have a constant sign (either always positive or always negative) within each interval determined by the real zeros of P and Q plotted on a number line. If neither P nor Q have real zeros, then the rational expression P/Q is either positive over the whole line or negative over the whole line.

To solve a polynomial inequality:

Step 1: Write the polynomial inequality in standard form.

Step 2: Find all real zeros of the polynomial.

Step 3: Plot the real zeros on a number line, dividing the number line into intervals.

Step 4: Choose a test number (that is easy to compute with) in each interval and evaluate the polynomial for each number.

Step 5: Using the results of Step 4 construct a sign chart, showing the sign of the polynomial in each interval.

Step 6: From the sign chart, write down the solution (and draw the graph, if required) of the original polynomial inequality.

A similar sequence of steps is required to solve a rational inequality.

1.
$$x^2 < 10 - 3x$$
$$x^2 + 3x - 10 < 0$$
$$(x + 5)(x - 2) < 0$$
Zeros: $-5, 2$

$x^2 + 3x - 10 = (x + 5)(x - 2)$			
Test Number	-6	0	3
Value of Polynomial for Test Number	8	-10	8
Sign of Polynomial in Interval	$+$	$-$	$+$
Interval	$(-\infty, -5)$	$(-5, 2)$	$(2, \infty)$

$x^2 + 3x - 10$ is negative within the interval $(-5, 2)$. $-5 < x < 2$.

3.
$$x^2 + 21 > 10x$$
$$x^2 - 10x + 21 > 0$$
$$(x - 3)(x - 7) > 0$$
Zeros: 3, 7

$x^2 - 10x + 21 = (x + 3)(x - 7)$			
Test Number	2	5	8
Value of Polynomial for Test Number	5	-4	5
Sign of Polynomial in Interval	+	$-$	+
Interval	$(-\infty, 3)$	$(3, 7)$	$(7, \infty)$

$x^2 - 10x + 21$ is positive within the intervals $(-\infty, 3)$ and $(7, \infty)$. $x < 3$ or $x > 7$

5.
$$x^2 \leq 8x$$
$$x^2 - 8x \leq 0$$
$$x(x - 8) \leq 0$$
Zeros: 0,8
0,8 are part of the solution set, so we use solid dots.

$x^2 - 8x = x(x - 8)$			
Test Number	-1	4	9
Value of Polynomial for Test Number	9	-16	9
Sign of Polynomial in Interval	+	$-$	+
Interval	$(-\infty, 0)$	$(0, 8)$	$(8, \infty)$

$x^2 - 8x$ is non-positive within the interval $[0,8]$. $0 \leq x \leq 8$

7.
$$x^2 + 5x \leq 0$$
$$x(x + 5) \leq 0$$
Zeros: $-5, 0$
$-5, 0$ are part of the solution set, so we use solid dots.

$x^2 + 5x = x(x + 5)$			
Test Number	-6	-1	1
Value of Polynomial for Test Number	6	-4	6
Sign of Polynomial in Interval	+	$-$	+
Interval	$(-\infty, -5)$	$(-5, 0)$	$(0, \infty)$

$x^2 + 5x$ is non-positive within the interval $[-5, 0]$. $-5 \leq x \leq 0$.

9.
$$x^2 > 4$$
$$x^2 - 4 > 0$$
$$(x+2)(x-2) > 0$$
Zeros: $-2, 2$

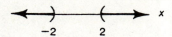

$x^2 - 4x = (x+2)(x-2)$			
Test Number	-3	0	3
Value of Polynomial for Test Number	5	-4	5
Sign of Polynomial in Interval	$+$	$-$	$+$
Interval	$(-\infty, -2)$	$(-2, 2)$	$(2, \infty)$

$x^2 - 4$ is positive within the interval $(-\infty, -2)$ and $(2, \infty)$. $x < -2$ or $x > 2$.

11. $\frac{x-2}{x+4} \le 0$

Zeros of $P, Q : -4, 2$

2 is part of the solution set so we use a solid dot there.

-4 is not part of the solid set, $\left(\frac{P}{Q} \text{ is not} \right.$

defined there$)$ so we use an open dot there.

$\frac{P}{Q} = \frac{x-2}{x+4}$			
Test Number	-5	0	3
Value of $\frac{P}{Q}$	7	$-\frac{1}{2}$	$\frac{1}{7}$
Sign of $\frac{P}{Q}$	$+$	$-$	$+$
Interval	$(-\infty, -4)$	$(-4, 2)$	$(2, \infty)$

$\frac{x-2}{x+4}$ is non-positive within the interval $(-4, 2]$. $-4 < x \le 2$.

13. $\frac{x+4}{1-x} \le 0$

Zeros of $P, Q : -4, 1$

-4 is part of the solution set, so we use a solid dot there.

1 is not part of the solid set, $\left(\frac{P}{Q} \text{ is not} \right.$

defined there$)$ so we use an open dot there.

$\frac{P}{Q} = \frac{x+4}{1-x}$			
Test Number	-5	0	2
Value of $\frac{P}{Q}$	$-\frac{1}{6}$	4	-6
Sign of $\frac{P}{Q}$	$-$	$+$	$-$
Interval	$(-\infty, -4)$	$(-4, 1)$	$(1, \infty)$

$\frac{x+4}{1-x}$ is non-positive within the intervals $(-\infty, -4)$ and $(1, \infty)$. $x \le -4$ or $x > 1$.

15. $\dfrac{x^2+5x}{x-3} \geq 0$

$\dfrac{x(x+5)}{x-3} \geq 0$

Zeros of $P, Q: -5, 0, 3$

-5 and 0 are part of the solution set, so we use solid dots there.

3 is not part of the solution set, $\left(\dfrac{P}{Q}\right.$ is not defined there) so we use an open dot there.

$\dfrac{P}{Q} = \dfrac{x(x+5)}{x-3}$				
Test Number	-6	-1	1	4
Value of $\frac{P}{Q}$	$-\frac{1}{9}$	1	-3	36
Sign of $\frac{P}{Q}$	$-$	$+$	$-$	$+$
Interval	$(-\infty,-5)$	$(-5,0)$	$(0,3)$	$(3,\infty)$

$\dfrac{x^2+5x}{x-3}$ is non-negative within the interval $[-5,0]$ and $(3, \infty)$. $-5 \leq x \leq 0$ or $3 < x$.

17. $\dfrac{(x+1)^2}{x^2+2x-3} \leq 0$

$\dfrac{(x+1)(x+1)}{(x-1)(x+3)} \leq 0$

Zeros of $P, Q: -1, 1, -3$

-1 is part of the solution set, so we use a solid dot there.

-3 and 1 are not part of the solution set, $\left(\dfrac{P}{Q}\right.$ is not defined there) so we use open dots there.

$\dfrac{P}{Q} = \dfrac{(x+1)(x+1)}{(x-1)(x+3)}$				
Test Number	-4	-2	0	2
Value of $\frac{P}{Q}$	$\frac{9}{5}$	$-\frac{1}{3}$	$-\frac{1}{3}$	$\frac{9}{5}$
Sign of $\frac{P}{Q}$	$+$	$-$	$-$	$+$
Interval	$(-\infty, -3)$	$(-3,-1)$	$(-1,1)$	$(1,\infty)$

$\dfrac{(x+1)^2}{x^2+2x-3}$ is non-positive within the intervals $(-3,-1)$ and $(-1,1)$ as well as at $x = -1$. More simply, we can write $-3 < x < 1$ or $(-3,1)$.

19. $\dfrac{1}{x} < 4$

$\dfrac{1}{x} - 4 < 0$

$\dfrac{1-4x}{x} < 0$

Zeros of $P, Q: 0, \frac{1}{4}$

$\dfrac{P}{Q} = \dfrac{1-4x}{x}$			
Test Number	-1	0.1	1
Value of $\frac{P}{Q}$	-5	6	-3
Sign of $\frac{P}{Q}$	$-$	$+$	$-$
Interval	$(-\infty, 0)$	$(0,\frac{1}{4})$	$(\frac{1}{4},\infty)$

$\dfrac{1-4x}{x} < 0$ and $\dfrac{1}{x} < 4$ within the interval $(-\infty,0)$ and $\left(\frac{1}{4},\infty\right)$. $x < 0$ or $\frac{1}{4} < x$.

21.
$$\frac{3x+1}{x+4} \leq 1$$
$$\frac{3x+1}{x+4} - 1 \leq 0$$
$$\frac{3x+1-(x+4)}{x+4} \leq 0$$

Common Error: $3x+1-x+4$ is an incorrect numerator.

$$\frac{2x-3}{x+4} \leq 0$$

Zeros of P, Q: $0, -4, \frac{3}{2}$

	$\frac{P}{Q} = \frac{2x-3}{x+4}$		
Test Number	-5	0	2
Value of $\frac{P}{Q}$	13	$-\frac{3}{4}$	$-\frac{1}{6}$
Sign of $\frac{P}{Q}$	$+$	$-$	$+$
Interval	$(-\infty, -4)$	$(-4, \frac{3}{2})$	$(\frac{3}{2}, \infty)$

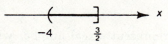

$\frac{3}{2}$ is part of the solution set so we use a solid dot there.

-4 is not part of the solution set ($\frac{P}{Q}$ is not defined there) so we use an open dot there.

$\frac{2x-3}{x+4} \leq 0$ and $\frac{3x+1}{x+4} \leq 1$ within the interval $(-4, \frac{3}{2}]. -4 < x \leq \frac{3}{2}$

23.
$$\frac{2}{x+1} \geq \frac{1}{x-2}$$
$$\frac{2}{x+1} - \frac{1}{x-2} \geq 0$$
$$\frac{2(x-2)-(x+1)}{(x+1)(x-2)} \geq 0$$
$$\frac{2x-4-x-1}{(x+1)(x-2)} \geq 0$$
$$\frac{x-5}{(x+1)(x-2)} \geq 0$$

	$\frac{P}{Q} = \frac{x-5}{(x+1)(x-2)}$			
Test Number	-2	0	3	6
Value of $\frac{P}{Q}$	$-\frac{7}{4}$	$\frac{5}{2}$	$-\frac{1}{2}$	$\frac{1}{28}$
Sign of $\frac{P}{Q}$	$-$	$+$	$-$	$+$
Interval	$(-\infty, -1)$	$(-1, 2)$	$(2, 5)$	$(5, \infty)$

Zeros of P, Q: $5, -1, 2$

5 is a part of the solution set, so we use a solid dot there.

-1 and 2 are not part of the solution set, ($\frac{P}{Q}$ is not defined there) so we use an open dot there.

$\frac{x-5}{(x-1)(x-2)} \geq 0$ and $\frac{2}{x+1} \geq \frac{1}{x-2}$ within the interval $(-1, 2)$ and $(5, \infty)$. $-1 < x < 2$ or $5 \leq x$.

25.
$$x^3 + 2x^2 \leq 8x$$
$$x^3 + 2x^2 - 8x \leq 0$$
$$x(x^2 + 2x - 8) \leq 0$$
$$x(x + 4)(x - 2) \leq 0$$

Zeros : $0, -4, 2$

$0, -4, 2$ are part of the solution set, so we use solid dots there.

$x^3 + 2x^2 - 8x = x(x+4)(x-2)$				
Test Number	-5	-1	1	3
Value of $\frac{P}{Q}$	-35	9	-5	21
Sign of $\frac{P}{Q}$	$-$	$+$	$-$	$+$
Interval	$(-\infty, -4)$	$(-4, 0)$	$(0, 2)$	$(2, \infty)$

$x(x-4)(x-2) \leq 0$ and $x^3 + 2x^2 \leq 8x$ within the intervals $(-\infty, -4]$ and $[0,2]$.
$x \leq -4$ or $0 \leq x \leq 2$.

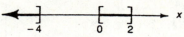

27. $\sqrt{x^2 - 9}$ will represent a
real number when $x^2 - 9 \geq 0$.
$x^2 - 9 \geq 0$
$(x+3)(x-3) \geq 0$
Zeros of $P, Q : -3, 3$
-3 and 3 are part of the
solution set, so we use solid dots.

$x^2 - 9 = (x+3)(x-3)$			
Test Number	-4	0	4
Value of Polynomial for Test Number	7	-9	7
Signal of Polynomial in Interval	$+$	$-$	$+$
Interval	$(-\infty, -3)$	$(-3, 3)$	$(3, \infty)$

$x^2 - 9 \geq 0$ and $\sqrt{x^2 - 9}$ will represent a real number within the intervals $(-\infty, -3]$ and $[3, \infty)$.
$x \leq -3$ or $x \geq 3$.

29. $\sqrt{2x^2 + x - 6}$ will represent a
real number when $2x^2 + x - 6 \geq 0$.
$2x^2 + x - 6 \geq 0$
$(2x-3)(x+2) \geq 0$
Zeros: $-2, \frac{3}{2}$
-2 and $\frac{3}{2}$ are part of the
solution set, so we use solid dots.

$2x^2 + x - 6 = (2x-3)(x+2)$			
Test Number	-3	0	2
Value of Polynomial for Test Number	9	-6	4
Sign of Polynomial in Interval	$+$	$-$	$+$
Interval	$(-\infty, -2)$	$(-2, \frac{3}{2})$	$(\frac{3}{2}, \infty)$

$2x^2 + x - 6 \geq 0$ and $\sqrt{2x^2 + x - 6}$ will represent a real number within the intervals $(-\infty, -2]$ and $[\frac{3}{2}, \infty)$. $x \leq -2$ or $x \geq \frac{3}{2}$.

31. $\sqrt{\frac{x+7}{3-x}}$ will represent a
real number when $\frac{x+7}{3-x} \geq 0$.
$\frac{x+7}{3-x} \geq 0$
Zeros of $\frac{P}{Q} : -7, 3$
-7 is part of the solution
set, so we use a solid dot there.
3 is not part of the solution
set ($\frac{P}{Q}$ is not
defined there) so we use an
open dot there.

$\frac{P}{Q} = \frac{x+7}{3-x}$			
Test Number	-8	0	4
Value of $\frac{P}{Q}$	$-\frac{1}{11}$	$\frac{7}{3}$	-11
Sign of $\frac{P}{Q}$	$-$	$+$	$-$
Interval	$(-\infty, -7)$	$(-7, 3)$	$(3, \infty)$

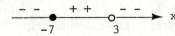

$\frac{x+7}{3-x} \geq 0$ and $\sqrt{\frac{x+7}{3-x}}$ will represent a real number within the interval $[-7,3)$. $-7 \leq x < 3$.

33.
$$\begin{aligned} x^2 + 1 &< 2x \\ x^2 - 2x + 1 &< 0 \\ (x-1)^2 &< 0 \end{aligned}$$

Since the square of no real number is negative, these statements are never true

for any real number x.　　　　　No solution (and no graph). ϕ is the solution set.

35.
$$\begin{aligned} x^2 &< 3x - 3 \\ x^2 - 3x + 3 &< 0 \end{aligned}$$

We attempt to find all real zeros of the polynomial.

$$\begin{aligned} x^2 - 3x + 3 &= 0 \\ x &= \frac{-b \pm \sqrt{b^2 - 4ac}}{2a} \\ a &= 1 \\ b &= -3 \\ c &= 3 \\ x &= \frac{-(-3) \pm \sqrt{(-3)^2 - 4(1)(3)}}{2(1)} \\ x &= \frac{3 \pm \sqrt{-3}}{2} \end{aligned}$$

The polynomial has no real zeros. Hence the statement is either true for all real x or for no real x.

To determine which, we choose a test number, say 0.

$$\begin{aligned} x^2 &\overset{?}{<} 3x \quad -3 \\ 0^2 &\overset{?}{<} 3(0) \quad -3 \\ 0 &< -3 \quad \text{False.} \end{aligned}$$

The statement is never true for any real number x.

No solution (and no graph). ϕ is the solution set.

37.
$$\begin{aligned} x^2 - 1 &\geq 4x \\ x^2 - 4x - 1 &\geq 0 \end{aligned}$$

Find all real zeros of the polynomial.

$$\begin{aligned} x^2 - 4x - 1 &= 0 \\ x &= \frac{-b \pm \sqrt{b^2 - 4ac}}{2a} \\ a &= 1 \\ b &= -4 \\ c &= -1 \\ x &= \frac{-(-4) \pm \sqrt{(-4)^2 - 4(1)(-1)}}{2(1)} \\ x &= \frac{4 \pm \sqrt{16 + 4}}{2} \\ x &= \frac{4 \pm \sqrt{20}}{2} \\ x &= 2 \pm \sqrt{5} \qquad\qquad Common Error: \\ &\approx -0.236, 4.236 \qquad x \neq 2 \pm \sqrt{20} \end{aligned}$$

Plot the real zeros on a number line.

$2-\sqrt{5}$ $2-\sqrt{5}$ x

Polynomial $x^2 - 4x - 1$			
Test Number	-1	0	5
Value of Polynomial for Test Number	4	-1	4
Sign of Polynomial in Interval	$+$	$-$	$+$
Interval	$(-\infty, 2-\sqrt{5})$	$(2-\sqrt{5}, 2+\sqrt{5})$	$(2+\sqrt{5}, \infty)$

$+$ $+$ $-$ $-$ $+$ $+$
$2-\sqrt{5}$ $2-\sqrt{5}$ x

$x^2 - 4x - 1 \geq 0$ and $x^2 - 1 \geq 4x$ within the intervals
$(-\infty, 2-\sqrt{5})$ and $(2+\sqrt{5}, \infty)$
$x \leq 2 - \sqrt{5}$ or $x \geq 2 + \sqrt{5}$

$2 - \sqrt{5}$ $2 + \sqrt{5}$ x

39. $$x^3 > 2x^2 + x$$
$$x^3 - 2x^2 - x > 0$$

Final all real zeros of the polynomial.
$$x^3 - 2x^2 - x = 0$$
$$x(x^2 - 2x - 1) = 0$$
$$x = 0 \quad \text{or} \quad x^2 - 2x - 1 = 0$$
$$x = \frac{-b \pm \sqrt{b^2 - 4ac}}{2a}$$
$$a = 1$$
$$b = -2$$
$$c = -1$$
$$x = \frac{-(-2) \pm \sqrt{(-2)^2 - 4(1)(-1)}}{2(1)}$$
$$x = \frac{2 \pm \sqrt{8}}{2}$$
$$x = 1 \pm \sqrt{2}$$
$$\approx -0.414, 2.414$$

Plot the real zeros on a number line.

$1-\sqrt{2}$ 0 $1-\sqrt{2}$ x

Polynomial $x^3 - 2x^2 - x$				
Test Number	-1	-0.1	1	3
Value of Polynomial for Test Number	-2	0.079	-2	6
Sign of Polynomial in Interval	$-$	$+$	$-$	$+$
Interval	$(-\infty, 1-\sqrt{2})$	$(1-\sqrt{2}, 0)$	$(0, 1+\sqrt{2})$	$(1+\sqrt{2}, \infty)$

$-$ $-$ $+$ $+$ $-$ $-$ $+$ $+$
$1-\sqrt{2}$ 0 $1-\sqrt{2}$ x

$x^3 - 2x^2 - x > 0$ and $x^3 > 2x^2 + x$ within the intervals

$(1 - \sqrt{2}, 0)$, and $(1 + \sqrt{2}, \infty)$
$1 - \sqrt{2} < x < 0$ or $x > 1 + \sqrt{2}$

41.
$$4x^4 + 4 \leq 17x^2$$
$$4x^4 - 17x^2 + 4 \leq 0$$
$$(4x^2 - 1)(x^2 - 4) \leq 0$$
$$(2x - 1)(2x + 1)(x - 2)(x + 2) \leq 0$$

Zeros: $-2, -\frac{1}{2}, \frac{1}{2}, 2$
$-2, -\frac{1}{2}, \frac{1}{2}$, and 2 are part of the solution set, so we use solid dots.

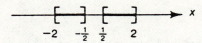

$4x^4 - 17x^2 + 4 = (2x - 1)(2x + 1)(x - 2)(x + 2)$					
Test Number	-3	-1	0	1	3
Value of Polynomial for Test Number	175	-9	4	-9	175
Sign of Polynomial in Interval	$+$	$-$	$+$	$-$	$+$
Interval	$(-\infty, -2)$	$(-2, -\frac{1}{2})$	$(-\frac{1}{2}, \frac{1}{2})$	$(\frac{1}{2}, 2)$	$(2, \infty)$

$4x^4 - 17x^2 + 4 \leq 0$ and $4x^4 + 4 \leq 17x^2$ within the intervals
$[-2, -\frac{1}{2}]$ and $[\frac{1}{2}, 2]$.
$-2 \leq x \leq -\frac{1}{2}$ or $\frac{1}{2} \leq x \leq 2$.

43. $\mid x^2 - 1 \mid \leq 3$ According to Theorem 5, Section 3-4, we must consider the equations
$x^2 - 1 = 3$ and $x^2 - 1 = -3$ to solve the equality part of the inequality statement,
and the double inequality $-3 < x^2 - 1 < 3$ to solve the inequality part.
Case 1:
$$x^2 - 1 = 3$$
$$x^2 = 4$$
There are two solutions, -2, and 2.
Case 2:
$$x^2 - 1 = -3$$
$$x^2 = -2$$
There are no real solutions.
Case 3:
$$-3 < x^2 - 1 < 3$$
To satisfy this relation, both statements $-3 < x^2 - 1$ and $x^2 - 1 < 3$ must hold.
But $-3 < x^2 - 1$ or $-4 < x^2$ holds for all real x, so we need only
examine $x^2 - 1 < 3$ or $x^2 - 4 < 0$. $(x - 2)(x + 2) < 0$
Zeros: $-2, 2$

$x^2 - 4 = (x-2)(x+2)$			
Test Number	-3	0	3
Value of Polynomial for Test Number	5	-4	5
Sign of Polynomial in Interval	$+$	$-$	$+$
Interval	$(-\infty, -2)$	$(-2, 2)$	$(2, \infty)$

$(x-2)(x+2)$ is negative within the interval $(-2, 2)$. Combining the solutions from the separate cases, $|\,x^2 - 1\,| \le 3$ when $-2 \le x \le 2. [-2, 2]$

45.   A. A profit will result
       if cost is less than revenue; that is, if

$$\begin{aligned} C &< R \\ 28 - 2p &< 9p - p^2 \\ p^2 - 11p + 28 &< 0 \\ (p-7)(p-4) &< 0 \end{aligned}$$

Zeros: 4, 7

$p^2 - 11p + 28 = (p-7)(p-4)$			
Test Number	3	5	8
Value of Polynomial for Test Number	4	-2	4
Sign of Polynomial in Interval	$+$	$-$	$+$
Interval	$(-\infty, 4)$	$(4, 7)$	$(7, \infty)$

$p^2 - 11p + 28 < 0$ and a profit will occur $(C < R)$, for $\$4 < p < \$7$ or $(\$4, \$7)$

B. A loss will result, if cost is greater than revenue;
that is, if

$$\begin{aligned} C &> R \\ 28 - 2p &> 9p - p^2 \\ p^2 - 11p + 28 &> 0 \end{aligned}$$

Referring to the sign chart in part (A), we see that $p^2 - 11p + 28 > 0$, and a loss will occur $(C > R)$, for $p < \$4$ or $p > \$7$.

Since a negative price doesn't make sense we delete any number to the left of 0.

Thus, a loss will occur for $\$\,0 \le p < \$4$ or $p > \$7$. $[\$0, \$4) \cup (\$7, \infty)$.

47.   The object will be 160 feet
       or higher while $160 \le d$.

$$\begin{aligned} 160 &\le 112t - 16t^2 \\ 16t^2 - 112t + 160 &\le 0 \\ t^2 - 7t + 10 &\le 0 \ (\text{dividing both} \\ & \qquad \text{sides by 16}) \\ (t-2)(t-5) &\le 0 \end{aligned}$$

Zeros: 2, 5

$t^2 - 7t + 10 = (t-2)(t-5)$			
Test Number	0	3	6
Value of Polynomial for Test Number	10	-2	4
Sign of Polynomial in Interval	$+$	$-$	$+$
Interval	$(-\infty, 2)$	$(2, 5)$	$(5, \infty)$

$t^2 - 7t + 10 \le 0$, and the object will be 160 feet or higher, for $2 \le t \le 5 \quad [2, 5]$

49. It will take the car more than
330 feet to stop when $d > 330$.

$$0.044v^2 + 1.1v > 330$$
$$0.044v^2 + 1.1v - 330 > 0$$
$$v^2 + 25v - 7,500 > 0$$

(dividing both sides by 0.044)

$$(v - 75)(v + 100) > 0$$

Zeros: $-100, 75$

$v^2 + 25v - 7,500 = (v - 75)(v + 100)$			
Test Number	-200	0	100
Value of Polynomial for Test Number	$27,500$	$-7,500$	$5,000$
Sign of Polynomial in Interval	$+$	$-$	$+$
Interval	$(-\infty, -100)$	$(-100, 75)$	$(75, \infty)$

$v^2 + 25v - 7,500 > 0$, and it will take the car more than 330 feet to stop, for
$v < -100$ or $v > 75$.
Since a negative speed doesn't make sense we delete any number to the left of 0.
Thus $v > 75$ miles per hour.

51. Sales will be 8,000 units or more when
$$8 \leq S$$
$$8 \leq \frac{200t}{t^2+100}$$
$$8 - \frac{200t}{t^2+100} \leq 0$$
$$\frac{8(t^2+100)-200t}{t^2+100} \leq 0$$
$$\frac{8t^2-200t+800}{t^2+100} \leq 0$$

Find all real zeros of P and Q,
$$\frac{P}{Q} = \frac{8t^2-200t+800}{t^2+100}.$$
$$8t^2 - 200t + 800 = 0$$
$$t^2 - 25t + 100 = 0$$
$$(t - 5)(t - 20) = 0$$
$$t = 5, 20 \text{ Zeros of P}$$
$t^2 + 100$ has no real zero.

$\frac{P}{Q} = \frac{8t^2-200t+800}{t^2+100}$			
Test Number	0	10	30
Value of $\frac{P}{Q}$	8	$-\frac{1}{50}$	2
Sign of $\frac{P}{Q}$	$+$	$-$	$+$
Interval	$(-\infty, 5)$	$(5, 20)$	$(20, \infty)$

$\frac{8t^2-200t+800}{t^2+100} \leq 0$ and sales will be 8 thousand units or more, for $5 \leq t \leq 20$.

Exercise 3-8

Key Ideas and Formulas

1. If both sides of an equation are squared, then the solution set of the original equation is a subset of the solution set of the new equation.

   Any new equation obtained by raising both members of an equation to the same power may have solutions (called *extraneous solutions*) that are not solutions of the original equation.

   Thus, every solution of the new equation must be checked in the original equation to eliminate extraneous solutions.

2. To solve an equation that is not quadratic, it may be possible to transform it to the form

$$au^2 + bu + c = 0$$

   where u is an expression in some other variable.

1.
$$\sqrt[3]{x+5} = 3$$
$$x+5 = 27$$
$$x = 22$$

   Check
$$\sqrt[3]{22+5} \overset{?}{=} 3$$
$$\sqrt[3]{27} \overset{?}{=} 3$$
$$3 \overset{\checkmark}{=} 3$$

   Solution: 22

3.
$$\sqrt{5n+9} = n-1$$
$$5n+9 = n^2 - 2n + 1$$
$$0 = n^2 - 7n - 8$$
$$n^2 - 7n - 8 = 0$$
$$(n-8)(n+1) = 0$$
$$n = 8, -1$$
$$\text{Check}\sqrt{5(8)+9} \overset{?}{=} 8-1$$
$$7 \overset{\checkmark}{=} 7$$
$$\sqrt{5(-1)+9} \overset{?}{=} -1-1$$
$$2 \neq -2$$

   Solution: 8

5.
$$\sqrt{x+5} + 7 = 0$$
$$\sqrt{x+5} = -7$$

   Since the left side is non-negative for real x, there is no solution. However, if we don't notice this, we proceed as follows:
$$x+5 = 49$$
$$x = 44$$

   But this is an extraneous solution, since checking gives
$$\sqrt{44+5} + 7 \overset{?}{=} 0$$
$$14 \neq 0$$

   No solution.

7.
$$\sqrt{3x+4} = 2 + \sqrt{x}$$
$$3x+4 = 4 + 4\sqrt{x} + x$$
$$2x = 4\sqrt{x}$$
$$x = 2\sqrt{x}$$
$$x^2 = 4x$$
$$x^2 - 4x = 0$$
$$x(x-4) = 0$$
$$x = 0, 4$$
$$\text{Check } \sqrt{3(0)+4} \overset{?}{=} 2 + \sqrt{0}$$
$$2 \overset{\checkmark}{=} 2$$
$$\sqrt{3(4)+4} \overset{?}{=} 2 + \sqrt{4}$$
$$4 \overset{\checkmark}{=} 4$$

   Solution: 0,4

9. $y^4 - 2y^2 - 8 = 0$
Let $u = y^2$, then
$$
\begin{aligned}
u^2 - 2u - 8 &= 0 \\
(u - 4)(u + 2) &= 0 \\
u &= 4, -2
\end{aligned}
$$
$$
\begin{array}{ll}
y^2 = 4 & y^2 = -2 \\
y = \pm 2 & y = \pm i\sqrt{2}
\end{array}
$$

13. $2x^{2/3} + 3x^{1/3} - 2 = 0$
Let $u = x^{1/3}$, then
$$
\begin{aligned}
2u^2 + 3u - 2 &= 0 \\
(2u - 1)(u + 2) &= 0 \\
u &= \tfrac{1}{2}, -2
\end{aligned}
$$
$$
\begin{array}{ll}
x^{1/3} = \tfrac{1}{2} & x^{1/3} = -2 \\
x = \tfrac{1}{8} & x = -8
\end{array}
$$

11. $x^{10} + 3x^5 - 10 = 0$
Let $u = x^5$, then
$$
\begin{aligned}
u^2 + 3u - 10 &= 0 \\
(u + 5)(u - 2) &= 0 \\
u &= -5, 2
\end{aligned}
$$
$$
\begin{array}{ll}
x^5 = -5 & x^5 = 2 \\
x = \sqrt[5]{-5} \text{ or } -\sqrt[5]{5} & x = \sqrt[5]{2}
\end{array}
$$

15. $(m^2 - m)^2 - 4(m^2 - m) = 12$
Let $u = m^2 - m$, then
$$
\begin{aligned}
u^2 - 4u &= 12 \\
u^2 - 4u - 12 &= 0 \\
(u - 6)(u + 2) &= 0 \\
u &= 6, -2
\end{aligned}
$$
$$
\begin{array}{ll}
m^2 - m = 6 & m^2 - m = -2 \\
m^2 - m - 6 = 0 & m^2 - m + 2 = 0 \\
(m - 3)(m + 2) = 0 & m = \dfrac{-(-1) \pm \sqrt{(-1)^2 - 4(1)(2)}}{2(1)} \\
m = 3, -2 & m = \dfrac{1 \pm \sqrt{-7}}{2} \text{ or } \dfrac{1 \pm i\sqrt{7}}{2} \\
& m = \tfrac{1}{2} \pm \tfrac{\sqrt{7}}{2}i
\end{array}
$$

17.
$$
\begin{aligned}
\sqrt{u - 2} &= 2 + \sqrt{2u + 3} \\
u - 2 &= 4 + 4\sqrt{2u + 3} + 2u + 3 \\
u - 2 &= 2u + 7 + 4\sqrt{2u + 3} \\
-u - 9 &= 4\sqrt{2u + 3} \\
u^2 + 18u + 81 &= 16(2u + 3) \\
u^2 + 18u + 81 &= 32u + 48 \\
u^2 - 14u + 33 &\doteq 0 \\
(u - 3)(u - 11) &= 0 \\
u &= 3, 11
\end{aligned}
$$

Common Error:
$u - 2 = 4 + 2u + 3$ is not an equivalent equation to the given equation.

$$(2 + \sqrt{2u + 3})^2 \neq 4 + 2u + 3$$

$$\text{Check } \sqrt{3-2} \ \overset{?}{=} \ 2\sqrt{2(3)+3}$$
$$1 \ne 5$$
$$\sqrt{11-2} \ \overset{?}{=} \ 2+\sqrt{2(11)+3}$$
$$3 \ne 7$$

No solution.

19.
$$\sqrt{3y-2} \ = \ 3-\sqrt{3y+1}$$
$$3y-2 \ = \ 9-6\sqrt{3y+1}+3y+1$$
$$3y-2 \ = \ 3y+10-6\sqrt{3y+1}$$
$$-12 \ = \ -6\sqrt{3y+1}$$
$$2 \ = \ \sqrt{3y+1}$$
$$4 \ = \ 3y+1$$
$$3 \ = \ 3y$$
$$y \ = \ 1$$

$$\text{Check } \sqrt{3(1)-2} \ \overset{?}{=} \ 3-\sqrt{3(1)+1}$$
$$1 \ \overset{\vee}{=} \ 1$$

Solution: 1

21.
$$\sqrt{7x-2}-\sqrt{x+1} \ = \ \sqrt{3}$$
$$\sqrt{7x-2} \ = \ \sqrt{x+1}+\sqrt{3}$$
$$7x-2 \ = \ x+1+2\sqrt{3}\sqrt{x+1}+3$$
$$7x-2 \ = \ x+4+2\sqrt{3(x+1)}$$
$$6x-6 \ = \ 2\sqrt{3(x+1)}$$
$$3x-3 \ = \ \sqrt{3x+3}$$
$$9x^2-18x+9 \ = \ 3x+3$$
$$9x^2-21x+6 \ = \ 0$$
$$3x^2-7x+2 \ = \ 0$$
$$(3x-1)(x-2) \ = \ 0$$
$$x \ = \ \tfrac{1}{3},2$$

$$\text{Check } \sqrt{7(\tfrac{1}{3})-2}-\sqrt{\tfrac{1}{3}+1} \ \overset{?}{=} \ \sqrt{3}$$
$$\sqrt{\tfrac{1}{3}}-\sqrt{\tfrac{4}{3}} \ \overset{?}{=} \ \sqrt{3}$$
$$-\tfrac{1}{3}\sqrt{3} \ \ne \ \sqrt{3}$$
$$\sqrt{7(2)-2}-\sqrt{2+1} \ \overset{?}{=} \ \sqrt{3}$$
$$\sqrt{12}-\sqrt{3} \ \overset{?}{=} \ \sqrt{3}$$
$$\sqrt{3} \ \overset{\vee}{=} \ \sqrt{3}$$

Solution: 2

23. $3n^{-2} - 11n^{-1} - 20 = 0$

Let $u = n^{-1}$, then

$$3u^2 - 11u - 20 = 0$$
$$(3u + 4)(u - 5) = 0$$
$$u = -\tfrac{4}{3}, 5$$

$$n^{-1} = -\tfrac{4}{3} \quad n^{-1} = 5$$
$$n = -\tfrac{3}{4} \quad n = \tfrac{1}{5}$$

25. $9y^{-4} - 10y^{-2} + 1 = 0$

Let $u = y^{-2}$, then

$$9u^2 - 10u + 1 = 0$$
$$(9u - 1)(u - 1) = 0$$
$$u = \tfrac{1}{9}, 1$$
$$y^{-2} = \tfrac{1}{9} \quad y^{-2} = 1$$
$$y^2 = 9 \quad y^2 = 1$$
$$y = \pm 3 \quad y = \pm 1$$

27. $y^{1/2} - 3y^{1/4} + 2 = 0$

Let $u = y^{1/4}$, then

$$u^2 - 3u + 2 = 0$$
$$(u - 1)(u - 2) = 0$$
$$u = 1, 2$$
$$y^{1/4} = 1 \quad y^{1/4} = 2$$
$$y = 1 \quad y = 16$$

Common Error:

$y \neq 2^{1/4}$

Both sides must be raised
to the fourth power to
eliminate the $y^{1/4}$.

29. $(m - 5)^4 + 36 = 13(m - 5)^2$

Let $u = (m - 5)^2$, then

$$u^2 + 36 = 13u$$
$$u^2 + 13u + 36 = 0$$
$$(u - 4)(u - 9) = 0$$
$$u = 4, 9$$

$$(m - 5)^2 = 4 \qquad (m - 5)^2 = 9$$
$$m - 5 = \pm 2 \qquad m - 5 = \pm 3$$
$$m = 5 \pm 2 \qquad m = 5 \pm 3$$
$$m = 3, 7 \qquad m = 2, 8$$

31.
$$\sqrt{5 - 2x} - \sqrt{x + 6} = \sqrt{x + 3}$$
$$\sqrt{5 - 2x} = \sqrt{x + 6} + \sqrt{x + 3}$$
$$5 - 2x = x + 6$$
$$+ 2\sqrt{x + 6}\sqrt{x + 3}$$
$$+ x + 3$$
$$5 - 2x = 2x + 9$$
$$+ 2\sqrt{x + 6}\sqrt{x + 3}$$
$$-4x - 4 = 2\sqrt{x + 6}\sqrt{x + 3}$$
$$-2x - 2 = \sqrt{(x + 6)(x + 3)}$$
$$4x^2 + 8x + 4 = (x + 6)(x + 3)$$
$$4x^2 + 8x + 4 = x^2 + 9x + 18$$
$$3x^2 - x - 14 = 0$$
$$(3x - 7)(x + 2) = 0$$
$$x = \tfrac{7}{3}, -2$$

Check

$$\sqrt{5 - 2(\tfrac{7}{3})} - \sqrt{\tfrac{7}{3} + 6} \overset{?}{=} \sqrt{\tfrac{7}{3}}$$
$$\sqrt{\tfrac{1}{3}} - \sqrt{\tfrac{25}{3}} \overset{?}{=} \sqrt{\tfrac{16}{3}}$$
$$-\tfrac{4\sqrt{3}}{3} \neq \tfrac{4\sqrt{3}}{3}$$
$$\sqrt{5 - 2(-2)} - \sqrt{-2 + 6} \overset{?}{=} \sqrt{-2 + 3}$$
$$1 \overset{\checkmark}{=} 1$$

Solution: -2

33.
$$2 + 3y^{-4} = 6y^{-2}$$
$$2y^4 + 3 = 6y^2 \text{ Multiply both}$$
$$\text{members by}$$
$$y^4 \quad y \neq 0$$
$$2y^4 - 6y^2 + 3 = 0$$
$$\text{Let } u = y^2, \text{ then}$$
$$2u^2 - 6u + 3 = 0$$
$$u = \frac{-b \pm \sqrt{b^2 - 4ac}}{2a}$$
$$a = 2$$
$$b = -6$$
$$c = 3$$
$$u = \frac{(-6) \pm \sqrt{(-6)^2 - 4(2)(3)}}{2(2)}$$
$$u = \frac{6 \pm \sqrt{36 - 24}}{4}$$
$$u = \frac{6 \pm \sqrt{12}}{4}$$
$$u = \frac{2(3 \pm \sqrt{3})}{4}$$
$$u = \frac{3 \pm \sqrt{3}}{2}$$
$$y^2 = \frac{3 \pm \sqrt{3}}{2}$$
$$y = \pm\sqrt{\frac{3 \pm \sqrt{3}}{2}} \text{ (four roots)}$$

35. **By squaring:**

$$m - 7\sqrt{m} + 12 = 0$$
$$m + 12 = 7\sqrt{m}$$
$$m^2 + 24m + 144 = 49m$$
$$m^2 - 25m + 144 = 0$$
$$(m-9)(m-16) = 0$$
$$m = 9, 16$$
$$0 \overset{\checkmark}{=} 0$$

Check

$$9 - 7\sqrt{9} + 12 \overset{?}{=} 0$$
$$0 \overset{\checkmark}{=} 0$$
$$16 - 7\sqrt{16} + 12 \overset{?}{=} 0$$
$$0 \overset{\checkmark}{=} 0$$

Solution: 9, 16

By substitution:

$$m - 7\sqrt{m} + 12 = 0$$

Let $u = \sqrt{m}$, then

$$u^2 - 7u + 12 = 0$$
$$(u-4)(u-3) = 0$$
$$u = 3, 4$$
$$\sqrt{m} = 3 \quad \sqrt{m} = 4$$
$$m = 9 \quad m = 16$$

These answers have already been checked.

37. **By squaring:**

$$t - 11\sqrt{t} + 18 = 0$$
$$t + 18 = 11\sqrt{t}$$
$$t^2 + 36t + 324 = 121t$$
$$t^2 - 85t + 324 = 0$$
$$(t-4)(t-81) = 0$$
$$t = 4, 81$$

Check

$$4 - 11\sqrt{4} + 18 \overset{?}{=} 0$$
$$0 \overset{\checkmark}{=} 0$$
$$81 - 11\sqrt{81} + 18 \overset{?}{=} 0$$
$$0 \overset{\checkmark}{=} 0$$

Solution: 4, 81

By substitution:

$$t - 11\sqrt{t} + 12 = 0$$

Let $u = \sqrt{t}$, then

$$u^2 - 11u + 18 = 0$$
$$(u-9)(u-2) = 0$$
$$u = 2, 9$$
$$u = 2 \quad u = 9$$
$$\sqrt{t} = 2 \quad \sqrt{t} = 9$$
$$t = 4 \quad t = 81$$

These answers have already been checked.

Exercise 3-9

CHAPTER REVIEW

1. $\begin{aligned} 0.05x + 0.25(30 - x) &= 3.3 \\ 0.05x + 7.5 - 0.25x &= 3.3 \\ -0.2x + 7.5 &= 3.3 \\ -0.2x &= -4.2 \\ x &= \frac{-4.2}{-0.2} \\ x &= 21 \quad (3-1) \end{aligned}$

2. $\begin{aligned} \frac{5x}{3} - \frac{4+x}{2} &= \frac{x-2}{4} + 1 \\ 12\frac{5x}{3} - 12\frac{(4+x)}{2} &= 12\frac{(x-2)}{4} + 12 \\ 20x - 6(4 + x) &= 3(x - 2) + 12 \\ 20x - 24 - 6x &= 3x - 6 + 15 \\ 14x - 24 &= 3x + 9 \\ 11x &= 33 \\ x &= 3 \qquad (3-1) \end{aligned}$

3. $\begin{aligned} 3(2 - x) - 2 &\leq 2x - 1 \\ 6 - 3x - 2 &\leq 2x - 1 \\ -3x + 4 &\leq 2x - 1 \\ -5x &\leq -5 \\ x &\geq 1 \\ [1, \infty) \quad (3-3) \end{aligned}$

4. $\begin{aligned} |y + 9| &< 5 \\ -5 &< y + 9 < 5 \\ -14 &< y < -4 \end{aligned}$

$(-14, 4) \quad (3-4)$

5. $\begin{aligned} |3 - 2x| &\leq 5 \\ -5 &\leq 3 - 2x \leq 5 \\ -8 &\leq -2x \leq 2 \\ 4 &\geq x \geq -1 \\ -1 &\leq x \leq 4 \end{aligned}$

$[-1, 4] \quad (3-4)$

6. $\begin{aligned} x^2 + x &< 20 \\ x^2 + x - 20 &< 0 \\ (x + 5)(x - 4) &< 0 \end{aligned}$
Zeros: $-5, 4$

$x^2 + x - 20 = (x + 5)(x - 4)$			
Test Number	-6	0	5
Value of Polynomial for Test Number	10	-20	10
Sign of Polynomial in Interval	$+$	$-$	$+$
Interval	$(-\infty, -5)$	$(-5, 4)$	$(4, \infty)$

$x^2 + x - 20$ is negative within the interval $(-5, 4)$. $-5 < x < 4$. $(3-7)$

7. $\begin{aligned} x^2 &\geq 4x + 21 \\ x^2 - 4x - 21 &\geq 0 \\ (x + 3)(x - 7) &\geq 0 \end{aligned}$
Zeros: $-3, 7$
$-3, 7$ are part of the solution set, so we use solid dots.

$x^2 - 4x - 21 = (x + 3)(x - 7)$			
Test Number	-4	0	8
Value of Polynomial for Test Number	11	-21	11
Sign of Polynomial in Interval	$+$	$-$	$+$
Interval	$(-\infty, -3)$	$(-3, 7)$	$(7, \infty)$

$x^2 - 4x - 21 \geq 0$ is negative within the intervals $(-\infty, -3]$ and $[7, \infty)$. $x \leq -3$ or $x \geq 7$. $(3-7)$

8.   A. $(-3 + 2i) + (6 - 8i) = -3 + 2i + 6 - 8i = 3 - 6i$

B. $(3 - 3i)(2 + 3i) = 6 + 3i - 9i^2 = 6 + 3i + 9 = 15 + 3i$

C. $\frac{13-i}{5-3i} = \frac{(13-i)}{(5-3i)}\frac{5+3i}{5+3i} = \frac{65+34i-3i^2}{25-9i^2} = \frac{65+34i+3}{25+9} = \frac{68+34i}{34} = 2 + i$ $(3-5)$

9. $\begin{aligned} 2x^2 - 7 &= 0 \\ 2x^2 &= 7 \\ x^2 &= \frac{7}{2} \\ x &= \pm\sqrt{\frac{7}{2}} \\ x &= \pm\frac{\sqrt{14}}{2} \quad (3-6) \end{aligned}$

10. $\begin{aligned} 2x^2 &= 4 \\ 2x^2 - 4x &= 0 \\ 2x(x - 2) &= 0 \\ 2x = 0 \quad x - 2 &= 0 \\ x = 0 \quad x &= 2 \quad (3-6) \end{aligned}$

11. $\begin{aligned} 2x^2 &= 7x - 3 \\ 2x^2 - 7x + 3 &= 0 \\ (2x - 1)(x - 3) &= 0 \\ 2x - 1 = 0 \quad x - 3 &= 0 \\ x = \tfrac{1}{2} \quad x &= 3 \quad (3-6) \end{aligned}$

12. $\begin{aligned} m^2 + m + 1 &= 0 \\ m &= \frac{-b \pm \sqrt{b^2 - 4ac}}{2a} \\ a &= 1 \\ b &= 1 \\ c &= 1 \\ m &= \frac{-1 \pm \sqrt{(1)^2 - 4(1)(1)}}{2(1)} \\ m &= \frac{-1 \pm \sqrt{-3}}{2} \\ m &= \frac{-1 \pm i\sqrt{3}}{2} \\ m &= -\tfrac{1}{2} \pm \tfrac{\sqrt{3}}{2}i \quad (3-6) \end{aligned}$

13. $\begin{aligned} y^2 &= \tfrac{3}{2}(y + 1) \\ 2y^2 &= 3(y + 1) \\ 2y^2 &= 3y + 3 \\ 2y^2 - 3y - 3 &= 0 \\ y &= \frac{-b \pm \sqrt{b^2 - 4ac}}{2a} \\ a &= 2 \\ b &= -3 \\ c &= -3 \\ y &= \frac{-(-3) \pm \sqrt{(-3)^2 - 4(2)(-3)}}{2(2)} \\ y &= \frac{3 \pm \sqrt{33}}{4} \quad (3-6) \end{aligned}$

14. $\begin{aligned} \sqrt{5x - 6} - x &= 0 \\ \sqrt{5x - 6} &= x \\ 5x - 6 &= x^2 \\ 0 &= x^2 - 5x + 6 \\ x^2 - 5x + 6 &= 0 \\ (x - 3)(x - 2) &= 0 \\ x &= 2, 3 \end{aligned}$

Check

$\sqrt{5(2) - 6} - 2 \overset{?}{=} 0$

$0 \overset{\checkmark}{=} 0$

$\sqrt{5(3) - 6} - 3 \overset{?}{=} 0$

$0 \overset{\checkmark}{=} 0$

Solution: 2,3 $(3-8)$

15. $\sqrt{3 - 5x}$ represents a real number exactly when $3 - 5x$ is positive or zero. We can write this as an inequality statement and solve for x.

$\begin{aligned} 3 - 5x &\geq 0 \\ -5x &\geq -3 \\ x &\leq \tfrac{3}{5} \quad (3-3) \end{aligned}$

16.
$$\frac{7}{2-x} = \frac{10-4x}{x^2+3x-10}$$

$$\frac{7}{2-x} = \frac{10-4x}{(x-2)(x+5)} \text{ Excluded values} : x \neq 2, -5$$

$$\overset{-1}{(x-2)}(x+5)\frac{7}{2-x} = (x-2)(x+5)\frac{10-4x}{(x-2)(x+5)}$$

$$-7(x+5) = 10-4x$$

$$-7x-35 = 10-4x$$

$$-3x = 45$$

$$x = -15 \quad (3-1)$$

17.
$$\frac{u-3}{2u-2} = \frac{1}{6} - \frac{1-u}{3u-3}$$

$$\frac{u-3}{2(u-1)} = \frac{1}{6} - \frac{1-u}{3(u-1)} \text{ Excluded value} : u \neq 1$$

$$6(u-1)\frac{(u-3)}{2(u-1)} = 6(u-1)\frac{1}{6} - 6(u-1)\frac{(1-u)}{3(u-1)}$$

$$3(u-3) = u-1-2(1-u)$$

$$3u-9 = u-1-2+2u$$

$$3u-9 = 3u-3$$

$$-9 = -3 \quad (3-1)$$

No solution

18.
$$\frac{x+3}{8} \leq 5 - \frac{2-x}{3}$$

$$24\frac{(x+3)}{8} \leq 120 - 24\frac{(2-x)}{3}$$

$$3(x+3) \leq 120 - 8(2-x)$$

$$3x+9 \leq 120 - 16 + 8x$$

$$3x+9 \leq 8x+104$$

$$-5x \leq 95$$

$$x \geq -19 \quad (3-3)$$

$$[-19, \infty)$$

19.
$$|3x-8| > 2$$

$$3x-8 < -2 \quad \text{or} \quad 3x-8 > 2$$

$$3x < 6 \qquad\qquad 3x > 10$$

$$x < 2 \quad \text{or} \quad x > \frac{10}{3}$$

$$(-\infty, 2) \cup (10/3, \infty) \quad (3-4)$$

20.
$$\frac{1}{x} < 2$$

$$\frac{1}{x} - 2 < 0$$

$$\frac{1-2x}{x} < 0$$

Zeros of P, Q: $0, \frac{1}{2}$

	$\frac{P}{Q} = \frac{1-2x}{x}$		
Test Number	-1	0.1	1
Value of $\frac{P}{Q}$	-3	8	-1
Sign of $\frac{P}{Q}$	$-$	$+$	$-$
Interval	$(-\infty, 0)$	$(0, \frac{1}{2})$	$(\frac{1}{2}, \infty)$

$\frac{1-2x}{x} < 0$ and $\frac{1}{x} < 2$ within the intervals $(-\infty, 0)$ and $(\frac{1}{2}, \infty)$. $x < 0$ or $x > \frac{1}{2}$. $\quad (3-7)$

21.
$$\frac{3}{x-4} \leq \frac{2}{x-3}$$

$$\frac{3}{x-4} - \frac{2}{x-3} \leq 0$$

$$\frac{3(x-3)-2(x-4)}{(x-4)(x-3)} \leq 0$$

$$\frac{3x-9-2x+8}{(x-4)(x-3)} \leq 0$$

$$\frac{x-1}{(x-4)(x-3)} \leq 0$$

$\frac{P}{Q} = \frac{x-1}{(x-4)(x-3)}$				
Test Number	0	2	3.5	5
Value of $\frac{P}{Q}$	$-\frac{1}{12}$	$\frac{1}{2}$	-10	2
Sign of $\frac{P}{Q}$	$-$	$+$	$-$	$+$
Interval	$(-\infty, 1)$	$(1, 3)$	$(3,4)$	$(4,\infty)$

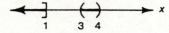

Zeros of P, Q: 1,3,4

1 is part of the solution set so we use a solid dot there. 3 and 4 are not part of the solution set ($\frac{P}{Q}$ is not defined there) so we use open dots there.

$\frac{x-1}{(x-4)(x-3)} \leq 0$ and $\frac{3}{x-4} \leq \frac{2}{x-3}$ within the intervals $(-\infty,1]$ and $(3,4)$. $x \leq 1$ or $3 < x < 4$ (3-7)

22.
$$\sqrt{(1-2m)^2} \leq 3$$
$$|1-2m| \leq 3$$
$$-3 \leq 1-2m \leq 3$$
$$-4 \leq -2m \leq 2$$
$$2 \geq m \geq -1$$
$$-1 \leq m \leq 2 \quad [-1,2] \quad (3-4)$$

23. $\sqrt{\frac{x+4}{2-x}}$ will represent
a real number when $\frac{x+4}{2-x} \geq 0$.
$\frac{x+4}{2-x} \geq 0$
Zeros of P, Q: $-4, 2$

$\frac{P}{Q} = \frac{x+4}{2-x}$			
Test Number	-5	0	3.
Value of $\frac{P}{Q}$	$-\frac{1}{7}$	2	-7
Sign of $\frac{P}{Q}$	$-$	$+$	$-$
Interval	$(-\infty,-4)$	$(-4,2)$	$(2,\infty)$

-4 is part of the solution set
so we use a solid dot there.
2 is not part of the solution
set ($\frac{P}{Q}$ is not defined there) so we use an open dot there.

$\frac{x+4}{2-x} \geq 0$ and $\sqrt{\frac{x+4}{2-x}}$ will represent a real number within the interval $[-4,2)$. $-4 \leq x < 2$.   (3-7)

24.   A. $d(A,B) = |-2-(-8)| = |6| = 6$

      B. $d(B,A) = |-8-(-2)| = |-6| = 6$ (3-4)

25.　　A. $(3+i)^2 - 2(3+i) + 3 \;=\; 9 + 6i + i^2 - 6 - 2i + 3$

$$= 9 + 6i - 1 - 6 - 2i + 3$$

$$= 5 + 4i$$

B. $i^{27} = i^{26}i = (i^2)^{13}i = (-1)^{13}i = (-1)i = -i$　(3 - 5)

26.　　A. $(2 - \sqrt{-4}) - (3 - \sqrt{-9}) \;=\; (2 - i\sqrt{4}) - (3 - i\sqrt{9}) = (2 - 2i) - (3 - 3i)$

$$= 2 - 2i - 3 + 3i = -1 + i$$

B. $\dfrac{2-\sqrt{-1}}{3+\sqrt{-4}} \;=\; \dfrac{2-i\sqrt{1}}{3+i\sqrt{4}} = \dfrac{2-i}{3+2i} = \dfrac{(2-i)}{(3+2i)}\dfrac{(3-2i)}{(3-2i)} = \dfrac{6-7i+2i^2}{9-4i^2} = \dfrac{6-7i-2}{9+4}$

$$= \dfrac{4-7i}{13} = \dfrac{4}{13} - \dfrac{7}{13}i$$

C. $\dfrac{4+\sqrt{-25}}{\sqrt{-4}} = \dfrac{4+i\sqrt{25}}{i\sqrt{4}} = \dfrac{4+5i}{2i} = \dfrac{4+5i}{2i}\dfrac{i}{i} = \dfrac{4i+5i^2}{2i^2} = \dfrac{4i-5}{-2} = \dfrac{5}{2} - 2i$　(3 - 5)

27.　$\left(u + \dfrac{5}{2}\right)^2 \;=\; \dfrac{5}{4}$

$$u + \dfrac{5}{2} \;=\; \pm\sqrt{\dfrac{5}{4}}$$

$$u + \dfrac{5}{2} \;=\; \pm\dfrac{\sqrt{5}}{2}$$

$$u \;=\; -\dfrac{5}{2} \pm \dfrac{\sqrt{5}}{2}$$

$$u \;=\; \dfrac{-5\pm\sqrt{5}}{2}$$　(3 - 6)

28.　$1 + \dfrac{3}{u^2} \;=\; \dfrac{2}{u}$　Excluded value : $u \neq 0$

$$u^2 + 3 \;=\; 2u$$

$$u^2 - 2u \;=\; -3$$

$$u^2 - 2u + 1 \;=\; -2$$

$$(u - 1)^2 \;=\; -2$$

$$u - 1 \;=\; \pm\sqrt{-2}$$

$$u \;=\; 1 \pm \sqrt{-2}$$

$$u \;=\; 2 \pm i\sqrt{2}$$　(3 - 6)

29.

$$\dfrac{x}{x^2-x-6} - \dfrac{2}{x-3} \;=\; 3$$

$$\dfrac{x}{(x-3)(x+2)} - \dfrac{2}{x-3} \;=\; 3 \quad \text{Excluded values} : x \neq 3, -2$$

$$(x-3)(x+2)\dfrac{x}{(x-3)(x+2)} - (x-3)(x+2)\dfrac{2}{x-3} \;=\; 3(x-3)(x+2)$$

$$x - 2(x+2) \;=\; 3(x-3)(x+2)$$

$$x - 2x - 4 \;=\; 3(x^2 - x - 6)$$

$$-x - 4 \;=\; 3x^2 - 3x - 18$$

$$0 \;=\; 3x^2 - 2x - 14$$

$$3x^2 - 2x - 14 \;=\; 0$$

$$x \;=\; \dfrac{-b\pm\sqrt{b^2-4ac}}{2a}$$

$$a = 3$$
$$b = -2$$
$$c = -14$$

$$x \;=\; \dfrac{-(-2)\pm\sqrt{(-2)^2-4(3)(-14)}}{2(3)}$$

$$x \;=\; \dfrac{2\pm\sqrt{172}}{6}$$

$$x \;=\; \dfrac{2\pm2\sqrt{43}}{6}$$

$$x \;=\; \dfrac{1\pm\sqrt{43}}{3}$$　(3 - 6)

30. $2x^{2/3} - 5x^{1/3} - 12 = 0$
Let $u = x^{1/3}$, then
$$2u^2 - 5u - 12 = 0$$
$$(2u + 3)(u - 4) = 0$$
$$u = -\tfrac{3}{2}, 4$$
$$x^{1/3} = -\tfrac{3}{2} \quad x^{1/3} = 4$$
$$x = -\tfrac{27}{8} \quad x = 64 \qquad (3-8)$$

31. $m^4 + 5m^2 - 36 = 0$
Let $u = m^2$, then
$$u^2 + 5u - 36 = 0$$
$$(u + 9)(u - 4) = 0$$
$$u = -9, 4$$
$$m^2 = -9 \quad m^2 = 4$$
$$m = \pm 3i \quad m = \pm 2 \quad (3-8)$$

32. $$\sqrt{y-2} - \sqrt{5y+1} = -3$$
$$-\sqrt{5y+1} = -3 - \sqrt{y-2}$$
$$5y+1 = 9 + 6\sqrt{y-2} + y - 2$$
$$5y+1 = y + 7 + 6\sqrt{y-2}$$
$$4y - 6 = 6\sqrt{y-2}$$
$$2y - 3 = 3\sqrt{y-2}$$
$$4y^2 - 12y + 9 = 9(y-2)$$
$$4y^2 - 12y + 9 = 9y - 18$$
$$4y^2 - 21y + 27 = 0$$
$$(4y - 9)(y - 3) = 0$$
$$y = \tfrac{9}{4}, 3$$

Check
$$\sqrt{\tfrac{9}{4} - 2} - \sqrt{5(\tfrac{9}{4}) + 1} \overset{?}{=} -3$$
$$\sqrt{\tfrac{1}{4}} - \sqrt{\tfrac{49}{4}} \overset{?}{=} -3$$
$$-3 \overset{\checkmark}{=} -3$$
$$\sqrt{3 - 2} - \sqrt{5(3) + 1} \overset{?}{=} -3$$
$$-3 \overset{\checkmark}{=} -3$$

Solution: $\tfrac{9}{4}, 3$

Common Error:
$y - 2 - 5y + 1 = 9$
is not equivalent to the
equation formed by
squaring both members
of the given equation.   (3-8)

33. $$P = M - Mdt$$
$$M - Mdt = P$$
$$M(1 - dt) = P$$
$$M = \tfrac{P}{1-dt} \quad (3-1)$$

34. $$P = EI - RI^2$$
$$RI^2 - EI + P = 0$$
$$I = \tfrac{-b \pm \sqrt{b^2 - 4ac}}{2a}$$
$$a = R$$
$$b = -E$$
$$c = P$$
$$I = \tfrac{-(-E) \pm \sqrt{(-E)^2 - 4(R)(P)}}{2(R)}$$
$$I = \tfrac{E \pm \sqrt{E^2 - 4PR}}{2R} \qquad (3-6)$$

35. $$x = \tfrac{4y+5}{2y+1}$$
$$x(2y+1) = (2y+1)\tfrac{4y+5}{2y+1}$$
$$2xy + x = 4y + 5$$
$$2xy + x - 4y = 5$$
$$2xy - 4y = 5 - x$$
$$y(2x - 4) = 5 - x$$
$$y = \tfrac{5-x}{2x-4} \quad (3-1)$$

36.   The given inequality
$a + b < b - a$
is equivalent to, successively,
$$a < -a$$
$$2a < 0$$
$$a < 0$$
Thus its truth is
independent of the value
of b, and dependent on
on a being negative.
True for all real b and
all negative a.   (3-3)

37.   If $a > b$ and b is negative,
then $\tfrac{a}{b} < \tfrac{b}{b}$, that is $\tfrac{a}{b} < 1$,
since dividing both sides
by b reverses the order
of the inequality.
$\tfrac{a}{b}$ is less than 1.   (3-3)

38.
$$y = \frac{1}{1-\frac{1}{1-x}}$$

$$y = \frac{1(1-x)}{(1-x)1-(1-x)\frac{1}{1-x}}$$

$$y = \frac{1-x}{1-x-1}$$

$$y = \frac{1-x}{-x}$$

$$-xy = 1-x$$

$$x - xy = 1$$

$$x(1-y) = 1$$

$$x = \frac{1}{1-y} \quad (3-1)$$

39. $0 < |\,x-6\,| < d$ means:
the distance between x and 6 is less
than d but $x \neq 6$.

$$-d < x-6 < d \text{ except } x \neq 6$$

$$6-d < x < 6+d \text{ but } x \neq 6$$

$$6-d < x < 6 \text{ or } 6 < x < 6+d$$

$$(6-d,6) \ \bigcup \ (6,6+d) \qquad (3-4)$$

40.
$$2x^2 = \sqrt{3}x - \tfrac{1}{2}$$

$$4x^2 = 2\sqrt{3}x - 1$$

$$4x^2 - 2\sqrt{3}x + 1 = 0$$

$$x = \frac{-b \pm \sqrt{b^2-4ac}}{2a}$$

$$a = 4$$

$$b = -2\sqrt{3}$$

$$c = 1$$

$$x = \frac{-(-2\sqrt{3}) \pm \sqrt{(-2\sqrt{3})^2 - 4(4)(1)}}{2(4)}$$

$$x = \frac{2\sqrt{3} \pm \sqrt{-4}}{8}$$

$$x = \frac{2\sqrt{3} \pm 2i}{8}$$

$$x = \frac{\sqrt{3} \pm i}{4} \text{ or } \frac{\sqrt{3}}{4} \pm \tfrac{1}{4}i \quad (3-6)$$

41.
$$4 = 8x^{-2} + x^{-4}$$

$$4x^4 = 8x^2 + 1$$

Multiply both
members by x^4 $x \neq 0$

$$4x^4 - 8x^2 - 1 = 0$$

Let $u = x^2$, then

$$4u^2 - 8u - 1 = 0$$

$$u = \frac{-b \pm \sqrt{b^2 - 4ac}}{2a}$$

$$a = 4$$

$$b = -8$$

$$c = -1$$

$$u = \frac{-(-8) \pm \sqrt{(-8)^2 - 4(4)(-1)}}{2(4)}$$

$$u = \frac{8 \pm \sqrt{64 + 16}}{8}$$

$$u = \frac{8 \pm \sqrt{80}}{8}$$

$$u = \frac{8 \pm 4\sqrt{5}}{8} \quad CommonError:$$

It is incorrect to
"cancel" the 8's
at this point.

$$u = \frac{4(2 \pm \sqrt{5})}{8}$$

$$u = \frac{2 \pm \sqrt{5}}{2}$$

$$x^2 = \frac{2 \pm \sqrt{5}}{2}$$

$$x = \pm\sqrt{\frac{2 \pm \sqrt{5}}{2}}$$

(four roots) $(3-8)$

42.
$$(a+bi)\{\frac{a}{a^2+b^2} - \frac{b}{a^2+b^2}i\} =$$

$$\frac{(a+bi)}{1}\{\frac{a}{a^2+b^2} - \frac{bi}{a^2+b^2}\} =$$

$$\frac{a(a+bi)}{a^2+b^2} - \frac{bi(a+bi)}{a^2+b^2} =$$

$$\frac{a^2 + abi - abi - b^2i^2}{a^2+b^2}$$

$$= \frac{a^2+b^2}{a^2+b^2} = 1 \quad (3-5)$$

43.
$$2x > \frac{x^2}{5} + 5$$

$$10x > x^2 + 25$$

$$0 > x^2 - 10x + 25$$

$$0 > (x-5)^2$$

Since the right side is never
negative for x a real number,
this statement is satisfied by
no real number. $(3-7)$

44.
$$\frac{x^2}{4} + 4 \geq 2x$$

$$x^2 + 16 \geq 8x$$

$$x^2 - 8x + 16 \geq 0$$

$$(x-4)^2 \geq 0$$

Since the left side is
always positive or zero for
x a real number, this
statement is satisfied by
all real numbers x. $(3-7)$

45. $\left| x - \dfrac{8}{x} \right| \geq 2$

There are two ways for this statement to hold. The solution set is the union of the solution sets for:

$x - \dfrac{8}{x} \leq -2$ and

$x - \dfrac{8}{x} \geq 2$

Case I:

$$x - \frac{8}{x} \leq -2$$

$$x - \frac{8}{x} + 2 \leq 0$$

$$\frac{x^2 - 8 + 2x}{x} \leq 0$$

$$\frac{x^2 + 2x - 8}{x} \leq 0$$

$$\frac{(x-2)(x+4)}{x} \leq 0$$

	$\dfrac{P}{Q} = \dfrac{(x-2)(x+4)}{x}$			
Test Number	-5	-1	1	3
Value of $\frac{P}{Q}$	$-\frac{7}{5}$	9	-5	$\frac{7}{3}$
Sign of $\frac{P}{Q}$	$-$	$+$	$-$	$+$
Interval	$(-\infty, -4)$	$(-4, 0)$	$(0,2)$	$(2, \infty)$

Zeros of $P, Q : 2, -4, 0$

2 and -4 are part of the solution set, so we use solid dots there.

0 is not part of the solution set, so we use an open dot there.

$\dfrac{(x-2)(x+4)}{x} \leq 0$ and $x - \dfrac{8}{x} \leq -2$ within the intervals $(-\infty, -4]$ and $(0,2]$.

Case II:

$$x - \frac{8}{x} \geq 2$$

$$x - \frac{8}{x} - 2 \geq 0$$

$$\frac{x^2 - 8 - 2x}{x} \geq 0$$

$$\frac{x^2 - 2x - 8}{x} \geq 0$$

$$\frac{(x-4)(x+2)}{x} \geq 0$$

	$\dfrac{P}{Q} = \dfrac{(x-4)(x+2)}{x}$			
Test Number	-3	-1	1	5
Value of $\frac{P}{Q}$	$-\frac{7}{3}$	5	-9	$\frac{7}{5}$
Sign of $\frac{P}{Q}$	$-$	$+$	$-$	$+$
Interval	$(-\infty, -2)$	$(-2, 0)$	$(0,4)$	$(4, \infty)$

Zeros of $P, Q : -2, 4, 0. -2$ and 4 are part of the solution set, so we use solid dots there.

0 is not part of the solution set, so we use an open dot there.

$\dfrac{(x-4)(x+2)}{x} \geq 0$ and $x - \dfrac{8}{x} \geq 2$ within the intervals $[-2, 0)$ and $[4, \infty)$.

Combining Case I and Case II, we have:

$x \leq -4$ or $-2 \leq x < 0$ or $0 < x \leq 2$ or $x \geq 4$

$(-\infty, -4] \cup [-2, 0) \cup (0, 2] \cup [4, \infty)$ $(3-7)$

46.
$$2.15x - 3.73(x - 0.93) = 6.11x$$
$$2.15x - 3.73x + 3.4689 = 6.11x$$
$$-1.58x + 3.4689 = 6.11x$$
$$3.4689 = 7.69x$$
$$x = 0.45 \quad (3-1)$$

47.
$$-1.52 \leq 0.77 - 2.04x \leq 5.33$$
$$-2.29 \leq -2.04x \leq 4.56$$
$$1.12 \geq x \geq -2.24$$
$$-2.24 \leq x \leq 1.12 \text{ or } [-2.24, 1.12] \quad (3-3)$$

48.
$$\frac{3.77 - 8.47i}{6.82 - 7.06i} = \frac{(3.77 - 8.47i)\,(6.82 + 7.06i)}{(6.82 - 7.06i)\,(6.82 + 7.06i)}$$

$$= \frac{(3.77)(6.82) + (3.77)(7.06i) - (6.82)(8.47i) - (8.47)(7.06)i^2}{(6.82)^2 + (7.06)^2}$$

$$= \frac{25.7114 + 26.6162i - 57.7654i + 59.7982}{46.5124 + 49.8436}$$

$$= \frac{85.5096 - 31.1492i}{96.356}$$

$$= 0.89 - 032i \quad (3-5)$$

49.
$$6.09x^2 + 4.57x - 8.86 = 0$$
$$x = \frac{-b \pm \sqrt{b^2 - 4ac}}{2a}$$
$$a = 6.09$$
$$b = 4.57$$
$$c = -8.86$$
$$x = \frac{-4.57 \pm \sqrt{(4.57)^2 - 4(6.09)(-8.86)}}{2(6.09)}$$
$$= \frac{-4.57 \pm \sqrt{236.7145}}{12.18}$$
$$= \frac{-4.57 \pm 15.3855}{12.18}$$
$$= -1.64, 0.89 \quad (3-6)$$

50. Let x = the number
$\frac{1}{x}$ = its reciprocal
Then
$$x - \frac{1}{x} = \frac{16}{15} \quad \text{Excluded value} : x \neq 0$$
$$15x^2 - 15 = 16x$$
$$15x^2 - 16x - 15 = 0$$
$$(5x + 3)(3x - 5) = 0$$
$$5x + 3 = 0 \quad 3x - 5 = 0$$
$$x = -\frac{3}{5} \quad x = \frac{5}{3} \quad (3-6)$$

51. A. $H = 0.7(220 - A)$
B. We are to find H when $A = 20$.
$H = 0.7\,(220\text{-}20)$
$H = 140$ beats per minute.
C. We are to find A when $H = 126$.
$126 = 0.7(220 - A)$

$$126 = 154 - 0.7A$$
$$-28 = -0.7A$$
$$A = 40 \text{ years old.} \quad (3-2)$$

52. Let x = amount of 80% solution
     Then $50 - x$ = amount of 30% solution
     since 50 = amount of 60% solution

acid in		acid in		acid in
80% solution	$+$	30% solution	$=$	60% solution

$$0.8(x) + 0.3(50 - x) = 0.6(50)$$
$$0.8x + 15 - 0.3x = 30$$
$$15 + 0.5x = 30$$
$$0.5x = 15$$
$$x = \frac{15}{0.5}$$
$$x = 30 \text{ milliliters of 80% solution}$$
$$50 - x = 20 \text{ milliliters of 30% solution} \quad (3-2)$$

53. Let x = the rate of the current
     Then $12 - x$ = the rate of the boat upstream
     $12 + x$ = the rate of the boat downstream

Solving $d = rt$ for t, we have $t = \frac{d}{r}$. We use this formula, together with

time upstream $= 2 +$ time downstream time upstream

time upstream $= \frac{\text{distance upstream}}{\text{rate upstream}} = \frac{45}{12-x}$

time downstream $= \frac{\text{distance downstream}}{\text{rate downstream}} = \frac{45}{12+x}$

So,
$$\frac{45}{12-x} = 2 + \frac{45}{12+x}$$

Excluded values : $x \neq -12, 12$

$$45(12 + x) = (12 + x)(12 - x)2 + 45(12 - x)$$
$$540 + 45x = 288 - 2x^2 + 540 - 45x$$
$$2x^2 + 90x - 288 = 0$$
$$2(x + 48)(x - 3) = 0$$
$$x = -48, 3$$

Discarding the negative answer, we have rate = 3 miles per hour $(3-6)$

54.   A. Apply the given formula with $C = 15$.

$$15 = x^2 - 10x + 31$$
$$0 = x^2 - 10x + 16$$
$$x^2 - 10x + 16 = 0$$
$$(x - 8)(x - 2) = 0$$
$$x = 2 \text{ or } 8$$

Thus the output could be either 2000 or 8000 units. $(3 - 6)$

B. Apply the given formula with $C = 6$.

$$6 = x^2 - 10x = 31$$
$$0 = x^2 - 10x + 25$$
$$x^2 - 10x + 25 = 0$$
$$(x - 5)^2 = 0$$
$$x = 5$$

Thus the output must be 5000 units.

55.   The break-even points are defined by $C = R$ (cost = revenue).

Applying the formulas in this problem and the previous one, we have

$$x^2 - 10x + 31 = 3x$$
$$x^2 - 13x + 31 = 0$$
$$x = \frac{-b \pm \sqrt{b^2 - 4ac}}{2a}$$
$$a = 1$$
$$b = -13$$
$$c = 31$$
$$x = \frac{-(-13) \pm \sqrt{(-13)^2 - 4(1)(31)}}{2(1)}$$
$$x = \frac{13 \pm \sqrt{45}}{2} \quad \text{thousand}$$

or approximately 3,146 and 9,854 units $(3 - 6)$

56.   A profit will result if $R > C$, that is, if

$$3x > x^2 - 10x + 31$$
$$x^2 - 10x + 31 < 3x$$
$$x^2 - 13x + 31 < 0$$

The zeros of this polynomial were determined in problem 55 to be

$\frac{13 \pm \sqrt{45}}{2}$ (in thousands)

$x^2 - 13x + 31$			
Test Number	0	5	10
Value of Polynomial for Test Number	31	−9	1
Sign of Polynomial in Interval	+	−	+
Interval	$\left(-\infty, \frac{13-\sqrt{45}}{2}\right)$	$\left(\frac{13-\sqrt{45}}{2}, \frac{13+\sqrt{45}}{2}\right)$	$\left(\frac{13+\sqrt{45}}{2}, \infty\right)$
approximately	$(-\infty, 3.146)$	$(3.146, 9.854)$	$(9.854, \infty)$

$x^2 - 13x + 31 < 0$, and a profit will result, if $\frac{13-\sqrt{45}}{2} < x < \frac{13+\sqrt{45}}{2}$, or, approximately, $3.146 < x < 9.854$, in thousands. $(3-7)$

57.   The distance of T from 110 must be no greater than 5.
$|T - 110| \leq 5$ $(3-4)$

CHAPTER 4

Exercise 4-1

Key Ideas and Formulas

There is a one-to-one correspondence between the points in a plane and the elements in a set of all ordered pairs of real numbers.

The graph of an equation in two variables is the graph of its solution set, the set of all ordered pairs of real numbers (a, b) that satisfy the equation.

To sketch a graph of an equation, we include enough points from its solution set so that the total graph is apparent. We may use point-to-point plotting and aids to graphing, such as symmetry.

A graph is symmetric with respect to the y axis if $(-a, b)$ is on the graph whenever (a, b) is on the graph.

A graph is symmetric with respect to the x axis if $(a, -b)$ is on the graph whenever (a, b) is on the graph.

A graph is symmetric with respect to the origin if $(-a, -b)$ is on the graph whenever (a, b) is on the graph.

Test for symmetry with respect to the	if equation is equivalent when
y axis	x is replaced with $-x$
x axis	y is replaced with $-y$
origin	x and y are replaced with $-x$ and $-y$.

Distance Formula:

Distance between $P_1(x_1, y_1)$ and $P_2(x_2, y_2): d(P_1, P_2) = \sqrt{(x_2 - x_1)^2 + (y_2 - y_1)^2}$

Standard Equations of a Circle
1. Circle with radius $r(r > 0)$ and center at $(h, k): (x - h)^2 + (y - k)^2 = r^2$
2. Circle with radius $r(r > 0)$ and center at $(0,0): x^2 + y^2 = r^2$

---

1. $y = 2x - 4$ y axis symmetry? $y = 2(-x) - 4$ $y = -2x - 4$ No, not equivalent
     x axis symmetry? $-y = 2x - 4$ $y = 4 - 2x$ No, not equivalent
     origin symmetry? $-y = 2(-x) - 4$ $-y = -2x - 4$
                                $y = 2x + 4$ No, not equivalent

x	y
0	-4
2	0
4	4

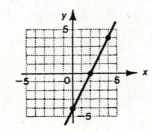

3. $y = |x|$ y axis symmetry? $y = |-x|$ $y = |x|$ Yes, equivalent

                x axis symmetry? $-y = |x|$ $y = -|x|$ No, not equivalent

                origin symmetry? $-y = |-x|$ $y = -|x|$ No, not equivalent

x	y
0	0
4	4

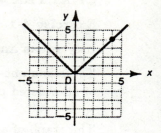

We reflect the portion of the graph in quadrant I through the y axis, using the y axis symmetry.

5. $|y| = x$ y axis symmetry? $|y| = -x$ No, not equivalent

                x axis symmetry? $|-y| = x$ $|y| = x,$ Yes, equivalent

                origin symmetry? $|-y| = -x$ $|y| = -x$ No, not equivalent

x	y
0	0
4	4

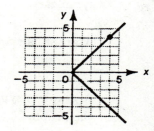

We reflect the portion of the graph in quadrant I through the x axis, using the x axis symmetry.

7. $|x| = |y|$ y axis symmetry? $|-x| = |y|$ $|x| = |y|$ Yes, equivalent

                x axis symmetry? $|x| = |-y|$ $|x| = |y|$ Yes, equivalent

                origin symmetry follows

                automatically.

x	y
0	0
4	4

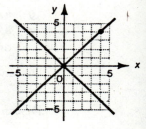

We reflect the portion of the graph in quadrant I through the y axis, and the x axis, and the origin, using all three symmetries.

9. $d = \sqrt{[3 - (-6)]^2 + [4 - (-4)]^2}$

     $= \sqrt{(9)^2 + (8)^2} = \sqrt{145}$

     Common Error: not 9+8.

11. $d = \sqrt{[(4) - (6)]^2 + [(-2) - (6)]^2}$

      $= \sqrt{(-2)^2 + (-8)^2} = \sqrt{68}$

13. $(x - 0)^2 + (y - 0)^2 = 7^2$

               $x^2 + y^2 = 49$

15. $(x - 2)^2 + (y - 3)^2 = 6^2$

      $(x - 2)^2 + (y - 3)^2 = 36$

17. $[x - (-4)]^2 + (y - 1)^2 = (\sqrt{7})^2$

       $(x + 4)^2 + (y - 1)^2 = 7$

      Common Error: not $(x - 4)^2$

19. $[x - (-3)]^2 + [y - (-4)]^2 = (\sqrt{2})^2$

       $(x + 3)^2 + (y + 4)^2 = 2$

21. $y^2 = x + 2$ y axis symmetry? $y^2 = (-x) + 2$ $y^2 = 2 - x$ No, not equivalent
                       x axis symmetry? $(-y)^2 = x + 2$ $y^2 = x + 2$ Yes, equivalent
                       origin symmetry? $(-y)^2 = (-x) + 2$ $y^2 = 2 - x$ No, not equivalent

x	y
-2	0
-1	± 1
2	± 2

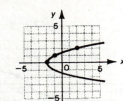

Automatically reflected across x axis.

23. $y = x^2 + 1$ y axis symmetry? $y = (-x)^2 + 1$ $y = x^2 + 1$ Yes, equivalent
                       x axis symmetry? $-y = x^2 + 1$ $y = -x^2 - 1$ No, not equivalent
                       origin symmetry? $-y = (-x)^2 + 1$ $y = -x^2 - 1$ No, equivalent

x	y
0	1
1	2
2	5

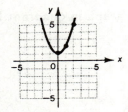

We reflect the portion of the graph in quadrant I across the y axis, using the y axis symmetry.

25. $9x^2 + y^2 = 9$ y axis symmetry? $9(-x)^2 + y^2 = 9$ $9x^2 + y^2 = 9$ Yes, equivalent
                          x axis symmetry? $9x^2 + (-y)^2 = 9$ $9x^2 + y^2 = 9$ Yes, equivalent
                          origin symmetry
                          follows
                          automatically

x	y
0	0
$\pm\frac{1}{2}$	$\pm\frac{\sqrt{27}}{2} = 2.6$
± 1	0

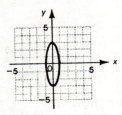

Reflections follow automatically.

27. $x^2 + 4y = 4$ y axis symmetry? $(-x)^2 + 4y^2 = 4$ $x^2 + 4y^2 = 4$ Yes, equivalent
                        x axis symmetry? $x^2 + 4(-y)^2 = 4$ $x^2 + 4y^2 = 4$ Yes, equivalent
                        origin symmetry
                        follows automatically

x	y
0	± 1
± 1	$\pm\frac{\sqrt{3}}{2} = .9$
± 2	0

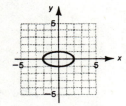

Reflections follow automatically.

29. $y^3 = x$ **y axis symmetry?** $y^3 = -x$ No, not equivalent
      x axis symmetry? $(-y)^3 = x$ $y^3 = -x$ No, not equivalent
      origin symmetry? $(-y)^3 = -x$ $y^3 = x$ Yes, equivalent

x	y
0	0
1	1
8	2

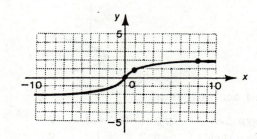

We reflect the portion of the graph in quadrant I through the origin, using the origin symmetry.

31. The lengths of the three sides are found from the distance formula to be:

$$= \frac{\sqrt{[1-(-3)]^2 + [(-2)-(2)]^2}}{\sqrt{4^2 + (-4)^2} = \sqrt{32}}$$

$$= \frac{\sqrt{[8-1]^2 + [5-(-2)]^2}}{\sqrt{7^2 + 7^2} = \sqrt{98}}$$

$$= \frac{\sqrt{[8-(-3)]^2 + [5-2]^2}}{\sqrt{11^2 + 3^2} = \sqrt{130}}$$

Since $(\sqrt{130})^2 = (\sqrt{32})^2$

$+ (\sqrt{98})^2$, the triangle is a right triangle.

Common Error: Confusing these true statements with the *false* statement $\sqrt{130} = \sqrt{32} + \sqrt{98}$.

33. The distance formula requires that
$$\sqrt{(x-2)^2 + [8-(-4)]^2} = 13$$
Solving, we have

$$\sqrt{(x-2)^2 + (12)^2} = 13$$
$$(x-2)^2 + (12)^2 = (13)^2$$
$$(x-2)^2 + 144 = 169$$
$$(x-2)^2 = 25$$
$$x - 2 = \pm 5$$
$$x = 2 \pm 5$$
$$x = -3, 7$$

35. $(x-3)^2 + (y-5)^2 = 49$
      $(x-3)^2 + (y-5)^2 = 7^2$
      Center $(3,5)$; Radius $= 7$

37. $(x+4)^2 + (y-2)^2 = 7$
      $[x-(-4)]^2 + (y-2)^2 = (\sqrt{7})^2$
      Center $(-4,2)$; Radius $= \sqrt{7}$

39.
$$x^2 + y^2 - 6x - 4y = 36$$
$$x^2 - 6x + y^2 - 4y = 36$$
$$x^2 - 6x + 9 + y^2 - 4y + 4 = 36 + 9 + 4$$
$$(x-3)^2 + (y-2)^2 = 49$$
Center $(3,2)$; Radius $= 7$

41.
$$x^2 + y^2 + 8x - 6y + 8 = 0$$
$$x^2 + 8x + y^2 - 6y = -8$$
$$x^2 + 8x + 16 + y^2 - 6y + 9 = -8 + 16 + 9$$
$$(x+4)^2 + (y-3)^2 = 17$$
Center $(-4,3)$; Radius $= \sqrt{17}$

43. $y^3 = |x|$ **y axis symmetry?** $y^3 = |-x|$ $y^3 = |x|$ Yes, equivalent
      x axis symmetry? $(-y)^3 = |x|$ $-y^3 = |x|$ No, not equivalent
      origin symmetry? $(-y)^3 = |-x|$ $-y^3 = |x|$ No, no equivalent

x	y
0	0
1	1
8	2

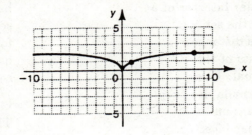

We reflect the portion of the graph in quadrant I across the y axis, using the y axis symmetry.

45. $xy = 1$ **y axis symmetry?** $-xy = 1$ No, not equivalent
      x axis symmtry? $x(-y) = 1$ $-xy = 1$ No, not equivalent
      origin symmetry? $(-x)(-y) = 1$ $xy = 1$ Yes, equivalent.

x	y
1	1
2	$\frac{1}{2}$
3	$\frac{1}{3}$
$\frac{1}{2}$	2
$\frac{1}{3}$	3

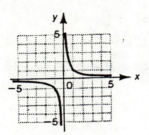

We reflect the portion of the graph in quadrant I through the origin, using the origin symmetry.

47. $y = 6x - x^2$ **y axis symmetry?** $y = 6(-x) - (-x)^2$ $y = -6x - x^2$ No, not equivalent
      x axis symmetry? $-y = 6x - x^2$ $y = x^2 - 6x$ No, not equivalent
      origin symmetry? $-y = 6(-x) - (-x)^2$ $y = x^2 + 6x$ No, not equivalent

x	y
−1	−7
0	0
1	5
2	8
3	9
4	8
5	5
6	0
7	−7

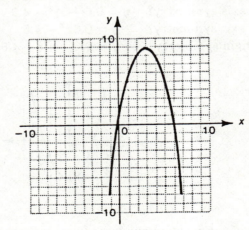

A larger table of values is needed since we have no symmetry information.

49. From geometry we know that the perpendicular bisector of a line segment is the set of points whose distances to the end points of the segment are equal. Using the distance formula, let (x, y) be a typical point on the perpendicular bisector. Then the distance from (x, y) to $(-6, -2)$ equals the distance from (x, y) to $(4, 4)$.

$$\sqrt{[x - (-6)]^2 + [y - (-2)]^2}$$
$$= \sqrt{(x - 4)^2 + (y - 4)^2}$$
$$(x + 6)^2 + (y + 2)^2$$
$$= (x - 4)^2 + (y - 4)^2$$
$$x^2 + 12x + 36 + y^2 + 4y + 4$$
$$= x^2 - 8x + 16 + y^2 - 8y + 16$$
$$20x + 12y = -8$$
$$5x + 3y = -2$$

51. The center of the circle must be the midpoint of the diameter segment. Using the midpoint formula from problem 50, we have

center = midpoint has coordinates
$$\left(\frac{7 + 1}{2}, \frac{-3 + 7}{2}\right) = (4, 2)$$

The radius of the circle must the distance from the center to one of the endpoints of the diameter. It does not matter which endpoint we choose. For example, $r = \sqrt{(7 - 4)^2 + [-3 - 2]^2}$ $= \sqrt{3^2 + (-5)^2} = \sqrt{34}$.

The equation of the circle is
$$(x - 4)^2 + (y - 2)^2 = (\sqrt{34})^2$$
or $(x - 4)^2 + (y - 2)^2 = 34$.

53. Perimeter
$$= \text{sum of lengths of all three sides}$$
$$= \sqrt{[1 - (-3)]^2 + [(-2) - 1]^2}$$
$$+ \sqrt{(4 - 1)^2 + [3 - (-2)]^2}$$
$$+ \sqrt{[4 - (-3)]^2 + (3 - 1)^2}$$
$$= \sqrt{16 + 9} + \sqrt{9 + 25} + \sqrt{49 + 4}$$
$$= \sqrt{25} + \sqrt{34} + \sqrt{53}$$
$$= 18.11$$

55. $y = 0.6x^2 - 4.5$

y axis symmetry?	$y = 0.6(-x)^2 - 4.5$	$y = 0.6x^2 - 4.5$	Yes, equivalent
x axis symmetry?	$-y = 0.6x^2 - 4.5$	$y = -0.6x^2 + 4.5$	No, not equivalent
origin symmetry?	$-y = 0.6(-x^2) - 4.5$	$y = -0.6x^2 + 4.5$	No, not equivalent

x	y
0	-4.5
1	-3.9
2	-2.1
3	0.9
4	5.1

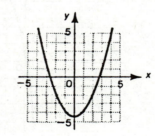

We reflect the portion of the graph in quadrant I across the y axis, using the y axis symmetry.

57. $y = \sqrt{17 - x^2}$

y axis symmetry?	$y = \sqrt{17 - (-x)^2}$	$y = \sqrt{17 - x^2}$	Yes, equivalent
x axis symmetry?	$-y = \sqrt{17 - x^2}$	$y = -\sqrt{17 - x^2}$	No, not equivalent
origin symmetry?	$-y = \sqrt{17 - (-x)^2}$	$y = -\sqrt{17 - x^2}$	No, not equivalent

x	y
0	$\sqrt{17} = 4.1$
1	$\sqrt{16} = 4$
2	$\sqrt{13} = 3.6$
3	$\sqrt{8} = 2.8$
4	$\sqrt{1} = 1$
$\sqrt{17}$	0

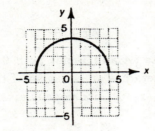

We reflect the portion of the graph in quadrant I across the y axis, using the y axis symmetry.

59. $y = x^{2/3}$

y axis symmetry?	$y = (-x)^{2/3}$	$y = x^{2/3}$	Yes, equivalent
x axis symmetry?	$-y = x^{2/3}$	$y = -(x^{2/3})$	No, not equivalent
origin symmetry?	$-y = (-x)^{2/3}$	$y = -(x^{2/3})$	No, not equivalent

x	y
0	0
1	1
2	$2^{2/3} = 1.6$
3	$3^{2/3} = 2.1$

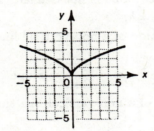

We reflect the portion of the graph in quadrant I across the y axis, using the y axis symmetry.

61. $R = (10 - p)\,p, 5 \le p \le 10$. Since the indicated values of p (and the corresponding values of R) are positive, there is no information to be gained from testing for symmetry. We make a larger table of values.

p	R
5	25
6	24
7	21
8	16
9	9
10	0

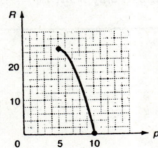

63. Using the hint, we note that $(2, r - 1)$ must satisfy $x^2 + y^2 = r^2$, that is

$$
\begin{aligned}
2^2 + (r - 1)^2 &= r^2 \\
4 + r^2 - 2r + 1 &= r^2 \\
-2r + 5 &= 0 \\
r &= \tfrac{5}{2} \text{ or } 2.5 \text{ ft.}
\end{aligned}
$$

Exercise 4-2

Key Ideas and Formulas

The graph of the equation

$$Ax + By = C$$

where $A, B,$ and C are constants (A and B not both 0), is a straight line. Any straight line in a rectangular coordinate system has an equation of this form. If a line passes through two distinct points $P_1(x_1, y_1)$ and $P_2(x_2, y_2)$, its slope is given by

$$m = \frac{y_2 - y_1}{x_2 - x_1} \quad x_2 \neq x_1$$
$$= \frac{\text{vertical change}}{\text{horizontal change}} = \frac{\text{rise}}{\text{run}}.$$

For a vertical line, $x_2 = x_1$, hence slope is not defined. Equation: $x = a$. For a horizontal line, $y_2 = y_1$, slope $= 0$. Equation: $y = b$.

Slope-intercept form of an equation of a non-vertical line

$$y = mx + b \quad m = \frac{\text{Rise}}{\text{Run}} = \text{Slope} \quad b = y \text{ intercept}$$

Point-slope form of an equation of a non-vertical line through (x_1, y_1) with slope m

$$y - y_1 = m(x - x_1)$$

Given two nonvertical lines L_1 and L_2 with slopes m_1 and m_2, respectively, then

$$L_1 \parallel L_2 \text{ if and only if } m_1 = m_2$$

$$L_1 \perp L_2 \text{ if and only if } m_1 m_2 = -1 \left(m_1 = -\frac{1}{m_2} \right)$$

---

1. $y = -\frac{3}{5}x + 4$ slope $-\frac{3}{5}$

x	y
0	4
5	1
-5	7

3. $y = -\frac{3}{4}x$ slope $-\frac{3}{4}$

x	y
0	0
4	-3
-4	3

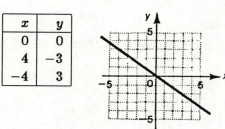

5. $\quad 2x - 3y = 15$

$\quad\quad\quad -3y = -2x + 15$

$\quad\quad\quad\quad y = \dfrac{2}{3}x - 5 \quad$ slope $\dfrac{2}{3}$

x	y
0	-5
3	-3
-3	-7

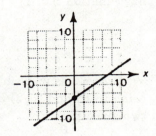

7. $\quad 4x - 5y = -24$

$\quad\quad\quad -5y = -4x - 24$

$\quad\quad\quad\quad y = \dfrac{4}{5}x + \dfrac{24}{5} \quad$ slope $\dfrac{4}{5}$

x	y
-1	4
4	8
9	12

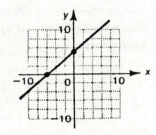

9. $\quad \dfrac{y}{8} - \dfrac{x}{4} = 1$

$\quad\quad\quad \dfrac{y}{8} = \dfrac{x}{4} + 1$

$\quad\quad\quad\quad y = 2x + 8 \quad$ slope 2

x	y
0	8
-4	0
-8	-8

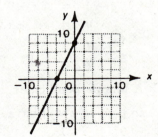

11. $x = -3$ slope not defined
vertical line

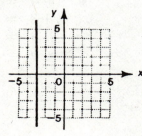

13. $y = 3.5$ slope 0
horizontal line

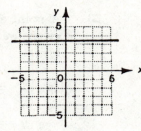

15. Slope and y intercept are given; we use slope-intercept form.

$\quad y = 1x + 0$

$\quad y = x$

17. Slope and y intercept are given; we use slope-intercept form.

$\quad y = -\dfrac{2}{3}x + (-4)$

$\quad y = -\dfrac{2}{3}x - 4$

19. A point and the slope are given; we use point-slope form.

$\quad y - 4 = -3(x - 0)$

$\quad y - 4 = -3x$

$\quad\quad\quad y = -3x + 4$

21. A point and the slope are given; we use point-slope form.

$\quad y - 4 = -\dfrac{2}{5}[x - (-5)]$

$\quad y - 4 = -\dfrac{2}{5}[x + 5]$

$\quad y - 4 = -\dfrac{2}{5}x - 2$

$\quad\quad\quad y = -\dfrac{2}{5}x + 2$

23. A point and the slope are given; we use point-slope form.

$$y - 5 = 0(x - 5)$$
$$y - 5 = 0$$
$$y = 5$$

25. Two points are given; we first find slope, then use point-slope form.

$$m = \frac{-2 - 6}{5 - 1} = \frac{-8}{4} = -2$$
$$y - 6 = (-2)(x - 1)$$
$$y - 6 = -2x + 2$$
$$y = -2x + 8$$

or

$$y - (-2) = (-2)(x - 5)$$
$$y + 2 = -2x + 10$$
$$y = -2x + 8$$

Thus, it does not matter which point is chosen in substituting into the point-slope form; both points must give rise to the same equation.

27. We proceed as in problem 25.

$$m = \frac{8 - 0}{-4 - 2} = \frac{8}{-6} = -\frac{4}{3}$$
$$y - 0 = -\frac{4}{3}(x - 2)$$
$$y = -\frac{4}{3}x + \frac{8}{3}$$

29. We proceed as in problem 25.

$$m = \frac{4 - 4}{5 - (-3)} = \frac{0}{8} = 0$$
$$y - 4 = 0(x - 5)$$
$$y - 4 = 0$$
$$y = 4$$

31. We proceed as in problem 25.

$$m = \frac{-3 - 6}{4 - 4} = \frac{-9}{0}$$

slope is undefined.
A vertical line through $(4,6)$ has equation $x = 4$.

33. We proceed as in problem 25, using $(6,0)$ and $(0,2)$ as the two given points.

$$m = \frac{0 - 2}{6 - 0} = \frac{-2}{6} = -\frac{1}{3}$$
$$y - 0 = -\frac{1}{3}(x - 6)$$
$$y = -\frac{1}{3}x + 2$$

35. We proceed as in problem 25, using $(-4,0)$ and $(0,3)$ as the two given points.

$$m = \frac{3 - 0}{0 - (4)} = \frac{3}{4}$$
$$y - 0 = \frac{3}{4}[x - (-4)]$$
$$y = \frac{3}{4}x + 3$$

37. A line parallel to $y = 3x - 5$ will have the same slope, namely 3. We now use the point slope form.

$$y - 4 = 3[x - (-3)]$$
$$y - 4 = 3x + 9$$
$$y = 3x + 13$$
$$-3x + y = 13$$
$$3x - y = -13$$

39. A line perpendicular to $y = -\frac{1}{3}x$ will have slope satisfying $-\frac{1}{3}m = -1$, or $m = 3$. We use the point-slope form.

$$y - (-3) = 3(x - 2)$$
$$y + 3 = 3x - 6$$
$$9 = 3x - y$$
$$3x - y = 9$$

41. The equation of the vertical line through $(2,5)$ is $x = 2$.

43. The equation of the vertical line through $(3,-2)$ is $x = 3$.

45. A line parallel to $3x - 2y = 4$ will have the same slope. The slope of $3x - 2y = 4$, or $2y = 3x - 4$, or $y = \frac{3}{2}x - 4$ is $\frac{3}{2}$. We use the point-slope form,

$$
\begin{aligned}
y - 0 &= \frac{3}{2}(x - 5) \\
y &= \frac{3}{2}x - \frac{15}{2} \\
\frac{15}{2} &= \frac{3}{2}x - y \\
15 &= 3x - 2y \\
3x - 2y &= 15
\end{aligned}
$$

(Alternatively, we could notice that a line parallel to $3x - 2y = 4$ will have an equation of the form $3x - 2y = C$. Since the required line must contain $(5,0)$, $(5,0)$ must satisfy its equation. Therefore $3(5) - 2(0) = C$. Since $15 = C$, the equation desired is $3x - 2y = 15$.)

47. A line perpendicular to $x + 3y = 9$, which has slope $-\frac{1}{3}$, will have slope satisfying $-\frac{1}{3}m = -1$, or $m = 3$. We use the point-slope form.

$$
\begin{aligned}
y - (-4) &= 3(x - 0) \\
y + 4 &= 3x \\
4 &= 3x - y \\
3x - y &= 4
\end{aligned}
$$

49.

x	$y = 2x + 2$	$y = \frac{1}{2}x + 2$	$y = 0x + 2$	$y = -\frac{1}{2}x + 2$	$y = -2x + 2$
0	2	2	2	2	2
2	6	3	2	1	-2
-2	-2	1	2	3	6

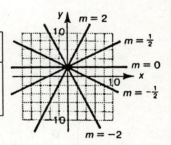

51. slope of $AB = \dfrac{-1 - 2}{4 - 0} = -\dfrac{3}{4}$

slope of $DC = \dfrac{-5 - (-2)}{1 - (-3)} = -\dfrac{3}{4}$

Therefore $AB \parallel DC$.

53. slope of $AB = -\dfrac{3}{4}$.

slope of $BC = \dfrac{-5 - (-1)}{1 - 4} = \dfrac{4}{3}$

(slope AB) (slope BC) $= (-\dfrac{3}{4})(\dfrac{4}{3}) = -1$

Therefore $AB \perp BC$.

55.

$$\text{midpoint of } AD = (\frac{0+(-3)}{2}, \frac{2+(-2)}{2}) = (-\frac{3}{2}, 0)$$

$$\text{slope of } AD = \frac{-2-2}{-3-0} = \frac{4}{3}$$

We require the equation of a line, through the midpoint of AD, which is perpendicular to AD. Its slope will satisfy $\frac{4}{3}m = -1$, or $m = -\frac{3}{4}$. We use the point-slope form.

$$y - 0 = -\frac{3}{4}[x - (-\frac{3}{2})]$$

$$y = -\frac{3}{4}(x + \frac{3}{2})$$

$$y = -\frac{3}{4}x - \frac{9}{8}$$

$$8y = -6x - 9$$

$$6x + 8y = -9$$

57. $y = |x|$ y axis symmetry? $y = |-x|$ $y = |x|$ Yes, equivalent

                 x axis symmetry? $-y = |x|$ $y = -|x|$ No, not equivalent

                 origin symmetry? $-y = |-x|$ $y = -|x|$ No, not equivalent

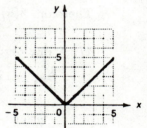

x	y
0	0
4	4

We reflect the portion of the graph in quadrant I through the y axis, using the y axis symmetry.

59. $y = |x - 1|$ y axis symmetry? $y = |-x - 1|$ No, not equivalent

                   x axis symmetry? $-y = |x - 1|$ $y = -|x - 1|$ No, not equivalent

                   origin symmetry? $y = -|-x - 1|$ $-y = -|-x - 1|$ No, not equivalent

x	y
-4	5
-3	4
-2	3
-1	2
0	1
1	0
2	1
3	2
4	3
5	4

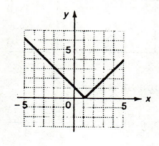

A larger table of values is needed since we have no symmetry information.

61. $x^2 - y^2 = 0$ y axis symmetry? $(-x)^2 - y^2 = 0$ $x^2 - y^2 = 0$ Yes, equivalent
         x axis symmetry? $x^2 - (-y)^2 = 0$ $x^2 - y^2 = 0$ Yes, equivalent
         origin symmetry   follows automatically

x	y
0	0
4	4

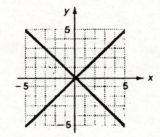

We reflect the portion of the graph in   quadrant I through the y axis,
and the x axis, and the origin, using all three symmetries.

63.   *Case I:* $B \neq 0$

$Ax + By = C$ can be rewritten $By = -Ax + C$, $y = -\frac{A}{B}x + \frac{C}{B}$

$Ax + By = D$ can be rewritten $By = -Ax + D$, $y = -\frac{A}{B}x + \frac{D}{B}$

Since both lines have the same slope, $-\frac{A}{B}$, they are parallel.
Case II: $B = 0$

$Ax + By = C$ can be rewritten $Ax = C$
$Ax + By = C$ can be rewritten $Ax = D$

Both lines are vertical, hence they are parallel.

65.   Two points given; we first find slope, then use point-slope form.

$$m = \frac{0-b}{a-0} = -\frac{b}{a} \quad a \neq 0$$
$$y - b = -\frac{b}{a}(x - 0)$$
$$y - b = -\frac{bx}{a}$$

Divide both sides by b, then $(b \neq 0)$

$$\frac{y-b}{b} = -\frac{x}{a}$$
$$\frac{y}{b} - 1 = -\frac{x}{a}$$
$$\frac{y}{b} = 1 - \frac{x}{a}$$
$$\frac{x}{a} + \frac{y}{b} = 1$$

67. A. If F is linearly related to C_1 then we are looking for an equation whose graph passes
    through $(C_1, F_1) = (0, 32)$ and $(C_2, F_2) = (100, 212)$. We find the slope, and then use the
    point-slope form to find the equation.

$$m = \frac{F_2 - F_1}{C_2 - C_1} = \frac{212 - 32}{100 - 0} = \frac{180}{100} = \frac{9}{5}$$

$$F - F_1 = m(C - C_1)$$

$$F - 32 = \frac{9}{5}(C - 0)$$

$$F - 32 = \frac{9}{5}C$$

$$F = \frac{9}{5}C + 32$$

B. We are asked for F when $C = 20$.

$$F = \frac{9}{5}(20) + 32 = 36 + 32 = 68°$$

We are then asked for C when $F = 86°$

$$86 = \frac{9}{5}C + 32$$

$$430 = 9C + 160$$

$$270 = 9C$$

$$C = 30°$$

C. The slope is $m = \frac{9}{5}$.

69. A. If V is linearly related to t, then we are looking for an equation whose graph passes through $(t_1, V_1) = (0, 8,000)$ and $(t_2, V_2) = (5, 0)$. We find the slope, and then we use the point-slope form to find the equation.

$$m = \frac{V_2 - V_1}{t_2 - t_1} = \frac{0 - 8,000}{5 - 0} = \frac{-8,000}{5} = -1,600$$

$$V - V_1 = m(t - t_1)$$

$$V - 8,000 = -1,600(t - 0)$$

$$V - 8,000 = -1,600t$$

$$V = -1,600t + 8,000$$

$$V = -1,600t + 8,000$$

B. We are asked for V when $t = 3$

$$V = -1,600(3) + 8,000$$

$$V = -4,800 + 8,000$$

$$V = 3,200$$

C. The slope is $m = -1,600$

71. A. If T is linearly related to A, then we are looking for an equation whose graph passes through $(A_1, T_1) = (0, 70)$ and $(A_2, T_2) = (18, -20)$. We find the slope, and then we use the point-slope form to find the equation.

$$m = \frac{T_2 - T_1}{A_2 - A_1} = \frac{(-20) - 70}{18 - 0} = \frac{-90}{18} = -5$$

$$T - T_1 = m(A - A_1)$$

$$T - 70 = -5(A - 0)$$

$$T - 70 = -5A$$

$$T = -5A + 70 \quad A \geq 0$$

B. We are asked for A when $T = 0$.

$$0 = -5A + 70$$

$$5A = 70$$

$$A = 14 \text{ thousand feet}$$

C. The slope is $m = -5$; the temperature changes $-5°F$ for each 1,000 foot rise in altitude.

73. A. If h is linearly related to t, then we are looking for an equation whose graph passes through $(t_1, h_1) = (9, 23)$ and $(t_2, h_2) = (24, 40)$. We find the slope, and then we use the point-slope form to find the equation.

$$m = \frac{h_2 - h_1}{t_2 - t_1} = \frac{40 - 23}{24 - 9} = \frac{17}{15} \approx 1.13$$

$$h - h_1 = m(t - t_1)$$

$$h - 23 = \frac{17}{15}(t - 9)$$

$$h - 23 = \frac{17}{15}t - \frac{51}{5}$$

$$h = \frac{17}{15}t - \frac{51}{5} + 23$$

$$h = 1.13t + 12.8$$

B. We are asked for t when $h = 50$.

$$50 = 1.13t + 12.8$$

$$37.2 = 1.13t$$

$$t = 32.9 \text{ hours}$$

75. A. If R is linearly related to C, then we are looking for an equation whose graph passes through $(C_1, R_1) = (210, 0.160)$ and $(C_2, R_2) = (231, 0.192)$. We find the slope, and then we use the point-slope form to find the equation.

$$m = \frac{R_2 - R_1}{C_2 - C_1} = \frac{0.192 - 0.160}{231 - 210} = \frac{0.032}{21} = 0.00152$$

$$R - R_1 = m(C - C_1)$$

$$R - 0.160 = 0.00152(C - 210)$$

$$R - 0.160 = 0.00152C - 0.319$$

$$R = 0.00152C - 0.159, \quad C \geq 210$$

B. We are asked for R when $C = 260$.

$$R = 0.00152(260) - 0.159$$

$$R = 0.236$$

C. The slope is $m = 0.00152$; coronary risk increases 0.00152 per unit increase in cholesterol above the 210 cholesterol level.

Exercise 4-3

Key Ideas and Formulas

A function is a rule (process or method) that produces a correspondence between two sets of elements such that to each element in the first set there corresponds one and only element in the second set. The first set is called the domain and the second set is called the range.

In an equation in two variables, if to each value of the independent variable (input) there corresponds exactly one value of the dependent variable (output), then the equation specifies a function. If there is any value of the independent variable to which there corresponds more than one value of the dependent variable, then the equation does not specify a function.

An equation specifies a function if each vertical line in the coordinate system passes through at most one point on the graph of the equation. If any vertical line passes through two or more points on the graph of an equation, then the equation does not specify a function.

If a function is specified by an equation and the domain is not indicated, we assume that the domain is the set of all real number replacements of the independent variable (inputs) that produce real values for the dependent variable (outputs).

If x represents an element in the domain of a function f, we frequently use the symbol $f(x)$ to designate the number in the range of the function to which f is paired.

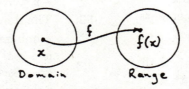

---

1. A function

3. Not a function (two range values correspond to some domain values)

5. A function

7. A function

9. Not a function (two range values correspond to some domain values)

11. A function

13. A function

15. Not a function (fails vertical line test)

17. Not a function (fails vertical line test)

19. $f(-1) = 3(-1) - 5 = -8$

21. $G(-2) = (-2) - (-2)^2 = -6$

23. $F(-1) + f(3) = 3(-1)^2 + 3(3) - 5 = 7$

25. $2F(-2) - G(-1) = 2 \cdot 3(-2)^2$
$- [(-1) - (-1)^2] = 26$

27. $\dfrac{f(0) \cdot g(-2)}{F(-3)} = \dfrac{[3(0) - 5] \cdot [4 - (-2)]}{3(-3)^2}$
$= \dfrac{(-5)(6)}{27} = \dfrac{-30}{27} = \dfrac{-10}{9}$

29. $\dfrac{f(4) - f(2)}{2} = \dfrac{[3(4) - 5] - [3(2) - 5]}{2}$
$= \dfrac{7 - 1}{2} = \dfrac{6}{2} = 3$

31. A function

33. A function

35. Not a function (two values of y correspond to any value of x greater than -2).

37. A function

39. Not a function (two values of y correspond to any value of x, $-9 < x < 9$)

41. A function

43. A function

45. A function

47. A function

49. The domain is R, since $3x + 8$ represents a real number for all replacements of x by real numbers.

51. The domain is the set of all real numbers x such that $\sqrt{x + 2}$ is a real number – that is, such that $x + 2 \geq 0$, or $x \geq -2$.

53. $\dfrac{2 + 3x}{4 - x}$

represents a real number for all replacements of x by real numbers except for $x = 4$. All R except 4. (*Common Error:* $3x + 2 = 0$, $x = -\dfrac{2}{3}$ is *not* excluded from the domain. Zero *can* be divided by a non-zero number.)

55. $\dfrac{x^2 - 2x + 9}{x^2 - 2x - 8} = \dfrac{x^2 - 2x + 9}{(x - 4)(x + 2)}$.

This represents a real number for all replacements of x by real numbers except -2 and 4 (division by zero is not defined). All R except -2 and 4.

57. The domain is the set of all real numbers x such that $\sqrt{4 - x^2}$ is a real number, that is, such that $4 - x^2 \geq 0$. Solving by the methods of Section 3-7, we obtain the sign chart:

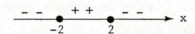

Thus $4 - x^2$ is non-negative and $\sqrt{4 - x^2}$ is real for $-2 \leq x \leq 2$.

59. The domain is the set of all real numbers x such that $\sqrt{x^2 - 3x - 4}$ is a real number, that is, such that $x^2 - 3x - 4 = (x - 4)(x + 1) \geq 0$. Solving by the methods of Section 3-7, we obtain the sign chart:

Thus $x^2 - 3x - 4$ is non-negative and $\sqrt{x^2 - 3x - 4}$ is real for $x \leq -1$ or $x \geq 4$

61. The domain is the set of all real numbers x such that $\sqrt{\frac{5-x}{x-2}}$ is a real number, that is, such that $\frac{5-x}{x-2} \geq 0$. Solving by the methods of Section 3-7, we obtain the sign chart:

$$\xleftarrow{\hspace{1cm}} \underset{2}{\circ} \xrightarrow{\hspace{0.3cm} + + \hspace{0.3cm}} \underset{5}{\bullet} \xrightarrow{\hspace{0.3cm} - - \hspace{0.3cm}} x$$

Thus $\dfrac{5 - x}{x - 2}$ is non-negative and $\sqrt{\dfrac{5 - x}{x - 2}}$ is real for $2 < x \leq 5$

63. The domain is the set of all real numbers x such that $\dfrac{1}{\sqrt[3]{x^2-1}}$ is a real number. Since $\sqrt[3]{a}$ exists for any real number a, this will represent a real number unless $\sqrt[3]{x^2-1}=0$ or $x^2-1=0$. The solutions of this equation are $x=1$ and $x=-1$, thus $\dfrac{1}{\sqrt[3]{x^2-1}}$ represents a real number for x any real number except -1 and 1. All R except -1 and 1.

65.
$$F(s) = 3s + 15$$
$$F(2+h) = 3(2+h) + 15$$
$$F(2) = 3(2) + 15$$
$$\frac{F(2+h) - F(2)}{h} = \frac{[3(2+h)+15]-[3(2)+15]}{h} = \frac{[6+3h+15]-[21]}{h}$$
$$= \frac{3h+21-21}{h} = \frac{3h}{h} = 3$$

67.
$$g(x) = 2 - x^2$$
$$g(3+h) = 2 - (3+h)^2$$
$$g(3) = 2 - (3)^2$$
$$\frac{g(3+h) - g(3)}{h} = \frac{[2-(3+h)]^2 - [2-(3)^2]}{h} = \frac{[2-9-6h-h^2]-[-7]}{h}$$
$$= \frac{-6h-h^2}{h} = \frac{h(-6-h)}{h} = -6 - h$$

69.
$$Q(t) = t^2 - 2t + 1$$
$$Q(1+h) = (1+h)^2 - 2(1+h) + 1$$
$$Q(1) = 1^2 - 2(1) + 1$$
$$\frac{Q(1+h) - Q(1)}{h} = \frac{[(1+h)^2 - 2(1+h) + 1] - [1^2 - 2(1) + 1]}{h}$$
$$= \frac{[1+2h+h^2-2-2h+1]-0}{h} = \frac{h^2}{h} = h$$

71. A.
$$f(x) = 3x - 4$$
$$f(x+h) = 3(x+h) - 4$$
$$\frac{f(x+h) - f(x)}{h} = \frac{[3(x+h)-4]-[3x-4]}{h}$$
$$= \frac{3x+3h-4-3x+4}{h}$$
$$= \frac{3h}{4}$$
$$= 3$$

B.
$$f(x) = 3x - 4$$
$$f(a) = 3a - 4$$
$$\frac{f(x) - f(a)}{x - a} = \frac{(3x - 4) - (3a - 4)}{x - a}$$
$$= \frac{3x - 4 - 3a + 4}{x - a}$$
$$= \frac{3x - 3a}{x - a}$$
$$= \frac{3(x - a)}{x - a}$$
$$= 3$$

73. A.
$$f(x) = x^2 - 1$$
$$f(x + h) = (x + h)^2 - 1$$
$$\frac{f(x + h) - f(x)}{h} = \frac{[(x + h)^2 - 1] - [x^2 - 1]}{h}$$
$$= \frac{x^2 + 2xh + h^2 - 1 - x^2 + 1}{h}$$
$$= \frac{2xh + h^2}{h}$$
$$= \frac{h(2x + h)}{h}$$
$$= 2x + h$$

B.
$$f(x) = x^2 - 1$$
$$f(a) = a^2 - 1$$
$$\frac{f(x) - f(a)}{x - a} = \frac{(x^2 - 1) - (a^2 - 1)}{x - a}$$
$$= \frac{x^2 - 1 - a^2 + 1}{x - a}$$
$$= \frac{x^2 - a^2}{x - a}$$
$$= \frac{(x - a)(x + a)}{x - a}$$
$$= x + a$$

75. A.
$$f(x) = x^3$$
$$f(x + h) = (x + h)^3$$
$$\frac{f(x + h) - f(x)}{h} = \frac{(x + h)^3 - x^3}{h}$$
$$= \frac{x^3 + 3x^2h + 3xh^2 + h^3 - x^3}{h}$$
$$= \frac{3x^2h + 3xh^2 + h^3}{h}$$
$$= \frac{h(3x^2 + 3xh + h^2)}{h}$$
$$= 3x^2 + 3xh + h^2$$

B.
$$f(x) = x^3$$
$$f(a) = a^3$$
$$\frac{f(x) - f(a)}{x - a} = \frac{x^3 - a^3}{(x - a)}$$
$$= \frac{(x - a)(x^2 + xa + a^2)}{x - a}$$
$$= x^2 + ax + a^2$$

77. Given $w =$ width and Area $= 64$, we use $A = \ell w$ to write $\ell = \dfrac{A}{w} = \dfrac{64}{w}$.

   Then $P = 2w + 2\ell = 2w + 2\left(\dfrac{64}{w}\right) = 2w + \dfrac{128}{w}$.

79.

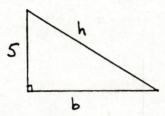

Using the given letters, the Pythagorem theorem gives

$$h^2 = b^2 + 5^2$$
$$h^2 = b^2 + 25$$
$$h = \sqrt{b^2 + 25} \text{ (since } h \text{ is positive)}$$

81.
$$\text{Daily cost} = \text{fixed cost} + \text{variable cost}$$
$$c(x) = \$300 + (\$1.75 \text{ per dozen doughnuts}) \times (\text{number of dozen doughnuts})$$
$$c(x) = 300 + 1.75x$$

83.

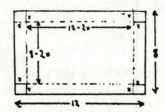 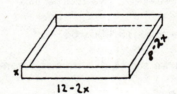

From the above figures it should be clear that
$V =$ length $\times$ width $\times$ height $= (12 - 2x)(8 - 2x)x$.
Since all distances must be positive, $x > 0, 8 - 2x > 0, 0, 12 - 2x > 0$.
Thus, $0 < x, 4 > x, 6 > x$, or $0 < x < 4$ (the last condition, $6 > x$,
will be automatically satisfied if $x < 4$). Domain: $0 < x < 4$.

85. From the text diagram, since each pen must have area 50 square feet, we see

   Area $=$ (length)(width) or $50 =$ (length) x. Thus the length of each pen is $\dfrac{50}{x}$ feet.

The total amount of fencing $= 4(width) + 5(length) + 4(width - gatewidth)$

$$F(x) = 4x + 5\left(\frac{50}{x}\right) + 4(x - 3)$$

$$F(x) = 4x + \frac{250}{x} + 4x - 12$$

$$F(x) = 8x + \frac{250}{x} - 12$$

Then $F(4) = 8(4) + \dfrac{250}{4} - 12 = 82.5$

$F(5) = 8(5) + \dfrac{250}{5} - 12 = 78$

$F(6) = 8(6) + \dfrac{250}{6} - 12 = 77.7$

$F(7) = 8(7) + \dfrac{250}{7} - 12 = 79.7$

87.

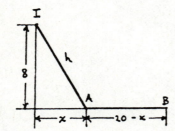

We note that the pipeline consists of the lake section IA, and shore section AB. The shore

section has length $20 - x$. The lake section has length h, where $h^2 = 8^2 + x^2$, thus $h = \sqrt{64 + x^2}$. The cost of all the pipeline $=$

$$\left(\begin{smallmatrix}\text{cost of shore} \\ \text{section per mile}\end{smallmatrix}\right) \left(\begin{smallmatrix}\text{number of} \\ \text{shore miles}\end{smallmatrix}\right) + \left(\begin{smallmatrix}\text{cost of lake} \\ \text{section per mile}\end{smallmatrix}\right) \left(\begin{smallmatrix}\text{number of} \\ \text{lake miles}\end{smallmatrix}\right)$$

$$c(x) = 10,000(20 - x) + 15,000\sqrt{64 + x^2}$$

From the diagram we see that x must be non-negative, but no more than 20.

Domain: $0 \le x \le 20$

89.   Cost $=$ (cost per hour) $\times$ number of hours
       $=$ (Fuel cost per hour $+$ Fixed cost per hour) $\times$ number of hours.
       For a 500 mile trip, since $D = rt, t = D/r = 500/v$ hours.
       Hence cost $= \left(\dfrac{v^2}{5} + 400\right)\left(\dfrac{500}{v}\right)$

$$c(v) = \frac{v^2}{5} \cdot \frac{500}{v} + 400 \cdot \frac{500}{v}$$

$$= 100v + \frac{200,000}{v}$$

91.   A. $s(t) = 16t^2 \quad s(0) = 16(0)^2 = 0 \quad s(1) = 16(1)^2 = 16$
            $s(2) = 16(2)^2 = 64 \quad s(3) = 16(3)^2 = 144$

B.
$$s(2+h) = 16(2+h)^2$$
$$s(2) = 64 \text{ from A.}$$
$$\frac{s(2+h) - s(2)}{h} = \frac{16(2+h)^2 - 64}{h} = \frac{16(4+4h+h^2) - 64}{h}$$
$$= \frac{64 + 64h + 16h^2 - 64}{h} = \frac{64h + 16h^2}{h} = \frac{h(64h + 16h)}{h} = 64 + 16h$$

C. As h tends to 0, $16h$ tends to 0, so $64 + 16h$ tends to 64.

If we interpret $s(2+h)$ as the distance fallen after $2 + h$ seconds, and $s(2)$ as the distance fallen after 2 seconds, then we can think of $s(2+h) - s(2)$ as the distance fallen in h seconds, starting from $t = 2$. Then

$$\frac{s(2+h) - s(2)}{h} = \frac{\text{distance fallen in } h \text{ seconds}}{h \text{ seconds}}$$

$$= \text{speed during } h \text{ seconds after } t = 2$$

As h tends to zero, this appears to tend to the speed of the object at the end of 2 seconds.

Exercise 4-4

Key Ideas and Formulas

A function f is a linear function if $f(x) = mx + b \quad m \neq 0$ where m and b are real numbers.

A function f is a quadratic function if $f(x) = ax^2 + bx + c \quad a \neq 0$ where $a, b,$ and c are real numbers.

Properties of $f(x) = ax^2 + bx + c, a \neq 0$

1.   Axis (of symmetry): $x = -\dfrac{b}{2a}$

2.   Vertex: $(-\dfrac{b}{2a}, f(-\dfrac{b}{2a}))$

3.   Maximum or minimum value of $f(x)$:

$$f(-\frac{b}{2a}) \begin{cases} \text{Minimum if } a > 0 \\ \text{Maximum if } a < 0 \end{cases}$$

4.   Graph (a parabola):

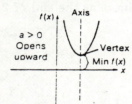

Domain: all real numbers

Range: $[f(-\frac{b}{2a}), \infty)$

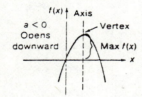

Domain: all real numbers

Range: $(-\infty, f(-\frac{b}{2a})]$

Functions whose definitions involve more than one formula are called piecewise-defined functions.

---

1.   A. $(-\infty, \infty)$ B. $(-4, \infty)$
    C. $-3, 1$ D. -3
    E. $[-1, \infty)$ F. $(-\infty, -1]$
    G. None   H. None

3.   A. $(-\infty, 2) \cup (2, \infty)$
    (The function is not defined at $x = 2$.)
    B. $(-\infty, -1) \cup [1, \infty)$ C. None
    D. 1   E. None
    F. $(-\infty, -2] \cup (2, \infty)$ G. $[-2, 2)$
    H. $x = 2$

5. $f(x) = \underbrace{2}_{\text{slope}} x + 4$

The y intercept is $f(0) = 4$, and the slope is 2. To find the x intercept, we solve the equation $f(x) = 0$ for x.

$$f(x) = 0$$
$$2x + 4 = 0$$
$$2x = -4$$
$$x = -2$$

x intercept is -2.

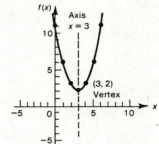

7. $f(x) = -\underbrace{\dfrac{1}{2}}_{\text{slope}} x - \dfrac{5}{3}$

The y intercept is $f(0) = -\dfrac{5}{3}$, and the slope is $-\dfrac{1}{2}$. To find the x intercept, we solve the equation $f(x) = 0$ for x.

$$f(x) = 0$$
$$-\tfrac{1}{2}x - \tfrac{5}{3} = 0$$
$$-\tfrac{1}{2}x = \tfrac{5}{3}$$
$$x = (-2)\tfrac{5}{3}$$
$$x = \tfrac{-10}{3}$$

x intercept is $\dfrac{-10}{3}$

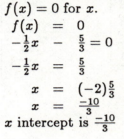

9. $f(x) = -3x + 4$

11.   We first find the slope using the given points, then use the point-slope form of the equation of a line.

$$\text{slope} = m = \frac{y_2 - y_1}{x_2 - x_1} = \frac{2 - 5}{4 - (-2)}$$
$$= \frac{-3}{6} = -\frac{1}{2}$$
$$y - y_1 = m(x - x_1)$$
$$y - 5 = -\frac{1}{2}[x(-2)]$$
$$y - 5 = -\frac{1}{2}[x + 2]$$
$$y - 5 = -\frac{1}{2}x - 1$$
$$y = -\frac{1}{2}x + 4$$
$$f(x) = -\frac{1}{2}x + 4$$

13. $f(x) = (x - 3)^2 + 2$

Minimum value of $f(x)$ (since $(x - 3)^2$ is never negative)

$\text{Min} f(x) = f(3) = 2$

Range: $[2, \infty)$

Vertex: $(3, f(3)) = (3, 2)$

Axis: $x = 3$

Graph: Locate axis and vertex, then plot several points on either side of the axis

x	$f(x)$
0	11
1	6
2	3
3	2
4	3
5	6

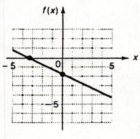

15. $f(x) = -(x + 3)^2 - 2$

Maximum value of $f(x)$ (since $-(x + 3)^2$ is never positive)

$\text{Max } f(x) = f(-3) = -2$

Range: $(-\infty, -2]$

Vertex: $(-3, f(-3)) = (-3, -2)$

Axis: $x = -3$

Graph: Locate axis and vertex, then plot several points on either side of the axis

x	$f(x)$
-5	-6
-4	-3
-3	-2
-2	-3
-1	-6
0	-11

17. $f(x) = x^2 - 4x - 5$. Note that
$a = 1, b = -4$, and $c = -5$.
Axis: $x = -\dfrac{b}{2a} = -\dfrac{-4}{2(1)} = 2$
Vertex: $(-\dfrac{b}{2a}, f(-\dfrac{b}{2a})) = (2, f(2)) = (2, -9)$
y intercept: $f(0) = -5$
x intercepts: We find the values
of x for which $f(x) = 0$.

$$x^2 - 4x - 5 = 0$$
$$(x - 5)(x + 1) = 0$$
$$x - 5 = 0 \quad x + 1 = 0$$
$$x = 5 \qquad x = -1$$

x intercepts : $5, -1$
Graph:

x	$f(x)$
-1	0
0	-5
1	-8
2	-9
3	-8
4	-5
5	0

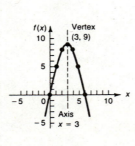

21. $f(x) = x^2 + 6x + 11$. Note
that $a = 1, b = 6$, and $c = 11$.
Axis: $x = -\dfrac{b}{2a} = -\dfrac{6}{2(1)} = -3$
Vertex: $(-\dfrac{b}{2a}, f(-\dfrac{b}{2a}))$
$= (-3, f(-3)) = (-3, 2)$
Since $a > 0$ f is decreasing
to the left of $x = -3$, on
the interval $(-\infty, -3]$
and f is increasing to the
right of $x = -3$, on the
interval $[-3, \infty)$.

x	$f(x)$
-5	6
-4	3
-3	2
-2	3
-1	6
0	11

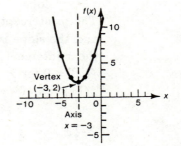

19. $f(x) = -x^2 + 6x$.
Note that $a = -1, b = 6$, and $c = 0$
Axis: $x = -\dfrac{b}{2a} = -\dfrac{6}{2(-1)} = 3$
Vertex: $(-\dfrac{b}{2a}, f(-\dfrac{b}{2a})) = (3, f(3)) = (3, 9)$
y intercept: $f(0) = 0$
x intercepts: We find the values of
x for which $f(x) = 0$

$$-x^2 + 6x = 0$$
$$x(-x + 6) = 0$$
$$x = 0 \quad -x + 6 = 0$$
$$x = 6$$

x − intercepts : $0, 6$
Graph:

x	$f(x)$
0	0
1	5
2	8
3	9
4	8
5	5
6	0

23. $f(x) = -x^2 + 6x - 6$. Note that
$a = -1, b = 6$, and $c = -6$.
Axis: $x = -\dfrac{b}{2a} = -\dfrac{6}{2(-1)} = 3$
Vertex: $(-\dfrac{b}{2a}, f(-\dfrac{b}{2a}))$
$= (3, f(3)) = (3, 3)$
Since $a < 0$, f is increasing to
the left of $x = 3$, on the
interval $(-\infty, 3]$
and f is decreasing to the right
of $x = 3$, on the interval $[3, \infty)$.
Graph:

x	$f(x)$
0	-6
1	-1
2	2
3	3
4	2
5	-1

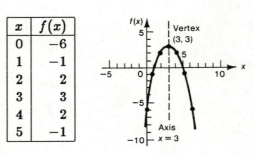

25.

x	$y = x + 1$	x	$y = -x + 1$
-1	0	0	1
$-\frac{1}{2}$	$\frac{1}{2}$	$\frac{1}{2}$	$\frac{1}{2}$
		1	0

Domain: $[-1, 1]$
Range: $[0, 1]$

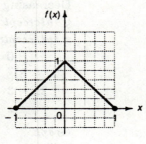

27.

x	$y = x^2 + 1$	x	$y = -x^2 - 1$
-2	5	1	-2
-1	2	2	-5

Domain: $(-\infty, 0) \cup (0, \infty)$
We graph using open dots
at $(0, 1)$, and $(0, -1)$ to
indicate that these points
do not belong to the
graph of g. f is discontinuous at $x = 0$.
Range: $(-\infty, -1) \cup (1, \infty)$

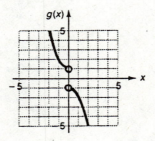

29.

x	$y = 1$	x	$y = x + 1$	x	$y = 2$
-1	1	0	1	2	2
-2	1	1	2	3	2

Domain: R
We graph using an open dot at $(2, 3)$
and a solid dot at $(2, 2)$ to show
that only the latter belongs to the
graph of g.
Range: $[1, 3)$ Discontinuous: at $x = 2$

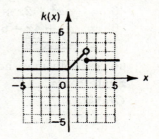

31.

x	$y = 0$	x	$y = 4 - x^2$	x	$y = 0$
-1	0	0	4	2	0
-2	0	1	3	3	0

Domain: R
We graph using an open dot at $(0, 0)$
and a solid dot at $(0, 4)$ to show that
only the latter belongs to the graph
of g.
Range: $[0, 4]$ Discontinuous: at $x = 0$

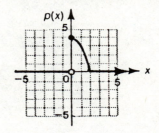

33. $f(x) = \frac{1}{2}x^2 + 2x + 3.$
Note that $a = \frac{1}{2}$, $b = 2$, and $c = 3$
Axis: $x = -\frac{b}{2a} = -\frac{2}{2(1/2)} = -2$
Vertex: $\left(-\frac{b}{2a}, f\left(-\frac{b}{2a}\right)\right)$
$\qquad = (-2, f(-2)) = (-2, 1)$
Minimum value of $f(x)$ (since $a > 0$):
Min $f(x) = f\left(-\frac{b}{2a}\right) = f(-2) = 1$
Increasing on $[-2, \infty)$
Decreasing on $(-\infty, -2]$
y intercept: $f(0) = 3$
x intercept: Solutions of
$$0 = \frac{1}{2}x^2 + 2x + 3$$
$$0 = x^2 + 4x + 6$$
$$x = \frac{-4 \pm \sqrt{-8}}{2}$$
no real solutions, hence no
x intercepts
Graph:

x	$f(x)$
-5	5.5
-4	3
-3	1.5
-2	1
-1	1.5
0	3
1	5.5

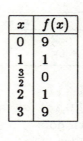

35. $f(x) = 4x^2 - 12x + 9.$ Note that $a = 4$,
$b = -12$, and $c = 9$.
Axis: $x = -\frac{b}{2a} = -\frac{-12}{2(4)} = \frac{3}{2}$
Vertex: $\left(-\frac{b}{2a}, f\left(-\frac{b}{2a}\right)\right) = \left(\frac{3}{2}, f\left(\frac{3}{2}\right)\right) = \left(\frac{3}{2}, 0\right)$
Minimum value of $f(x)$ (since $a > 0$) :
Min $f(x) = f\left(-\frac{b}{2a}\right) = f\left(\frac{3}{2}\right) = 0$
Increasing on $\left[\frac{3}{2}, \infty\right)$
Decreasing on $\left(-\infty, \frac{3}{2}\right]$
y intercept: $f(0) = 9$
x intercept: Solutions of
$$0 = 4x^2 - 12x + 9$$
$$(2x - 3)^2 = 0$$
$$2x - 3 = 0$$
$$x = \frac{3}{2}$$

Graph:

x	$f(x)$
0	9
1	1
$\frac{3}{2}$	0
2	1
3	9

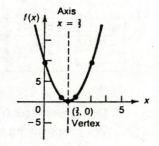

37. $f(x) = -2x^2 - 8x - 2.$

Note that $a = -2$,
$b = -8$, and $c = -2$.

Axis: $x = -\frac{b}{2a} = -\frac{-8}{2(-2)} = -2$

Vertex: $\left(-\frac{b}{2a}, f\left(-\frac{b}{2a}\right)\right)$
$\quad = (-2, f(-2)) = (-2, 6)$

Maximum value of $f(x)$
(Since $a < 0$) :

Max $f(x) = f(-2) = 6$

Decreasing on $[-2, \infty)$

Increasing on $[-\infty, -2]$

y *intercept:* $f(0) = -2$

x *intercept:* Solutions of
$$0 = -2x^2 - 8x - 2$$
$$0 = x^2 + 4x + 1$$
$$x = \frac{-4 \pm \sqrt{12}}{2} = -2 \pm \sqrt{3}$$

Graph:

x	$f(x)$
-4	-2
-3	4
-2	6
-1	4
0	-2

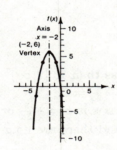

39. $f(x) = \frac{|x|}{x}$

If $x < 0$ then $|x| = -x$ and
$f(x) = \frac{|x|}{x} = \frac{-x}{x} = -1$

If $x = 0$, then f is not defined. f
is discontinuous at $x = 0$

If $x > 0$, then $|x| = x$ and
$f(x) = \frac{|x|}{x} = \frac{x}{x} = 1$

Domain: $(-\infty, 0) \bigcup (0, \infty)$

Range: $\{-1, 1\}$, From the graph
we see that only two values of
f are possible.

Piecewise definition for f :
$$f(x) = \begin{cases} -1 \text{ if } x < 0 \\ 1 \text{ if } x > 0 \end{cases}$$

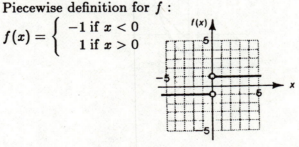

41. $f(x) = x + \frac{|x-1|}{x-1}$

If $x < 1$, then $|x - 1| = -(x - 1)$
and
$f(x) = x + \frac{-(x-1)}{x-1} = x - 1$

If $x = 1$, then $f(x)$ is not defined.

f is discontinuous at $x = 1$.

If $x > 1$, then $|x - 1| = x - 1$ and
$f(x) = x + \frac{x-1}{x-1} = x + 1$

x	$y = x - 1$	x	$y = x + 1$
-1	-2	2	3
0	-1	3	4

Domain: $(-\infty, 1) \bigcup (1, \infty)$

Piecewise definition for f:
$$f(x) = \begin{cases} x - 1 & \text{if } x < 1 \\ x + 1 & \text{if } x > 1 \end{cases}$$

Range: From the graph, we see that y is
less than 0 or y is greater than 2:
$(-\infty, 0) \bigcup (2, \infty)$

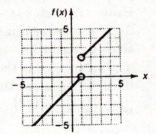

43. $f(x) = |x| + |x - 2|$
      If $x < 0$, then $|x| = -x$ and $|x - 2| = -(x - 2) = 2 - x$ and
      $f(x) = |x| + |x - 2| = -x + (2 - x) = 2 - 2x$
      If $0 \leq x < 2$, then $|x| = x$ but $|x - 2| = -(x - 2) = 2 - x$ and
      $f(x) = |x| + |x - 2| = x + (2 - x) = 2$
      If $2 \leq x$, then $|x| = x$ and $|x - 2| = x - 2$ and
      $f(x) = |x| + |x - 2| = x + (x - 2) = 2x - 2$
      Domain: R
      Piecewise definition for f :
      $$f(x) = \begin{cases} 2 - 2x & x < 0 \\ 2 & 0 \leq x < 2 \\ 2x - 2 & x \geq 2 \end{cases}$$

      There are no discontinuities for f.

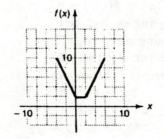

x	$y = 2 - 2x$	x	$y = 2$	x	$y = 2x - 2$
-2	6	0	2	2	2
-1	4	1	2	3	4

Range: From the graph we see that y is never less than 2. $[2, \infty)$

45. $f(x) = ax^2 + bx + c$ Since Min $f(x) = f(2) = 4$, $-\frac{b}{2a} = 2$
      Hence the vertex is $\left(-\frac{b}{2a}, f\left(-\frac{b}{2a}\right)\right) = (2, 4)$. The axis is $x = -\frac{b}{2a}$ or $x = 2$.
      Since y is never less than 4, the range is $[4, \infty)$ and there are no x-intercepts
      since y cannot be equal to 0.

47.   The secant line is the line through (-1, 3) and (3,5). To find its equation we find
      its slope, then use the point- slope form of the equation of a line.
      $$m = \frac{f(x_2) - f(x_1)}{x_2 - x_1} = \frac{5 - (-3)}{3 - (-1)} = \frac{8}{4} = 2$$
      $y - f(x_1) = m(x - x_1)$
      $y - (-3) = 2[x - (-1)]$
      $y + 3 = 2[x + 1]$
      $y + 3 = 2x + 2$
      $y = 2x - 1$
      Graph:

x	$y = 2x - 1$	$y = x^2 - 4$
-2	-5	0
-1	-3	-3
0	-1	-4
1	1	-3
2	3	0
3	5	5

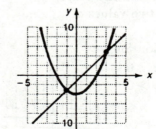

49.   A. The slope of the secant line is given by

$$m = \frac{f(2+h) - f(2)}{(2+h) - 2} = \frac{f(2+h) - f(2)}{h}$$

Since $f(2+h) = (2+h)^2 - 3(2+h) + 5 = 4 + 4h + h^2 - 6 - 3h + 5 = 3 + h + h^2$
and $f(2) = (2)^2 - 3(2) + 5 = 4 - 6 + 5 = 3$
We have

$$m = \frac{f(2+h) - f(2)}{h} = \frac{(3 + h + h^2) - (3)}{h} = \frac{h + h^2}{h} = \frac{h(1+h)}{h} = 1 + h$$

B.

h	$slope = 1 + h$
1	2
0.1	1.1
0.01	1.01
0.001	1.001

The slope seems to be approaching 1.

51.   A. Using the hint, we have $f(10) = 1$ and $f(0) = 0$. Since $f(w) = mw + b$,
we must have $1 = f(10) = m(10) + b$ and $0 = f(0) = m(0) + b$.
Therefore $b = 0$ and $m = \frac{1}{10}$. So $s = f(w) = \frac{1}{10}$.
B. $f(w) = \frac{w}{10}$. Therefore $f(15) = \frac{15}{10} = 1.5$ inches and $f(30) = \frac{30}{10} = 3$ inches.
C. The slope is the coefficient of w, or $\frac{1}{10}$.
D.

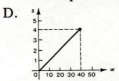

53.   If $0 \le x \le 3,000$, $E(x) = 200$

		Base Salary	+	Commission on Sales over \$3,000
If \$3,000 < x < 8,000, E(x)	=	200	+	0.04(x - 3,000)
	=	200	+	0.04x - 120
	=	80	+	0.04x

Common Error:
Commission is *not* 0.04x
There is a point of discontinuity at $x = 8,000$.

		Salary	+	Bonus
If x ≥ 8,000, E(x)	=	80 + 0.04x	+	100
	=	180	+	0.04x

$$E(x) = \begin{cases} 200 & \text{if } 0 \le x \le 3{,}000 \\ 80 + 0.04x & \text{if } 3{,}000 < x < 8{,}000 \\ 180 + 0.04x & \text{if } 8{,}000 \le x \end{cases}$$

Summarizing,

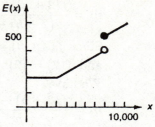

$$E(5{,}750) = 80 + 0.04(5{,}750) = \$310$$
$$E(9{,}200) = 180 + 0.04(9{,}200) = \$548$$

x	$y = 200$	x	$y = 80 + 0.04x$	x	$y = 180 + 0.04x$
0	200	3,000	200	8,000	500
2,000	200	7,000	360	10,000	580

55.   A. Let $x = $ width of pen

$$\text{Then } 2x + 2\ell = 100$$
$$\text{so } 2\ell = 100 - 2x$$
$$\text{Therefore } \ell = 50 - x = \text{length of pen}$$
$$A(x) = x(50 - x) = 50x - x^2$$

B. Since all distances must be positive, $x > 0$ and $50 - x > 0$.
    Therefore $0 < x < 50$ or $(0, 50)$ is the domain of $A(x)$.

C. Note: 0, 50 were excluded from the domain for geometrical reasons; but can be used to help draw the graph since the *polynomial* $50x - x^2$ has domain including these values.

x	$A(x)$
0	0
10	400
20	600
30	600
40	400
50	0

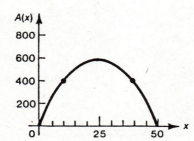

D. From the graph, the maximum value of $A(x)$ seems to occur when $x = 25$.
    Checking by completing the square, we see

$$\begin{aligned} A(x) &= 50x - x^2 \\ &= -(x^2 - 50x + ?) \\ &= -(x^2 - 50x + 625) + 625 \\ &= -(x - 25)^2 + 625 \end{aligned}$$

Indeed, the maximum value of $A(x)$ occurs when $x = 25$. Then $50 - x = 25$, and the dimensions are 25 feet by 25 feet.

57.   Let $x = $ rate per car per day.

$$\begin{aligned} \text{Then (number of cars rented)} &= 300 - 5(\text{increase of rate over } \$40) \\ &= 300 - 5(x - 40) \\ &= 300 - 5x + 200 \\ &= 500 - 5x \end{aligned}$$

$$\text{Income} \quad = \quad I(x) \quad = \quad \begin{pmatrix}\text{Number of}\\\text{Cars Rented}\end{pmatrix}\begin{pmatrix}\text{Rate Per}\\\text{Car Per Day}\end{pmatrix}$$
$$I(x) \quad = \quad (500 - 5x)x$$
$$I(x) \quad = \quad 500x - 5x^2$$

This is a quadratic function with $a = -5$, $b = 500$, and $c = 0$.

The maximum value of this function occurs when $x = -\dfrac{b}{2a} = -\dfrac{500}{2(-5)} = 50$

Hence the rate should be \$50 a day. The resulting maximum income is

$$I(50) \quad = \quad 500(50) - 5(50)^2$$
$$= \quad 25,000 - 12,500$$
$$= \quad \$12,500$$

59. The given function $f(x) = \frac{1}{4}x - \left(\frac{16}{v^2}\right)x^2$ is a quadratic function for every positive value of v. In the coordinate system given, this function has x intercepts 0 and 80, and maximum value when $x = 40$. Thus $-\frac{b}{2a} = 40$. If the given function, $a = -\frac{16}{v^2}$ and $b = \frac{1}{4}$, thus

$$-\frac{\frac{1}{4}}{2(-\frac{16}{v^2})} \quad = \quad 40$$

$$-\frac{4v^2 \cdot \frac{1}{4}}{4v^2 \cdot 2(-\frac{16}{v^2})} \quad = \quad 40$$

$$-\frac{v^2}{-128} \quad = \quad 40$$

$$v^2 \quad = \quad 5120$$

$$v \quad = \quad \sqrt{5120} \text{ (discarding the negative solution)}$$

$$v \quad = \quad 32\sqrt{5} \approx 71.55 \text{ ft/sec}$$

B. The maximum height of the cycle above the ground is

$$10 + \text{Max } f(x) \quad = \quad 10 + f\left(-\frac{b}{2a}\right)$$

$$= \quad 10 + f(40)$$

$$= \quad 10 + \frac{1}{4}(40) - \frac{16}{(32\sqrt{5})^2}(40)^2$$

$$= \quad 10 + 10 - \frac{16}{5120}(1600)$$

$$= \quad 10 + 10 - 5$$

$$= \quad 15 \quad \text{feet}$$

Exercise 4-5

Key Ideas and Formulas

The intersection of sets A and B is the set of all elements in A that are also in B. The sum, difference, product and quotient of the functions f and g are the functions defined by:

$$(f+g)(x) \ = \ f(x) + g(x) \quad \text{Sum function}$$

$$(f-g)(x) \ = \ f(x) - g(x) \quad \text{Difference function}$$

$$(fg)(x) \ = \ f(x)g(x) \qquad \text{Product function}$$

$$\left(\tfrac{f}{g}\right)(x) \ = \ \tfrac{f(x)}{g(x)} \qquad \text{Quotient function}$$

Each function is defined on the intersection of the domains of f and g, with the exception that the values of x where $g(x) = 0$ must be excluded from the domain of the quotient function.

Given functions f and g, then fog is called their composite and is defined by the equation

$$(fog)(x) = f[g(x)]$$

The domain of fog is the set of all numbers x such that x is in the domain of g and $g(x)$ is the domain of f.

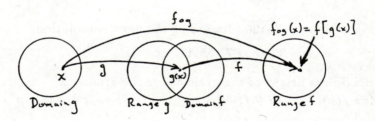

---

1. $\quad (f+g)(x) \ = \ f(x) + g(x) = 4x + x + 1 = 5x + 1 \quad$ Domain : R

    $\quad (f-g)(x) \ = \ f(x) - g(x) = 4x - (x+1) = 3x - 1 \quad$ Domain : R

    $\quad (fg)(x) \ = \ f(x)g(x) = 4x(x+1) = 4x^2 + 4x \qquad$ Domain : R

    $\quad \left(\tfrac{f}{g}\right)(x) \ = \ \tfrac{f(x)}{g(x)} = \tfrac{4x}{x+1} \qquad\qquad\qquad$ Domain : $\{x \mid x \neq 1\}$, or $(-\infty, 1) \cup (1, \infty)$

    Common Error:

    $f(x) - g(x) \neq 4x - x + 1$. The parentheses are necessary.

3. $\quad (f+g)(x) \ = \ f(x) + g(x) = 2x^2 + x^2 + 1 = 3x^2 + 1 \quad$ Domain : R

    $\quad (f-g)(x) \ = \ f(x) - g(x) = 2x^2 - (x^2 + 1) = x^2 - 1 \quad$ Domain : R

    $\quad (fg)(x) \ = \ f(x)g(x) = 2x^2(x^2+1) = 2x^4 + 2x^2 \qquad$ Domain : R

    $\quad \left(\tfrac{f}{g}\right)(x) \ = \ \tfrac{f(x)}{g(x)} = \tfrac{2x^2}{x^2+2} \qquad\qquad\qquad$ Domain : R

5. $(f+g)(x) = f(x) + g(x) = 3x + 5 + x^2 - 1$ Domain : R

$\qquad\qquad\qquad = x^2 + 3x + 4$

$\qquad (f-g)(x) = f(x) - g(x) = 3x + 5 - (x^2 - 1)$

$\qquad\qquad\qquad = 3x + 5 - x^2 + 1 = -x^2 + 3x + 6$ Domain : R

$\qquad (fg)(x) = f(x)g(x) = (3x + 5)(x^2 - 1)$

$\qquad\qquad\qquad = 3x^3 - 3x + 5x^2 - 5 = 3x^3 + 5x^2 - 3x - 5$ Domain : R

$\qquad \left(\frac{f}{g}\right)(x) = \frac{f(x)}{g(x)} = \frac{3x+5}{x^2-1}$ Domain : $\{x \mid x \neq -1\}$,

$\qquad\qquad\qquad\qquad\qquad\qquad\qquad\qquad$ or $(-\infty, -1) \cup (-1, 1) \cup (1, \infty)$

7. $(f \circ g)(x) = f[g(x)] = [g(x)]^3 = (x^2 - x + 1)^3$ Domain : R

$\qquad (g \circ f)(x) = g[f(x)] = [f(x)]^2 - [f(x)] + 1 = (x^3)^2 - x^3 + 1 = x^6 - x^3 + 1$ Domain : R

9. $(f \circ g)(x) = f[g(x)] = \mid g(x) + 1 \mid = \mid 2x + 3 + 1 \mid = \mid 2x + 4 \mid$ Domain : R

$\qquad (g \circ f)(x) = g[f(x)] = 2f(x) + 3 = 2 \mid x + 1 \mid + 3$ Domain : R

11. $(f \circ g)(x) = f[g(x)] = [g(x)]^{1/3} = (2x^3 + 4)^{1/3}$ Domain : R

$\qquad (g \circ f)(x) = g[f(x)] = 2[f(x)]^3 + 4 = 2(x^{1/3})^3 + 4 = 2x + 4$ Domain : R

13. $(f+g)(x) = f(x) + g(x) = \sqrt{4 - x^2} + \sqrt{4 + x^2}$

$\qquad (f-g)(x) = f(x) - g(x) = \sqrt{4 - x^2} - \sqrt{4 + x^2}$

$\qquad (fg)(x) = f(x)g(x) = \sqrt{4 - x^2}\sqrt{4 + x^2} = \sqrt{(4 - x^2)(4 + x^2)} = \sqrt{16 - x^4}$

$\qquad \left(\frac{f}{g}\right)(x) = \frac{f(x)}{g(x)} = \frac{\sqrt{4-x^2}}{\sqrt{4+x^2}} = \sqrt{\frac{4-x^2}{4+x^2}}$

The domains of f and g are:

Domain of $f = \{x \mid 4 - x^2 \geq 0\} = [-2, 2]$

Domain of $g = R$

The intersection of these domains is $[-2, 2]$.

This is the domain of the functions $f + g$, $f - g$, fg, and $\frac{f}{g}$ (since $g(x)$ is never 0).

15. $(f+g)(x) = f(x) + g(x) = \sqrt{2 - x} + \sqrt{x + 3}$

$\qquad (f-g)(x) = f(x) - g(x) = \sqrt{2 - x} - \sqrt{x + 3}$

$\qquad (fg)(x) = f(x)g(x) = \sqrt{2 - x}\ \sqrt{x + 3} = \sqrt{(2 - x)(3 + x)} = \sqrt{6 - x - x^2}$

$\qquad \left(\frac{f}{g}\right)(x) = \frac{f(x)}{g(x)} = \frac{\sqrt{2-x}}{\sqrt{x+3}} = \sqrt{\frac{2-x}{x+3}}$

The domains of f and g are

Domain of $f = \{x \mid 2 - x \geq 0\} = [-\infty, 2]$

Domain of $g = \{x \mid x + 3 \geq 0\} = [-3, \infty)$

The intersection of these domains is $[-3, 2]$.

This is the domain of the functions $f + g$, $f - g$, and fg.

Since $g(-3) = 0, x = -3$ must be excluded from the domains of $\frac{f}{g}$, so its domain is $(-3, 2]$.

17.
$$(f + g)(x) = f(x) + g(x) = \sqrt{x} + 2 + \sqrt{x} - 4 = 2\sqrt{x} - 2$$
$$(f - g)(x) = f(x) - g(x) = \sqrt{x} + 2 - (\sqrt{x} - 4) = \sqrt{x} + 2 - \sqrt{x} + 4 = 6$$
$$(fg)(x) = f(x)g(x) = (\sqrt{x} + 2)(\sqrt{x} - 4) = x - 2\sqrt{x} - 8$$
$$\left(\frac{f}{g}\right)(x) = \frac{f(x)}{f(x)} = \frac{\sqrt{x} + 2}{\sqrt{x} - 4}$$

The domains of f and g are both

$\{x \mid x \geq 0\} = [0, \infty)$

This is the domain of $f + g$, $f - g$, and fg. We note that in the domain of $\frac{f}{g}, g(x) \neq 0$.

Thus $\sqrt{x} - 4 \neq 0$. To solve this, we solve

$$\sqrt{x} - 4 = 0$$
$$\sqrt{x} = 4 \quad Common \quad Error : x \neq \sqrt{4}$$
$$x = 16$$

Hence, 16 must be excluded from $\{x \mid x \geq 0\}$ to find the domain of $\frac{f}{g}$.

Domain of $\frac{f}{g} = \{x \mid x \geq 0, x \neq 16\} = [0, 16) \cup (16, \infty)$.

19.
$$(f \circ g)(x) = f[g(x)] = [g(x) - 1]^3 = (1 + x^{1/3} - 1)^3 = (x^{1/3})^3 = x. \qquad \text{Domain} : R$$
$$(g \circ f)(x) = g[f(x)] = 1 + [f(x)]^{1/3} = 1 + [(x - 1)^3]^{1/3} = 1 + x - 1 = x. \quad \text{Domain} : R$$

21.
$$(f \circ g)(x) = f[g(x)] = \sqrt{g(x)} = \sqrt{x - 4} \qquad \text{Domain} : \{x \mid x \geq 4\} \text{ or } [4, \infty)$$
$$(g \circ f)(x) = g[f(x)] = f(x) - 4 = \sqrt{x} - 4 \quad \text{Domain} : \{x \mid x \geq 0\} \text{ or } (0, \infty)$$

23.
$$(f \circ g)(x) = f[g(x)] = g(x) + 2 = \frac{1}{x} + 2 \quad \text{Domain} : \{x \mid x \neq 0\} \text{ or} (-\infty, 0) \cup (0, \infty)$$
$$(g \circ f)(x) = g[f(x)] = \frac{1}{f(x)} = \frac{1}{x + 2} \qquad \text{Domain} : \{x \mid x \neq -2\} \text{ or } (-\infty, -2) \cup (-2, \infty)$$

25.
$$(f \circ g)(x) = f[g(x)] = \mid g(x) \mid = \mid \frac{1}{x - 1} \mid = \frac{|1|}{|x - 1|} = \frac{1}{|x - 1|} \quad \text{Domain} : \{x \mid x \neq 1\} \text{ or } (-\infty, 1) \cup (1, \infty)$$
$$(g \circ f)(x) = g[f(x)] = \frac{1}{f(x) - 1} = \frac{1}{|x| - 1}$$

To find the domain we must exclude $\mid x \mid -1 = 0$ from the domain of f.

$$\mid x \mid -1 = 0$$
$$\mid x \mid = 1$$
$$x = -1 \text{ or } 1$$

Domain of $g \circ f$: $\{x \mid x \neq -1 \text{ or } 1\} = (-\infty, -1) \cup (-1, 1) \cup (1, \infty)$

27. If we let $g(x) = 2x - 7$, then

$$h(x) = [g(x)]^4$$

Now if we let $f(x) = x^4$, we have

$$h(x) = [g(x)]^4 = f[g(x)] = (f \circ g)(x)$$

29. If we let $g(x) = 4 + 2x$, then

$$h(x) = \sqrt{g(x)}$$

Now if we let $f(x) = x^{1/2}$, we have

$$h(x) = \sqrt{g(x)} = [g(x)]^{1/2} = f[g(x)] = (f \circ g)(x).$$

31. If we let $f(x) = x^7$, then

$$h(x) = 3f(x) - 5$$

Now if we let $g(x) = 3x - 5$, we have

$$h(x) = 3f(x) - 5 = g[f(x)] = (g \circ f)(x)$$

33. If we let $f(x) = x^{-1/2}$, then

$$h(x) = 4f(x) + 3$$

Now if we let $g(x) = 4x + 3$, we have

$$h(x) = 4f(x) + 3 = g[f(x)] = (g \circ f)(x)$$

35.

$$
\begin{aligned}
(f+g)(x) &= f(x) + g(x) = x + \tfrac{1}{x} + x - \tfrac{1}{x} = 2x \quad \textit{Common Error}: \\
(f-g)(x) &= f(x) - g(x) = x + \tfrac{1}{x} - (x - \tfrac{1}{x}) = \tfrac{2}{x} \quad \text{Domain is not } R. \text{ See below.} \\
(fg)(x) &= f(x)g(x) = (x + \tfrac{1}{x})(x - \tfrac{1}{x}) = x^2 - \tfrac{1}{x^2} \\
(\tfrac{f}{g})(x) &= \tfrac{f(x)}{g(x)} = \tfrac{x + \frac{1}{x}}{x - \frac{1}{x}} = \tfrac{x^2+1}{x^2-1}
\end{aligned}
$$

The domains of f and g are both

$$\{x \mid x \neq 0\} = (-\infty, 0) \cup (0, \infty)$$

This is therefore the domain of $f + g$, $f - g$, and fg. To find the domain of $\frac{f}{g}$, we must exclude from this domain the set of values of x for which $g(x) = 0$.

$$
\begin{aligned}
x - \tfrac{1}{x} &= 0 \\
x^2 - 1 &= 0 \\
x^2 &= 1 \\
x &= -1, 1
\end{aligned}
$$

Hence, the domain of $\frac{f}{g}$ is $\{x \mid x \neq 0, -1, \text{ or } 1\}$ or

$$(-\infty, -1) \cup (-1, 0) \cup (0, 1) \cup (1, \infty).$$

37.

$$
\begin{aligned}
(f+g)(x) &= f(x) + g(x) = x + \tfrac{8}{\sqrt{x}} + x - \tfrac{8}{\sqrt{x}} = 2x \\
(f-g)(x) &= f(x) - g(x) = x + \tfrac{8}{\sqrt{x}} - (x - \tfrac{8}{\sqrt{x}}) = x + \tfrac{8}{\sqrt{x}} - x + \tfrac{8}{\sqrt{x}} = \tfrac{16}{\sqrt{x}} \\
(fg)(x) &= f(x)g(x) = (x + \tfrac{8}{\sqrt{x}})(x - \tfrac{8}{\sqrt{x}}) = x^2 - (\tfrac{8}{\sqrt{x}})^2 = x^2 - \tfrac{64}{x} \\
(\tfrac{f}{g})(x) &= \tfrac{f(x)}{g(x)} = \tfrac{x + \frac{8}{\sqrt{x}}}{x - \frac{8}{\sqrt{x}}} = \tfrac{x\sqrt{x}+8}{x\sqrt{x}-8} \text{ or } \tfrac{x^{3/2}+8}{x^{3/2}-8}
\end{aligned}
$$

The domains of f and g are both

$$\{x \mid x > 0\} = (0, \infty)$$

This is therefore the domain of $f + g$, $f - g$ and fg. To find the domain of $\frac{f}{g}$,

we must exclude from this domain the set of values of x for which $g(x) = 0$

$$
\begin{aligned}
x - \frac{8}{\sqrt{x}} &= 0 \\
x\sqrt{x} - 8 &= 0 \\
x^{3/2} - 8 &= 0 \\
x^{3/2} &= 8 \\
x &= 8^{2/3} \text{ or } 4
\end{aligned}
$$

Common Error:
$x \neq 8^{3/2}$

Hence the domain of $\frac{f}{g}$ is $\{x \mid x > 0, x \neq 4\}$ or

$(0, 4) \cup (4, \infty)$.

39.
$$(f + g)(x) = f(x) + g(x) = 1 - \frac{x}{|x|} + 1 + \frac{x}{|x|} = 2$$
$$(f - g)(x) = f(x) - g(x) = 1 - \frac{x}{|x|} - \left(1 - \frac{x}{|x|}\right) = 1 - \frac{x}{|x|} - 1 - \frac{x}{|x|} = \frac{-2x}{|x|}$$
$$(fg)(x) = f(x)g(x) = \left(1 - \frac{x}{|x|}\right)\left(1 - \frac{x}{|x|}\right) = (1)^2 - \left(\frac{x}{|x|}\right)^2 = 1 - \frac{x^2}{|x|^2} = 1 - \frac{x^2}{x^2} = 1 - 1 = 0$$
$$\left(\frac{f}{g}\right)(x) = \frac{f(x)}{g(x)} = \frac{1 - \frac{x}{|x|}}{1 + \frac{x}{|x|}} = \frac{|x| - x}{|x| + x}. \text{This can be further simplified}$$

however, when we examine the domain of $\frac{f}{g}$ below.

The domains of f and g are both

$$\{x \mid x \neq 0\} = (-\infty, 0) \cup (0, \infty)$$

This is therefore the domain of $f + g$, $f - g$, and fg. To find the domain of $\frac{f}{g}$

we must exclude from this domain the set of values of x for which $g(x) = 0$

$$
\begin{aligned}
1 + \frac{x}{|x|} &= 0 \\
|x| + x &= 0 \\
|x| &= -x
\end{aligned}
$$

x is negative.

Thus the domain of $\frac{f}{g}$ is the positive numbers, $(0, \infty)$. On this domain,

$|x| = x$, hence

$$\left(\frac{f}{g}\right)(x) = \frac{|x| - x}{|x| + x} = \frac{x - x}{x + x} = \frac{0}{2x} = 0$$

41. $(f \circ g)(x) = f[g(x)] = \sqrt{4 - g(x)} = \sqrt{4 - x^2}$

The domain of f is $(-\infty, 4]$. The domain of g is R. Hence the domain of $f \circ g$

is the set of those real numbers x for which $g(x)$ is in $(-\infty, 4]$,

that is, for which $x^2 \leq 4$, or $-2 \leq x \leq 2$.

Domain of $f \circ g = \{x \mid -2 \leq x \leq 2\} = [-2, 2]$

$(gof)(x) = g[f(x)] = [f(x)]^2 = (\sqrt{4-x})^2 = 4 - x$

The domain of gof is the set of those numbers x in $(-\infty, 4]$ for which $f(x)$ is in R, that is, $(-\infty, 4]$.

43.　$(fog)(x) = f[g(x)] = \sqrt{g(x)} = \sqrt{x^2} = |x|$

The domain of f is $[0, \infty)$. The domain of g is R. Hence the domain of fog

is the set of those real numbers x for which $g(x)$ is in $[0, \infty)$,

that is, $x^2 \geq 0$. This condition is satisfied for all $x \epsilon R$, hence

Domain $fog = R$.

$(gof)(x) = g[f(x)] = [f(x)]^2 = (\sqrt{x})^2 = x$

The domain of fog is the set of those numbers x in $[0, \infty)$ for which $f(x)$ is real, that is $x \geq 0$. Hence, the domain of gof is $[0, \infty)$.

45.　$(fog)(x) = f[g(x)] = \frac{g(x)+5}{g(x)} = \frac{\frac{x}{x-2}+5}{\frac{x}{x-2}} = \frac{x+5(x-2)}{x} = \frac{x+5x-10}{x} = \frac{6x-10}{x}$

The domain of f is $\{x \mid \neq 0\}$. The domain of g is $\{x \mid x \neq 2\}$.

Hence the domain of fog is the set of those numbers in $\{x \mid x \neq 2\}$ for

which $g(x)$ is in $\{x \mid x \neq 0\}$. Thus we must exclude from $\{x \mid x \neq 2\}$

those numbers x for which $\frac{x}{x-2} = 0$, or $x = 0$. Hence the domain of fog is

$\{x \mid x \neq 0, x \neq 2\}$, or $(-\infty, 0) \bigcup (0, 2) \bigcup (2, \infty)$.

$(gof)(x) = g[f(x)] = \frac{f(x)}{f(x)-2} = \frac{\frac{x+5}{x}}{\frac{x+5}{x}-2} = \frac{x+5}{x+5-2x} = \frac{x+5}{5-x}$

The domain of gof is the set of those numbers in $\{x \mid x \neq 0\}$ for

which $f(x)$ is in $\{x \mid x \neq 2\}$. Thus we must exclude from $\{x \mid x \neq 0\}$

those numbers x for which $\frac{x+5}{x} = 2$, or $x + 5 = 2x$, or $x = 5$.

Hence the domain of gof is $\{x \mid x \neq 0, x \neq 5\}$ or $(-\infty, 0) \bigcup (0, 5) \bigcup (5, \infty)$.

47.　$(fog)(x) = f[g(x)] = \sqrt{25 - [g(x)]^2} = \sqrt{25 - (\sqrt{9+x^2})^2} = \sqrt{25 - (9+x^2)} = \sqrt{16 - x^2}$

The domain of f is $[-5, 5]$. The domain of g is R.

Hence the domain of fog is the set of those real numbers x for which

$g(x)$ is in $[-5, 5]$, that is, $\sqrt{9+x^2} \leq 5$, or $9 + x^2 \leq 25$, or $x^2 \leq 16$,

or $-4 \leq x \leq 4$. Hence the domain of fog is $\{x \mid -4 \leq x \leq 4\}$ or $[-4, 4]$.

$(gof)(x) = g[f(x)] = \sqrt{9 + [f(x)]^2} = \sqrt{9 + (\sqrt{25-x^2})^2} = \sqrt{9 + 25 - x^2} = \sqrt{34 - x^2}$

The domain of gof is the set of those numbers x in $[-5, 5]$ for which $g(x)$ is real.

Since $g(x)$ is real for all x, the domain of gof is $[-5, 5]$.

Common error:
The domain of gof is not evident from the final form $\sqrt{34 - x^2}$.

It is not $[-\sqrt{34}, \sqrt{34}]$.

49.   The profit function is the difference of the revenue and the cost functions.

$$P = R - C$$

Hence

$$
\begin{aligned}
P(x) &= R(x) - C(x) \\
&= (20x - \tfrac{1}{200}x^2) - (10x + 30,000) \\
&= 20x - \tfrac{1}{200}x^2 - 10x - 30,000 \\
&= 10x - \tfrac{1}{200}x^2 - 30,000
\end{aligned}
$$

Next we use composition to express P as a function of the price p.

$$
\begin{aligned}
(Pof)(p) &= P[f(p)] \\
&= 10f(p) - \tfrac{1}{200}[f(p)]^2 - 30,000 \\
&= 10(4,000 - 200p) - \tfrac{1}{200}(4,000 - 200\,p)^2 - 30,000 \\
&= 40,000 - 2,000\,p - \tfrac{1}{200}(16,000,000 - 1,600,000\,p + 40,000\,p^2) - 30,000 \\
&= 40,000 - 2,000\,p - 80,000 + 8,000\,p - 200\,p^2 - 30,000 \\
&= -70,000 + 6,000\,p - 200\,p^2
\end{aligned}
$$

51.   We are given $V(r) = 0.1A(r) = 0.1\pi r^2$ and $r(t) = 0.4t^{1/3}$.

Hence we use composition to express V as a function of the time.

$$
\begin{aligned}
(Vor)(t) &= V[r(t)] \\
&= 0.1\pi[r(t)]^2 \\
&= 0.1\pi[0.4t^{1/3}]^2 \\
&= 0.1\pi[0.16t^{2/3}] \\
&= 0.016\pi t^{2/3}
\end{aligned}
$$

We write $V(t) = 0.016\pi t^{2/3}$

53.

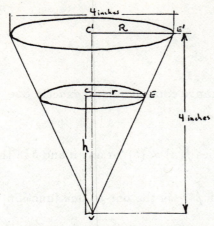

A. We note: In the figure, triangles VCE and $VC'E'$ are similar.

Moreover $R =$ radius of cup $= \frac{1}{2}$ diameter of cup $= \frac{1}{2}(4) = 2$ inches.

Hence $\frac{r}{2} = \frac{h}{4}$ or $r = \frac{1}{2}h$. We write $r(h) = \frac{1}{2}h$.

B. Since $V = \frac{1}{3}\pi r^2 h$ and $r = \frac{1}{2}h$, $V = \frac{1}{3}\pi(\frac{1}{2}h)^2 h = \frac{1}{3}\pi\frac{1}{4}h^2 h = \frac{1}{12}\pi h^3$.

We write $V(h) = \frac{1}{12}\pi h^3$.

C. Since $V(h) = \frac{1}{12}\pi h^3$ and $h(t) = 0.5\sqrt{t}$, we use composition to express V

as a function of t.

$$
\begin{aligned}
(V \circ h)(t) &= V[h(t)] \\
&= \frac{1}{12}\pi[h(t)]^3 \\
&= \frac{1}{12}\pi[0.5\sqrt{t}]^3 \\
&= \frac{1}{12}\pi(0.125)(t^{1/2})^3 \\
&= \frac{0.125}{12}\pi t^{3/2}
\end{aligned}
$$

We write $V(t) = \frac{0.125}{12}\pi t^{3/2}$

Exercise 4-6

Key Ideas and Formulas

A function is one-to-one if each element in the range corresponds to exactly one element in the domain.

A function is one-to-one if and only if $a \neq b$ implies $f(a) \neq (b)$ for any a and b in the domain of f.

The inverse of a one-to-one function f, denoted f^{-1}, is the one-to-one function formed by reversing all the ordered pairs in the function f.

$f^{-1} = \{(y, x) \mid y = f(x), y \epsilon$ Range of $f, x\epsilon$ Domain of f $\}$

$x = f^{-1}(y)$ if and only if $y = f(x)$.

Range of f^{-1} = Domain of f.   Domain of f^{-1} = Range of f.

$(f^{-1}of)(x) = x$ for all x in the domain of f.

$(fof^{-1})(y) = y$ for all y in the domain of f^{-1}.

To find the inverse of a function f

1.   Find the domain of f and verify that f is one-to-one.
        If f is not one-to-one, then stop $-f^{-1}$ does not exist.
2.   Solve the equation $y = f(x)$ for x. The result is an equation of the form $x = f^{-1}(y)$.
3.   Interchange x and y in the equation found in Step 2.
        The result is an equation of the form $y = f^{-1}(x)$.
4.   State the inverse function and its domain.
5.   Check by showing that
        $(f^{-1}of)(x) \quad = \quad x \quad$ for all x in the domain of f
        $(fof^{-1})(x) \quad = \quad x \quad$ for all x in in the domain of f^{-1}

---

1.   One-to-one
3.   The range element 7 corresponds to more than one domain element. Not one-to-one.
5.   One-to-one
7.   Some range elements, (0, for example) correspond to more than one domain element. Not one-to-one
9.   One-to-one

11. $F(x) = \frac{1}{2}x + 2.$

$$\text{Since } a \;\neq\; b$$
$$\text{implies } \tfrac{1}{2}a \;\neq\; \tfrac{1}{2}b$$
$$\text{implies } \tfrac{1}{2}a + 1 \;\neq\; \tfrac{1}{2}b + 1$$
$$\text{or } F(a) \;\neq\; F(b),$$

F is one-to-one.

13. $H(x) = 4 - x^2$

Since $H(1) = 4 - 1^2 = 3$ and $H(-1) = 4 - (-1)^2 = 4 - 1 = 3$, both $(1,3)$ and $(-1,3)$ belong to H and H is not one-to-one.

15. $M(x) = \sqrt{x+1}$

$$\text{Since } a \;\neq\; b$$
$$\text{implies } a + 1 \;\neq\; b + 1$$
$$\text{imples } \sqrt{a+1} \;\neq\; \sqrt{b+1}$$
$$\text{or } M(a) \;\neq\; M(b),$$

M is one-to-one.

17.   From the given graph of f, we see
Domain of $f = [-4,4]$ Range of $f = [1,5]$
Hence
Range of $f^{-1} = [-4,4]$ Domain of $f^{-1} = [1,5]$
The graph of f^{-1} is drawn by reflecting the given graph of f in the line $y = x$.

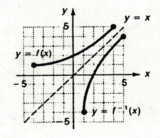

19.   From the given graph of f, we see
Domain of $f = [-5,3]$ Range of $f = [-3,5]$
Hence
Range of $f^{-1} = [-5,3]$ Domain of $f^{-1} = [-3,5]$
The graph of f^{-1} is drawn by reflecting the given graph of f in the line $y = x$.

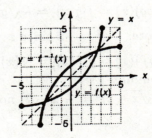

21. $f(x) = 3x + 6$ $g(x) = \frac{1}{3}x - 2$

$(g \circ f)(x)\ =\ g[f(x)] = \frac{1}{3}f(x) - 2 = \frac{1}{3}(3x + 6) - 2 = x + 2 - 2 = x$

$(f \circ g)(x)\ =\ f[g(x)] = 3g(x) + 6 = 3(\frac{1}{3}x - 2) + 6 = x - 6 + 6 = x$

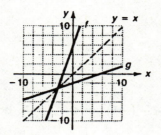

23. $f(x) = 4 + x^2$ $x \geq 0$ $g(x) = \sqrt{x - 4}$

$(g \circ f)(x)\ =\ g[f(x)] = \sqrt{f(x) - 4} = \sqrt{4 + x^2 - 4} = \sqrt{x^2} = x$ since $x \geq 0$

$(f \circ g)(x)\ =\ f[g(x)] = 4 + [g(x)]^2 = 4 + (\sqrt{x - 4})^2 = 4 + x - 4 = x$

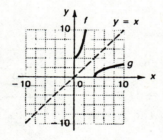

25. $f(x) = -\sqrt{x - 2}$ $g(x) = x^2 + 2,$ $x \leq 0$

$(f \circ g)(x) = f[g(x)] = -\sqrt{g(x) - 2} = -\sqrt{x^2 + 2 - 2} = -\sqrt{x^2} = -(-x) = x$

(Note: $\sqrt{x^2} = -x$ since $x \leq 0$)

$(g \circ f)(x) = g[f(x)] = [f(x)]^2 + 2 = (-\sqrt{x - 2})^2 + 2 = x - 2 + 2 = x$

Common Error:

$(-\sqrt{x - 2})^2 \neq -(x - 2)$ nor $-x - 2$

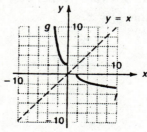

27. $f(x) = 4x - 3$
Solve $y = f(x)$ for x :
$$y = 4x - 3$$
$$y + 3 = 4x$$
$$x = \tfrac{1}{4}y + \tfrac{3}{4}$$
Interchange x and y :

$$y = \tfrac{1}{4}x + \tfrac{3}{4}$$

State the inverse function
of its domain:
$$f^{-1}(x) = \tfrac{1}{4}x + \tfrac{3}{4}$$

Domain: R
Check:
$$(f^{-1}of)(x) = \tfrac{1}{4}(4x - 3) + \tfrac{3}{4}$$
$$= x - \tfrac{3}{4} + \tfrac{3}{4}$$
$$= x$$
$$(fof^{-1})(x) = 4(\tfrac{1}{4}x + \tfrac{3}{4}) - 3$$
$$= x + 3 - 3$$
$$= x$$
$$f^{-1}(x) = \tfrac{1}{4}x + \tfrac{3}{4}$$

29. $f(x) = \frac{x}{x+2}$
Solve $y = f(x)$ for x :
$$y = \frac{x}{x+2}$$
$$y(x + 2) = x$$
$$xy + 2y = x$$
$$2y = x - xy$$
$$2y = x(1 - y)$$
$$x = \frac{2y}{1-y}$$
Interchange x and y:

$$y = \frac{2x}{1-x}$$

State the inverse function
and its domain:
$$f^{-1}(x) = \frac{2x}{1-x}$$

Domain:$(-\infty, 1) \cup (1, -\infty)$
Check:

$$(f^{-1}of)(x) = \frac{2\frac{x}{x+2}}{1-\frac{x}{x+2}}$$
$$= \frac{2x}{x+2-x}$$
$$= \frac{2x}{2}$$
$$= x$$
$$(fof^{-1})(x) = \frac{\frac{2x}{1-x}}{\frac{2x}{1-x}+2}$$
$$= \frac{2x}{2x+2(1-x)}$$
$$= \frac{2x}{2x+2-2x}$$
$$= \frac{2x}{2}$$
$$= x$$
$$f^{-1}(x) = \frac{2x}{1-x}$$

31. $f(x) = x^3 + 1$

Solve $y = f(x)$ for x :

$$y = x^3 + 1$$
$$y - 1 = x^3$$
$$x = \sqrt[3]{y - 1}$$

Interchange x and y :

$$y = \sqrt[3]{x - 1}$$

State the inverse function and its domain:

$$f^{-1}(x) = \sqrt[3]{x - 1}$$

Domain: R

Check:

$$(f^{-1}of)(x) = \sqrt[3]{x^3 + 1 - 1} = \sqrt[3]{x^3} = x$$
$$(fof^{-1})(x) = (\sqrt[3]{x - 1})^3 + 1 = x - 1 + 1 = x$$
$$f^{-1}(x) = \sqrt[3]{x - 1}$$

33. $f(x) = 4 - \sqrt[5]{x + 2}$

Solve $y = f(x)$ for x :

$$y = 4 - \sqrt[5]{x + 2}$$
$$y - 4 = -\sqrt[5]{x + 2}$$
$$4 - y = \sqrt[5]{x + 2}$$
$$(4 - y)^5 = x + 2$$
$$x = (4 - y)^5 - 2$$

Interchange x and y :

$$y = (4 - x)^5 - 2$$

State the inverse function and its domain:

$$f^{-1}(x) = (4 - x)^5 - 2$$

Domain: R

Check:

$$(f^{-1}of) = (x)[4 - (4 - \sqrt[5]{x + 2})]^5 - 2$$
$$= [4 - 4 + \sqrt[5]{x + 2}]^5 - 2$$
$$= [\sqrt[5]{x + 2}]^5 - 2$$
$$= x + 2 - 2$$
$$= x$$
$$(fof^{-1})x = 4 - \sqrt[5]{(4 - x)^5 - 2 + 2}$$
$$= 4 - \sqrt[5]{(4 - x)^5}$$
$$= 4 - (4 - x)$$
$$= 4 - 4 + x$$
$$= x$$

35. $f(x) = \frac{1}{2}\sqrt{16 - x}$

Solve $y = f(x)$ for x :

$y = \frac{1}{2}\sqrt{16 - x}$

$\left. \begin{array}{rcl} 2y & = & \sqrt{16 - x} \\[6pt] 4y^2 & = & 16 - x \end{array} \right\}$ These are equivalent only if $y \geq 0$

$x = 16 - 4y^2 \quad y \geq 0$

State the inverse function and its domain:

$f^{-1}(x) = 16 - 4x^2, x \geq 0$

Check:

$$
\begin{aligned}
(f^{-1}of)(x) &= 16 - 4[\tfrac{1}{2}\sqrt{16 - x}]^2 \\
&= 16 - 4[\tfrac{1}{4}(16 - x)] \\
&= 16 - (16 - x) \\
&= 16 - 16 + x \\
&= x \\
(fof^{-1})(x) &= \tfrac{1}{2}\sqrt{16 - (16 - 4x^2)} \\
&= \tfrac{1}{2}\sqrt{16 - 16 + 4x^2} \\
&= \tfrac{1}{2}\sqrt{4x^2} \quad \sqrt{4x^2} = 2x \text{ since } x \geq 0 \\
&= \tfrac{1}{2}(2x) \\
&= x \\
f^{-1}(x) &= 16 - 4x^2, x \geq 0
\end{aligned}
$$

37. $f(x) = 3 - \sqrt{x - 2}$

Solve $y = f(x)$ for x :

$$
\begin{aligned}
y &= 3 - \sqrt{x - 2} \\
y - 3 &= -\sqrt{x - 2}
\end{aligned}
$$

$\left. \begin{array}{rcl} 3 - y & = & \sqrt{x - 2} \\[6pt] (3 - y)^2 & = & x - 2 \end{array} \right\}$ these are equivalent only if $y \leq 3$

$x = (3 - y)^2 + 2 \quad y \leq 3$

Interchange x and y

$y = (3 - x)^2 + 2 \quad x \leq 3$

State the inverse function and its domain:

$f^{-1}(x) = (3 - x)^2 + 2 \quad x \leq 3.$

Check:

$$(f^{-1} \circ f)(x) = [3 - (3 - \sqrt{x-2})]^2 + 2$$
$$= (\sqrt{x-2})^2 + 2$$
$$= x - 2 + 2$$
$$= x$$
$$(f \circ f^{-1})(x) = 3 - \sqrt{(3-x)^2 + 2 - 2}$$
$$= 3 - \sqrt{(3-x)^2} \quad \sqrt{(3-x)^2} = 3 - x \text{ since } x \le 3$$
$$= 3 - (3 - x)$$
$$= 3 - 3 + x$$
$$= x$$
$$f^{-1}(x) = (3-x)^2 + 2, \quad x \le 3$$

39. $f(x) = (x-1)^2 + 2 \quad x \ge 1$

Solve $y = f(x)$ for x:
$$y = (x-1)^2 + 2 \quad x \ge 1$$
$$y - 2 = (x-1)^2 \quad x \ge 1$$
$$\sqrt{y-2} = \sqrt{(x-1)^2} \text{ since } x \ge 1, \sqrt{(x-1)^2} = x - 1$$
$$\sqrt{y-2} = x - 1$$
$$x = 1 + \sqrt{y-2}$$

Interchange x and y:
$$y = 1 + \sqrt{x-2}$$

State the inverse function and its domain:
$$f^{-1}(x) = 1 + \sqrt{x-2} \quad x \ge 2$$

Check:
$$(f^{-1} \circ f)(x) = 1 + \sqrt{(x-1)^2 + 2 - 2}$$
$$= 1 + \sqrt{(x-1)^2}$$
$$\sqrt{(x-1)^2} = x - 1 \text{ since } x \ge 1$$
$$= 1 + x - 1$$
$$= x$$
$$(f \circ f^{-1})(x) = (1 + \sqrt{x-2} - 1)^2 + 2$$
$$= (\sqrt{x-2})^2 + 2$$
$$= x - 2 + 2$$
$$= x$$
$$f^{-1}(x) = 1 + \sqrt{x-2}, x \ge 2$$

41. $f(x) = x^2 + 2x - 2 \quad x \le -1$

Solve $y = f(x)$ for x :
$$y = x^2 + 2x - 2 \qquad x \le -1$$
$$y + 3 = x^2 + 2x + 1 \qquad x \le -1$$
$$y + 3 = (x+1)^2 \qquad x \le -1$$
$$\sqrt{y+3} = \sqrt{(x+1)^2} \text{ since } x \le -1 \quad \sqrt{(x+1)^2} = -(x+1)$$
$$\sqrt{y+3} = -(x+1)$$
$$-\sqrt{y+3} = x + 1$$
$$x = -1 - \sqrt{y+3}$$

Interchange x and y:

$y = -1 - \sqrt{x+3}$

State the inverse function and its domain

$f^{-1}(x) = -1 - \sqrt{x+3} \quad x \geq -3$

Check:

$$\begin{aligned}
(f^{-1} \circ f)(x) &= -1 - \sqrt{x^2 + 2x - 2 + 3} & x &\leq -1 \\
&= -1 - \sqrt{x^2 + 2x + 1} & x &\leq -1 \\
&= -1 - \sqrt{(x+1)^2} & x &\leq -1 \\
& \quad \sqrt{(x+1)^2} = -(x+1) \text{ since} & x &\leq -1 \\
&= -1 - [-(x+1)] \\
&= -1 + x + 1 \\
&= x \\
(f \circ f^{-1})(x) &= (-1 - \sqrt{x+3})^2 + 2(-1 - \sqrt{x+3}) - 2 \\
&= 1 + 2\sqrt{x+3} + x + 3 - 2 - 2\sqrt{x+3} - 2 \\
&= 4 + x - 4 \\
&= x
\end{aligned}$$

In the following solutions we have omitted the verification that f is one-to-one for lack of space. The verification is analogous to that in Example 30 of the text.

43. $f(x) = -\sqrt{9 - x^2} \quad 0 \leq x \leq 3$

Solve $y = f(x)$ for x :

$$\begin{aligned}
y &= -\sqrt{9 - x^2} & 0 &\leq x \leq 3 \\
y^2 &= 9 - x^2 & y &\leq 0 \\
y^2 - 9 &= -x^2 & y &\leq 0 \\
9 - y^2 &= x^2 & y &\leq 0 \\
x &= \underbrace{\sqrt{9 - y^2}} & y &\leq 0
\end{aligned}$$

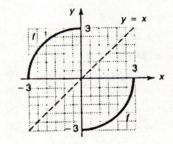

positive square root only because $0 \leq x$

Interchange x and y :

$y = \sqrt{9 - x^2} \quad x \leq 0$

State the inverse function and its domain:

$f^{-1}(x) = \sqrt{9 - x^2}, -3 \leq x \leq 0$

Check:

$$(f^{-1} \circ f)(x) = \sqrt{9 - (-\sqrt{9 - x^2})^2}$$
$$= \sqrt{9 - (9 - x^2)}$$
$$= \sqrt{9 - 9 + x^2}$$
$$= \sqrt{x^2} \quad \sqrt{x^2} = x \text{ since } x \geq 0 \text{ in the domain of } f$$
$$= x$$

$$(f \circ f^{-1})(x) = -\sqrt{9 - (\sqrt{9 - x^2})^2}$$
$$= -\sqrt{9 - (9 - x^2)}$$
$$= -\sqrt{9 - 9 + x^2}$$
$$= -\sqrt{x^2} \quad \sqrt{x^2} = -x \text{ since } x \leq 0 \text{ in the domain of } f^{-1}$$
$$= -(-x)$$
$$= x$$

$f^{-1}(x) = \sqrt{9 - x^2}$ Domain of $f^{-1} = [-3, 0]$
Range of $f^{-1} = $ Domain of $f = [0, 3]$

45. $f(x) = \sqrt{9 - x^2} \quad -3 \leq x \leq 0$
Solve $y = f(x)$ for x

$$y = \sqrt{9 - x^2} \quad -3 \leq x \leq 0$$
$$y^2 = 9 - x^2 \quad y \geq 0$$
$$y^2 - 9 = -x^2 \quad y \geq 0$$
$$9 - y^2 = x^2 \quad y \geq 0$$
$$x = -\underbrace{\sqrt{9 - y^2}}_{} \quad y \geq 0$$

negative square root only because $x \leq 0$
Interchange x and y:
$y = -\sqrt{9 - x^2} \quad x \geq 0$
State the inverse function and its domain:
$f^{-1}(x) = -\sqrt{9 - x^2}, 0 \leq x \leq 3$
Check:

$$(f^{-1} \circ f)(x) = -\sqrt{9 - (-\sqrt{9 - x^2})^2}$$
$$= -\sqrt{9 - (9 - x^2)}$$
$$= -\sqrt{9 - 9 + x^2}$$
$$= -\sqrt{x^2}$$
$$\sqrt{x^2} = -x \text{ since } x \leq 0 \text{ in the domain of } f$$
$$= -(-x)$$
$$= x$$

$$(f \circ f^{-1})(x) = \sqrt{9 - \left(-\sqrt{9 - x^2}\right)^2}$$
$$= \sqrt{9 - (9 - x^2)}$$
$$= \sqrt{9 - 9 + x^2}$$
$$= \sqrt{x^2}$$

$\sqrt{x^2} = x$ since $x \geq 0$

in the domain of f^{-1}

$$= x$$

$f^{-1}(x) = -\sqrt{9 - x^2}$

Domain $f^{-1} = [0, 3]$

Range of $f^{-1} =$

Domain of $f = [-3, 0]$

47. $f(x) = 1 + \sqrt{1 - x^2}$ $0 \leq x \leq 1$

Solve $y = f(x)$ for x:

$$y = 1 + \sqrt{1 - x^2} \quad 0 \leq x \leq 1$$
$$y - 1 = \sqrt{1 - x^2}$$
$$(y - 1)^2 = 1 - x^2 \qquad y \geq 1$$
$$y^2 - 2y + 1 = 1 - x^2$$
$$y^2 - 2y = -x^2$$
$$2y - y^2 = x^2$$
$$x = \underbrace{\sqrt{2y - y^2}}_{} \qquad y \geq 1$$

positive square root only because $0 \leq x$

Interchange x and y :

$y = \sqrt{2x - x^2} \quad x \geq 1$

State the inverse function and its domain:

$f^{-1}(x) = \sqrt{2x - x^2} \quad 1 \leq x \leq 2$

Check:

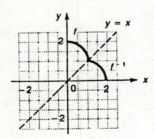

$$(f^{-1} of)(x) = 1 + \sqrt{1 - (\sqrt{2x - x^2})^2}$$
$$= 1 + \sqrt{1 - (2x - x^2)}$$
$$= 1 + \sqrt{1 - 2x + x^2}$$
$$= 1 + \sqrt{(1 - x)^2}$$
$$= 1 + [-(1 - x)] \text{ because } 1 \leq x \text{ } in \text{ } the \text{ } domain \text{ } of \text{ } f$$
$$= 1 - 1 + x$$
$$= x$$

$$(f of^{-1})(x) = \sqrt{2(1 + \sqrt{1 - x^2}) - (1 + \sqrt{1 - x^2})^2}$$
$$= \sqrt{2 + 2\sqrt{1 - x^2} - (1 + 2\sqrt{1 - x^2} + 1 - x^2)}$$
$$= \sqrt{2 + 2\sqrt{1 - x^2} - 1 - 2\sqrt{1 - x^2} - 1 + x^2}$$
$$= \sqrt{x^2}$$
$$= x$$

because $x \geq 0$ *in the domain of* f^{-1}

$f^{-1}(x) = \sqrt{2x - x^2}$ Domain of $f^{-1} = [1, 2]$
Range of $f^{-1} =$ Domain of $f = [0, 1]$

49. $f(x) = 1 - \sqrt{1 - x^2} \quad -1 \leq x \leq 0$

Solve $y = f(x)$ for x :

$$y = 1 - \sqrt{1 - x^2} \quad -1 \leq x \leq 0$$
$$y - 1 = -\sqrt{1 - x^2}$$
$$1 - y = \sqrt{1 - x^2}$$
$$(1 - y)^2 = 1 - x^2 \qquad y \leq 1$$
$$1 - 2y + y^2 = 1 - x^2$$
$$-2y + y^2 = -x^2$$
$$2y - y^2 = x^2$$
$$x = \underbrace{-\sqrt{2y - y^2}}_{} \quad y \leq 1$$

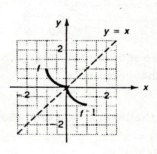

negative square root only because $x \leq 0$
Interchange x and y :
$y = -\sqrt{2x - x^2} \quad x \leq 1$
State the inverse function and its domain:
$f^{-1}(x) = -\sqrt{2x - x^2} \qquad 0 \leq x \leq 1$
Check:

$$(f^{-1}of)(x) \;=\; 1 - \sqrt{1 - \left(-\sqrt{2x - x^2}\right)^2}$$

$$= \; 1 - \sqrt{1 - (2x - x^2)}$$

$$= \; 1 - \sqrt{1 - 2x + x^2}$$

$$= \; 1 - \sqrt{(1 - x)^2}$$

$$= \; 1 - (1 - x) \text{ because } x \le 1 \text{ in the domain of } f$$

$$= \; 1 - 1 + x$$

$$= \; x$$

$$(fof^{-1})(x) \;=\; - \sqrt{2(1 - \sqrt{1 - x^2}) - (1 - \sqrt{1 - x^2})^2}$$

$$= \; -\sqrt{2 - 2\sqrt{1 - x^2} - (1 - 2\sqrt{1 - x^2} + 1 - x^2)}$$

$$= \; -\sqrt{2 - 2\sqrt{1 - x^2} - 1 + 2\sqrt{1 - x^2} - 1 + x^2}$$

$$= \; -\sqrt{x^2}$$

$$= \; -(-x) \text{ because } x \le 0 \text{ in the domain of } f^{-1}$$

$$= \; x$$

$f^{-1}(x) = -\sqrt{2x - x^2}$

Domain of $f^{-1} = [0, 1]$

Range of f^{-1} = Domain of $f = [-1, 0]$

51. $f(x) = ax + b \quad a \ne 0$

Solve $y = f(x)$ for x:

$$y \;=\; ax + b$$
$$y - b \;=\; ax$$
$$x \;=\; \frac{y - b}{a}$$

Interchange x and y :

$y = \frac{x - b}{a}$

State the inverse function and its domain

$f^{-1}(x) = \frac{x - b}{a}$

Domain: R

Check:

$$(f^{-1}of)(x) \;=\; \frac{ax + b - b}{a} = \frac{ax}{a} = x$$
$$(fof^{-1})(x) \;=\; a\frac{x - b}{a} + b = x - b + b = x$$
$$f^{-1}(x) \;=\; \frac{x - b}{a}$$

53.   If f is increasing on I, then $a < b$ implies $f(a) < f(b)$.

Interchanging a and b restates this:

$b < a$

or

$a > b$ implies $f(b) < f(a)$.

Thus $a \ne b$ implies $f(a) \ne f(b)$.

This is precisely the statement that f is one-to-one.

55. $f(x) = (2 - x)^2$

A. $x \leq 2$

Solve $y = f(x)$ for x:

$$\begin{aligned} y &= (2 - x)^2 \quad x \leq 2 \\ \sqrt{y} &= 2 - x \text{ since } 2 - x \geq 0 \\ \sqrt{y} - 2 &= -x \\ x &= 2 - \sqrt{y} \end{aligned}$$

Interchange x and y:

$y = 2 - \sqrt{x}$

State the inverse function and its domain:

$f^{-1}(x) = 2 - \sqrt{x} \qquad x \geq 0$

Check:

$$\begin{aligned} (f^{-1} o f)(x) &= 2 - \sqrt{(2 - x)^2} \\ &= 2 - (2 - x) \text{ since } 2 - x \geq 0 \\ &\quad \textit{in the domain of } f \\ &= 2 - 2 + x \\ &= x \\ (f o f^{-1})(x) &= [2 - (2 - \sqrt{x})]^2 \\ &= [2 - 2 + \sqrt{x}]^2 \\ &= [\sqrt{x}]^2 \\ &= x \\ f^{-1}(x) &= 2 - \sqrt{x} \end{aligned}$$

B. $x \geq 2$

Solve $y = f(x)$ for x:

$$\begin{aligned} y &= (2 - x)^2 \qquad x \geq 2 \\ \sqrt{y} &= -(2 - x) \text{ since } \quad 2 - x \leq 0 \\ \sqrt{y} &= -2 + x \\ x &= 2 + \sqrt{y} \end{aligned}$$

Interchange x and y:

$y = 2 + \sqrt{x}$

State the inverse function and its domain:

$f^{-1}(x) = 2 + \sqrt{x} \quad x \geq 0$

Check:

$$(f^{-1}of)(x) \;=\; 2 + \sqrt{(2-x)^2}$$

$$\;=\; 2 + [-(2-x)] \text{ since } 2 - x \le 0$$
in the domain of f

$$\;=\; 2 - 2 + x$$

$$\;=\; x$$

$$(fof^{-1})(x) \;=\; [2 - (2 + \sqrt{x})]^2$$

$$\;=\; [2 - 2 - \sqrt{x}]^2$$

$$\;=\; [-\sqrt{x}]^2$$

$$\;=\; x$$

$$f^{-1}(x) = 2 + \sqrt{x}$$

57. $\quad f(x) = \sqrt{4x - x^2}$

A. $0 \le x \le 2$

Solve $y = f(x)$ for x:

$$y \;=\; \sqrt{4x - x^2}$$

$$y^2 \;=\; 4x - x^2 \qquad y \ge 0$$

$$-y^2 \;=\; x^2 - 4x$$

$$4 - y^2 \;=\; x^2 - 4x + 4$$

$$4 - y^2 \;=\; (x - 2)^2$$

$$\underbrace{-\sqrt{4 - y^2}}\;=\; x - 2$$

negative square root only because $x \le 2$

$x = 2 - \sqrt{4 - y^2} \quad y \ge 0$

Interchange x and y:

$y = 2 - \sqrt{4 - x^2} \quad x \ge 0$

State the inverse function and its domain:

$$f^{-1}(x) = 2 - \sqrt{4 - x^2} \qquad 0 \le x \le 2$$

Check:

$$(f^{-1}of)(x) \;=\; 2 - \sqrt{4 - (\sqrt{4x - x^2})^2}$$

$$\;=\; 2 - \sqrt{4 - (4x - x^2)}$$

$$\;=\; 2 - \sqrt{4 - 4x + x^2}$$

$$\;=\; 2 - \sqrt{(2 - x)^2}$$

$$\;=\; 2 - (2 - x) \text{ since } 2 - x \ge 0$$

$$\;=\; x$$

$$(fof^{-1})(x) \;=\; \sqrt{4(2 - \sqrt{4 - x^2}) - (2 - \sqrt{4 - x^2})^2}$$

$$\;=\; \sqrt{8 - 4\sqrt{4 - x^2} - (4 - 4\sqrt{4 - x^2} + 4 - x^2)}$$

$$\;=\; \sqrt{8 - 4\sqrt{4 - x^2} - 4 + 4\sqrt{4 - x^2} - 4 + x^2}$$

$$\;=\; \sqrt{x^2}$$

$$\;=\; x \text{ since } 0 \le x$$

$$f^{-1}(x) = 2 - \sqrt{4 - x^2}, 0 \le x \le 2$$

B. $2 \leq x \leq 4$

Solve $y = f(x)$ for x:

$$
\begin{aligned}
y &= \sqrt{4x - x^2} \\
y^2 &= 4x - x^2 \quad y \geq 0 \\
-y^2 &= x^2 - 4x \\
4 - y^2 &= x^2 - 4x + 4 \\
4 - y^2 &= (x - 2)^2 \\
\underbrace{\sqrt{4 - y^2}}_{} &= x - 2
\end{aligned}
$$

positive square root only because $x \geq 2$

$x = 2 + \sqrt{4 - y^2} \quad y \geq 0$

Interchange x and y:

$y = 2 + \sqrt{4 - x^2} \quad x \geq 0$

State the inverse function and its domain:

$$f^{-1}(x) = 2 + \sqrt{4 - x^2} \quad 0 \leq x \leq 2$$

$$
\begin{aligned}
(f^{-1}of)(x) &= 2 + \sqrt{4 - \left(\sqrt{4x - x^2}\right)^2} \\
&= 2 + \sqrt{4 - (4x - x^2)} \\
&= 2 + \sqrt{4 - 4x + x^2} \\
&= 2 + \sqrt{(2 - x)^2} \\
&= 2 + [-(2 - x)] \text{ since } 2 \leq x \\
&\quad \textit{in the domain of } f \\
&= 2 - 2 + x \\
&= x \\
(fof^{-1})(x) &= \sqrt{4(2 - \sqrt{4 - x^2}) - (2 - \sqrt{4 - x^2})^2} \\
&= \sqrt{8 - 4\sqrt{4 - x^2} - (4 - 4\sqrt{4 - x^2} + 4 - x^2)} \\
&= \sqrt{8 - 4\sqrt{4 - x^2} - 4 + 4\sqrt{4 - x^2} - 4 + x^2} \\
&= \sqrt{x^2} \\
&= x \text{ since } 0 \leq x \\
&\quad \textit{in the domain of } f^{-1}
\end{aligned}
$$

$f^{-1}(x) = 2 + \sqrt{4 - x^2} \quad 0 \leq x \leq 2$

Exercise 4-7
CHAPTER REVIEW

1. A. $d(A,B) = \sqrt{[4-(-2)]^2 + (0-3)^2} = \sqrt{36+9} = \sqrt{45}$
    B. $m = \frac{0-3}{4-(-2)} = \frac{-3}{6} = -\frac{1}{2}$
    C. The slope m_1 of a line perpendicular to AB must satisfy $m_1(-\frac{1}{2}) = -1$.
    Therefore $m_1 = 2.$ $(4-1, 4-2)$

2. A. Center at $(0,0)$ and radius $\sqrt{7}$ B. Center at $(3,-2)$ and radius $\sqrt{7}$
    $$x^2 + y^2 = r^2$$ $$(h,k) = (3,-2) \; r = \sqrt{7}$$
    $$x^2 + y^2 = (\sqrt{7})^2$$ $$(x-h)^2 + (y-k)^2 = r^2$$
    $$x^2 + y^2 = 7$$ $$(x-3)^2 + [y-(-2)]^2 = (\sqrt{7})^2$$
    $$(x-3)^2 + (y+2)^2 = 7 \quad (4-1)$$

3. $$(x+3)^2 + (y-2)^2 = 5$$
    $$[x-(-3)]^2 + (y-2)^2 = (\sqrt{5})^2$$
    $$\text{Center}: C(h,k) = (-3,2)$$
    $$\text{Radius}: r = \sqrt{5} \quad (4-1)$$

4.

x	y
0	$\frac{9}{2}$
3	0
1	3

$$3x + 2y = 9$$
$$2y = -3x + 9$$
$$y = -\frac{3}{2}x + \frac{9}{2}$$
$$\text{Slope}: -\frac{3}{2}$$

$(4-2)$

5. The line passes through the two given points, $(6,0)$ and $(0,4)$. Thus, its slope is given by
    $m = \frac{0-4}{6-0} = \frac{-4}{6} = -\frac{2}{3}$
    The equation of the line is, therefore, using the point-slope form
    $y - 0 = -\frac{2}{3}(x-6)$ or $3y = -2(x-6)$ or $3y = -2x + 12$ $2x + 3y = 12$ $(4-2)$

6. $y = mx + b \quad m = -\frac{2}{3} \quad b = 2$
    $y = -\frac{2}{3}x + 2 \quad (4-2)$

7. vertical : $x = -3$, slope not defined
    horizontal : $y = 4$, slope $= 0$ $(4-2)$

8. A. Function.
    B. Not a function – two range elements correspond to some domain
    elements; for example 2 and -2 correspond to 4.
    C. Function.

D. Not a function — two range elements correspond to some domain elements; for example 2 and -2 correspond to 2.   (4 – 3)

9. $f(2) = 3(2) + 5 = 11$
$g(-2) = 4 - (-2)^2 = 0$
$k(0) = 5$
Therefore $f(2) + g(-2) + k(0) = 11 + 0 + 5 = 16$ (4 – 3)

10. $m(-2) = 2\,|-2\,|-1 = 3$
$g(2) = 4 - (2)^2 = 0$
Therefore $\frac{m(-2)+1}{g(2)+4} = \frac{3+1}{0+4} = 1$ (4 – 3)

11. $\dfrac{f(2+h)-f(2)}{h} = \dfrac{[3(2+h)+5]-[3(2)+5]}{h}$
$= \dfrac{6+3h+5-11}{h}$
$= \dfrac{3h}{h}$
$= 3$ (4 – 3)

12. $\dfrac{g(a+h)-g(a)}{h} = \dfrac{[4-(a+h)^2]-[4-a^2]}{h}$
$= \dfrac{4-a^2-2ah-h^2-4+a^2}{h}$
$= \dfrac{-2ah-h^2}{h}$
$= \dfrac{h(-2a-h)}{h}$
$= -2a - h$ (4 – 3)

13. $(f+g)(x) = f(x) + g(x) = 3x + 5 + 4 - x^2 = 9 + 3x - x^2$ (4 – 5)

14. $(f-g)(x) = f(x) - g(x) = 3x + 5 - (4 - x^2) = 3x + 5 - 4 + x^2 = x^2 + 3x + 1$ (4 – 5)

15. $(fg)(x) = f(x)g(x) = (3x+5)(4-x^2) = 12x - 3x^3 + 20 - 5x^2 = 20 + 12x - 5x^2 - 3x^3$ (4 – 5)

16. $\left(\frac{f}{g}\right)(x) = \frac{f(x)}{g(x)} = \frac{3x+5}{4-x^2}$ Domain: $\{x \mid 4 - x^2 \neq 0\}$ or $\{x \mid x \neq \pm 2\}$ (4 – 5)

17. $(f \circ g)(x) = f[g(x)] = 3g(x) + 5 = 3(4 - x^2) + 5 = 12 - 3x^2 + 5 = 17 - 3x^2$ (4 – 5)

18. $(g \circ f)(x) + \cdot g[f(x)] = 4 - [f(x)]^2 = 4 - (3x+5)^2 = 4 - (9x^2 + 30x + 25) = 4 - 9x^2 - 30x - 25$
$= -21 - 30x - 9x^2$ (4 – 5)

19. $f(x) = x^2 - 6x + 11$. Note that $a = 1, b = -6$, and $c = 11$.
Since $a > 0$, the function has a minimum where $x = -\frac{b}{2a} = -\frac{-6}{2(1)} = 3$
Min $f(x)$: $f(-\frac{b}{2a}) = f(3) = 3^2 - 6(3) + 11 = 9 - 18 + 11 - 2.$
Vertex: $(-\frac{b}{2a}, f(-\frac{b}{2a})) = (3, f(3)) = (3, 2)$ (4 – 4)

20.   Domain : $= R$
Range : $= (-3, \infty)$ (4 – 4)

21. $[-2, -1], [1, \infty)$ (4 – 4)

22. $[-1, 1)$ (4 – 4)

23. $(-\infty, -2)$ (4 – 4)

24. $x = -2, x = 1$ (4 – 4)

25.   A. Since two points are given, we find the slope, then apply the point-slope form.

$$m = \frac{-3-3}{0-(-4)} = \frac{-6}{4} = -\frac{3}{2}$$

$$y - 3 = -\frac{3}{2}[x - (-4)]$$

$$2(y - 3) = -3(x + 4)$$

$$2y - 6 = -3x - 12$$

$$3x + 2y = -6$$

B. $d(P, Q) = \sqrt{(-3 - 3)^2 + [0 - (-4)]^2} = \sqrt{36 + 16} = \sqrt{52}$ $(4 - 1, 4 - 2)$

26.   The line $6x + 3y = 5$, or $3y = -6x + 5$, or $y = -2x + \frac{5}{3}$, has slope -2.

A. We require a line through $(-2, 1)$, with slope -2. Applying the point-slope form, we have

$$y - 1 = -2[x - (-2)]$$
$$y - 1 = -2x - 4$$
$$y = -2x - 3$$

B. We require a line with slope m satisfying $-2m = -1$, or $m = \frac{1}{2}$.
Again applying the point-slope form, we have

$$y - 1 = \frac{1}{2}[x - (-2)]$$
$$y - 1 = \frac{1}{2}x + 1$$
$$y = \frac{1}{2}x + 2 \quad (4 - 2)$$

27.   Since the equation is unchanged by any substitution of $-x$ for x or $-y$ for y, or both, the graph must be symmetric with respect to all three. $(4 - 1)$

28.   The domain is the set of all real numbers x such that $\frac{1}{\sqrt{3-x}}$ is a real number – that is, such that $3 - x > 0$, or $x < 3$.
Domain: $(-\infty, 3)$ $(4 - 3)$

29. $f(x) = x^2 - 6x + 5$. Note that $a = 1, b = -6$, and $c = 5$.
Axis: $x = -\frac{b}{2a} = -\frac{-6}{2(1)} = 3$
Vertex: $(-\frac{b}{2a}, f(-\frac{b}{2a})) = (3, f(3)) = (3, -4)$
Minimum value of $f(x)$ (since $a > 0$):
Min $f(x) = f(-\frac{b}{2a}) = f(3) = -4$
y intercept: $f(0) = 5$
x intercepts: Solutions of:

$$0 = x^2 - 6x + 5$$
$$0 = (x - 1)(x - 5)$$

$$x - 1 = 0 \quad \text{or} \quad x - 5 = 0$$
$$x = 1 \quad \text{or} \quad x = 5$$

Graph:

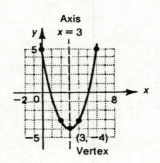

x	$f(x)$
0	5
1	0
2	-3
3	-4
4	-3
5	0

30. $f(x) = \sqrt{x} - 8 \quad g(x) = |x|$

   A. $(f \circ g)(x) = f[g(x)] = \sqrt{g(x)} - 8 = \sqrt{|x|} - 8$
   $(g \circ f)(x) = g[f(x)] = |f(x)| = |\sqrt{x} - 8|$

   B. The domain of f is $\{x \mid x \geq 0\}$. The domain of g is R.

   Hence the domain of $f \circ g$ is the set of all those numbers in R for which $g(x)$ is non-negative, that is, R. The domain of $g \circ f$ is the set of all those numbers in $\{x \mid x \geq 0\}$ for which $f(x)$ is in R, that is, $\{x \mid x \geq 0\}$ or $[0, \infty)$. $(4-5)$

31.   A. $f(x) = x^3$. $a \neq b$ implies $a^3 \neq b^3$. f is one $-$ to $-$ one.

   B. $g(x) = (x-2)^2$. Since $g(3) = g(1) = 1, g$ is not one $-$ to $-$ one.

   C. $h(x) = 2x - 3$. $a \neq b$ implies $2a \neq 2b$, which implies $2a - 3 \neq 2b - 3$. h is one $-$ to $-$ one.

   D. $F(x) = (x+3)^2, x \geq -3 \quad a \neq b, a \geq -3$ and $b \geq -3$,
      imply $a + 3 \neq b + 3$, $a + 3$ and $b + 3$ both non $-$ negative.

   Since different *non-negative* numbers will have different squares, $(a+3)^2 \neq (b+3)^2$, hence F is one-to-one. $(4-6)$

32.   A. $f(x) = 3x - 7$ f is one-to-one, since $a \neq b$ implies $3a \neq 3b$, which implies $3a - 7 \neq 3b - 7$.

   Solve $y = f(x)$
   for x :

   $$y = 3x - 7$$
   $$y + 7 = 3x$$
   $$x = \tfrac{1}{3}y + \tfrac{7}{3}$$

   Interchange x and y :
   $y = \tfrac{1}{3}x + \tfrac{7}{3}$
   State the inverse function and its domain:
   $f^{-1}(x) = \tfrac{1}{3}x + \tfrac{7}{3}$ Domain: R
   Check:
   $$(f^{-1} \circ f)(x) = \tfrac{1}{3}(3x - 7) + \tfrac{7}{3} = x - \tfrac{7}{3} + \tfrac{7}{3} = x$$
   $$(f \circ f^{-1})(x) = 3(\tfrac{1}{3}x + \tfrac{7}{3}) - 7 = x + 7 - 7 = x$$
   $$f^{-1}(x) = \tfrac{1}{3}x + \tfrac{7}{3} = \tfrac{x+7}{3}$$

   B. $f^{-1}(5) = \tfrac{5+7}{3} = 4$
   C. $(f^{-1} \circ f)(x) = x$ (See part A)
   D. Since $a < b$ implies $3a < 3b$, which implies $3a - 7 < 3b - 7$, or $f(a) < f(b)$, f is increasing. $(4-6)$

33.

x	$y = 2 - x$	x	$y = x^2$
-1	3	0	0
$-\frac{1}{2}$	$2\frac{1}{2}$	$\frac{1}{2}$	$\frac{1}{4}$
		1	1

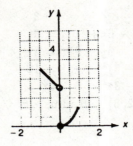

Domain: $[-1, 1]$

Range: $[0, 1] \cup (2, 3]$ (see graph)

Discontinuous at $x = 0$ $(4 - 4)$

34. $f(x) = \sqrt{x - 1}$ f is one-to-one, since $a \neq b$ implies

$a - 1 \neq b - 1$, hence $\sqrt{a - 1} \neq \sqrt{b - 1}$

A. Solve $y = f(x)$ for x :

$$y = \sqrt{x - 1}$$
$$y^2 = x - 1 \quad y \geq 0$$
$$x = 1 + y^2 \quad y \geq 0$$

Interchange x and y :

$$y = 1 + x^2 \quad x \geq 0$$

State the inverse function and its domain:

$$f^{-1}(x) = 1 + x^2 \quad x \geq 0$$

Check:

$$(f^{-1} \circ f)(x) = 1 + (\sqrt{x - 1})^2 = 1 + x - 1 = x$$
$$(f \circ f^{-1})(x) = \sqrt{1 + x^2 - 1} = \sqrt{x^2} = x \quad \text{Since } x \geq 0 \text{ in the domain of } f^{-1}$$

B. Domain of $f = [1, \infty) = $ Range of f^{-1}

Range of $f = [0, \infty) = $ Domain of f^{-1}

C.

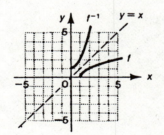

$(4 - 6)$

35.   We are given $C(h, k) = (3, 0)$. To find r, we use the distance formula.

$$\begin{aligned} r &= \text{distance from center to } (-1, 4) \\ &= \sqrt{[(-1) - 3]^2 + (0 - 4)^2} \\ &= \sqrt{16 + 16} \\ &= \sqrt{32} \end{aligned}$$

Then the equation of the circle is

$$\begin{aligned} (x - h)^2 + (y - k)^2 &= r^2 \\ (x - 3)^2 + (y - 0)^2 &= (\sqrt{32})^2 \\ (x - 3)^2 + y^2 &= 32 \quad (4 - 1) \end{aligned}$$

36.
$$x^2 + y^2 + 4x - 6y = 3$$
$$(x^2 + 4x + ?) + (y^2 - 6y + ?) = 3$$
$$(x^2 + 4x + 4) + (y^2 - 6y + 9) = 3 + 4 + 9$$
$$(x + 2)^2 + (y - 3)^2 = 16$$
$$[x - (-2)]^2 + (y - 3)^2 \quad 4^2$$

Center: $C(h, k) = C(-2, 3)$ Radius $r = \sqrt{16} = 4$ $(4 - 1)$

37. $xy = 4$ y axis symmetry? $(-x)y = 4$ $-xy = 4$ No, not equivalent

   x axis symmetry? $x(-y) = 4$ $-xy = 4$ No, not equivalent

   origin symmetry? $(-x)(-y) = 4$ $xy = 4$ Yes, equivalent

x	y
1	4
2	2
3	$\frac{4}{3}$
4	1

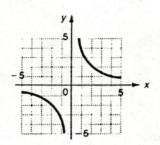

$(4 - 1)$

We reflect the portion of the graph in quadrant I through the origin, using origin symmetry.

38.   decreasing $(4 - 2, 4 - 3, 4 - 4)$

39.   A. Domain of $f = [0, \infty) =$ Range of f^{-1}

Since $x^2 \geq 0, x^2 - 1 \geq -1$, so

Range of $f = [-1, \infty) =$ Domain of f^{-1}

B. $f(x) = x^2 - 1$ $x \geq 0$ f is one-to-one on its domain (steps omitted)

Solve $y = f(x)$ for x:
$$y = x^2 - 1 \quad x \geq 0$$
$$x^2 = y + 1 \quad x \geq 0$$

positive square root since
$$x = \sqrt{y + 1} \quad x \geq 0$$

Interchange x and y:
$$y = \sqrt{x + 1} \quad y \geq 0$$

State the inverse function and its domain:
$$f^{-1}(x) = \sqrt{x + 1} \quad x \geq -1$$

Check:

$$(f^{-1} \circ f)(x) = \sqrt{x^2 - 1 + 1} \quad x \geq 0$$
$$= \sqrt{x^2} \quad x \geq 0$$
$$= x \text{ since } x \geq 0$$
$$(f \circ f^{-1})(x) = (\sqrt{x+1})^2 - 1$$
$$= x + 1 - 1$$
$$= x \quad (4-7)$$

C. $f^{-1}(3) = \sqrt{3+1} = 2$

D. $(f^{-1} \circ f)(4) = 4$

E. $(f^{-1} \circ f)(x) = x$

40.

x	$|x+1|$	$y = f(x)$
-5	4	-5
-4	3	-4
-3	2	-3
-2	1	-2
-1	0	-1
0	1	-2
1	2	-3
2	3	-4
3	4	-5

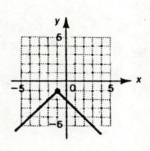

$$(4-4)$$

41.   The domain is the set of all real numbers x such that
$\sqrt{25 - x^2}$ is a real number – that is, such that $25 - x^2 \geq 0$.
Solving by the methods of Section 3-7, we obtain $-5 \leq x \leq 5$ or $[-5, 5]$ $(3-7, 4-3)$

42.   A. $(fg)(x) = f(x)g(x) = x^2 \sqrt{1-x}$
The domain of f is R. The domain of g is $(-\infty, 1]$.
Hence the domain of fg is the intersection of these sets, that is, $(-\infty, 1]$.

B. $\left(\frac{f}{g}\right)(x) = \frac{f(x)}{g(x)} = \frac{x^2}{\sqrt{1-x}}$
To find the domain of $\frac{f}{g}$, we exclude from $(-\infty, 1]$ the set
of values of x for which $g(x) = 0$

$$\sqrt{1-x} = 0$$
$$1 - x = 0$$
$$x = 1$$

Thus the domain of $\frac{f}{g}$ is $(-\infty, 1)$.

C. $(f \circ g)(x) = f[g(x)] = [g(x)]^2 = [\sqrt{1-x}]^2 = 1 - x$.
The domain of $f \circ g$ is the set of those numbers in $(-\infty, 1]$ for which
$g(x)$ is real, that is, $(-\infty, 1]$.

D. $(g \circ f)(x) = f[f(x)] = \sqrt{1 - f(x)} = \sqrt{1 - x^2}$
The domain of $g \circ f$ is the set of those real numbers for which $f(x)$ is
in $(-\infty, 1]$, that is, $x^2 \leq 1$, or $-1 \leq x \leq 1$. $[-1, 1]$ $(4-5)$

43.   A. $f(x) = \frac{x+2}{x-3}$

Solve $\qquad y = f(x)$ for x :

$$y = \frac{x+2}{x-3}$$

$$y(x-3) = x+2$$

$$xy - 3y = x+2$$

$$xy - x = 3y+2$$

$$x(y-1) = 3y+2$$

$$x = \frac{3y+2}{y-1}$$

Interchange x and y:

$y = \frac{3x+2}{x-1}$

State the inverse function and its domain

$f^{-1}(x) = \frac{3x+2}{x-1} \quad x \neq 1$

Check:

$$(f^{-1}of)(x) = \frac{3\frac{x+2}{x-3}+2}{\frac{x+2}{x-3}-1}$$

$$= \frac{3(x+2)+2(x-3)}{(x+2)-(x-3)}$$

$$= \frac{3x+6+2x-6}{x+2-x+3}$$

$$= \frac{5x}{5}$$

$$= x$$

$$(fof^{-1})(x) = \frac{\frac{3x+2}{x-1}+2}{\frac{3x+2}{x-1}-3}$$

$$= \frac{3x+2+2(x-1)}{3x+2-3(x-1)}$$

$$= \frac{3x+2+2x-2}{3x+2-3x+3}$$

$$= \frac{5x}{5}$$

$$= x$$

B. $f^{-1}(3) = \frac{3(3)+2}{3-1} = \frac{11}{2}$

C. $(f^{-1}of)(x) = x$ (See Part A) $(4-6)$

44. $f(x) = |x+1| - |x-1|$

If $x < -1$, then $|x+1| = -(x+1)$ and $|x-1| = -(x-1)$, hence

$$f(x) = -(x+1) - [-(x-1)]$$

$$= -x - 1 + x - 1$$

$$= -2$$

If $-1 \leq x < 1$, then $|x+1| = x+1$ but $|x-1| = -(x-1)$, hence

$$f(x) = x+1 - [-(x-1)]$$

$$= x+1+x-1$$

$$= 2x$$

If $x \geq 1$, then $|x+1| = x+1$ and $|x-1| = x-1$, hence

$$\begin{aligned} f(x) &= x+1-(x-1) \\ &= x+1-x+1 \\ &= 2 \end{aligned}$$

Domain: R

Piecewise definition for f : $f(x) = \begin{cases} -2 \text{ if } x < -1 \\ 2x \text{ if } -1 \le x < 1 \\ 2 \text{ if } x \ge 1 \end{cases}$

Range: Since if $-1 \le x < 1$, then $-2 \le 2x < 2$, $f(x)$ is always between -2 and 2. The range if $[-2,2]$ $(4-4)$

45. Let (x,y) be a point equidistant from $(3,3)$ and $(6,0)$. Then
$$\begin{aligned} \sqrt{(x-3)^2+(y-3)^2} &= \sqrt{(x-6)^2+(y-0)^2} \\ (x-3)^2+(y-3)^2 &= (x-6)^2+y^2 \\ x^2-6x+9+y^2-6y+9 &= x^2-12x+36+y^2 \\ -6x-6y+18 &= -12x+36 \\ 6x-6y &= 18 \\ x-y &= 3 \end{aligned}$$
This is the equation of a line. $(4-1)$

46. We will show separately:

A. If two non-vertical lines are parallel, then they have the same slope.

B. If two lines have the same slope they are non-vertical and parallel.

A. Let $y = m_1 x + b_1$ and $y = m_2 x + b_2$ be parallel. Then there is no point with coordinates that satisfy both equations. But for any y,
$$m_1 x + b_1 = m_2 x + b_2$$
Then (x,y) will satisfy both equations unless this equation has no solution. But this equation will have a solution
$$\begin{aligned} m_1 x - m_2 x &= b_2 - b_1 \\ (m_1 - m_2)x &= b_2 - b_1, \end{aligned}$$
that is, $x = \frac{b_2-b_1}{m_1-m_2}$
unless $m_1 - m_2 = 0$. So $m_1 = m_2$. So the lines have the same slope.

B. Assume the two lines have equations $y = mx + b$, and $y = mx + b_2$. Then both have slopes, hence are non-vertical. For any y a point that lies on both lines must have coordinates that satisfy $mx + b_1 = mx + b_2$, or $b_1 = b_2$. So unless the lines are the same line there is no point which lies on both of them. Hence the lines do not intersect, that is, they are parallel. $(4-2)$

47. Let f be an even function. Then the graph of f is symmetric with respect to the y axis, that is, when $(x, f(x))$ is on the graph, $(-x, f(x))$ is on the graph. Since $(-x, f(-x))$ is on the graph, and there is exactly one range element for each domain element, $f(-x) = f(x)$. $(4-1, 4-4)$

48. Let f be an odd function. Then the graph of f is symmetric with respect to the origin, that is, when $(x, f(x))$ is on the graph, $(-x, -f(x))$ is on the graph. Since $(-x, f(-x))$ is on the graph, and there is exactly one range element for each domain element, $f(-x) = -f(x)$. $(4-1, 4-4)$

49. A. $f(x) = x^2$ $f(-x) = (-x)^2 = x^2 = f(x)$. Even
     B. $g(x) = x^3$ $g(-x) = (-x)^3 = -x^3 = -g(x)$. Odd
     C. $h(x) = x + 1$ $h(-x) = -x + 1$.
     Since $h(-x) \neq h(x)$ and $h(-x) \neq -h(x)$, h is neither
     even nor odd. $(4-3)$

50. A. If V is linearly related to t, then we are looking for an equation
     whose graph passes through $(t_1, V_1) = (0, 12,000)$ and $(t_2, V_2) = (8, 2,000)$.
     We find the slope, and then we use the point-slope for m to find the equation.

$$m = \frac{V_2 - V_1}{t_2 - t_1} = \frac{2,000 - 12,000}{8 - 0} = \frac{-10,000}{8} = -1,250$$
$$V - V_1 = m(t - t_1)$$
$$V - 12,000 = -1,250(t - 0)$$
$$V - 12,000 = -1,250t$$
$$V = -1,250t + 12,000$$

     B. We are asked for V when $t = 5$
$$V = -1,250(5) + 12,000$$
$$V = -6,250 + 12,000$$
$$V = \$5,750 \quad (4-2)$$

51. A. If R is linearly related to C, then we are looking for an
     equation whose graph passes through $(C_1, R_1) = (30, 48)$ and $(C_2, R_2) = (20, 32)$.

     We find the slope, and then we use the point-slope form to find the equation.
$$m = \frac{R_2 - R_1}{C_2 - C_1} = \frac{32 - 48}{20 - 30} = \frac{-16}{-10} = 1.6$$

$$R - R_1 = m(C - C_1)$$
$$R - 48 = 1.6(C - 32)$$
$$R - 48 = 1.6C - 48$$
$$R = 1.6C$$

     B. We are asked for R when $C = 105$.
$$R = 1.6(105)$$
$$R = \$168 \quad (4-2)$$

52. If $0 \leq x \leq 3,000$, $E(x) = 200$
     If

	(Base Salary)	+	(Commission on Sales over \$3,000)
$x > 3,000, E(x) =$	200	+	$0.1(x - 3,000)$
$=$	200	+	$0.1x - 300$
$=$	$0.1x - 100$		

$$\text{Summarizing, } E(x) = \begin{cases} 200 \text{ if } 0 \leq x \leq 3,000 \\ 0.1x - 100 \text{ if } x > 3,000 \end{cases}$$

$$E(2,000) = 200$$

$$E(5,000) = 0.1(5,000) - 100 = 500 - 100 = 400 \quad (4-4)$$

53. The profit function is the difference of the revenue and the cost functions.

$P = R - C$

Hence,
$$
\begin{aligned}
P(x) &= R(x) - C(x) \\
&= (50x - \tfrac{1}{10}x^2) - (20x + 4,000) \\
&= 50x - \tfrac{1}{10}x^2 - 20x - 4,000 \\
&= 30x - \tfrac{1}{10}x^2 - 4,000
\end{aligned}
$$

Next we use composition to express P as a function of the price p.

$$
\begin{aligned}
(P \circ f)(p) &= P[f(p)] \\
&= 30f(p) - \tfrac{1}{10}[f(p)]^2 - 4,000 \\
&= 30(500 - 10p) - \tfrac{1}{10}(500 - 10p)^2 - 4,000 \\
&= 15,000 - 300p - \tfrac{1}{10}(250,000 - 10,000p + 100p^2) - 4,000 \\
&= 15,000 - 300p - 25,000 + 1,000p - 10p^2 - 4,000 \\
&= -14,000 + 700p - 10p^2 \quad (4-5)
\end{aligned}
$$

54.

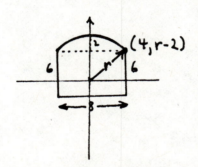

In the above sketch, we note that the point $(4, r-2)$ is on the circle with equation $x^2 + y^2 = r^2$, hence $(4, r-2)$ must satisfy this equation

$$
\begin{aligned}
4^2 + (r-2)^2 &= r^2 \\
16 + r^2 - 4r + 4 &= r^2 \\
-4r + 20 &= 0 \\
r &= 5 \text{ feet} \quad (4-1)
\end{aligned}
$$

55. A. From the figure, we see that $A = x(y + y) = 2xy$. Since the fence consists of four pieces of length y, and three pieces of length x, we have

$3x + 4y = 120$.

Hence:
$$
\begin{aligned}
4y &= 120 - 3x \\
y &= 30 - \tfrac{3}{4}x \\
A &= 2x(30 - \tfrac{3}{4}x) \\
A(x) &= 60x - \tfrac{3}{2}x^2
\end{aligned}
$$

B. Since both x and y must be positive, we have

$$
\begin{aligned}
x &> 0 \\
30 - \tfrac{3}{4}x &> 0 \text{ or } -\tfrac{3}{4}x > -30 \text{ or } x < 40
\end{aligned}
$$

Hence: $0 < x < 40$ is the domain of A.

C. The function A is a quadratic function with $a = -\frac{3}{2}, b = 60, c = 0$.
Hence its maximum value occurs when

$$x = -\frac{b}{2a}$$
$$x = -\frac{60}{2(-\frac{3}{2})}$$
$$= \frac{-60}{-3}$$
$$= 20$$

Thus the total area will be maximum when

$x = 20$ and $y = 30 - \frac{3}{4}x = 30 - \frac{3}{4}(20) = 15$ (4 − 4)

CHAPTER 5

Exercise 5-1

Key Ideas and Formulas

The equation $f(x) = b^x$ $b > 0, b \neq 1$ defines an exponential function for each different constant b, called the base. Domain: R Range: $(0, \infty)$

Basic Properties of the graph of $f(x) = b^x$ $b > 0, \quad b \neq 1$

1. All graphs pass through $(0,1)$
2. All graphs are continuous with no holes or jumps.
3. The x-axis is a horizontal asymptote.
4. If $b > 1$, then b^x increases as x increases.
5. If $0 < b < 1$, then b^x decreases as x increases.
6. The function f is one-to-one.

Additional Exponential Function Properties.

1. Exponent laws
   $a^x a^y = a^{x+y}$ $(a^x)^y = a^{xy}$ $(ab)^x = a^x b^x$ $(a/b)^x = a^x/b^x$ $a^x/a^y = a^{x-y}$
2. $a^x = a^y$ if and only if $x = y$
3. For $x \neq 0$, then $a^x = b^x$ if and only if $a = b$.

Doubling Time Growth Model

$P = P_0 2^{t/d}$ where

P = Population at time t
P_0 = Population at time $t = 0$
d = doubling time

Half-Life Decay Model

$$A = A_0\left(\tfrac{1}{2}\right)^{t/h}$$
$$= A_0 2^{-t/h}$$

where

A = amount at time t
A_0 = amount at time $t = 0$
h = half $-$ life

Compound Interest:

If a principal P is invested at an annual rate r compounded n times a year, then the amount A in the account at the end of t years is given by

$A = P(1 + \tfrac{r}{n})^{nt}$

1.

x	y
-3	0.04
-2	0.11
-1	0.33
0	1.00
1	3.00
2	9.00
3	27.00

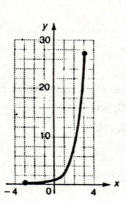

5.

x	$y = g(x)$
-3	-27.00
-2	-9.00
-1	-3.00
0	-1.00
1	-0.33
2	-0.11
3	-0.04

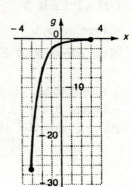

3.

x	y
-3	27.00
-2	9.00
-1	3.00
0	1.00
1	0.33
2	0.11
3	0.04

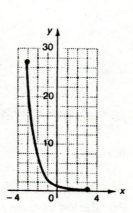

7.

x	$y = h(x)$
-3	0.19
-2	0.55
-1	1.67
0	5.00
1	15.00
2	45.00
3	135.00

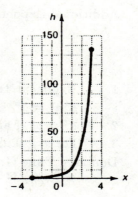

9.

x	y
-6	-4.96
-5	-4.89
-4	-4.67
-3	-4.00
-2	-2.00
-1	4.00
0	22.00

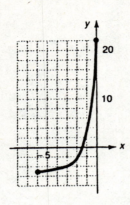

11. $10^{3x-1}10^{4-x} = 10^{3x-1+4-x} = 10^{2x+3}$

13. $\dfrac{3^x}{3^{1-x}} = 3^{x-(1-x)} = 3^{x-1+x} = 3^{2x+1}$

15. $\left(\dfrac{4^x}{5^y}\right)^{3z} = \dfrac{4^{3xz}}{5^{3yz}}$

17.
$$
\begin{aligned}
5^{3x} &= 5^{4x-2} \text{ if and only if} \\
3x &= 4x - 2 \\
-x &= -2 \\
x &= 2
\end{aligned}
$$

19.
$$
\begin{aligned}
7^{x^2} &= 7^{2x+3} \text{ if and only if} \\
x^2 &= 2x + 3 \\
x^2 - 2x - 3 &= 0 \\
(x-3)(x+1) &= 0 \\
x &= -1, 3
\end{aligned}
$$

21.
$$
\begin{aligned}
(1-x)^5 &= (2x-1)^5 \text{ if and only if} \\
1 - x &= 2x - 1 \\
-3x &= -2 \\
x &= \tfrac{2}{3}
\end{aligned}
$$

23. $2^x = 4^{x+1}$

$2^x = (2^2)^{x+1}$

$2^x = 2^{2(x+1)}$ if and only if

$x = 2(x+1)$

$x = 2x+2$

$-x = 2$

$x = -2$

25.

t	G(t)
−200	0.11
−150	0.19
−100	0.33
−50	0.58
0	1.00
50	1.73
100	3.00
150	5.20
200	9.00

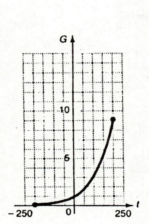

27.

x	y
−9	1543.3
−8	891.0
−7	514.4
−6	297.0
−5	171.5
−4	99
−3	57.2
−2	33
0	11
3	2.1
6	0.4
9	0.1

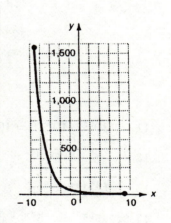

29.

x	$y = g(x)$
-3	0.13
-2	0.25
-1	0.5
0	1.0
1	0.5
2	0.25
3	0.13

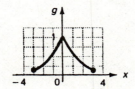

31.

x	y
0	1,000
1	1,080
2	1,166
3	1,260
4	1,360
5	1,469
6	1,587
7	1,714
8	1,851
9	1,999
10	2,159

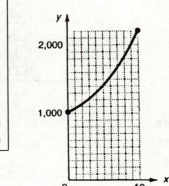

33.

x	y
-2	0.06
-1.5	0.21
-1	0.50
-0.5	0.84
0	1.00
0.5	0.84
1	0.50
1.5	0.21
2	0.06

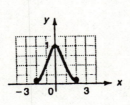

35. $\quad (6^x + 6^{-x})(6^x - 6^{-x}) \; = \; (6^x)^2 - (6^{-x})^2$

$\qquad\qquad\qquad\qquad\qquad\quad = \; 6^{2x} - 6^{-2x}$

(think: $(a+b)(a-b) = a^2 - b^2$)

Common Errors:

$(6^x)^2 \neq 6^{x^2}$

$6^{2x} - 6^{-2x} \neq 6^{4x}$

37. $\quad (6^x + 6^{-x})^2 - (6^x - 6^{-x})^2 \; = \; (6^x)^2 + 2(6^x)(6^{-x}) + (6^{-x})^2$

$\qquad\qquad\qquad\qquad\qquad\qquad - \; [(6^x)^2 - 2(6^x)(6^{-x}) + (6^{-x})^2]$

$\qquad\qquad\qquad\qquad\qquad\; = \; 6^{2x} + 2 + 6^{-2x} - [6^{2x} - 2 + 6^{-2x}]$

$\qquad\qquad\qquad\qquad\qquad\; = \; 6^{2x} + 2 + 6^{-2x} - 6^{2x} + 2 - 6^{-2x}$

$\qquad\qquad\qquad\qquad\qquad\; = \; 4$

39.

x	$y = m(x)$
0	0
0.5	0.29
1	0.33
1.5	0.29
2	0.22
2.5	0.16
3	0.11

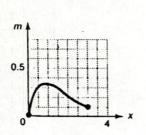

41.

x	$y = f(x)$
-3	4.06
-2	2.13
-1	1.25
0	1.00
1	1.25
2	2.13
3	4.06

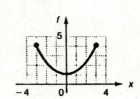

43.   16.24

45.   5.047

47.   4.469

49.

n	L
1	2
2	4
3	8
4	16
5	32
6	64
7	128
8	256
9	512
10	1,024

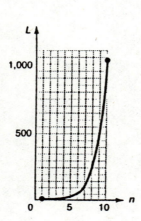

51.   We use the Doubling Time Growth Model:

$P = P_0 2^{t/d}$

Substituting $P_0 = 10$ and $d = 2.4$, we have

$P = 10(2^{t/2.4})$

A. $t = 7$, hence $P = 10(2^{7/2.4})$

$\qquad\qquad\qquad = 75.5$

76 flies

B. $t = 14$, hence $P = 10(2^{14/2.4})$

$\qquad\qquad\qquad\quad = 570.2$

570 flies

53.   We use the Half-Life Decay Model

$A = A_0 \left(\frac{1}{2}\right)^{t/h} = A_0 2^{-t/h}$

Substituting $A_0 = 25$ and $h = 12$, we have

$A = 25(2^{-t/12})$

A. $t = 5$, hence $A = 25(2^{-5/12}) = 19$ pounds

B. $t = 20$, hence $A = 25(2^{-20/12}) = 7.9$ pounds

55.   We use the compound interest formula

$$A = P\left(1+\frac{r}{n}\right)^{nt}$$

$$n = \frac{365}{7} P = 4,000 \; r = 0.11$$

$$A = 4,000\left(1+\frac{0.11}{365/7}\right)^{365t/7}$$

$$= 4,000\left(1+\frac{0.77}{365}\right)^{365t/7}$$

A. $t = 0.5$, hence $A = 4,000\left(1+\frac{0.77}{365}\right)^{365(0.5)/7}$

$$= \$4,225.92$$

B. $t = 10$, hence $A = 4,000\left(1+\frac{0.77}{365}\right)^{365(10)/7}$

$$= \$12,002.75$$

57.   We use the compound interest formula

$A = P\left(1+\frac{r}{n}\right)^{nt}$ to find $P : P = \frac{A}{\left(1+\frac{r}{n}\right)^{nt}}$

$$n = 365 \; r = 0.0825 \; A = 40,000 \; n = 17$$

$$P\frac{40,000}{\left(1+\frac{0.0825}{365}\right)^{365\cdot17}} = \$9,841$$

Exercise 5-2

Key Ideas and Formulas

As m increases without bound, the value of $(1 + \frac{1}{m})^m$ approaches an irrational number called e. To twelve decimal places, $e = 2.718\,281\,828\,459$.

For x a real number, the equation $f(x) = e^x$ defines the exponential function with base e.

Exponential Growth and Decay

Description	Equation	Graph	Uses
Unlimited Growth	$y = ce^{kt}$ $c, k > 0$		Short term population growth (people, bacteria etc.) Growth of money at continuous compound interest.$(A = Pe^{rt})$
Exponential Decay	$y = ce^{-kt}$ $c, k > 0$		Radioactive decay. Light absorption in water, glass, etc. Atmosphere pressure. Electric circuits.
Limited Growth	$y = c(1 - e^{-kt})$ $c, k > 0$		Learning skills. Sales fads. Company growth. Electric Circuits.
Logistic Growth	$y = \frac{M}{1 + ce^{-kt}}$ $c, k, M > 0$		Long-term population growth. Epidemics. Sales of new products. Company growth.

1.

x	y
-3	-0.05
-2	-0.13
-1	-0.37
0	-1.00
1	-2.72
2	-7.39
3	-20.09

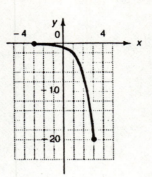

7.

x	y
-1	2.05
0	2.14
1	2.37
2	3.00
3	4.72
4	9.39
5	22.09

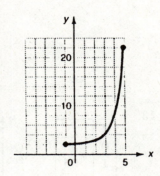

3.

x	y
-10	1.35
-8	2.02
-6	3.01
-4	4.49
-2	6.70
0	10.00
2	14.92
4	22.26
6	33.20
8	49.53
10	73.89

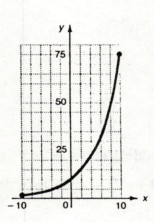

9.

x	y
-3	0.05
-2	0.14
-1	0.37
0	1.00
1	0.37
2	0.14
3	0.05

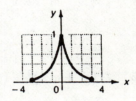

5.

t	f(t)
-5	165
-4	149
-3	135
-2	122
-1	111
0	100
1	90
2	82
3	74
4	67
5	61

11.

x	$y = M(x)$
-5	12.3
-4	7.5
-3	4.7
-2	3.1
-1	2.3
0	2.0
1	2.3
2	3.1
3	4.7
4	7.5
5	12.3

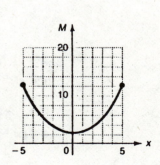

13.

t	N
0	50
1	95
2	142
3	174
4	190
5	196

15.
$$(e^x + e^{-x})^2 + (e^x - e^{-x})^2 = (e^x)^2 + 2(e^x)(e^{-x}) + (e^{-x})^2 + (e^x)^2 - 2(e^x)(e^{-x}) + (e^{-x})^2$$
$$= e^{2x} + 2 + e^{-2x} + e^{2x} - 2 + e^{-2x}$$
$$= 2e^{2x} + 2e^{-2x}$$

Common Errors:
$$(e^x)^2 \neq e^{x^2}$$
$$e^{2x} + e^{2x} \neq e^{4x}$$

17.
$$\frac{e^{-x}(e^x - e^{-x}) + e^{-x}(e^x + e^{-x})}{e^{-2x}} = \frac{e^{-x}e^x - e^{-x}e^{-x} + e^{-x}e^x + e^{-x}e^{-x}}{e^{-2x}}$$
$$= \frac{1 - e^{-2x} + 1 + e^{-2x}}{e^{-2x}}$$
$$= \frac{2}{e^{-2x}}$$
$$= 2e^{2x}$$

19. $2xe^{-x} = 0$ if $2x = 0$ or $e^{-x} = 0$.

Since e^{-x} is never 0, the only solution is $x = 0$.

21.

$$x^2 e^x - 5xe^x = 0$$
$$xe^x(x - 5) = 0$$
$$x = 0 \text{ or } \quad e^x = 0 \text{ or } \quad x - 5 = 0$$
$$\text{never} \quad\quad x = 5$$
$$x = 0, 5$$

23.

x	$f(x)$
-2	0.02
-1.5	0.11
-1	0.37
-0.5	0.78
0	1.00
0.5	0.78
1	0.37
1.5	0.11
2	0.02

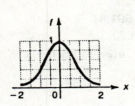

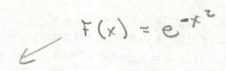

$F(x) = e^{-x^2}$

25. A.

s	$f(s)$
-0.5	4.0000
-0.2	3.0518
-0.1	2.8680
-0.01	2.7320
-0.001	2.7196
-0.0001	2.7184

s	$f(s)$
0.5	2.2500
0.2	2.4883
0.1	2.5937
0.01	2.7048
0.001	2.7169
0.0001	2.7181

   B. Both tables are "closing in" on 2.7182 ... or e.

27.   We use the continuous compounding formula

$A = Pe^{rt}$

$P = 5$ billion $r = 0.017$ $t = 0$ in 1988, hence $t = 12$ in 2000

$A = (5 \text{ billion}) e^{(0.017)(12)}$

$A = 6.1$ billion

29.

t	P
0	75
10	72
20	70
30	68
40	65
50	63
60	61
70	59
80	57
90	55
100	53

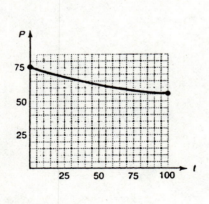

31. $I = I_0 e^{-0.00942d}$

A. $d = 50$ $I = I_0 e^{-0.00942(50)} = 0.62 I_0$ 62%

B. $d = 100$ $I = I_0 e^{-0.00942(100)} = 0.39 I_0$ 39%

33.   We use the Continuous Compound Interest Formula

$$A \ = \ Pe^{rt}$$
$$P \ = \ 5,250 \quad r = 0.1138 \quad A = 5,250e^{0.1138t}$$
A. $t \ = \ 6.25 \quad A = 5,250e^{0.1138(6.25)} = \$10,691.81$
B. $t \ = \ 17 \quad A = 5,250e^{0.1138(17)} = \$36,336.69$

35. Gill Savings: Use the Continuous Compound Interest Formula
$$A \ = \ Pe^{rt} \quad P = 1,000 \quad r = 0.083 \quad t = 2.5$$
$$A \ = \ 1,000e^{(0.083)(2.5)}$$
$$A \ = \ \$1,230.60$$

Richardson $S\&L$: Use the Compound Interest Formula
$$A \ = \ P\left(1 + \tfrac{r}{n}\right)^{nt} \quad P = 1,000 \quad r = 0.084 \quad n = 4 \quad t = 2.5$$
$$A \ = \ 1,000\left(1 + \tfrac{0.084}{4}\right)^{(4)(2.5)}$$
$$A \ = \ \$1,231.00$$

U.S.A. Savings: Use the Compound Interest Formula
$$A \ = \ P\left(1 + \tfrac{r}{n}\right)^{nt} \quad P = 1,000 \quad r = 0.0825 \quad n = 365 \quad t = 2.5$$
$$A \ = \ 1,000\left(1 + \tfrac{0.0825}{365}\right)^{(365)(2.5)}$$
$$A \ = \ \$1,229.03$$

37. We use the Continuous Compound Interest Formula
$$A \ = \ Pe^{rt}$$
$$P \ = \ \tfrac{A}{e^{rt}} \text{ or } P = Ae^{-rt}$$
$$A \ = \ 30,000 \quad r = 0.09 \quad t = 10$$
$$P \ = \ 30,000e^{(-0.09)(10)}$$
$$P \ = \ \$12,197.09$$

39. We use the Continuous Compounding Formula
$$A \ = \ Pe^{rt}$$
$$P \ = \ 45,000 \quad r = 0.38$$
$$A \ = \ 45,000e^{0.38t}$$
A. In 1987, $t = 0$, hence in 1992, $t = 5$
$$A \ = \ 45,000e^{(0.38)(5)}$$
$$A \ = \ 30,000 \text{ cases}$$
B. In 2000, $t = 13$
$$A \ = \ 45,000e^{(0.38)(13)}$$
$$A \ = \ 6,300,000 \text{ cases}$$

41.

t	N
0	0
5	18
10	28
15	33
20	36
25	38
30	39

As t increases without bound, $e^{-0.12t}$ approaches 0, hence
$N = 40(1 - e^{-0.12t})$ approaches
40. Hence 40 boards is the maximum number of boards an average

person could be expected to produce in one day.

43. $\quad T \;=\; T_m + (T_0 - T_m)e^{-kt}$
$\quad T_m \;=\; 40° \;\; T_0 = 72° \;\; k = 0.4 \;\; t = 3$
$\quad T \;=\; 40 + (72 - 40)e^{-0.4(3)}$
$\quad T \;=\; 50°$

45.

t	q
0	0
1	0.00016
2	0.00020
3	0.00041
4	0.00050
5	0.00057
6	0.00063
7	0.00068
8	0.00072
9	0.00075
10	0.00078

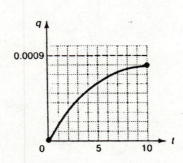

As t increases without bound, $e^{-0.2t}$ approaches 0, hence
$q = 0.0009(1 - e^{-0.2t})$
approaches 0.0009. Hence 0.0009 is the
maximum charge on the capacitor.

47.

t	N
0	20
5	33
10	50
15	67
20	80
25	89
30	94

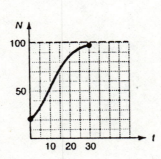

As t increases without bound, $e^{-0.14t}$ approaches 0, hence
$N = \frac{100}{1+4e^{-0.14t}}$ approaches 100. Hence 100 is the number of deer the island can support.

49. $\quad y \;=\; \dfrac{e^{0.25x}+e^{-0.25x}}{2(0.25)}$
$\quad\quad =\; \dfrac{e^{0.25x}+e^{-0.25x}}{0.5}$

x	y
−5	7.6
−4	6.2
−3	5.2
−2	4.5
−1	4.1
0	4.0
1	4.1
2	4.5
3	5.2
4	6.2
5	7.6

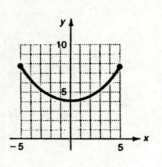

Exercise 5-3

Key Ideas and Formulas

Definition of Logarithmic Function:

For $b > 0$ and $b \neq 1$

      logarithmic form   exponential form

      $y = \log_b x$ is equivalent to $x = b^y$

The log to the base b of x is the exponent to which b must be raised to obtain x. A logarithm is, therefore, an exponent.

Typical Logarithmic Curves

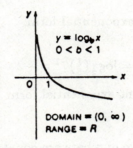
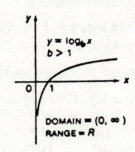

Properties of Logarithmic Functions:

If $b, M,$ and N are positive real numbers, $b \neq 1$, and p and x are real numbers, then:

1. $\log_b 1 = 0$
2. $\log_b b = 1$
3. $\log_b b^x = x$
4. $b^{\log_b x} = x$
5. $\log_b MN = \log_b M + \log_b N$
6. $\log_b \frac{M}{N} = \log_b M - \log_b N$
7. $\log_b M^p = p\log_b M$
8. $\log_b M = \log_b N$ if and only if $M = N$

---

1. $81 = 3^4$
3. $0.001 = 10^{-3}$
5. $3 = 81^{1/4}$
7. $16 = (\frac{1}{2})^{-4}$
9. $\log_{10} 0.0001 = -4$
11. $\log_4 8 = \frac{3}{2}$
13. $\log_{32} \frac{1}{2} = -\frac{1}{5}$
15. $7 = \sqrt{49}$ is rewritten $7 = 49^{\frac{1}{2}}$.

In equivalent logarithmic form this becomes $\log_{49} 7 = \frac{1}{2}$.

17. 0

19. 1
21. 4
23. $\log_{10} 0.01 = \log_{10} 10^{-2} = -2$
25. $\log_5 \sqrt[3]{5} = \log_5 5^{\frac{1}{3}} = \frac{1}{3}$
27. $\sqrt{x}$
29. $e^{2\log_e x} = e^{\log_e x^2} = x^2$

    Common Error: $e^{2\log_e x} \neq 2x$

31. Write $\log_2 x = 2$ in equivalent exponential form.

    $x = 2^2 = 4$

33. $\log_4 16 = \log_4 4^2 = 2$

                 $y = 2$

35. Write $\log_b 16 = 2$ in equivalent exponential form.

$$16 = b^2$$
$$b^2 = 16$$
$$b = 4 \text{ since bases are required to be positive}$$

37. Write $\log_b 1 = 0$ in equivalent exponential form.

$$1 = b^0$$

This statement is true if b is any real number except 0. However, bases are required to be positive and 1 is not allowed, so the original statement is true if b is any positive real number except 1.

39. Write $\log_4 x = \frac{1}{2}$ in equivalent exponential form.

$$x = 4^{\frac{1}{2}} = 2$$

41. $\log_{1/3} 9 = \log_{1/3} 3^2 = \log_{\frac{1}{3}} \frac{1}{(\frac{1}{3})^2} = \log_{\frac{1}{3}} (\frac{1}{3})^{-2} = -2$

43. Write $\log_b 1000 = \frac{3}{2}$ in equivalent exponential form

$$1000 = b^{3/2}$$
$$10^3 = b^{3/2}$$
$$(10^3)^{2/3} = (b^{3/2})^{2/3} \quad \text{(If two numbers are equal the results are equal if they are raised to the same exponent.)}$$
$$10^{3(\frac{2}{3})} = b^{\frac{3}{2}(\frac{2}{3})}$$
$$10^2 = b$$
$$b = 100$$

45. $\log_b u^2 v^7 = \log_b u^2 + \log_b v^7 = 2\log_b u + 7\log_b v$

47. $\log_b \frac{m^{2/3}}{n^{1/2}} = \log_b m^{2/3} - \log_b n^{1/2} = \frac{2}{3}\log_b m - \frac{1}{2}\log_b n$

49.
$$\log_b \frac{u}{vw} = \log_b u - \log_b vw$$
$$= \log_b u - (\log_b v + \log_b w)$$
$$= \log_b u - \log_b v - \log_b w$$
$$= \log_b u - \log_b v - \log_b w$$

Common Error:

Forgetting the parentheses. $-\log_b v + \log_b w$ is incorrect

51. $\log_b \frac{1}{a^2} = \log_b a^{-2} = -2\log_b a$

53. $\log_b \sqrt[3]{x^2 - y^2} = \log_b (x^2 - y^2)^{1/3} = \frac{1}{3}\log_b(x^2 - y^2)$

Common Error:

$\log_b(x^2 - y^2)$ is not $\log_b x^2 - \log_b y^2$.

55.
$$\log_b \frac{\sqrt[3]{N}}{p^2 q^3} = \log_b \sqrt[3]{N} - \log_b p^2 q^3$$
$$= \log_b N^{1/3} - (\log_b p^2 + \log_b q^3)$$
$$= \frac{1}{3}\log_b N - 2\log_b p - 3\log_b q$$

57.
$$\log_b \sqrt[4]{\frac{x^2 y^3}{\sqrt{z}}} = \log_b \left(\frac{x^2 y^3}{z^{1/2}}\right)^{1/4} = \frac{1}{4}\log_b \frac{x^2 y^3}{z^{1/2}} = \frac{1}{4}(\log_b x^2 + \log_b y^3 - \log_b z^{1/2})$$
$$= \frac{1}{4}(2\log_b x + 3\log_b y - \frac{1}{2}\log_b z)$$

59. $2\log_b x - \log_b y = \log_b x^2 - \log_b y = \log_b \frac{x^2}{y}$

61. $\log_b w - \log_b x - \log_b y = \log_b \frac{w}{x} - \log_b y$

$$= \log_b(\frac{w}{x} \div y)$$

$$= \log_b(\frac{w}{x} \cdot \frac{1}{y})$$

$$= \log_b \frac{w}{xy}$$

63. $3\log_b x + 2\log_b y - \frac{1}{4}\log_b z = \log_b x^3 + \log_b y^2 - \log_b z^{1/4} = \log_b x^3 y^2 - \log_b z^{1/4}$

$$= \log_b \frac{x^3 y^2}{z^{1/4}}$$

65. $5(\frac{1}{2}\log_b u - 2\log_b v) = 5(\log_b u^{1/2} - \log_b v^2) = 5 \log_b \frac{u^{1/2}}{v^2} = \log_b(\frac{u^{1/2}}{v^2})^5$

67. $\frac{1}{5}(2\log_b x + 3\log_b y) = \frac{1}{5}(\log_b x^2 \log_b y^3) = \frac{1}{5}\log_b x^2 y^3 = \log_b(x^2 y^3)^{1/5}$

$$= \log_b \sqrt[5]{x^2 y^3}$$

69.

			Check:
$\log_2(x+5)$	$=$	$2\log_2 3$	
$\log_2(x+5)$	$=$	$\log_2 3^2$	$\log_2(x+5) = 2\log_2 3$
$x+5$	$=$	3^2	$\log_2(4+5) \overset{?}{=} 2\log_2 3$
$x+5$	$=$	9	$\log_2 9 \overset{\checkmark}{=} \log_2 9$
x	$=$	4	

71.

$2\log_5 x = \log_5(x^2 - 6x + 2)$ Check: $2\log_5 x = \log_5(x^2 - 6x + 2)$

$\log_5 x^2 = \log_5(x^2 - 6x + 2)$ $2\log_5 \frac{1}{3} \overset{?}{=} \log_5[(\frac{1}{3})^2 - 6(\frac{1}{3}) + 2)]$

$x^2 = x^2 - 6x + 2$ $\log_5 \frac{1}{9} \overset{\checkmark}{=} \log_5(\frac{1}{9})$

$6x = 2$

$x = \frac{1}{3}$

73.

$\log_e(x+8) - \log_e x = 3\log_e 2$ Check: $\log_e(x+8) - \log_e x = 3\log_e 2$

$\log_e \frac{x+8}{x} = \log_e 2^3$ $\log_e(\frac{8}{7}+8) - \log_e \frac{8}{7} \overset{?}{=} 3\log_e 2$

$\frac{x+8}{x} = 2^3$ $\log_e(\frac{64}{7}) - \log_e \frac{8}{7} \overset{?}{=} \log_e 8$

$x+8 = 8x$ $\log_e(\frac{64}{7} \div \frac{8}{7}) \overset{?}{=} \log_e 8$

$8 = 7x$ $\log_e 8 \overset{\checkmark}{=} \log_e 8$

$x = \frac{8}{7}$

75.

$2\log_3 x = \log_3 2 + \log_3(4 - x)$ Check: $2\log_3 x = \log_3 2 + \log_3(4 - x)$

$\log_3 x^2 = \log_3 2(4 - x)$ $x = -4$

$x^2 = 2(4 - x)$ $2\log_3(-4) \overset{?}{=} \log_3 2 + \log_3[4 - (-4)]$

$x^2 = 8 - 2x$ False. -4 is not in the domain

$x^2 + 2x - 8 = 0$ of $\log_3 x$. -4 is not a solution.

$(x+4)(x-2) = 0$

$x = -4$ or $x = 2$ $x = 2$

$2\log_3 2 \overset{?}{=} \log_3 2 + \log_3(4 - 2)$

$2\log_3 2 \overset{\checkmark}{=} 2\log_3 2$

Solution: 2

77. $3 \log_b 2 + \frac{1}{2} \log_b 25 - \log_b 20 \;=\; \log_b x$ Check: $3 \log_b 2 + \frac{1}{2} \log_b 25 - \log_b 20 \;=\; \log_b x$

$\log_b 2^3 + \log_b 25^{1/2} - \log_b 20 \;=\; \log_b x$ $3 \log_b 2 + \frac{1}{2} \log_b 25 - \log_b 20 \overset{?}{=} \log_b 2$

$\log_b \frac{2^3 \cdot 25^{1/2}}{20} \;=\; \log_b x$ $\log_b 8 + \log_b 5 - \log_b 20 \overset{?}{=} \log_b 2$

$\frac{2^3 \cdot 25^{1/2}}{20} \;=\; x$ $\log_b \frac{40}{20} \overset{?}{=} \log_b 2$

$x \;=\; \frac{8 \cdot 5}{20}$ $\log_b 2 \overset{\vee}{=} \log_b 2$

$x \;=\; 2$

79. $\log_b(30) = \log_b(2 \cdot 3 \cdot 5) = \log_b 2 + \log_b 3 + \log_b 5 = 0.69 + 1.10 + 1.61 = 3.40$

81. $\log_b \frac{2}{5} = \log_b 2 - \log_b 5 = 0.69 - 1.61 = -0.92$

83. $\log_b 27 = \log_b 3^3 = 3 \log_b 3 = 3(1.10) = 3.30$

85. $\log_b \sqrt[3]{2} = \log_b 2^{\frac{1}{3}} = \frac{1}{3} \log_b 2 = \frac{1}{3}(0.69) = 0.23$

87. $\log_b \sqrt{0.9} \;=\; \log_b(\frac{9}{10})^{1/2} = \frac{1}{2} \log_b \frac{3^2}{2 \cdot 5} = \frac{1}{2}[\log_b 3^2 - \log_b 2 - \log_b 5]$

$= \; \frac{1}{2}[2 \log_b 3 - \log_b 2 - \log_b 5] = \frac{1}{2}[2.20 - 0.69 - 1.61] = \frac{1}{2}(-0.10)$

$= \; -0.05$

89.

x	y
10	3
6	2
4	1
3	0
$2\frac{1}{2}$	-1
$2\frac{1}{4}$	-2

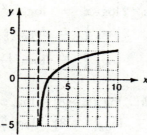

91.

x	y
8	1
4	0
2	-1
1	-2
$\frac{1}{2}$	-3
$\frac{1}{4}$	-4

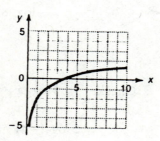

93.   A.

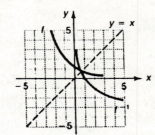

B. Domain $f = R =$ Range f^{-1}
Range $f = (0, \infty) =$ Domain f^{-1}
C. $f^{-1}(x) = \log_{1/2} x = -\log_2 x$

95. $f(x) = 5^{3x-1} + 4$
Solve $y = f(x)$ for x

$$
\begin{aligned}
y &= 5^{3x-1} + 4 \\
y - 4 &= 5^{3x-1} \\
\log_5(y - 4) &= 3x - 1 \\
3x - 1 &= \log_5(y - 4) \\
3x &= 1 + \log_5(y - 4) \\
x &= \tfrac{1}{3}[1 + \log_5(y - 4)]
\end{aligned}
$$

Interchange x and y
$y = \tfrac{1}{3}[1 + \log_5(x - 4)]$
State the inverse function
and its domain:
$f^{-1}(x) = \tfrac{1}{3}[1 + \log_5(x - 4)] \quad x > 4$
(Check omitted for lack of space)

97. $g(x) = 3\log_e(5x - 2)$
Solve $y = g(x)$ for x

$$
\begin{aligned}
y &= 3\log_e(5x - 2) \\
\tfrac{y}{3} &= \log_e(5x - 2) \\
5x - 2 &= e^{y/3} \\
5x &= e^{y/3} + 2 \\
x &= \tfrac{1}{5}(e^{y/3} + 2)
\end{aligned}
$$

Interchange x and y
$y = \tfrac{1}{5}(e^{x/3} + 2)$
State the inverse function
and its domain:
$g^{-1}(x) = \tfrac{1}{5}(e^{x/3} + 2)$ domain: R
(Check omitted for lack of space)

99. $\log_e x - \log_e 100 = -0.08t$
$\log_e \tfrac{x}{100} = -0.08t$
Write this in equivalent
exponential form
$\tfrac{x}{100} = e^{-0.08t}$
$x = 100e^{-0.08t}$

101.   Let $u = \log_b M$ and $v = \log_b N$;
then $M = b^u$ and $N = b^v$.
Thus, $\log_b M/N = \log_b b^u/b^v =$
$\log_b b^{u-v} = u - v = \log_b M - \log_b N$.

Exercise 5-4

Key Ideas and Formulas

Logarithmic Notation

$$\log x = \log_{10} x \quad \text{Common logarithm}$$
$$\ln x = \log_e x \quad \text{Natural logarithm}$$
$$\log x = y \quad \text{is equivalent to } x = 10^y$$
$$\ln x = y \quad \text{is equivalent to } x = e^y$$

Sound Intensity	Earthquake Intensity	Rocket Flight Velocity
$D = 10 \log \frac{I}{I_0}$ $D =$ decibel level $I =$ intensity of sound $I_0 =$ threshold of hearing	$M = \frac{2}{3} \log \frac{E}{E_0}$ $M =$ Magnitude on Richter Scale $E =$ Energy released by earthquake $E_0 =$ Energy released by small reference earthquake	$v = c \ln \frac{W_t}{W_b}$ $v =$ Velocity at fuel burnout level $c =$ exhaust velocity $W_t =$ take off weight $W_b =$ burnout weight

1.	4.9177	13.	47.73
3.	-2.8419	15.	0. 6760
5.	3.7623	17.	4.959
7.	-2.5128	19.	7.861
9.	200,800	21.	3.301
11.	$6.648 \times 10^{-4} = 0.000648$	23.	4.561

25. $x = \log(5.3147 \times 10^{12}) = \log 5.3147 + \log 10^{12} = 0.725 + 12 = 12.725$
    (Round off error is significantly reduced if we enter 5.3147 rather than 5.3147×10^{12})

27. -25.715

29. $\log x = 32.068523 = .068523 + 32 = \log 1.1709 + \log 10^{32} = \log(1.1709 \times 10^{32})$
    Hence $x = 1.1709 \times 10^{32}$

31. 4.2672×10^{-7}

33.

x	y
0.5	-0.7
1	0
2	0.7
3	1.1
4	1.4
5	1.6

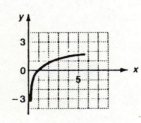

35.

x	y
0.25	1.4
0.5	0.7
0.75	0.3
1	0
2	0.7
3	1.1
4	1.4
5	1.6

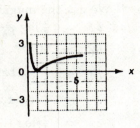

37.

x	y
−1.5	−1.4
−1	0
−0.5	0.8
0	1.4
1	2.2
2	2.8
3	3.2

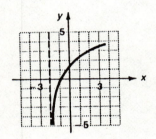

39.

x	y
0.5	−5.8
1	−3.0
2	−0.2
3	1.4
4	2.5
5	3.4
6	4.2
7	4.8
8	5.3
9	5.8
10	6.2

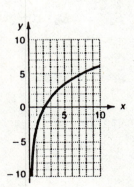

41. The inequality sign in the last step reverses because $\log \frac{1}{3}$ is negative.

43. We use the decibel formula

$$D = 10 \log \frac{I}{I_0}$$

A. $I = I_0$

$$D = 10 \log \frac{I_0}{I_0}$$
$$D = 10 \log 1$$
$$D = 0 \text{ decibels}$$

B. $I_0 = 1.0 \times 1.0^{-12}$ $I = 1.0$

$$D = 10 \log \frac{1.0}{1.0 \times 1.0^{-12}}$$
$$D = 120 \text{ decibels}$$

45. We use the decibel formula

$$D = 10 \log \frac{I}{I_0}$$
$$I_2 = 1000 I_1$$
$$D_1 = 10 \log \frac{I_1}{I_0} \qquad D_2 = 10 \log \frac{I_2}{I_0}$$
$$D_2 - D_1 = 10 \log \frac{I_2}{I_0} - 10 \log \frac{I_1}{I_0}$$
$$= 10 \log \left(\frac{I_2}{I_0} \div \frac{I_1}{I_0} \right)$$
$$= 10 \log \frac{I_2}{I_1}$$
$$= 10 \log \frac{1000 I_1}{I_1}$$
$$= 10 \log 1000$$
$$= 30 \text{ decibels}$$

47. We use the magnitude formula

$$M = \frac{2}{3} \log \frac{E}{E_0}$$
$$E_0 = 10^{4.40} \quad E = 1.99 \times 10^{17}$$
$$M = \frac{2}{3} \log \frac{1.99 \times 10^{17}}{10^{4.40}}$$
$$= \frac{2}{3} \log(1.99 \times 10^{12.6})$$
$$= \frac{2}{3}(\log 1.99 + \log 10^{12.6})$$
$$= \frac{2}{3}(0.299 + 12.6)$$
$$= 8.6$$

49. We use the magnitude formula

$$M = \frac{2}{3} \log \frac{E}{E_0}$$

For the Long Beach earthquake,

$$6.3 = \frac{2}{3} \log \frac{E_1}{E_0}$$
$$9.45 = \log \frac{E_1}{E_0}$$
$$\frac{E_1}{E_0} = 10^{9.45}$$
$$E_1 = E_0 \cdot 10^{9.45}$$

For the Anchorage earthquake,

$$8.3 = \frac{2}{3} \log \frac{E_2}{E_0}$$
$$12.45 = \log \frac{E_2}{E_0}$$
$$\frac{E_2}{E_0} = 10^{12.45}$$
$$E_2 = E_0 \cdot 10^{12.45}$$

Hence, the Anchorage earthquake compares to the Long Beach earthquake as follows:

$$\frac{E_2}{E_1} = \frac{E_0 \cdot 10^{12.45}}{E_0 \cdot 10^{9.45}} = 10^3$$
$$E_2 = 10^3 E, \text{ or } 1000 \text{ times as powerful}$$

51. We use the rocket equation.

$$v = c \ln \frac{W_t}{W_b}$$

$$v = 2.57 \ln (19.8)$$

$$v = 7.67 \text{ km/s}$$

53. A. $pH = -\log[H^+] = -\log(4.63 \times 10^{-9}) = 8.3$.

Since this is greater than 7, the substance is basic.

B. $pH = -\log[H^+] = -\log(9.32 \times 10^{-4}) = 3.0$

Since this is less than 7, the substance is acidic.

55. Since $pH = -\log[H^+]$, we have

$$5.2 = -\log[H^+], \text{ or}$$

$$[H^+] = 10^{-5.2} = 6.3 \times 10^{-6} \text{ moles per liter}$$

Exercise 5-5

Key Ideas and Formulas

To solve equations in which the variable appears only in the exponent, use logarithms. If $a^x = b$, then $x \log a = \log b$.

To solve equations in which the variable appears only in the argument of a logarithm, change to equivalent exponential form.

If $\log_a x = b$, then $x = a^b$

Change-of-Base Formula: $\log_b N = \dfrac{\log_a N}{\log_a b}$

---

1. $10^{-x} = 0.0347$
    $-x = \log_{10} 0.0347$
    $x = -\log_{10} 0.0347$
    $x = 1.46$

3. $10^{3x+1} = 92$
    $3x + 1 = \log_{10} 92$
    $3x = \log_{10} 92 - 1$
    $x = \dfrac{\log_{10} 92 - 1}{3}$
    $x = 0.321$

5. $e^x = 3.65$
    $x = \ln 3.65$
    $x = 1.29$

7. $e^{2x-1} = 405$
    $2x - 1 = \ln 405$
    $2x - 1 = 1 + \ln 405$
    $x = \dfrac{1 + \ln 405}{2}$
    $x = 3.50$

9. $5^x = 18$
    $x \ln 5 = \ln 18$
    $x = \dfrac{\ln 18}{\ln 5}$
    $x = 1.80$

11. $2^{-x} = 0.238$
    $-x \ln 2 = \ln 0.238$
    $-x = \dfrac{\ln 0.238}{\ln 2}$
    $x = -\dfrac{\ln 0.238}{\ln 2}$
    $x = 2.07$

13. $\log 5 + \log x = 2$
    $\log(5x) = 2$
    $5x = 10^2$
    $5x = 100$
    $x = 20$

15. $\log x + \log(x - 3) = 1$
    $\log[x(x - 3)] = 1$
    $x(x - 3) = 10^1$
    $x^2 - 3x = 10$
    $x^2 - 3x - 10 = 0$
    $(x - 5)(x + 2) = 0$
    $x = 5 \text{ or } -2$

Common Error: $\log(x - 3) \neq \log x - \log 3$

Check:

$\log 5 + \log(5 - 3) \overset{\vee}{=} 1$
$\log(-2) + \log(-2 - 3)$ is not defined.
$x = 5$

17. $\log(x+1) - \log(x-1) = 1$

$$\log \tfrac{x+1}{x-1} = 1$$

$$\tfrac{x+1}{x-1} = 10^1$$

$$\tfrac{x+1}{x-1} = 10$$

$$x+1 = 10(x-1)$$

$$x+1 = 10x - 10$$

$$11 = 9x$$

$$x = \tfrac{11}{9}$$

Common Error:
$$\tfrac{x+1}{x-1} \neq \log 1$$

Check:

$$\log(\tfrac{11}{9} + 1) - \log(\tfrac{11}{9} - 1) \overset{?}{=} 1$$

$$\log \tfrac{20}{9} - \log \tfrac{2}{9} \overset{?}{=} 1$$

$$\log 10 \overset{\checkmark}{=} 1$$

19.
$$2 = 1.05^x$$

$$\ln 2 = x \ln 1.05$$

$$\tfrac{\ln 2}{\ln 1.05} = x$$

$$x = 14.2$$

21.
$$e^{-1.4x} = 13$$

$$-1.4x = \ln 13$$

$$x = \tfrac{\ln 13}{-1.4}$$

$$x = -1.83$$

23.
$$123 = 500e^{-0.12x}$$

$$\tfrac{123}{500} = e^{-0.12x}$$

$$\ln \left(\tfrac{123}{500}\right) = -0.12x$$

$$\tfrac{\ln \left(\tfrac{123}{500}\right)}{-0.12} = x$$

$$x = 11.7$$

25.
$$e^{-x^2} = 0.23$$

$$-x^2 = \ln 0.23$$

$$x^2 = -\ln 0.23$$

$$x = \pm\sqrt{-\ln 0.23}$$

$$x = \pm 1.21$$

27.
$$\log x - \log 5 = \log 2 - \log(x-3)$$

$$\log \tfrac{x}{5} = \log \tfrac{2}{x-3}$$

$$\tfrac{x}{5} = \tfrac{2}{x-3}$$

Excluded value: $x \neq 3$

$$5(x-3)\tfrac{x}{5} = 5(x-3)\tfrac{2}{x-3}$$

$$(x-3)x = 10$$

$$x^2 - 3x = 10$$

$$x^2 - 3x - 10 = 0$$

$$(x-5)(x+2) = 0$$

$$x = 5, -2$$

Check: $\log 5 - \log 5 \overset{\checkmark}{=} \log 2 - \log 2$

$\log(-2)$ is not defined

Solution: 5

29.
$$\ln x = \ln(2x-1) - \ln(x-2)$$

$$\ln x = \ln \tfrac{2x-1}{x-2}$$

$$x = \tfrac{2x-1}{x-2}$$

Excluded value: $x \neq 2$

$$x(x-2) = (x-2)\tfrac{2x-1}{x-2}$$

$$x(x-2) = 2x-1$$

$$x^2 - 2x = 2x - 1$$

$$x^2 - 4x + 1 = 0$$

$$x = \tfrac{-b \pm \sqrt{b^2 - 4ac}}{2a}$$

$$a = 1$$

$$b = -4$$

$$c = 1$$

$$x = \tfrac{-(-4) \pm \sqrt{(-4)^2 - 4(1)(1)}}{2(1)}$$

$$x = \tfrac{4 \pm \sqrt{12}}{2}$$

$$x = 2 \pm \sqrt{3}$$

Check: $\ln(2+\sqrt{3}) \overset{?}{=} \ln[2(2+\sqrt{3})-1]$
$$-\ln[(2+\sqrt{3})-2]$$

$$\ln(2+\sqrt{3}) \overset{?}{=} \ln(3+2\sqrt{3}) - \ln\sqrt{3}$$

$$\ln(2+\sqrt{3}) \overset{?}{=} \ln\left(\frac{3+2\sqrt{3}}{\sqrt{3}}\right)$$

$$\ln(2+\sqrt{3}) \overset{\checkmark}{=} \ln(\sqrt{3}+2)$$

$\ln(x-2)$ is not defined if $x = 2 - \sqrt{3}$

Solution: $2 + \sqrt{3}$

31.
$$\log(2x+1) = 1 - \log(x-1)$$
$$\log(2x+1) + \log(x-1) = 1$$
$$\log[(2x+1)(x-1)] = 1$$
$$(2x+1)(x-1) = 10$$
$$2x^2 - x - 1 = 10$$
$$2x^2 - x - 11 = 0$$
$$x = \frac{-b \pm \sqrt{b^2-4ac}}{2a}$$

$$a = 2$$
$$b = -1$$
$$c = -11$$
$$x = \frac{-(-1) \pm \sqrt{(-1)^2 - 4(2)(-11)}}{2(2)}$$

$$x = \frac{1 \pm \sqrt{89}}{4}$$

Check:

$$\log\left(2\frac{1+\sqrt{89}}{4} + 1\right) \overset{?}{=} 1 - \log\left(\frac{1+\sqrt{89}}{4} - 1\right)$$

$$\log\left(\frac{1+\sqrt{89}+2}{2}\right) \overset{?}{=} 1 - \log\left(\frac{1+\sqrt{89}-4}{4}\right)$$

$$\log\left(\frac{3+\sqrt{89}}{2}\right) \overset{?}{=} 1 - \log\left(\frac{\sqrt{89}-3}{4}\right)$$

$$\log\left(\frac{3+\sqrt{89}}{2}\right) \overset{?}{=} \log 10 - \log\left(\frac{\sqrt{89}-3}{4}\right)$$

$$\overset{?}{=} \log\left(\frac{40}{\sqrt{89}-3}\right)$$

$$\overset{?}{=} \log\left[\frac{40(\sqrt{89}+3)}{89-9}\right]$$

$$\overset{\checkmark}{=} \log\left(\frac{\sqrt{89}+3}{2}\right)$$

$\log(x-1)$ is not defined if $x = \frac{1-\sqrt{89}}{4}$

Solution: $\frac{1+\sqrt{89}}{4}$

33.
$$(\ln x)^3 = \ln x^4$$
$$(\ln x)^3 = 4\ln x$$
$$(\ln x)^3 - 4\ln x = 0$$
$$\ln x[(\ln x)^2 - 4] = 0$$
$$\ln x(\ln x - 2)(\ln x + 2) = 0$$

$$\ln x = 0 \quad \ln x - 2 = 0 \quad \ln x + 2 = 0$$
$$x = 1 \quad\quad \ln x = 2 \quad\quad \ln x = -2$$
$$\quad\quad\quad\quad x = e^2 \quad\quad x = e^{-2}$$

Check:

$$(\ln 1)^3 \overset{?}{=} \ln 1^4 \quad (\ln e^2)^3 \overset{?}{=} \ln(e^{-2})^4 \quad (\ln e^{-2})^3 \overset{?}{=} \ln(e^{-2})^4$$
$$0 \overset{\checkmark}{=} 0 \quad\quad 8 \overset{\checkmark}{=} -8 \quad\quad -8 \overset{\checkmark}{=} -8$$

Solution: $1, e^2, e^{-2}$

35.
$$\ln(\ln x) = 1$$
$$\ln x = e^1$$
$$\ln x = e$$
$$x = e^e$$

37. $x^{\log x} = 100x$

We start by taking logarithms of both sides.
$$\log(x^{\log x}) = \log 100x$$
$$\log x \log x = \log 100 + \log x$$
$$(\log x)^2 = 2 + \log x$$
$$(\log x)^2 - \log x - 2 = 0$$
$$(\log x - 2)(\log x - 1) = 0$$

$$\log x - 2 = 0 \quad\quad \log x + 1 = 0$$
$$\log x = 2 \quad\quad\quad \log x = -1$$
$$x = 10^2 \quad\quad\quad x = 10^{-1}$$
$$x = 100 \quad\quad\quad x = 0.1$$

Check:

$$100^{\log 100} \overset{?}{=} 100 \cdot 100 \quad\quad (0.1)^{\log(0.1)} \overset{?}{=} 100(0.1)$$
$$\quad\quad\quad\quad\quad\quad\quad\quad\quad 0.1^{-1} \overset{?}{=} 10$$
$$10^4 \overset{\checkmark}{=} 10^4 \quad\quad\quad\quad\quad 10 \overset{\checkmark}{=} 10$$

39. $\log_5 372 = \dfrac{\ln 372}{\ln 5} = 3.6776$ or $\log_5 372 = \dfrac{\log_{10} 372}{\log_{10} 5} = 3.6776$

41. $\log_8 0.0352 = \dfrac{\ln 0.0352}{\ln 8} = -1.6094$

43. $\log_3 0.1483 = \dfrac{\ln 0.1483}{\ln 3} = -1.7372$

45.
$$A = Pe^{rt}$$
$$\frac{A}{P} = e^{rt}$$
$$\ln \frac{A}{P} = rt$$
$$\frac{1}{t}\ln \frac{A}{P} = r$$
$$r = \frac{1}{t}\ln \frac{A}{P}$$

47.
$$D = 10\log\frac{I}{I_0}$$
$$\frac{D}{10} = \log\frac{I}{I_0}$$
$$\frac{I}{I_0} = 10^{D/10}$$
$$I = I_0 10^{D/10}$$

49.
$$M = 6 - 2.5\log\frac{I}{I_0}$$
$$M = 6 - 2.5\log\frac{I}{I_0}$$
$$6 - M = 2.5\log\frac{I}{I_0}$$
$$\frac{6-M}{2.5} = \log\frac{I}{I_0}$$
$$\frac{I}{I_0} = 10^{(6-M)/2.5}$$
$$I = I_0 10^{(6-M)/2.5}$$

51.
$$I = \frac{E}{R}(1 - e^{-Rt/L})$$
$$RI = E(1 - e^{-Rt/L})$$
$$\frac{RI}{E} = 1 - e^{-\frac{Rt}{L}}$$
$$\frac{RI}{E} - 1 = -e^{-\frac{Rt}{L}}$$
$$-\left(\frac{RI}{E} - 1\right) = e^{\frac{-Rt}{L}}$$
$$-\frac{RI}{E} + 1 = e^{-\frac{Rt}{L}}$$
$$1 - \frac{RI}{E} = e^{-\frac{Rt}{L}}$$
$$\ln\left(1 - \frac{RI}{E}\right) = -\frac{Rt}{L}$$
$$-\frac{L}{R}\ln\left(1 - \frac{RI}{E}\right) = t$$
$$t = -\frac{L}{R}\ln\left(1 - \frac{RI}{E}\right)$$

53.
$$y = \frac{e^x + e^{-x}}{2}$$
$$2y = e^x + e^{-x}$$
$$2y = e^x + \frac{1}{e^x}$$
$$2ye^x = (e^x)^2 + 1$$
$$0 = (e^x)^2 - 2ye^x + 1$$

This equation is quadratic in e^x

$$e^x = \frac{-b \pm \sqrt{b^2 - 4ac}}{2a}$$
$$a = 1$$
$$b = -2y$$
$$c = 1$$
$$e^x = \frac{-(-2y) \pm \sqrt{(-2y)^2 - 4(1)(1)}}{2(1)}$$
$$e^x = \frac{2y \pm \sqrt{4y^2 - 4}}{2}$$
$$e^x = \frac{2(y \pm \sqrt{y^2 - 1})}{2}$$
$$e^x = y \pm \sqrt{y^2 - 1}$$
$$x = \ln\left(y \pm \sqrt{y^2 - 1}\right)$$

55.
$$y = \frac{e^x - e^{-x}}{e^x + e^{-x}}$$
$$y = \frac{e^x - \frac{1}{e^x}}{e^x + \frac{1}{e^x}}$$
$$y = \frac{e^x e^x - \frac{1}{e^x}e^x}{e^x e^x + \frac{1}{e^x}e^x}$$
$$y = \frac{e^{2x} - 1}{e^{2x} + 1}$$
$$y(e^{2x} + 1) = e^{2x} - 1$$
$$ye^{2x} + y = e^{2x} - 1$$
$$1 + y = e^{2x} - ye^{2x}$$
$$1 + y = (1 - y)e^{2x}$$
$$e^{2x} = \frac{1+y}{1-y}$$
$$2x = \ln\frac{1+y}{1-y}$$
$$x = \frac{1}{2}\ln\frac{1+y}{1-y}$$

57. To find the doubling time we replace A in $A = P(1 + 0.15)^n$ with $2P$ and solve for n.

$$2P = P(1.15)^n$$
$$2 = (1.15)^n$$
$$\ln 2 = n \ln 1.15$$
$$n = \frac{\ln 2}{\ln 1.15}$$

$$n = 5 \text{ years to the nearest year}$$

59. We solve $A = Pe^{rt}$ for r, with $A = 2,500, P = 1,000, t = 10$

$$2,500 = 1,000 e^{r(10)}$$
$$2.5 = e^{10r}$$
$$10r = \ln(2.5)$$
$$r = \tfrac{1}{10}\ln 2.5$$

$$r = 0.0916 \text{ or } 9.16\%$$

61. $m = 6 - 2.5 \log \frac{L}{L_0}$

A. We find m when $L = L_0$
$$m = 6 - 2.5 \log \frac{L_0}{L_0}$$

$$m = 6 - 2.5 \log 1$$
$$m = 6$$

B. We compare L_1 for $m = 1$ with L_2 for $m = 6$

$1 = 6 - 2.5 \log \frac{L_1}{L_0}$	$6 = 6 - 2.5 \log \frac{L_2}{L_0}$
$-5 = -2.5 \log \frac{L_1}{L_0}$	$0 = -2.5 \log \frac{L_2}{L_0}$
$2 = \log \frac{L_1}{L_0}$	$0 = \log \frac{L_2}{L_0}$
$\frac{L_1}{L_0} = 10^2$	$\frac{L_2}{L_0} = 1$
$L_1 = 100 L_0$	$L_2 = L_0$

Hence $\frac{L_1}{L_2} = \frac{100 L_0}{L_0} = 100$. The star of magnitude 1 is 100 times brighter.

63. We solve $P = P_0 e^{rt}$ for t with $P = 2P_0, r = 0.02$.

$$2P_0 = P_o e^{0.02t}$$
$$2 = e^{0.02t}$$
$$\ln 2 = 0.02t$$
$$\frac{\ln 2}{0.02} = t$$

$$t = 35 \text{ years to the nearest year}$$

65. We solve $A = A_0 e^{-0.000124t}$ for t with $A = 0.1 A_0$

$$0.1 A_0 = A_0 e^{-0.000124t}$$
$$0.1 = e^{-0.000124t}$$
$$\ln 0.1 = -0.000124t$$
$$t = \frac{\ln 0.1}{-0.000124}$$

$$t = 18,600 \text{ years old}$$

67. We solve $q = 0.0009(1 - e^{-0.2t})$ for t with $q = 0.0007$

$$0.0007 = 0.0009(1 - e^{-0.2t})$$

$$\frac{0.0007}{0.0009} = 1 - e^{-0.2t}$$

$$\frac{7}{9} = 1 - e^{-0.2t}$$

$$-\frac{2}{9} = -e^{-0.2t}$$

$$\frac{2}{9} = e^{-0.2t}$$

$$\ln \frac{2}{9} = -0.2t$$

$$t = \frac{\ln \frac{2}{9}}{-0.2}$$

$$t = 7.52 \text{ seconds}$$

69.   First, we solve $T = T_m + (T_0 - T_m)e^{-kt}$ for k, with $T = 61.5°, T_m = 40°, T_0 = 72°, t = 1$

$$61.5 = 40 + (72 - 40)e^{-k(1)}$$

$$21.5 = 32e^{-k}$$

$$\frac{21.5}{32} = e^{-k}$$

$$\ln \frac{21.5}{32} = -k$$

$$k = -\ln \frac{21.5}{32}$$

$$k = 0.40$$

Now we solve $T = T_m + (T_0 - T_m)e^{-0.40t}$ for t, with

$$T = 50°, T_m = 40°, T_0 = 72°$$

$$50 = 40 + (72 - 50)e^{-0.40t}$$

$$10 = 32e^{-0.40t}$$

$$\frac{10}{32} = e^{-0.40t}$$

$$\ln \frac{10}{32} = -0.40t$$

$$t = \frac{\ln 10/32}{-0.40}$$

$$t = 2.9 \text{ hours}$$

71.   We solve $N = \frac{100}{1 + 4e^{-0.14t}}$ for t, with $N = 50$

$$50 = \frac{100}{1 + 4e^{-0.14t}}$$

$$\frac{1}{50} = \frac{1 + 4e^{-0.14t}}{100}$$

$$2 = 1 + 4e^{-0.14t}$$

$$1 = 4e^{-0.14t}$$

$$0.25 = e^{-0.14t}$$

$$\ln 0.25 = -0.14t$$

$$t = \frac{\ln 0.25}{-0.14}$$

$$t = 10 \text{ years}$$

Exercise 5-6

CHAPTER REVIEW

1. $\log m = n$ $(5-3)$

2. $\ln x = y$ $(5-3)$

3. $x = 10^y$ $(5-3)$

4. $y = e^x$ $(5-3)$

5. $\dfrac{7^{x+2}}{7^{2-x}} = 7^{(x+2)-(2-x)}$
   $= 7^{x+2-2+x}$
   $= 7^{2x}$ $(5-1)$

6. $\left(\dfrac{e^x}{e^{-x}}\right)^x = [e^{x-(-x)}]^x$
   $= (e^{2x})^x = e^{2x \cdot x}$
   $= e^{2x^2}$ $(5-1)$

7. $\log x = 3$
   $x = 2^3$
   $x = 8$ $(5-3)$

8. $\log_x 25 = 2$
   $25 = x^2$
   $x = 5$
   since bases are
   restricted positive $(5-3)$

9. $\log_3 27 = x$
   $\log_3 3^3 = x$
   $3 = x$ $(5-3)$

10. $10^x = 17.5$
    $x = \log_{10} 17.5$
    $x = 1.24$ $(5-3)$

11. $e^x = 143{,}000$
    $x = \ln 143{,}000$
    $x = 11.9$ $(5-3)$

12. $\ln x = -0.01573$
    $x = e^{-0.01573}$
    $x = 0.984$ $(5-3)$

13. $\log x = 2.013$
    $x = 10^{2.013}$
    $x = 103$ $(5-3)$

14. $\ln(2x-1) = \ln(x+3)$
    $2x-1 = x+3$
    $x = 4$
    Check
    $\ln(2\cdot 4 - 1) \overset{?}{=} \ln(4+3)$
    $\ln 7 \overset{\checkmark}{=} \ln 7$ $(5-3)$

15. $\log(x^2-3) = 2\log(x-1)$
    $\log(x^2-3) = \log(x-1)^2$
    $x^2-3 = (x-1)^2$
    $x^2-3 = x^2-2x+1$
    $-3 = -2x+1$
    $-4 = -2x$
    $x = 2$
    Check:
    $\log(2^2-3) \overset{?}{=} \log(2-1)$
    $\log 1 \overset{?}{=} 2\log 1$
    $0 \overset{?}{=} 0$ $(5-3)$

16. $e^{x^2-3} = e^{2x}$
    $x^2-3 = 2x$
    $x^2-2x-3 = 0$
    $(x-3)(x+1) = 0$
    $x = 3, -1$ $(5-2)$

17. $4^{x-1} = 2^{1-x}$
    $(2^2)^{x-1} = 2^{1-x}$
    $2^{2(x-1)} = 2^{1-x}$
    $2(x-1) = 1-x$
    $2x-2 = 1-x$
    $3x = 3$
    $x = 1$ $(5-1)$

18. $2x^2 e^{-x} = 18 e^{-x}$
    $2x^2 e^{-x} - 18 e^{-x} = 0$
    $2e^{-x}(x^2-9) = 0$
    $2e^{-x}(x-3)(x+3) = 0$
    $2e^{-x} = 0 \quad x-3 = 0 \quad x+3 = 0$
    never $\qquad x = 3 \qquad x = -3$
    Solution: $3, -3$ $(5-2)$

19. $\log_{1/4} 16 = x$
    $\log_{1/4} 4^2 = x$
    $\log_{1/4}\left(\tfrac{1}{4}\right)^{-2} = x$
    $-2 = x$ $(5-3)$

20. $\log_x 9 = -2$

$x^{-2} = 9$

$\frac{1}{x^2} = 9$

$1 = 9x^2$

$\frac{1}{9} = x^2$

$x = \pm\sqrt{\frac{1}{9}}$

$x = \frac{1}{3}$

since bases are restricted positive $(5-3)$

21. $\log_{16} x = \frac{3}{2}$

$16^{3/2} = x$

$64 = x$

$x = 64 \quad (5-3)$

22. $\log_x e^5 = 5$

$e^5 = x^5$

$x = e \quad (5-3)$

23. $10^{\log_{10} x} = 33$

$\log_{10} x = \log_{10} 33$

$x = 33 \quad (5-3)$

24. $\ln x = 0$

$e^0 = x$

$x = 1 \quad (5-3)$

25. $x = 2(10^{1.32})$

$x = 41.8 \quad (5-1)$

26. $x = \log_5 23$

$x = \frac{\log 23}{\log 5} \text{ or } \frac{\ln 23}{\ln 5}$

$x = 1.95 \quad (5-3)$

27. $\ln x = -3.218$

$x = e^{-3.218}$

$x = 0.0400 \quad (5-3)$

28. $x = \log(2.156 \times 10^{-7})$

$x = \log 2.156 + \log 10^{-7}$

$x = \log 2.156 - 7$

$x = -6.67 \quad (5-3)$

29. $x = \frac{\ln 4}{\ln 2.31}$

$x = 1.66 \quad (5-3)$

30. $25 = 5(2)^x$

$\frac{25}{5} = 2^x$

$5 = 2^x$

$\ln 5 = x \ln 2$

$\frac{\ln 5}{\ln 2} = x$

$x = 2.32 \quad (5-5)$

31. $4,000 = 2,500e^{0.12x}$

$\frac{4,000}{2,500} = e^{0.12x}$

$0.12x = \ln \frac{4,000}{2,500}$

$x = \frac{1}{0.12}\ln \frac{4,000}{2,500}$

$x = 3.92 \quad (5-5)$

32. $0.01 = e^{-0.05x}$

$-0.05x = \ln 0.01$

$x = \frac{\ln 0.01}{-0.05}$

$x = 92.1 \quad (5-5)$

33. $5^{2x-3} = 7.08$

$(2x-3)\log 5 = \log 7.08$

$2x-3 = \frac{\log 7.08}{\log 5}$

$2x-3 = \frac{\log 7.08}{\log 5}$

$x = \frac{1}{2}[3 + \frac{\log 7.08}{\log 5}]$

$x = 2.11 \quad (5-5)$

34. $\frac{e^x - e^{-x}}{2} = 1$

$e^x - e^{-x} = 2$

$e^x - \frac{1}{e^x} = 2$

$e^x e^x - e^x(\frac{1}{e^x}) = 2e^x$

$(e^x)^2 - 1 = 2e^x$

$(e^x)^2 - 2e^x - 1 = 0$

34. (Continued)

This equation is quadratic in e^x

$$e^x = \frac{-b \pm \sqrt{b^2 - 4ac}}{2a}$$
$$a = 1$$
$$b = -2$$
$$c = -1$$
$$e^x = \frac{-(-2) \pm \sqrt{(-2)^2 - 4(1)(-1)}}{2}$$
$$e^x = \frac{2 \pm \sqrt{8}}{2}$$
$$e^x = 1 \pm \sqrt{2}$$
$$x = \ln(1 \pm \sqrt{2})$$

$1 - \sqrt{2}$ is negative, hence not in the domain of the logarithm function.

$$x = \ln(1 + \sqrt{2})$$
$$x = 0.881 \quad (5-5)$$

35.
$$\log 3x^2 - \log 9x = 2$$
$$\log \frac{3x^2}{9x} = 2$$
$$\frac{3x^2}{9x} = 10^2$$
$$\frac{x}{3} = 100$$
$$x = 300$$

Check:
$$\log(3 \cdot 300^2) - \log(9 \cdot 300) \overset{?}{=} 2$$
$$\log(270,000) - \log(2,700) \overset{?}{=} 2$$
$$\log \frac{270,000}{2,700} \overset{?}{=} 2$$
$$\log 100 \overset{\checkmark}{=} 2 \quad (5-5)$$

36.
$$\log x - \log 3 = \log 4 - \log(x+4)$$
$$\log \frac{x}{3} = \log \frac{4}{x+4}$$
$$\frac{x}{3} = \frac{4}{x+4}$$

excluded value : $x \neq -4$

$$3(x+4)\frac{x}{3} = 3(x+4)\frac{4}{x+4}$$
$$(x+4)x = 12$$
$$x^2 + 4x = 12$$
$$x^2 + 4x = 12$$
$$x^2 + 4x - 12 = 0$$
$$(x+6)(x-2) = 0$$
$$x = -6 \quad x = 2$$

Check: $\log(-6)$ is not defined
$$\log 2 - \log 3 \overset{?}{=} \log 4 - \log(2+4)$$
$$\log \frac{2}{3} \overset{?}{=} \log \frac{4}{6}$$
$$\log \frac{2}{3} \overset{\checkmark}{=} \log \frac{2}{3}$$

Solution : 2 $(5-5)$

37.
$$\ln(x+3) - \ln x = 2\ln 2$$
$$\ln \frac{x+3}{x} = \ln 2^2$$
$$\frac{x+3}{x} = 2^2$$
$$\frac{x+3}{x} = 4$$
$$x + 3 = 4x$$
$$3 = 3x$$
$$x = 1$$

Check:
$$\ln(1+3) - \ln 1 \overset{?}{=} 2\ln 2$$
$$\ln 4 - 0 \overset{?}{=} 2\ln 2$$
$$\ln 4 \overset{\checkmark}{=} \ln 4 \quad (5-5)$$

38.
$$\ln(2x+1) - \ln(x-1) = \ln x$$
$$\ln \frac{2x+1}{x-1} = \ln x$$
$$\frac{2x+1}{x-1} = x$$

Excluded value $x \neq 1$
$$(x-1)\frac{2x+1}{x-1} = x(x-1)$$
$$2x + 1 = x^2 - x$$
$$0 = x^2 - 3x - 1$$
$$x = \frac{-b \pm \sqrt{b^2 - 4ac}}{2a}$$
$$a = 1$$
$$b = -3$$
$$c = -1$$
$$x = \frac{(-3) \pm \sqrt{(-3)^2 - 4(1)(-1)}}{2(1)}$$
$$x = \frac{3 \pm \sqrt{13}}{2}$$

$\ln\left(\frac{3-\sqrt{13}}{2}\right)$ is not defined

Solution: $\frac{3+\sqrt{13}}{2}$

Check:

$$\ln\left(2 \cdot \frac{3+\sqrt{13}}{2} + 1\right) - \ln\left(\frac{3+\sqrt{13}}{2} - 1\right) \overset{?}{=} \ln\left(\frac{3+\sqrt{13}}{2}\right)$$
$$\ln(3 + \sqrt{13} + 1) - \ln\left(\frac{3+\sqrt{13}-2}{2}\right) \overset{?}{=} \ln\left(\frac{3+\sqrt{13}}{2}\right)$$
$$\ln(4 + \sqrt{13}) - \ln\left(\frac{1+\sqrt{13}}{2}\right) \overset{?}{=} \ln\left(\frac{3+\sqrt{13}}{2}\right)$$
$$\ln\left(\frac{4+\sqrt{13}}{1} \cdot \frac{2}{1+\sqrt{13}}\right) \overset{?}{=} \ln\left(\frac{3+\sqrt{13}}{2}\right)$$
$$\ln\left(\frac{(4+\sqrt{13})2}{1+\sqrt{13}}\right) \overset{?}{=} \ln\left(\frac{3+\sqrt{13}}{2}\right)$$
$$\ln\left(\frac{(4+\sqrt{13})2(1-\sqrt{13})}{(1+\sqrt{13})(1-\sqrt{13})}\right) \overset{?}{=} \ln\left(\frac{3+\sqrt{13}}{2}\right)$$
$$\ln\left(\frac{2(4-3\sqrt{13}-13)}{1-13}\right) \overset{?}{=} \ln\left(\frac{3+\sqrt{13}}{2}\right)$$
$$\ln\left(\frac{-18-6\sqrt{13}}{-12}\right) \overset{?}{=} \ln\left(\frac{3+\sqrt{13}}{2}\right)$$
$$\ln\left(\frac{3+\sqrt{13}}{2}\right) \overset{\checkmark}{=} \ln\left(\frac{3+\sqrt{13}}{2}\right)$$

$(5-5)$

39.
$$(\log x)^3 = \log x^9$$
$$(\log x)^3 = 9\log x$$
$$(\log x)^3 - 9\log x = 0$$
$$\log x[(\log x)^2 - 9] = 0$$
$$\log x(\log x - 3)(\log x + 3) = 0$$
$$\log x = 0 \quad \log x - 3 = 0 \quad \log x + 3 = 0$$
$$x = 1 \quad\quad \log x = 3 \quad\quad \log x = -3$$
$$x = 10^3 \quad\quad x = 10^{-3}$$

Check:

$$(\log 1)^3 \overset{?}{=} \log 1^9$$
$$0 \overset{\checkmark}{=} 0$$
$$(\log 10^3) \overset{?}{=} \log(10^3)^9$$
$$27 \overset{\checkmark}{=} 27$$
$$(\log 10^{-3})^3 \overset{?}{=} \log(10^{-3})^9$$
$$-27 \overset{\checkmark}{=} -27$$
Solution: $1, 10^3, 10^{-3}$ $(5-5)$

40. $\ln(\log x) = 1$
$$\log x = e$$
$$x = 10^e \quad (5-5)$$

41.
$$(e^x + 1)(e^{-x} - 1) - e^x(e^{-x} - 1) = e^x e^{-x} - e^x + e^{-x} - 1 - e^x e^{-x} + e^x$$
$$= 1 - e^x + e^{-x} - 1 - 1 + e^x = e^{-x} - 1 \quad (5-2)$$

42. $(e^x + e^{-x})(e^x - e^{-x}) - (e^x - e^{-x})^2 = (e^x)^2 - (e^{-x})^2 - [(e^x)^2 - 2e^x e^{-x} + (e^{-x})^2]$
$$= e^{2x} - e^{-2x} - [e^{2x} - 2 + e^{-2x}]$$
$$= e^{2x} - e^{-2x} - e^{2x} + 2 - e^{-2x}$$
$$= 2 - 2e^{-2x} \quad (5-2)$$

43.

x	y
-2	0.13
-1	0.25
0	0.5
1	1
2	2
3	4
4	8

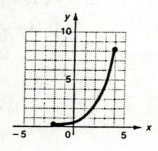

$(5-1)$

44.

t	f(t)
5	6.7
10	4.5
15	3.0
20	2.0
25	1.4
30	1.0
35	0.6
40	0.4

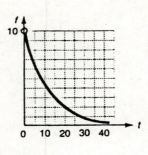

Note when $t = 0, 10e^{-0.08t} = 10$

This point is not in the domain of the specified function, but we indicate it with an open dot.   (5 − 2)

45.

x	y
-0.5	-0.7
0	0
1	0.7
2	1.1
4	1.6
6	1.9
8	2.2
10	2.4

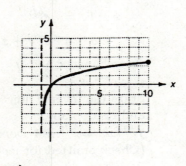

(5 − 3)

46.

t	N
0	25
1	48
2	71
3	87
4	95
5	98

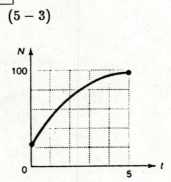

(5 − 2)

47.

$$D = 10 \log \frac{I}{I_0}$$

$$\frac{D}{10} = 10 \log \frac{I}{I_0}$$

$$10^{\frac{D}{10}} = \frac{I}{I_0}$$

$$I_0 10^{D/10} = I$$

$$I = I_0 10^{D/10} \quad (5 − 5)$$

48.
$$y = \frac{1}{\sqrt{2\pi}}e^{-x^2/2}$$

$$\sqrt{2\pi}y = e^{-x^2/2}$$

$$-\frac{x^2}{2} = \ln\left(\sqrt{2\pi}y\right)$$

$$x^2 = -2\ln\left(\sqrt{2\pi}y\right)$$

$$x = \pm\sqrt{-2\ln\left(\sqrt{2\pi}y\right)} \quad (5-5)$$

49.
$$x = -\frac{1}{k}\ln\frac{I}{I_0}$$

$$-kx = \ln\frac{I}{I_0}$$

$$\frac{I}{I_0} = e^{-kx}$$

$$I = I_0 e^{-kx} \quad (5-5)$$

50.
$$r = P\frac{i}{1-(1+i)^{-n}}$$

$$\frac{r}{P} = \frac{i}{1-(1+i)^{-n}}$$

$$\frac{P}{r} = \frac{1-(1+i)^{-n}}{i}$$

$$\frac{Pi}{r} = 1-(1+i)^{-n}$$

$$\frac{Pi}{r}-1 = -(1+i)^{-n}$$

$$1-\frac{Pi}{r} = (1+i)^{-n}$$

$$\ln\left(1-\frac{Pi}{r}\right) = -n\ln(1+i)$$

$$\frac{\ln\left(1-\frac{Pi}{r}\right)}{-\ln(1+i)} = n$$

$$n = -\frac{\ln\left(1-\frac{Pi}{r}\right)}{\ln(1+i)} \quad (5-5)$$

51.
$$f(x) = 2\ln(x-1)$$

Solve $y = f(x)$ for x

$$y = 2\ln(x-1)$$

$$\frac{y}{2} = \ln(x-1)$$

$$x-1 = e^{y/2}$$

$$x = e^{y/2}+1$$

Interchange x and y

$y = e^{x/2}+1$

State the inverse function
and its domain

$f^{-1}(x) = e^{x/2}+1$ Domain: R

(Check omitted for lack of space) $\quad (5-5, 4-6)$

52.
$$f(x) = \frac{e^x - e^{-x}}{2}$$

Solve $y = f(x)$ for x

$$y = \frac{e^x - e^{-x}}{2}$$

$$2y = e^x - e^{-x}$$

$$2y = e^x - \frac{1}{e^x}$$

$$2ye^x = e^x e^x - e^x\left(\frac{1}{e^x}\right)$$

$$2ye^x = (e^x)^2 - 1$$

$$0 = (e^x)^2 - 2ye^x - 1$$

This equation is quadratic in e^x

$$e^x = \frac{-b\pm\sqrt{b^2-4ac}}{2a}$$

$$a = 1$$

$$b = -2y$$

$$c = -1$$

$$e^x = \frac{-(-2y)\pm\sqrt{(-2y)^2-4(1)(-1)}}{2(1)}$$

$$e^x = \frac{(2y)\pm\sqrt{4y^2+4}}{2}$$

$$e^x = y \pm \sqrt{y^2+1}$$

Note: Since $0 < 1, y^2 < y^2 + 1, \sqrt{y^2} < \sqrt{y^2 + 1}$ and
$y < \sqrt{y^2 + 1}$ for all real y.
Hence $y - \sqrt{y^2 + 1}$ is always negative
Also, $y + \sqrt{y^2 + 1}$ is always positive.
$x = \ln\left(y + \sqrt{y^2 + 1}\right)$
Interchange x and y
$y = \ln\left(x + \sqrt{x^2 + 1}\right)$
State the inverse function and its domain
$f^{-1}(x) = \ln\left(x + \sqrt{x^2 + 1}\right)$
Domain: R (since $x + \sqrt{x^2 + 1}$ is always positive)
(Check omitted for lack of space) $(5 - 5, 4 - 6)$

53.
$$\begin{aligned}
\ln y &= -5t + \ln c \\
\ln y - \ln c &= -5t \\
\ln\left(\tfrac{y}{c}\right) &= -5t \\
\tfrac{y}{c} &= e^{-5t} \\
y &= ce^{-5t} \quad (5 - 3, 5 - 5)
\end{aligned}$$

54.

x	$y = \log_2 x$	$x = \log_2 y$	y
1	0	0	1
2	1	1	2
4	2	2	4
8	3	3	8

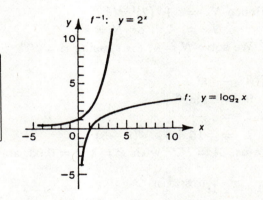

Domain $f = (0, \infty) =$ Range f^{-1}
Range $f = R =$ Domain f^{-1} $(5 - 3)$

55. If $\log_1 x = y$, then we would have to have $1^y = x$; that is, $1 = x$ for arbitrary positive x, which is impossible. $(5 - 3)$

56. Let $u = \log_b M$ and $v = \log_b N$; then $M = b^u$ and $N = b^v$.
Thus, $\log(M/N) = \log_b(b^u/b^v) = \log_b b^{u-v} = u - v = \log_b M - \log_b N$.

57. 1.145 $(5 - 3)$

58. Not defined $(5 - 3)$

59. 2.211 $(5 - 1)$

60. 11.59 $(5 - 2)$

61. We solve $P = P_0(1.03)^t$ for t, using $P = 2P_0$.

$$\begin{aligned}
2P_0 &= P_0(1.03)^t \\
2 &= (1.03)^t \\
\ln 2 &= \ln 1.03 \\
\tfrac{\ln 2}{\ln 1.03} &= t \\
t &= 23.4 \text{ years} \quad (5 - 5)
\end{aligned}$$

62. We solve $P = P_0 e^{0.03t}$ for t using $P = 2P_0$.

$$2P_0 = P_0 e^{0.03t}$$
$$2 = e^{0.03t}$$
$$\ln 2 = 0.03t$$
$$\frac{\ln 2}{0.03} = t$$
$$t = 23.1 \text{ years} \quad (5-5)$$

63.
$$A_0 = \text{original amount}$$
$$0.01A_0 = 1 \text{ percent of original amount}$$
We solve $A = A_0 e^{-0.000124t}$ for t, using $A = 0.01A_0$.

$$0.01A_0 = A_0 e^{-0.000124t}$$
$$0.01 = e^{-0.000124t}$$
$$\ln 0.01 = -0.000124t$$
$$\frac{\ln 0.01}{-0.000124} = t$$
$$t = 37,100 \text{ years} \quad (5-5)$$

64. A. When $t = 0$, $N = 1$. As t increases by $1/2$, N doubles.

$$\text{Hence } N = 1 \cdot (2)^{t \div 1/2}$$
$$N = 2^{2t} (\text{ or } N = 4^t)$$

B. We solve $N = 4^t$ for t, using $n = 10^9$

$$10^9 = 4^t$$
$$9 = t \log^4$$
$$t = \frac{9}{\log 4}$$
$$t = 15 \text{ days} \quad (5-5)$$

65. We use $A = Pe^{rt}$ with $P = 1$, $r = 0.03$, and $t = 2000$.

$$A = 1e^{0.03(2000)}$$
$$A = 1.1 \times 10^{26} \text{ dollars} \quad (5-2)$$

66. A.

t	P
0	1,000
5	670
10	450
15	301
20	202
25	135
30	91

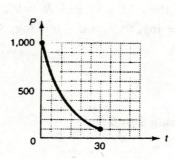

B. As t tends to infinity, P appears to tend to 0.   (5 − 2)

67. $M = \frac{2}{3} \log \frac{E}{E_0}$ $E_0 = 10^{4.40}$

We use $E = 1.99 \times 10^{14}$

$$M = \tfrac{2}{3} \log \frac{1.99 \times 10^{14}}{10^{4.40}}$$

$$M = \tfrac{2}{3} \log(1.99 \times 10^{9.6})$$

$$M = \tfrac{2}{3}(\log 1.99 + 9.6)$$

$$M = \tfrac{2}{3}(0.299 + 9.6)$$

$$M = 6.6 \quad (5-4)$$

68.　We solve $M = \tfrac{2}{3}\log\frac{I}{I_0}$ for I, using $I_0 = 10^{4.40}$, $M = 8.3$

$$8.3 = \tfrac{2}{3}\log\frac{I}{10^{4.40}}$$

$$\tfrac{3}{2}(8.3) = \log\frac{I}{10^{4.40}}$$

$$12.45 = \log\frac{I}{10^{4.40}}$$

$$\frac{I}{10^{4.40}} = 10^{12.45}$$

$$I = 10^{4.40} \cdot 10^{12.45}$$

$$I = 10^{16.85} \text{ or } 7.08 \times 10^{16} \text{ joules} \quad (5-4)$$

69.　We use the given formula twice, with

$$I_2 = 100,000 I_1$$

$$D_1 = 10\log\frac{I_1}{I_0} \quad D_2 = \log\frac{I_2}{I_0}$$

$$D_2 - D_1 = 10\log\frac{I_2}{I_0} - 10\log\frac{I_1}{I_0}$$

$$= 10\log\left(\frac{I_2}{I_0} \div \frac{I_1}{I_0}\right)$$

$$= 10\log\frac{I_2}{I_1}$$

$$= 10\log\frac{100,000 I_1}{I_1}$$

$$= 10\log 100,000$$

$$= 50 \text{ decibels}$$

The level of the louder sound is 50 decibels more.　　(5 − 4)

70.　$I = I_0 e^{-kd}$

To find k, we solve for k using $I = \tfrac{1}{2}I_0$ and $d = 73.6$

$$\tfrac{1}{2}I_0 = I_0 e^{-k(73.6)}$$

$$\tfrac{1}{2} = e^{-73.6k}$$

$$-73.6k = \ln\tfrac{1}{2}$$

$$k = \frac{\ln\tfrac{1}{2}}{-73.6}$$

$$k = 0.00942$$

We now find the depth at which 1% of the surface light remains. We solve

$$I = I_0 e^{-0.00942\, d} \text{ for } d \text{ with } I = 0.01\, I_0$$

$$0.01\, I_0 = I_0 e^{-0.00942\, d}$$

$$0.01 = e^{-0.00942\, d}$$

$$-0.00942\, d = \ln 0.01$$

$$d = \frac{\ln 0.01}{-0.00942}$$

$$d = 489 \text{ feet}$$

71.   We solve $N = \frac{30}{1+29e^{-1.35t}}$ for t with $N = 20$.

$$20 = \frac{30}{1+29e^{-1.35t}}$$

$$\frac{1}{20} = \frac{1+29e^{-1.35t}}{30}$$

$$1.5 = 1 + 29e^{-1.35t}$$

$$0.5 = 29e^{-1.35t}$$

$$\frac{0.5}{29} = e^{-1.35t}$$

$$-1.35t = \ln\frac{0.5}{29}$$

$$t = \frac{\ln\frac{0.5}{2.9}}{-1.35}$$

$$t = 3 \text{ years} \quad (5-5)$$

CHAPTER 6

Exercise 6-1

Key Ideas and Formulas

For the *nth* degree polynomial function P given by

$$p(x) = a_n x^n + a_{n-1} x^{n-1} + \cdots + a_1 x + a_0 \quad a_n \neq 0$$

the number r is said to be a zero of the function P, or a zero of the polynomial $P(x)$, or a solution or root of the equation $P(x) = 0$, if $P(r) = 0$.

The x intercepts of the graph of $y = P(x)$ are real zeros of P and $P(x)$ and real solutions or roots for the equation $P(x) = 0$.

Synthetic Division: To divide $P(x)$ by $x - r$ arrange the coefficients of $P(x)$ in order of descending powers of x (write 0 as the coefficient for each missing power).

Use the bring down, multiply, add scheme below:

$$
\begin{array}{cccccc}
 & a_n & & a_{n-1} & \cdots & a_1 \quad a_0 \\
\text{Bring down} & \downarrow & & ra_n & & \cdots \\
 & & \text{multiply} & \text{add} & & \\
\hline
r & a_n & & ra_n + a_{n-1} & & R
\end{array}
$$

The last number to the right in the third row of numbers is the remainder R; the other numbers in the third row are the coefficients of the quotient $Q(x)$, which is of degree one less than $P(x)$.

$$\frac{P(x)}{x - r} = Q(x) + \frac{R}{x - r}$$

1.

$$
\begin{array}{r}
2m + 1 \\
2m - 1 \overline{\big)\ 4m^2 + 0m - 1} \\
\underline{4m^2 - 2m} \\
2m - 1 \\
\underline{2m - 1} \\
0
\end{array}
$$

$2m + 1, R = 0$

3.

$$
\begin{array}{r}
4x \quad - \quad 5 \\
2x+1 \,\overline{\big)\, 8x^2 \;-\; 6x \;+\; 6} \\
\underline{8x^2 \;+\; 4x} \\
-\; 10x \;+\; 6 \\
\underline{-\; 10x \;-\; 5} \\
11
\end{array}
$$

$4x - 5,\; R = 11$

5.

$$
\begin{array}{r}
x^2 \;+\; x \;+\; 1 \\
x-1 \,\overline{\big)\, x^3 \;-\; 0x^2 \;+\; 0x \;-\; 1} \\
\underline{x^3 \;-\; x^2} \\
x^2 \;+\; 0x \\
\underline{x^2 \;-\; x} \\
x \;-\; 1 \\
\underline{x \;-\; 1} \\
0
\end{array}
$$

$x^2 + x + 1,\; R = 0$

7.

$$
\begin{array}{r}
2y^2 \;-\; 5y \;+\; 13 \\
y+2 \,\overline{\big)\, 2y^3 \;-\; y^2 \;+\; 3y \;-\; 1} \\
\underline{2y^3 \;+\; 4y^2} \\
-\; 5y^2 \;+\; 3y \\
\underline{-\; 5y^2 \;-\; 10y} \\
13y \;-\; 1 \\
\underline{13y \;+\; 26} \\
-\; 27
\end{array}
$$

$2y^2 - 5y + 13,\; R = -27$

9.

$$
\begin{array}{r}
1 \quad 3 \quad -7 \\
2 \quad 10 \\
2 \,\overline{\big)\, 1 \quad 5 \quad\;\; 3}
\end{array}
$$

$\dfrac{x^2+3x-7}{x-2} = x + 5 + \dfrac{3}{x-2}$

11.

$$
\begin{array}{r}
4 \quad\;\; 10 \quad -9 \\
-12 \quad\;\; 6 \\
-3 \,\overline{\big)\, 4 \quad -2 \quad -3}
\end{array}
$$

$\dfrac{4x^2+10x-9}{x+3} = 4x - 2 - \dfrac{3}{x+3}$

13.

$$\begin{array}{r} 2 \quad\ \ 0 \quad -3 \quad\ \ 1 \\ 4 \quad\ \ 8 \quad 10 \\ \hline 2\ \big|\ 2 \quad\ \ 4 \quad\ \ 5 \quad 11 \end{array}$$

$\frac{2x^3-3x+1}{x-2} = 2x^2 + 4x + 5 + \frac{11}{x-2}$

Common Error:

The first row is *not*

2   −3   1

The 0 must be inserted for the missing power.

15.

$$\begin{array}{r} 3 \quad\ \ 0 \quad\ \ 0 \quad -1 \quad -4 \\ -3 \quad\ \ 3 \quad -3 \quad\ \ 4 \\ \hline -1\ \big|\ 3 \quad -3 \quad\ \ 3 \quad -4 \quad\ \ 0 \end{array}$$

$3x^3 - 3x^2 + 3x - 4, R = 0$

17.

$$\begin{array}{r} 1 \quad\ \ 0 \quad\ \ 0 \quad\ \ 0 \quad\ \ 0 \quad\ \ 1 \\ -1 \quad\ \ 1 \quad -1 \quad\ \ 1 \quad -1 \\ \hline -1\ \big|\ 1 \quad -1 \quad\ \ 1 \quad -1 \quad\ \ 1 \quad\ \ 0 \end{array}$$

$x^4 - x^3 + x^2 - x + 1, R = 0$

19.

$$\begin{array}{r} 3 \quad\ \ 2 \quad\ \ 0 \quad -4 \quad -1 \\ -9 \quad 21 \quad -63 \quad 201 \\ \hline -3\ \big|\ 3 \quad -7 \quad 21 \quad -67 \quad 200 \end{array}$$

$3x^3 - 7x^2 + 21x - 67, R = 200$

21.

$$\begin{array}{r} 2 \quad -13 \quad\ \ 0 \quad\ \ 75 \quad 2 \quad\ \ 0 \quad -50 \\ 10 \quad -15 \quad -75 \quad 0 \quad 10 \quad\ \ 50 \\ \hline 5\ \big|\ 2 \quad -3 \quad -15 \quad\ \ 0 \quad 2 \quad 10 \quad\ \ 0 \end{array}$$

$2x^5 - 3x^4 - 15x^3 + 2x + 10, R = 0$

23.

$$\begin{array}{r} 4 \quad\ \ 2 \quad -6 \quad -5 \quad 1 \\ -2 \quad\ \ 0 \quad\ \ 3 \quad 1 \\ \hline -\frac{1}{2}\ \big|\ 4 \quad\ \ 0 \quad -6 \quad -2 \quad 2 \end{array}$$

$$4x^3 - 6x - 2, R = 2$$

25.

$$
\begin{array}{r}
4 \quad\quad 4 \quad -7 \quad -6 \\
\quad\quad -6 \quad\quad 3 \quad\quad 6 \\
\hline
-\tfrac{3}{2}\,|\quad 4 \quad -2 \quad -4 \quad\quad 0
\end{array}
$$

$$4x^2 - 2x - 4, R = 0$$

27.

$$
\begin{array}{r}
3 \quad -2 \quad\quad 2 \quad\quad -3 \quad\quad\quad 1 \\
1.2 \quad -0.32 \quad\quad 0.672 \quad -0.9312 \\
\hline
0.4\,|\quad 3 \quad -0.8 \quad 1.68 \quad -2.328 \quad\quad 0.0688
\end{array}
$$

$$3x^3 - 0.8x^2 + 1.68x - 2.328, R = 0.0688$$

29.

$$
\begin{array}{r}
3 \quad\quad 2 \quad\quad 5 \quad\quad 0 \quad\quad\quad -7 \quad\quad\quad -3 \\
-2.4 \quad 0.32 \quad -4.256 \quad\quad 3.4048 \quad\quad 2.87616 \\
\hline
-0.8\,|\quad 3 \quad -0.4 \quad 5.32 \quad -4.256 \quad -3.5952 \quad -0.12384
\end{array}
$$

$$3x^3 - 0.4x^3 + 5.32x^2 - 4.256x - 3.5952, R = -0.12384$$

31.

$$
\begin{array}{r}
2x^2 \quad -3x \quad +2 \\
3x^2 + 2x - 4\,|\ \overline{6x^4 \quad -5x^3 \quad -8x^2 \quad +16x \quad -8} \\
6x^4 \quad +4x^3 \quad -8x^2 \\
\hline
-9x^3 \quad +16x \\
-9x^3 \quad -6x^2 \quad +12x \\
\hline
6x^2 \quad +4x \quad -8 \\
6x^2 \quad +4x \quad -8 \\
\hline
0
\end{array}
$$

$$2x^2 - 3x + 2, R = 0$$

33.

$$
\begin{array}{r}
2.14 \quad -5.23 \quad -8.71 \quad\quad 6.85 \\
7.21 \quad\quad 6.68 \quad -6.85 \\
\hline
3.37\,|\quad 2.14 \quad\quad 1.98 \quad -2.03 \quad\quad 0.00
\end{array}
$$

$$2.14x^2 + 1.98x - 2.03, R = 0.00$$

35. $\quad\quad\quad$ 0.96 $\quad\quad$ 0 $\quad$ 4.09 $\quad\quad$ 9.44 $\quad$ $-$ 1.87

$\quad\quad\quad\quad\quad\quad\quad\quad$ $-$ 1.32 $\quad$ 1.80 $\quad$ $-$ 8.07 $\quad$ $-$ 1.87

$\quad\quad$ $-1.37\;\rceil\;$ 0.96 $\quad$ $-$ 1.32 $\quad$ 5.89 $\quad\quad$ 1.37 $\quad$ $-$ 3.74

$$0.96x^3 - 1.32x^2 + 5.89x + 1.37,\, R = -3.74$$

37. $\quad\quad$ 1 $\quad\quad$ -3 $\quad\quad\quad$ 1 $\quad$ $-$ 3

$\quad\quad\quad\quad\quad\quad$ i $\quad$ $-3i - 1$ $\quad\quad$ 3

$\quad\quad$ $i\;\rceil\;$ 1 $\quad$ $-3 + i$ $\quad\quad\quad$ $-3i$ $\quad\quad$ 0

$$x^2 + (-3 + i)x - 3i,\, R = 0$$

39. A.

$$\require{enclose}
\begin{array}{r}
a_2x \;\;+\;\; (a_1 + a_2r) \\
x - r\;\enclose{longdiv}{a_2x^2 \;\;+\;\; a_1x \quad\quad\quad + \quad a_0} \\
\underline{a_2x^2 \;\;-\;\; a_2rx} \\
(a_1 + a_2r)x \;\;+\;\; a_0 \\
\underline{(a_1 + a_2r)x \;\;-\;\; r(a_1 + a_2r)} \\
a_0 + r(a_1 + a_2r)
\end{array}$$

$\quad\quad$ a_2 $\quad$ a_1 $\quad\quad\quad\quad$ a_0

$\quad\quad\quad\quad$ a_2r $\quad\quad$ $(a_1 + a_2r)r$

$\quad$ $r\;\rceil\;$ a_2 $\quad$ $a_1 + a_2r$ $\quad$ $a_0 + (a_1 + a_2r)r$

In both cases the coefficient of x is a_2, the constant term $a_2r + a_1$,
and the remainder is $(a_2r + a_1)r + a_0$.

B. The remainder expanded is $a_2r^2 + a_1r + a_0 = P(r)$.

Exercise 6-2

Key Ideas and Formulas

Division Algorithm:

For each polynomial $P(x)$ of degree greater than 0 and each number r, there exists a unique polynomial $Q(x)$ of degree 1 less than $P(x)$ and a unique number R (which may be zero) such that

$$P(x) = (x - r)Q(x) + R$$

$Q(x)$ is called the quotient, $x - r$ the divisor, and R the remainder.

Remainder Theorem:

If R is the remainder after dividing $P(x)$ by $x - r$, then $P(r) = R$.

Factor Theorem:

If r is a zero of the polynomial $P(x)$, then $x - r$ is a factor of $P(x)$, conversely, if $x - r$ is a factor of $P(x)$, then r is a zero of $P(x)$.

A polynomial function with real coefficients is continuous everywhere. Its graph has no holes or gaps.

1.
$$
\begin{array}{r r r}
 & 3 & -1 & -10 \\
 & & -6 & 14 \\
\hline
-2 \,| & 3 & -7 & 4
\end{array}
$$

$$P(-2) = 4$$

3.
$$
\begin{array}{r r r r}
 & 2 & -5 & 7 & -7 \\
 & & 4 & -2 & 10 \\
\hline
2 \,| & 2 & -1 & 5 & 3
\end{array}
$$

$$P(2) = 3$$

5.
$$
\begin{array}{r r r r r}
 & 1 & 0 & -10 & 25 & -2 \\
 & & -4 & 16 & -24 & -4 \\
\hline
-4 \,| & 1 & -4 & 6 & 1 & -6
\end{array}
$$

$$P(-4) = -6$$

7. $3, -5$

9. $-\frac{1}{2}, 8, -2$

11. $x - 1$ will be a factor of $P(x)$ if $P(1) = 0$. Since $P(x) = x^{18} - 1$, $P(1) = 1^{18} - 1 = 0$. Therefore $x - 1$ is a factor of $x^{18} - 1$.

13. $x + 1$ will be a factor of $P(x)$ if $P(-1) = 0$. Since $P(x) = 3x^3 - 7x^2 - 8x + 2$, $P(-1) = 3(-1)^3 - 7(-1)^2 - 8(-1) + 2 = -3 - 7 + 8 + 2 = 0$. Therefore $x + 1$ is a factor of $3x^3 - 7x^2 - 8x + 2$.

15.

$$
\begin{array}{r}
\phantom{\tfrac{1}{2}|}\;\;4 \quad -8 \quad\; 5 \quad -4 \\
\phantom{\tfrac{1}{2}|}\;\;\quad\;\;\; 2 \quad -3 \quad\;\; 1 \\
\hline
\tfrac{1}{2}\,\big|\;\;4 \quad -6 \quad\; 2 \quad -3
\end{array}
$$

$$P\!\left(\tfrac{1}{2}\right) = -3$$

17.

$$
\begin{array}{r}
\;1 \quad\; 0 \quad\quad -2 \quad\quad\;\; 1 \\
\;\quad 0.3 \quad\; 0.09 \quad -0.573 \\
\hline
0.3\,\big|\;1 \quad 0.3 \quad -1.91 \quad\; 0.427
\end{array}
$$

$$P(0.3) = 0.427$$

19. We form a synthetic division table:

	1	−5	2	8	
−2	1	−7	16	−24	$= P(-2)$
−1	1	−6	8	0	$= P(-1)$
0	1	−5	2	8	$= P(0)$
1	1	−4	−2	6	$= P(1)$
2	1	−3	−4	0	$= P(2)$
3	1	−2	−4	−4	$= P(3)$
4	1	−1	−2	0	$= P(4)$
5	1	0	2	18	$= P(5)$

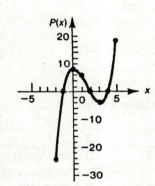

21. We form a synthetic division table:

	1	4	−1	−4	
−5	1	−1	4	16	$= P(-5)$
−4	1	0	−1	0	$= P(-4)$
−3	1	1	−4	8	$= P(-3)$
−2	1	2	−5	6	$= P(-2)$
−1	1	3	−4	0	$= P(-1)$
0	1	4	−1	−4	$= P(-0)$
1	1	5	4	0	$= P(1)$
2	1	6	11	18	$= P(2)$

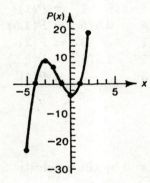

23. We form a synthetic division table:

	−1	2	0	−3	
−2	−1	4	−8	13	$= P(-2)$
−1	−1	3	−3	0	$= P(-1)$
0	−1	2	0	−3	$= P(0)$
1	−1	1	−1	−2	$= P(1)$
2	−1	0	0	−3	$= P(2)$
3	−1	−1	−3	−12	$= P(3)$

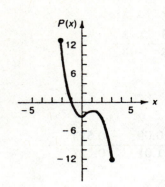

25. We form a synthetic division table:

	−1	3	−3	2	
−1	−1	4	−7	9	$= P(-1)$
−0.5	−1	3.5	−4.75	4.375	$= P(-0.5)$
0	−1	3	0	2	$= P(0)$
0.5	−1	2.5	−1.75	1.125	$= P(0.5)$
1	−1	2	−1	1	$= P(1)$
1.5	−1	1.5	−0.75	0.875	$= P(1.5)$
2	−1	1	−1	0	$= P(2)$
2.5	−1	0.5	−1.75	−2.375	$= P(2.5)$
3	−1	0	−3	−7	$= P(3)$

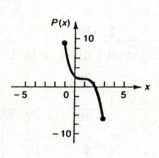

27. $-4, -8, 1$

29. $\frac{1}{8}, -\frac{3}{5}, -4$

31.   We solve $x^2 - 3x + 1 = 0$ by the quadratic formula with $a = 1, b = -3, c = 1$, to get
$$x = \frac{-(3) \pm \sqrt{(-3)^2 - 4(1)(1)}}{2(1)} = \frac{3 \pm \sqrt{5}}{2}$$
Therefore
$$P(x) = [x - (\tfrac{3+\sqrt{5}}{2})][x - (\tfrac{3-\sqrt{5}}{2})]$$

33.   We solve $x^2 - 6x + 10 = 0$ by the quadratic formula with $a = 1, b = -6, c = 10$, to get
$$x = \frac{-(-6) \pm \sqrt{(-6)^2 - 4(1)(10)}}{2(1)} = \frac{6 \pm \sqrt{-4}}{2} = \frac{6 \pm i\sqrt{4}}{2} = \frac{6 \pm 2i}{2} = 3 \pm i. \text{ Therefore}$$
$P(x) = [x - (3 + i)][x - (3 - i)]$.

35. $x - a$ will be a factor of $P(x)$ if $P(a) = 0$. Therefore, if $P(x) = x^n - a^n$, $P(a) = a^n - a^n = 0$
so $x - a$ is a factor of $x^n - a^n$.

37. $x - 1$ will be a factor of $P(x)$ if $P(1) = 0$.
Therefore, if $P(x) = 4x^7 - 2x^6 + x^2 + 2x + 5$, $P(1) = 4 - 2 + 1 + 2 + 5 = 10 \neq 0$,
so $x - 1$ is not a factor.

39. We form a synthetic division table:

	1	−2	−2	8	−8	
−3	1	−5	13	−31	85	$= P(-3)$
−2	1	−4	6	−4	0	$= P(-2)$
−1	1	−3	2	6	−14	$= P(-1)$
0	1	−2	−2	8	−8	$= P(0)$
1	1	−1	−3	5	−3	$= P(1)$
2	1	0	−2	4	0	$= P(2)$
3	1	1	1	11	41	$= P(3)$

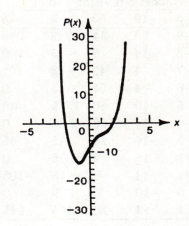

41. We form a synthetic division table:

	1	4	−1	−20	−20	
−4	1	0	−1	−16	44	$= P(-4)$
−3	1	1	−4	−8	4	$= P(-3)$
−2	1	2	−5	−10	0	$= P(-2)$
−1	1	3	−4	−16	−4	$= P(-1)$
0	1	4	−1	−20	−20	$= P(0)$
1	1	5	4	−16	−36	$= P(1)$
2	1	6	11	2	−16	$= P(2)$
3	1	7	20	40	100	$= P(3)$

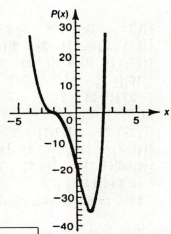

43. We form a synthetic division table:

	−1	2	0	1	20	
−3	−1	5	−15	46	−118	$= P(-3)$
−2	−1	4	−8	17	−34	$= P(-2)$
−1	−1	3	−3	4	16	$= P(-1)$
0	−1	2	0	1	20	$= P(0)$
1	−1	1	1	2	22	$= P(1)$
1.5	−1	0.5	0.75	2.125	23.1875	$= P(1.5)$
2	−1	0	0	1	22	$= P(2)$
3	−1	−1	−3	−8	−4	$= P(3)$
4	−1	−2	−8	−31	−104	$= P(4)$

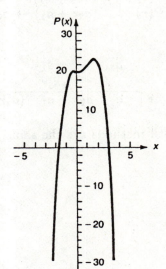

45. We form a synthetic division table:

	1	−5	−10	50	9	-45	
−4	1	−9	26	−54	225	−945	$= P(-4)$
−3	1	−8	14	8	−15	0	$= P(-3)$
−2	1	−7	4	42	−73	101	$= P(-2)$
−1	1	−6	−4	54	−45	0	$= P(-1)$
0	1	−5	−10	50	9	−45	$= P(0)$
1	1	−4	−14	36	45	0	$= P(1)$
2	1	−3	−16	18	45	45	$= P(2)$
3	1	−2	−16	2	15	0	$= P(3)$
4	1	−1	−14	−6	−15	−105	$= P(4)$
5	1	0	−10	0	9	0	$= P(5)$
6	1	1	−4	26	165	945	$= P(6)$

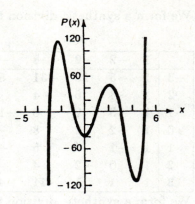

47.
$$
\begin{aligned}
P(x) &= \{[(2x - 3)x + 2]x - 5\}x + 7 \\
P(-2) &= \{[\{2(-2) - 3\}(-2) + 2](-2) - 5\}(-2) + 7 \\
&= \{[\{-7\}(-2) + 2](-2) - 5\}(-2) + 7 \\
&= \{[16](-2) - 5\}(-2) + 7 \\
&= \{-37\}(-2) + 7 \\
&= 81 \\
P(1.7) &= \{[\{2(1.7) - 3\}(1.7) + 2](1.7) - 5\}(1.7) + 7 \\
&= \{[\{0.4\}(1.7) + 2](1.7) - 5\}(1.7) + 7 \\
&= \{[2.68](1.7) - 5\}(1.7) + 7 \\
&= \{-0.444\}(1.7) + 7 \\
&= 6.2452 \text{ or } 6.2 \text{ to two significant digits.}
\end{aligned}
$$

49. A. $P(x) = (a_2x + a_1)x + a_0$
        $P(r) = (a_2r + a_1)r + a_0$

   B.

$$
\begin{array}{c|cccc}
 & a_2 & a_1 & a_0 & \\
 & & a_2r & (a_2r + a_1)r & \\
\hline
r & a_2 & a_2r + a_1 & (a_2r + a_1)r + a_0 & \quad P(r) = (a_2r + a_1)r + a_0
\end{array}
$$

Both methods are the same.

Exercise 6-3

Key Ideas and Formulas

Factor Theorem:

If r is a zero of the polynomial $P(x)$, then $x - r$ is a factor of $P(x)$; conversely, if $x - r$ is a factor of $P(x)$, then r is a zero of $P(x)$.

Fundamental Theorem of Algebra:

Every polynomial $P(x)$ of degree $n > 0$, with real or imaginary coefficients, has at least one real or imaginary zero.

n Zeros Theorem:

Every polynomial $P(x)$ of degree $n > 0$, with real or imaginary coefficients can be expressed as the product of n linear factors (hence, has exactly n zeros – not necessarily distinct.)

Imaginary Zeros Theorem:

Imaginary zeros of polynomials with real coefficients, if they exist, occur in conjugate pairs. A polynomial of odd degree with real coefficients always has at least one real zero.

---

1. -8 (multiplicity 3), 6 (multiplicity 2); degree of $P(x)$ is 5

3. -4 (multiplicity 3), 3 (multiplicity 2); -1; degree of $P(x)$ is 6.

5. $P(x) = (x - 3)^2(x + 4)$; degree 3

7. $P(x) = (x + 7)^3[x - (-3 + \sqrt{2})][x - (-3 - \sqrt{2})]$; degree 5

9. $P(x) = [x - (2 - 3i)][x - (2 + 3i)](x + 4)^2$; degree 4

11.   The degree of $P(x)$ is 3. Therefore, there could be as many as 3 real roots.
    If imaginary roots occur, they do so in pairs, so there could be a pair of
    imaginary roots, leaving one real root.

13.   The degree of $P(x)$ is 6. Therefore, there could be as many as 6 real roots.
    One pair of imaginary roots would lead to 2 imaginary, 4 real roots.
    Two pairs of imaginary roots would lead to 4 imaginary, 2 real roots.
    Three pairs of imaginary roots would lead to 6 imaginary roots.

15.   The degree of $P(x)$ is 5. Therefore, there could be as many as 5 real roots.
    One pair of imaginary roots would lead to 2 imaginary, 3 real roots.
    Two pairs of imaginary roots would lead to 4 imaginary, 1 real root.

17.   The degree of $P(x)$ is 4. Therefore, there could be as many as 4 real roots.
    One pair of imaginary roots would lead to 2 imaginary, 2 real roots.
    Two pairs of imaginary roots would lead to 4 imaginary roots.

19.   Since -1 is a zero of $P(x)$, we can write
$$P(x) = (x+1)Q(x),$$
and divide synthetically to find $Q(x)$.

$$
\begin{array}{r|rrrr}
 & 1 & 9 & 24 & 16 \\
 &   & -1 & -8 & -16 \\
\hline
-1 & 1 & 8 & 16 & 0 \\
\end{array}
$$

Since $Q(x) = x^2 + 8x + 16 = (x+4)^2$, we have
$$P(x) = (x+1)(x+4)^2$$

21.   Since -1 and 1 are zeros of $P(x)$, we can write
$$P(x) = (x+1)(x-1)Q(x) = (x^2-1)Q(x)$$
$Q(x)$ is clearly $x^2 + 1$ (check by division) or $x^2 - i^2 = (x-i)(x+i)$, so
$$P(x) = (x+1)(x-1)(x-i)(x+i)$$

23.   Since $\frac{1}{2}$ is a zero of $P(x)$, we can write
$$P(x) = (x - \tfrac{1}{2})Q(x) = (2x-1)\tfrac{1}{2}Q(x)$$
and divide synthetically to find $Q(x)$.

$$
\begin{array}{r|rrrr}
 & 2 & -17 & 90 & -41 \\
 &   & 1 & -8 & 41 \\
\hline
\frac{1}{2} & 2 & -16 & 82 & 0 \\
\end{array}
$$

So $Q(x) = 2x^2 - 16x + 82$, and $\frac{1}{2}Q(x) = x^2 - 8x + 41 = x^2 - 8x + 16 + 25$
$$\tfrac{1}{2}Q(x) = (x-4)^2 + 5^2 = (x-4)^2 - (5i)^2 = (x-4+5i)(x-4-5i)$$
So $P(x) = (2x-1)[x - (4-5i)][x - (4+5i)]$

25.   The degree of this polynomial is 4. Therefore, as explained in problem 17, there could be four real roots, two real and two imaginary roots, or four imaginary roots.

27.   The degree of this polynomial is 3. Therefore, as explained in problem 11, there could be three real roots, or one real and two imaginary roots.

29.   The degree of this polynomial is 6. Therefore, as explained in problem 13, there could be six real roots, two imaginary and four real roots, two real and four imaginary roots, or six imaginary roots.

31.   The degree of this polynomial is 5. Therefore, as explained in problem 15, there could be five real roots, three real and two imaginary roots, or one real and four imaginary roots.

33.
$$
\begin{aligned}
[x - (4-5i)][x - (4+5i)] &= [x - 4 + 5i][x - 4 - 5i] \\
&= [(x-4) + 5i][(x-4) - 5i] \\
&= (x-4)^2 - 25i^2 \\
&= x^2 - 8x + 16 + 25 \\
&= x^2 - 8x + 41
\end{aligned}
$$

35.
$$
\begin{aligned}
[x - (3 + 4i)][x - (3 - 4i)] &= [x - 3 - 4i][x - 3 + 4i] \\
&= [(x - 3) - 4i][(x - 3) + 4i] \\
&= (x - 3)^2 - 16i^2 \\
&= x^2 - 6x + 9 + 16 \\
&= x^2 - 6x + 25
\end{aligned}
$$

37.
$$
\begin{aligned}
[x - (a + bi)][x - (a - bi)] &= [x - a - bi][x - a + bi] \\
&= [(x - a) - bi][(x - a) + bi] \\
&= (x - a)^2 - b^2 i^2 \\
&= (x - a)^2 + b^2 \\
&= x^2 - 2ax + a^2 + b^2
\end{aligned}
$$

39.   If $3 - i$ is a zero, then $3 + i$ is a zero. So $Q(x) = [x - (3 - i)][x - (3 + i)]$ divides $P(x)$ evenly. Applying problem 37, $Q(x) = x^2 - 6x + 9 + 1 = x^2 - 6x + 10$. Dividing, we see

$$
\begin{array}{r}
x + \ 1 \\
x^2 - 6x + 10 \,\overline{\smash{\big)}\, x^3 - 5x^2 + 4x + 10} \\
\underline{x^3 - 6x^2 + 10x } \\
x^2 - 6x + 10 \\
\underline{x^2 - 6x + 10} \\
0
\end{array}
$$

So $P(x) = (x + 1)Q(x)$ and the two other zeros are -1 and $3 + i$.

41.   If $-5i$ is a zero, then so is $5i$. So $Q(x) = (x - 5i)(x + 5i)$ divides $P(x)$ evenly. $Q(x) = x^2 - 25i^2 = x^2 + 25$. Dividing, we see

$$
\begin{array}{r}
x - 3 \\
x^2 + 25 \,\overline{\smash{\big)}\, x^3 - 3x^2 + 25x - 75} \\
\underline{x^3 + 25x } \\
- 3x^2 - 75 \\
\underline{- 3x^2 - 75} \\
0
\end{array}
$$

So $P(x) = (x - 3)Q(x)$ and the two other zeros are $5i$ and 3.

43.   If $2 + i$ is a zero, then $2 - i$ is a zero. So $Q(x) = [x - (2 + i)][x - (2 - i)]$ divides $P(x)$ evenly. Applying problem 37, $Q(x) = x^2 - 4x + 4 + 1 = x^2 - 4x + 5$. Dividing, we see

$$
\begin{array}{r}
x^2 - 2 \\
x^2 - 4x + 5 \,\overline{\smash{\big)}\, x^4 - 4x^3 + 3x^2 + 8x - 10} \\
\underline{x^4 - 4x^3 + 5x^2 } \\
- 2x^2 + 8x - 10 \\
\underline{- 2x^2 + 8x - 10} \\
0
\end{array}
$$

So $P(x) = Q(x)(x^2 - 2)$. $x^2 - 2$ has two zeros: $\sqrt{2}$ and $-\sqrt{2}$. Summarizing, $P(x)$ has 4 zeros: $2 + i, 2 - i, \sqrt{2}, -\sqrt{2}$.

45.   A. Since there are 3 zeros of $x^3 - 1$, there are 3 cube roots of 1.

       B. $x^3 - 1 = (x - 1)(x^2 + x + 1)$. The other cube roots of 1 will be solutions to $x^2 + x + 1 = 0$. Applying the quadratic formula with $a = b = c = 1$, we have

$x = \frac{-1 \pm \sqrt{(1)^2 - 4(1)(1)}}{2(1)} = \frac{-1 \pm \sqrt{-3}}{2} = \frac{-1 \pm i\sqrt{3}}{2}$. Thus $-\frac{1}{2} + \frac{\sqrt{3}}{2}i$ and $-\frac{1}{2} + \frac{\sqrt{3}}{2}i$ are the other cube roots of 1.

47. $P(x)$ can have at most n and must have at least one real zero. Each zero of $P(x)$ represents a point where $P(x) = y = 0$ so the graph of $P(x)$ will cross the x-axis at and only at zeros of $P(x)$. Thus there can be a maximum of n axis crossings and there is a minimum of 1 axis crossing.

49. $$\begin{aligned} P(2+i) &= (2+i)^2 + 2i(2+i) - 5 \\ &= 4 + 4i + i^2 + 4i + 2i^2 - 5 \\ &= 4 + 4i - 1 + 4i - 2 - 5 \\ &= -4 + 8i \end{aligned}$$

So $P(2+i) \neq 0$ and $2+i$ is not a zero of $P(x)$. This does not contradict the theorem, since $P(x)$ is not a polynomial with real coefficients (the coefficient of x is the imaginary number $2i$).

Exercise 6-4

Key Ideas and Formulas

Descartes' Rule of Signs:

Given a polynomial $P(x)$ with real coefficients:

1. The number of positive zeros of $P(x)$ is never greater than the number of variations in sign in $P(x)$ and, if less, then always by an even number.

2. The number of negative zeros of $P(x)$ is never greater than the number of variations in sign in $P(-x)$ and, if less, then always by an even number.

Upper and Lower Bound of Real Zeros:

Given an nth degree polynomial $P(x)$ with real coefficients, $n > 0$, $a_n > 0$ and $P(x)$ divided by $x - r$ using synthetic division:

1. **Upper bound.** If $r > 0$ and all numbers in the quotient row of the synthetic division (including the remainder) are nonnegative, then r is an upper bound of the real zeros of $P(x)$.

2. **Lower Bound.** If $r < 0$ and all numbers in the quotient row of the synthetic division (including the remainder) alternate in sign, then r is a lower bound of the real zeros of $P(x)$.

Location Theorem:

If $P(x)$ is a polynomial with real coefficients, and if $P(a)$ and $P(b)$ are of opposite sign, then there is at least one real zero between a and b.

---

1. Apply Descartes' Rule of Signs: $P(x) = 2x^2 + x - 4$ 1 variation in sign

   $$P(-x) = 2x^2 - x - 4 \quad \text{1 variation in sign}$$

Common Errors:

Investigate $P(-x)$, *not* $-P(x)$ to determine possible negative zeros of $P(x)$.

Also, we are interested in sign variations *within* $P(x)$ and $P(-x)$.

Students often incorrectly count sign variations *between* $P(x)$ and $P(-x)$, which are irrelevant.

Possible Combinations of Zeros:

+	−	I
1	1	0

3.   Apply Descartes' Rule of Signs: $M(x) = 7x^2 + 2x + 4$ 0 variations in sign

$$M(-x) = 7x^2 - 2x + 4 \quad \text{2 variations in sign}$$

Possible Combinations of Zeros:

+	−	I
0	2	0
0	0	2

5. Apply Descartes' Rule of Signs: $Q(x) = 2x^3 - 4x^2 + x - 3$ 3   variations in sign

$$Q(x) = -2x^3 - 4x^2 - x - 3 \quad \text{0 variations in sign}$$

Possible Combinations of Zeros:

+	−	I
3	0	0
1	0	2

7.   We form a synthetic division table:

	1	−2	3		
1	1	−1	2		
2	1	0	3	→	This quotient row is non-negative; hence 2 is an upper bound.
−1	1	−3	6	→	This quotient row alternates in sign; hence, −1 is a lower bound.

9.   We form a synthetic division table:

	1	0	−3	5		
1	1	1	−2	3		
2	1	2	1	7	→	This quotient row is non-negative; hence 2 is an upper bound.
−1	1	−1	−2	7		
−2	1	−2	1	3		
−3	1	−3	6	−3	→	This quotient row alternates in sign; hence, −3 is a lower bound.

11.   We form a synthetic division table:

	1	0	−1	3	2	
1	1	1	0	3	2	→ This quotient row is non-negative hence, 1 is an upper bound.
−1	1	−1	0	3	−1	
−2	1	−2	3	−3	8	→ This quotient row alternates in sign, hence, −2 is a lower bound.

13. $P(3) = 3^2 - 3(3) - 2 = -2$

$P(4) = 4^2 - 3(4) - 2 = 2$

Since $P(3)$ and $P(4)$ have opposite signs, there is at least one real zero between 3 and 4.

15.   We form a synthetic division table:

	1	0	−3	5		
−3	1	−3	6	−13	=	$P(-3)$
−2	1	−2	1	3	=	$P(-2)$

Since $P(-3)$ and $P(-2)$ have opposite signs, there is at least one real zero between −3 and −2.

17.   We form a synthetic division table:

	1	−3	−3	9		
1	1	−2	−5	4	=	$P(1)$
2	1	−1	−5	−1	=	$P(2)$

Since $P(1)$ and $P(2)$ have opposite signs, there is at least one real zero between 1 and 2.

19.   A. Apply Descartes' Rule of Signs: $P(x) = x^3 - x^2 - 6x + 6$ 2 variations in sign

$P(-x) = -x^3 - x^2 + 6x + 6$ 1 variation in sign

Possible Combinations of Zeros:

+	−	I
2	1	0
0	1	2

B. We form a synthetic division table:

	1	-1	-6	6	
1	1	0	-6	0	$= P(1)$
2	1	1	-4	-2	$= P(2)$
3	1	2	0	6	$= P(3)$ → This quotient row is non-negative; hence, 3 is an upper bound.
0	1	-1	-6	6	$= P(0)$
-1	1	-2	-4	10	$= P(-1)$
-2	1	-3	0	6	$= P(-2)$
-3	1	-4	6	-12	$= P(-3)$ → This quotient row alternates in sign; hence, -3 is a lower bound.

C. The table shows clearly that 1 is a zero. Since $P(2)$ and $P(3)$ have opposite signs, there is one real zero in the interval $(2,3)$. Similarly, there is one real zero in the interval $(-3,-2)$.

21.   Apply Descartes' Rule of Signs: $P(x) \cong x^3 - 2x - 6$ 1 variation in sign

$$P(-x) = -x^3 + 2x - 6 \quad 2 \text{ variations in sign}$$

Possible combinations of zeros:

+	−	I
1	2	0
1	0	2

B. We form a synthetic division table:

	1	0	-2	-6	
1	1	1	-1	-7	$= P(1)$
2	1	2	2	-2	$= P(2)$
3	1	3	7	15	$= P(3)$ → This quotient row is non-negative; hence, 3 is an upper bound.
0	1	0	-2	-6	$= P(0)$
-1	1	-1	-1	-5	$= P(-1)$
-2	1	-2	2	-10	$= P(-2)$ → This quotient row alternates in sign; hence, -2 is a lower bound.

C. The table shows that $P(2)$ and $P(3)$ have opposite signs; hence, there is one real zero in the interval $(2,3)$.

23.   A. Apply Descartes' Rule of Signs: $P(x) = x^4 + 4x^3 - 2x^2 - 12x - 3$ 1 variation in sign

$$P(-x) = x^4 - 4x^3 - 2x^2 + 12x - 3 \quad 3 \text{ variations in sign}$$

Possible combinations of zeros:

+	−	I
1	3	0
1	1	2

B. We form a synthetic division table:

	1	4	2	−12	−3	
1	1	5	3	−9	−12	$= P(1)$
2	1	6	10	8	13	$= P(2)$ → This quotient row is non-negative; hence, 2 is an upper bound.
0	1	4	−2	−12	−3	$= P(0)$
−1	1	3	−5	−7	4	$= P(-1)$
−2	1	2	−6	0	−3	$= P(-2)$
−3	1	1	−5	3	−12	$= P(-3)$
−4	1	0	−2	−4	13	$= P(-4)$
−5	1	−1	3	−27	132	$= P(-5)$ → This quotient row alternates in sign; hence, −5 is a lower bound.

C. The table shows opposite signs for $P(1)$ and $P(2)$, $P(-1)$ and $P(0)$, $P(-2)$ and $P(-1)$, and $P(-4)$ and $P(-3)$. We conclude that there is one real zero in each of the four intervals $(1,2), (-1,0), (-2,-1),$ and $(-4,-3)$.

25. **A. Apply Descartes' Rule of Signs:** $P(x) = x^5 - 3x^3 + 2x - 5$ 3 variations in sign

$$P(-x) = -x^5 + 3x^3 - 2x - 5 \quad 2 \text{ variations in sign}$$

Possible combinations of zeros:

+	−	I
3	2	0
3	0	2
1	2	2
1	0	4

B. We form a synthetic division table:

	1	0	−3	0	2	−5	
1	1	1	−2	−2	0	−5	
2	1	2	1	2	6	7	→ This quotient row is non-negative; hence, 2 is an upper bound.
0	1	0	−3	0	2	−5	
−1	1	−1	−2	2	0	−5	
−2	1	−2	1	−2	6	−17	→ This quotient row alternates in sign; hence, −2 is a lower bound.

C. The table shows opposite signs for $P(1)$ and $P(2)$. We conclude that there is at least one real zero in the interval $(1,2)$.

27. A. Apply Descartes' Rule of Signs: $P(x) = x^5 + 2x^4 + 3x^2 - 2x - 8$ 1   variation in sign

$P(-x) = -x^5 + 2x^4 + 3x^2 + 2x - 8$ 2   variations in sign

Possible combinations of zeros:

+	−	I
1	2	2
1	0	4

B. We form a synthetic division table

	1	2	0	3	−2	−8	
1	1	3	3	6	4	−4	$= P(1)$
2	1	4	8	19	36	64	$= P(2)$ → This quotient row is non-negative; hence, 2 is an upper bound.
0	1	2	0	3	−2	−8	$= P(0)$
−1	1	1	−1	4	−6	−2	$= P(-1)$
−2	1	0	0	3	−8	8	$= P(-2)$
−3	1	−1	3	−6	16	−56	$-P(-3)$ → This quotient row alternates in sign; hence, −3 is a lower bound.

C. The table shows opposite signs for $P(1)$ and $P(2)$, $P(-2)$ and $P(-1)$, and $P(-3)$ and $P(-2)$.

We conclude that there is one real zero in each of the three intervals $(1,2), (-2,-1),$ and $(-3,-2)$.

29. A. Apply Descartes' Rule of Signs: $P(x) = x^6 - x^5 - 2x^3 + 3x - 5$ 3   variations in sign

$P(-x) = x^6 + x^5 + 2x^3 - 3x - 5$ 1   variation in sign

Possible combinations of zeros:

+	−	I
3	1	2
1	1	4

B. We form a synthetic division table:

	1	−1	0	−2	0	3	−5		
1	1	0	0	−2	−2	1	−4	$= P(1)$	
2	1	1	2	2	4	11	17	$= P(2)$	→ This quotient row is non-negative; hence, 2 is an upper bound.
0	1	−1	0	−2	0	3	−5	$= P(0)$	
−1	1	−2	2	−4	4	−1	−4	$= P(−1)$	
−2	1	−3	6	−14	28	−53	101	$= P(−2)$	→ This quotient row alternates in sign; hence, −2 is a lower bound.

C. The table shows that $P(1)$ and $P(2)$ have opposite signs; hence, there is at least one real zero in the interval $(1,2)$. $P(−2)$ and $P(−1)$ have opposite signs; hence there is at least one real zero in the interval $(−2,−1)$; since there can be only 1 negative zero, we conclude that there is exactly one zero in this interval.

31. $P(x) \;=\; x^4 + 3x^2 - x - 5$ 1   variation in sign.

   $P(-x) \;=\; x^4 + 3x^2 + x - 5$ 1   variation in sign.

   $P(0) \;=\; -5 \neq 0$

By Descartes' rule of signs, $P(x)$ has one positive and one negative real zero. By the fundamental theorem of algebra, $P(x)$ has four zeros. 0 is not a zero, so the two remaining zeros must be imaginary.

33. $P(x) \;=\; x^5 + 3x^3 + x$ no variations in sign

   $P(-x) \;=\; -x^5 - 3x^3 - x$ no variations in sign

   $P(0) \;=\; 0$

By Descartes' rule of signs, $P(x)$ has no positive or negative zeros. Since $P(0) = 0$ the graph crosses the x-axis at the origin, but nowhere else since it has no other real zeros.

Exercise 6-5

Key Ideas and Formulas

Rational Zero Theorem:

If the rational number $\frac{b}{c}$, in lowest terms, is a zero of the polynomial

$$P(x) = a_n x^n + a_{n-1} x^{n-1} + \cdots + a_1 x + a_0 \qquad a_n \neq 0$$

with integer coefficients, then b must be an integer factor of a_0 and c must be an integer factor of a_n.

Strategy for Finding Rational Zeros:

Assume $P(x)$ is a polynomial with integer coefficients and is of degree greater than two

Step 1. List possible rational zeros of $P(x)$ using the Rational Zero Theorem.

Step 2. Make a table of possible combinations of positive, negative, and imaginary zeros using Descartes' Rule of Signs and earlier theorems.
Use the results to reduce the size of the list from Step 1, if possible.

Step 3. Construct a synthetic division table.

    A.   If a rational zero is found at any time, stop, write
        $P(x) = (x - r)Q(x)$
        and proceed to find the rational zeros of $Q(x)$, the reduced polynomial.
        If $Q(x)$ is quadratic, find *all* its zeros.

    B.   If $P(x)$ changes sign, try the possible rational zeros from the list that are
        between the two integers producing the sign change.

    C.   If lower or upper bounds are found, modify the possible rational zero
        list appropriately.

Step 4. Test any untested numbers from the list of remaining possible zeros.
    If no rational number from the list is a zero, then $P(x)$ [or $Q(x)$] has no rational zeros.
    The remaining zeros are irrational or imaginary.

1.   A. Possible factors of 6 are $\pm1, \pm2, \pm3, \pm6$. Possible factors of 1 are ±1.
Therefore the possible rational zeros are $\pm1, \pm2, \pm3, \pm6$.

B. Descartes' rule of signs gives the following possible combinations of zeros:

+	−	I
2	1	0
0	1	2

Starting with 1, we find

	1	−2	−5	6
		1	−1	−6
1	1	−1	−6	0

$P(x) = (x - 1)(x^2 - x - 6)$
Since $x^2 - x - 6$ is factorable, we have
$P(x) = (x - 1)(x + 2)(x - 3)$
Therefore the rational zeros are $-2, 1, 3$.

3.   A. Possible factors of 4 are $\pm1, \pm2 \pm 4$. Possible factors of 3 are $\pm1, \pm3$.
Therefore the possible rational zeros are $\pm1, \pm2, \pm4, \pm\frac{1}{3}, \pm\frac{2}{3}, \pm\frac{4}{3}$.

B. Descartes' rule of signs gives the following possible combinations of zeros:

+	−	I
2	1	0
0	1	2

We form a synthetic division table:

	3	−11	8	4	
1	3	−8	0	7	
2	3	−5	−2	0	$= P(2)$

So $P(x) = (x - 2)(3x^2 - 5x - 2) = (x - 2)(3x + 1)(x - 2)$
Hence 2 is a double zero, and $-\frac{1}{3}$ is the last zero.

5.   A. Possible factors of 3 are $\pm1, \pm3$.
Possible factors of 12 are $\pm1, \pm2, \pm3, \pm4, \pm6, \pm12$.
Therefore the possible rational zeros are $\pm1, \pm3, \pm\frac{1}{2}, \pm\frac{3}{2}, \pm\frac{1}{3}, \pm\frac{1}{4}, \pm\frac{3}{4}, \pm\frac{1}{6}, \pm\frac{1}{12}$.

B. Descartes' rule of signs gives the following possible combinations of zeros:

+	−	I
2	1	0
0	1	2

We form a synthetic division table:

	12	−16	−5	3	
1	12	−4	−9	−6	
3	12	20	55	168	
−1	12	−28	23	−20	lower bound, eliminating −3 and −$\frac{3}{2}$
$\frac{1}{2}$	12	−10	−10	−2	
$\frac{3}{2}$	12	2	−2	0	a zero

$$\begin{aligned} \text{So } P(x) &= (x - \tfrac{3}{2})(12x^2 + 2x - 2) \\ &= (x - \tfrac{3}{2})2(6x^2 + 2x - 1) \\ &= (2x - 3)(3x - 1)(2x + 1) \end{aligned}$$

So the zeros of $P(x)$ are $\frac{3}{2}, \frac{1}{3}, -\frac{1}{2}$.

7. A. Possible factors of 3 are $\pm 1, \pm 3$. Possible factors of −4 are $\pm 1, \pm 2, \pm 4$.
Hence the possible rational zeros are $\pm 1, \pm 2, \pm 4, \pm \frac{1}{3}, \pm \frac{2}{3}, \pm \frac{4}{3}$.
B. Descartes' rule of signs gives the following possible combinations of zeros.

+	−	I
1	2	0
1	0	2

We form a synthetic division table:

	3	7	−10	−4	
1	3	10	0	−4	
2	3	13	16	28	2 is an upper bound, eliminating 4
$\frac{1}{3}$	3	8	$-10 + \frac{8}{3}$		abandon any row when fractions appear
$\frac{2}{3}$	3	9	−4	$-4 - \frac{8}{3}$	(unless merely trying to isolate zeros)
$\frac{4}{3}$	3	11	$-10 + \frac{44}{3}$		
$-\frac{1}{3}$	3	6	−12	0	$-\frac{1}{3}$ is a zero

$$\begin{aligned} \text{So } P(x) &= (x + \tfrac{1}{3})(3x^2 + 6x - 12) \\ &= (x + \tfrac{1}{3})3(x^2 + 2x - 4) \end{aligned}$$

Since $x^2 + 2x - 4$ cannot be factored in the integers, there are no more rational zeros.

9.   A. Possible factors of 6 are $\pm 1, \pm 2, \pm 3, \pm 6$; these are the possible
rational zeros.
B. Descartes' rule of signs gives the following possible combinations of zeros:

+	−	I
2	1	0
0	1	2

We form a synthetic division table:

	1	0	−3	6	
1	1	1	−2	4	
2	1	2	1	8	2 is an upper bound, eliminating 3 and 6
−1	1	−1	−2	8	
−2	1	−2	1	4	
−3	1	−3	6	−12	−3 is a lower bound, eliminating −6

Since all possible rational zeros have been eliminated, we conclude that the
polynomial has no rational zeros.

11.   A. Possible factors of −8 are $\pm 1, \pm 2, \pm 4, \pm 8$, these are the possible rational zeros.
B. Descartes' rule of signs gives the following possible combinations of zeros:

+	−	I
3	1	0
1	1	2

We form a synthetic division table:

	1	−2	−2	8	−8	
1	1	−1	−3	5	−3	
2	1	0	−2	4	0	2 is a zero

We now examine $x^3 - 2x + 4$. Only $\pm 2, \pm 4$ and -1 remain as possible rational zeros;
Descartes' rule of signs applied to this reduced polynomial gives the following
possible combinations of zeros:

+	−	I
2	1	0
0	1	2

We form a synthetic division table for the reduced polynomial:

	0	1	−2	4	
2	1	2	0	4	this quotient row is non-negative, eliminating 4
−1	1	−1	−3	7	
−2	1	−2	−4	0	−2 is a zero

So $P(x) = (x-2)(x+2)(x^2 - 2x - 4)$.

Since $x^2 - 2x - 4$ is not factorable in the integers, the only rational zeros are −2 and 2.

13. A. Possible factors of 6 are $\pm 1, \pm 2, \pm 3, \pm 6$. Possible factors of 3 are $\pm 1, \pm 3$. Hence the possible rational zeros are $\pm 1, \pm 2, \pm 3, \pm 6, +\frac{1}{3}, +\frac{2}{3}$.

B. Descartes' rule of signs gives the following possible combinations of zeros:

+	−	I
2	2	0
0	2	2
2	0	2
0	0	4

We form a synthetic division table:

	3	−8	−6	17	6	
1	3	−5	−11	6	12	
2	3	−2	−10	−3	0	2 is a zero

We now examine $3x^3 - 2x^2 - 10x - 3$. Only $-1, \pm 3, \pm \frac{1}{3}$ remain as possible rational zeros; Descartes' rule of signs applied to this reduced polynomial gives the following possible combinations of zeros:

+	−	I
1	2	0
1	0	2

We form a synthetic division table for the reduced polynomial:

	3	−2	−10	−3	
3	3	7	11	30	
−1	3	−5	−5	−2	
−3	3	−11	23	−70	
$\frac{1}{3}$	3	−1	$-10 - \frac{1}{3} \ldots$		
$-\frac{1}{3}$	3	−3	−9	0	$-\frac{1}{3}$ is a zero

So $P(x) = (x-2)(x+\frac{1}{3})(3x^2 - 3x - 9) = (x-2)(x+\frac{1}{3})3(x^2 - x - 3)$.

Since $x^2 - x - 3$ is not factorable in the integers, the only rational zeros are $2, -\frac{1}{3}$.

15. Let $P(x) = 2x^3 - 5x^2 + 1$. Possible rational zeros: $\pm 1, +\frac{1}{2}$.

Descartes' rule of signs applied gives the following possible combinations of zeros:

+	−	I
2	1	0
0	1	2

We form a synthetic division table:

	2	−5	0	1	
1	2	−3	−3	−2	
−1	2	−7	7	−6	
$\frac{1}{2}$	2	−4	−2	0	$\frac{1}{2}$ is a zero

$$
\begin{aligned}
\text{So } P(x) &= (x - \tfrac{1}{2})(2x^2 - 4x - 2) \\
&= (2x - 1)(x^2 - 2x - 1)
\end{aligned}
$$

To find the remaining zeros, we solve $x^2 - 2x - 1 = 0$, by completing the square:

$$
\begin{aligned}
x^2 - 2x &= 1 \\
x^2 - 2x + 1 &= 2 \\
(x - 1)^2 &= 2 \\
x - 1 &= \pm\sqrt{2} \\
x &= 1 \pm \sqrt{2}
\end{aligned}
$$

Hence the zeros are $\frac{1}{2}, 1 \pm \sqrt{2}$. These are the roots of the equation.

17. Let $P(x) = x^4 + 4x^3 - x^2 - 20x - 20$.

Possible rational zeros: $\pm 1, \pm 2, \pm 4, \pm 5, \pm 10, \pm 20$.

Descartes' rule of signs gives the following possible combinations of zeros:

+	−	I
1	3	0
1	1	2

We form a synthetic division table:

	1	4	−1	−20	−20	
1	1	5	4	−16	−36	
2	1	6	11	2	−16	
4	1	8	31	104	396	upper bound, eliminating 5, 10, 20
−1	1	3	−4	−16	−4	
−2	1	2	−5	−10	0	−2 is a zero

We now examine $x^3 + 2x^2 - 5x - 10$. The only remaining possible rational zeros are $-2, -5, -10$. Descartes' rule of signs applied to this reduced polynomial gives the following possible combinations of zeros:

+	−	I
1	2	0
1	0	2

We form a synthetic division table for the reduced polynomial:

	1	2	−5	−10	
−2	1	0	−5	0	−2 is a zero.

Common Error:
Students often miss double zeros. It is necessary to test whether -2 is again a zero of the reduced polynomial.
So $P(x) = (x+2)^2(x^2 - 5) = (x+2)^2(x - \sqrt{5})(x + \sqrt{5})$. So the zeros of the polynomial are -2 (double), $\pm\sqrt{5}$. These are the roots of the equation.

19.   Let $P(x) = x^4 - 2x^3 - 5x^2 + 8x + 4$. Possible rational zeros: $\pm 1, \pm 2, \pm 4$.
Descartes' rule of signs gives the following possible combinations of zeros:

+	−	I
2	2	0
0	2	2
2	0	2
0	0	4

We form a synthetic division table:

	1	−2	−5	8	4	
1	1	−1	−6	2	6	
2	1	0	−5	−2	0	2 is a zero.

We now examine $x^3 - 5x - 2$. The only remaining possible rational zeros are $-1, \pm 2$. Descartes' rule of signs applied to this reduced polynomial gives the following possible combinations of zeros:

+	−	I
1	2	0
1	0	2

We form a synthetic division table for the reduced polynomial:

	1	0	−5	−2	
2	1	2	−1	−4	
−1	1	−1	−4	2	
−2	1	−2	−1	0	−2 is a zero

So $P(x) = (x-2)(x+2)(x^2 - 2x - 1)$. The zeros of $x^2 - 2x - 1$ are $1 \pm \sqrt{2}$ (see problem 15). Hence the zeros of the polynomial are $\pm 2, 1 \pm \sqrt{2}$.
These are the roots of the equation.

21. Let $P(x) = 2x^5 - 3x^4 - 2x + 3$. The possible rational zeros are $\pm 1, \pm 3, \pm \frac{1}{2}, \pm \frac{3}{2}$.
Descartes' rule of signs gives the following possible combinations of zeros:

+	−	I
2	1	2
0	1	4

This indicates at least one pair of imaginary zeros.
We form a synthetic division table:

	2	−3	0	0	−2	3	
1	2	−1	−1	−1	−3	0	1 is a zero

We examine $2x^4 - x^3 - x^2 - x - 3$. The possible rational zeros are the same.
Descartes' rule of signs applied to this reduced polynomial, together with the
fact that the original polynomial has 2 imaginary zeros, leaves 1 positive, one negative,
and 2 imaginary zeros. We form a synthetic division table for the reduced polynomial.

	2	−1	−1	−1	−3	
1	2	1	0	−1	−4	
3	2	5	14	41	120	
−1	2	−3	2	−3	0	−1 is a zero

We examine $Q(x) = 2x^3 - 3x^2 + 2x - 3$. This clearly factors by grouping into $(2x - 3)(x^2 + 1)$,
but if we don't notice this we can proceed as before. Then the possible remaining
rational zeros are $-1, -3, \pm \frac{1}{2}, \pm \frac{3}{2}$. The negative possibilities are eliminated by
Descartes' rule of signs:
$Q(-x) = -2x^3 - 3x^2 - 2x - 3$ has no variations in sign.
We form a synthetic division table for $Q(x)$:

	2	−3	2	−3	
$\frac{1}{2}$	2	−2	1	$-\frac{5}{2}$	
$\frac{3}{2}$	2	0	2	0	$\frac{3}{2}$ is a zero

$$
\begin{aligned}
\text{So } P(x) &= (x-1)(x+1)(x-\tfrac{3}{2})(2x^2+2) \\
&= (x-1)(x+1)(x-\tfrac{3}{2})2(x^2+1) \\
&= (x-1)(x+1)(2x-3)(x^2+1) \\
&= (x-1)(x+1)(2x-3)(x-i)(x+i)
\end{aligned}
$$

The zeros are $\pm 1, \frac{3}{2}, \pm i$. These are the roots of the equation.

23.   The possible rational zeros are $\pm 1, \pm 2, \pm 4, \pm \frac{1}{2}, \pm \frac{1}{3}, \pm \frac{2}{3}, \pm \frac{4}{3}, \pm \frac{1}{6}$.
   Descartes' rule of signs gives the following possible combinations of zeros:

+	−	I
1	2	0
1	0	2

We form a synthetic division table:

	6	13	0	−4	
1	6	19	19	15	upper bound, eliminating 2, 4, $\frac{4}{3}$
−1	6	7	−7	3	
−2	6	1	−2	0	−2 is a zero

$$P(x) = (x+2)(6x^2+x-2) = (x+2)(3x+2)(2x-1)$$

25.   The possible rational zeros are $\pm 1, \pm 2, \pm 4$.
   Descartes' rule of signs gives the following possible combinations of zeros:

+	−	I
1	2	0
1	0	2

We form a synthetic division table:

	1	2	−9	−4	
1	1	3	−6	−10	
2	1	4	−1	−6	
4	1	6	15	56	
−1	1	1	−10	6	
−2	1	0	−9	14	
−4	1	−2	−1	0	−4 is a zero

So $P(x) = (x+4)(x^2 - 2x - 1)$.

The zeros of $x^2 - 2x - 1$ are $1 \pm \sqrt{2}$ (see problem 15).

Hence $P(x) = (x+4)[x - (1 + \sqrt{2})][x - (1 - \sqrt{2})]$

27. The possible rational zeros are $\pm 1, \pm 2, \pm\frac{1}{2}, \pm\frac{1}{4}$.

Descartes' rule of signs gives the following possible combinations of zeros:

+	−	I
2	2	0
2	0	2
0	2	2
0	0	4

We form a synthetic division table:

	4	−4	−9	1	2	
1	4	0	−9	−8	−6	
2	4	4	−1	−1	0	2 is a zero

So $P(x) = (x-2)(4x^3 + 4x^2 - x - 1)$. This clearly factors by grouping into
$(x-2)(x+1)(4x^2 - 1) = (x-2)(x+1)(2x-1)(2x+1)$, but if we don't notice
this we can proceed as before. Then we examine $Q(x) = 4x^3 + 4x^2 - x - 1$.

The possible remaining rational zeros are $-1, \pm\frac{1}{2} \pm\frac{1}{4}$.

Descartes' rule of signs applied to $Q(x)$ gives the following possible combinations of zeros:

+	−	I
1	−1	0
1	0	2

We form a synthetic division table for $Q(x)$:

	4	4	−1	−1	
−1	4	0	−1	0	−1 is a zero.

So $P(x) = (x-2)(x+1)(4x^2 - 1) = (x-2)(x+1)(2x+1)(2x-1)$.

29. $$x^2 \;\leq\; 4x - 1$$
$$x^2 - 4x + 1 \;\leq\; 0$$

We factor $x^2 - 4x + 1$ by solving $x^2 - 4x + 1 = 0$ by completing the square.

$$x^2 - 4x = -1$$
$$x^2 - 4x + 4 = 3$$
$$(x - 2)^2 = 3$$
$$x - 2 = \pm\sqrt{3}$$
$$x = 2 \pm \sqrt{3}$$

Hence $x^2 - 4x + 1 = [x - (2 + \sqrt{3})][x - (2 - \sqrt{3})]$. To solve $[x - (2 + \sqrt{3})][x - (2 - \sqrt{3})] \leq 0$ we form a sign chart, noting $2 - \sqrt{3} < 2 + \sqrt{3}$, hence $2 - \sqrt{3}$ is to the left of $2 + \sqrt{3}$.

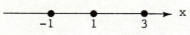

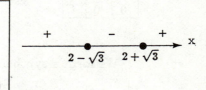

$x^2 - 4x + 1 = [x - (2 + \sqrt{3})][x - (2 - \sqrt{3})]$			
Test Number	0	2	4
Value of Polynomial for Test Number	1	−3	1
Sign of Polynomial in Interval	+	−	+
Interval	$(-\infty, 2 - \sqrt{3})$	$(2 - \sqrt{3}, 2 + \sqrt{3})$	$(2 + \sqrt{3}, \infty)$

$x^2 - 4x + 1 \leq 0$ and $x^2 \leq 4x - 1$ within the interval $[2 - \sqrt{3}, 2 + \sqrt{3}]$, or $2 - \sqrt{3} \leq x \leq 2 + \sqrt{3}$

31. $x^3 + 3 \leq 3x^2 + x$
$x^3 - 3x^2 - x + 3 \leq 0$
To factor $x^3 - 3x^2 - x + 3$, we can use factoring by grouping to obtain
$(x - 3)(x^2 - 1)$
or $(x - 3)(x - 1)(x + 1)$.
However, if we don't notice this, we can search for zeros by the methods of this section. A synthetic division table immediately gives:

	1	−3	−1	3
1	1	−2	−3	0

Hence $x^3 - 3x^2 - x + 3 = (x - 1)(x^2 - 2x - 3) = (x - 1)(x - 3)(x + 1)$.
We form a sign chart.
Zeros: $-1, 1, 3$

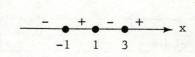

$x^3 - 3x^2 - x + 3 = (x - 1)(x - 3)(x + 1)$			
Test Number	−2	0	2	4
Value of Polynomial for Test Number	−15	3	−3	15
Sign of Polynomial in Interval	−	+	−	+
Interval	$(-\infty - 1)$	$(-1, 1)$	$(1, 3)$	$(3, \infty)$

$x^3 - 3x^2 - x + 3 \leq 0$ and $x^3 + 3 \leq 3x^2 + x$ within the intervals $(-\infty, -1]$ and $[1, 3]$, or $x \leq -1$ or $1 \leq x \leq 3$.

33. $\qquad 2x^3 + 6 \geq 13x - x^2$

$2x^3 + x^2 - 13x + 6 \geq 0$

To factor $2x^3 + x^2 - 13x + 6$, we search for zeros of the polynomial.

Possible rational zeros are $\pm 1, \pm 2, \pm 3, \pm 6, \pm \frac{1}{2}, \pm \frac{3}{2}$.

Descartes' rule of signs gives the following possible combinations of zeros:

+	−	I
2	1	0
0	1	2

We form a synthetic division table:

	2	1	−13	6	
1	2	3	−10	−4	
2	2	5	−3	0	2 is a zero

$2x^3 + x^2 - 13x + 6 = (x - 2)(2x^2 + 5x - 3) = (x - 2)(2x - 1)(x + 3)$.

Hence we must examine the sign behavior of $(x - 2)(2x - 1)(x + 3)$.

We form a sign chart.

Zeros: $-3, \frac{1}{2}, 2$

$2x^3 + x^2 - 13x + 6 = (x - 2)(2x - 1)(x + 3)$				
Test Number	−4	0	1	3
Value of Polynomial for Test Number	−54	6	−4	30
Sign of Polynomial in Interval	−	+	−	+
Interval	$(-\infty - 3)$	$(-3, \frac{1}{2})$	$(\frac{1}{2}, 2)$	$(2, \infty)$

$2x^3 + x^2 - 13x + 6 \geq 0$ and $2x^3 + 6 \geq 13x - x^2$ within the intervals $[-3, \frac{1}{2}]$ and $[2, \infty)$, or $-3 \leq x \leq \frac{1}{2}$ or $x \geq 2$.

35. $\dfrac{4}{2x^3 + 5x^2 - 2x - 5} \geq 0$

We need to form a sign chart for $\dfrac{P}{Q} = \dfrac{4}{2x^3 + 5x^2 - 2x - 5}$. We first locate the

zeros of P and Q. $P = 4$ has no zeros.

To find the zeros of $Q = 2x^3 + 5x^2 - 2x - 5$ we can factor by

grouping into $(2x + 5)(x^2 - 1) = (2x + 5)(x - 1)(x + 1)$.

However, if we don't notice this, we can search for zeros by the methods of this section.

A. synthetic division table immediately gives

	2	5	-2	-5
1	2	7	5	0

Hence $2x^3 + 5x^2 - 2x - 5 = (x-1)(2x^2 + 7x + 5) = (x-1)(x+1)(2x+5)$.
We form a sign chart.
Zeros of Q : $-\frac{5}{2}, -1, 1$

	$\frac{P}{Q} = \frac{4}{(x-1)(x+1)(2x+5)}$			
Test Number	-3	-2	0	2
Value of $\frac{P}{Q}$	$-\frac{1}{2}$	$\frac{4}{3}$	$-\frac{4}{5}$	$\frac{4}{27}$
Sign of $\frac{P}{Q}$	$-$	$+$	$-$	$+$
Interval	$(-\infty, -\frac{5}{2})$	$(-\frac{5}{2}, -1)$	$(-1, 1)$	$(1, \infty)$

$\frac{4}{2x^3 + 5x^2 - 2x - 5} \geq 0$ within the intervals $(-\frac{5}{2}, -1)$ and $(1, \infty)$,
or $-\frac{5}{2} < x < -1$ or $x > 1$.

37. $\frac{x^2 - 3x - 10}{x^3 - 4x^2 + x + 6} \leq 0$

We need to form a sign chart for $\frac{P}{Q} = \frac{x^2 - 3x - 10}{x^3 - 4x^2 + x + 6}$
We first locate the zeros of P and Q. $x^2 - 3x - 10 = (x-5)(x+2)$,
hence P has zeros at 5 and -2.
To find the zeros of Q, we note that the possible rational zeros are $\pm 1, \pm 2, \pm 3$, and ± 6.
We form a synthetic division table:

	1	-4	1	6	
1	1	-3	-2	4	
2	1	-2	-3	0	2 is a zero

Hence $Q = (x-2)(x^2 - 2x - 3) = (x-2)(x-3)(x+1)$. We form a sign chart.
Zeros of P, Q : $-2, -1, 2, 3, 5$
Open dots at zeros of Q
Solid dots at zeros of P

	$\frac{P}{Q} = \frac{(x-5)(x+2)}{(x-2)(x-3)(x+1)}$					
Test Number	-3	$-\frac{3}{2}$	0	$\frac{5}{2}$	4	6
Value of $\frac{P}{Q}$	$-\frac{2}{15}$	$\frac{26}{63}$	$-\frac{5}{3}$	$\frac{90}{7}$	$-\frac{3}{5}$	$\frac{2}{21}$
Sign of $\frac{P}{Q}$	$-$	$+$	$-$	$+$	$-$	$+$
Interval	$(-\infty,-2)$	$(-2,-1)$	$(-1,2)$	$(2,3)$	$(3,5)$	$(5,\infty)$

$\frac{x^2-3x-10}{x^3-4x^2+x+6} \leq 0$ within the intervals $(-\infty,-2]$ or $(-1,2)$ or $(3,5]$.

$x \leq -2$ or $-1 < x < 2$ or $3 < x \leq 5$

39. $\sqrt{6}$ is a root of $x^2 = 6$ or $x^2 - 6 = 0$. The possible rational roots of this equation are $\pm1, \pm2, \pm3, \pm6$. Since none of them satisfies $x^2 = 6$, there are no rational roots, hence $\sqrt{6}$ is not rational.

41. $\sqrt[3]{5}$ is a root of $x^3 = 5$ or $x^3 - 5 = 0$. The possible rational roots of this equation are $\pm1, \pm5$. Since none of them satisfies $x^3 = 5$, there are no rational roots, hence $\sqrt[3]{5}$ is not rational.

43. A. The possible rational zeros of $P(x) = x^3 + 2x - 4$ are $\pm1, \pm2$, and ±4. Forming a synthetic division table, we find $P(1)$ and $P(2)$.

	1	0	2	-4	
1	1	1	3	-1	$= P(1)$
2	1	2	6	8	$= P(2)$

By the location theorem there is a zero between 1 and 2.
There are no possible rational zeros between 1 and 2, hence the zero is irrational.
B. Since $|P(1)| < |P(2)|$ we can see that the zero probably lies closer to 1 than to 2.
We find $P(1.1), P(1.2), \ldots$ until a sign change occurs.

	1	0	2	-4	
1.1	1	1.1	3.21	-0.469	
1.2	1	1.2	3.44	0.128	the zero lies in the interval $[1.1, 1.2]$

C. Since $|P(1.1)| > |P(1.2)|$ we can see that the zero probably lies closer to 1.2 than to 1.1. Continuing with the interval divided into hundredths, we have

	1	0	2	-4
1.19	1	1.19	3.4161	0.0652
1.18	1	1.18	3.3924	0.0030
1.17	1	1.17	3.3689	-0.0584

The zero lies in the interval $[1.17, 1.18]$

45.   A. The possible rational zeros of $P(x) = x^3 + x - 1$ are ± 1.

We note that $P(1) = 1$ and $P(-1) = -3$ hence $P(x)$ has no rational zeros.

Since $P(0) = -1$, there is a zero between 0 and 1, which must then be irrational.

B. We find $P(0.1), P(0,2) \ldots$ until a sign change occurs.

	1	0	1	−1
0.1	1	0.1	1.01	−0.899
0.2	1	0.2	1.04	−0.796
0.3	1	0.3	1.09	−0.671
0.4	1	0.4	1.16	−0.536
0.5	1	0.5	1.25	−0.375
0.6	1	0.6	1.36	−0.184
0.7	1	0.7	1.49	0.043

The zero lies in the interval $[0.6, 0.7]$

C. Since $\mid P(0.6) \mid > \mid P(0.7) \mid$ we can see that the zero probably lies closer to 0.7 then to 0.6

We find $P(0.69), P(0.68), \ldots$

	1	0	1	−1
0.69	1	0.69	1.4761	0.0185
0.68	1	0.68	1.4624	−0.0056

The zero lies in the interval $[0.68, 0.69]$

47.   The possible rational zeros are $\pm 1, \pm 7$.

Descartes' rule of signs gives the following possible combinations of zeros:

+	−	I
1	1	2

The imaginary zeros are not required. We form a synthetic division table:

	1	0	0	−6	−7	
1	1	1	1	−5	−12	
2	1	2	4	2	−3	
3	1	3	9	21	56	Zero occurs between 2 and 3, probably much closer
						to 2; 3 is an upper bound, eliminating 7
−1	1	−1	1	−7	0	−1 is a zero

Hence $P(x) = (x + 1)(x^3 - x^2 - 7)$. We know that we must search for the one positive zero of $P(x)$, using $Q(x) = x^3 - x^2 + x - 7$, which will have the same zero, somewhere between 2 and 3. Forming a synthetic division table, we have:

	1	−1	1	−7	
2	1	1	3	−1	
3	1	2	7	20	zero is closer to 2 than to 3.
2.1	1	1.1	3.31	−0.049	
2.2	1	1.2	3.64	1.008	zero between 2.1 and 2.2, probably closer to 2.1.

Continuing with the interval subdivided into hundredths, we have;

2.11	1	1.11	3.3421	0.0518

So the zero is between 2.10 and 2.11, or 2.1 to one-place accuracy. Real zeros; $-1, 2.1$

49. The possible rational zeros are $\pm 1, \pm 3, \pm 9, \pm\frac{1}{2}, \pm\frac{3}{2}, \pm\frac{9}{2}$.

Descartes' rule of signs gives the following possible combinations of zeros:

+	−	I
2	3	0
0	3	2
2	1	2
0	1	4

We form a synthetic division table:

	2	−5	−7	4	21	9	
1	2	−3	−10	−6	15	24	
2	2	−1	−9	−14	−7	−5	zero between 1 and 2
3	2	−1	−4	−8	−3	0	3 is a zero

We have located the two possible positive zeros, so $\frac{1}{2}, \frac{9}{2}$ and 9 are eliminated.
We continue to examine $2x^4 + x^3 - 4x^2 - 8x - 3$. The possible rational zeros are $\pm 3, -1, \pm\frac{3}{2}, -\frac{1}{2}$.
Descartes' rule of signs applied to this polynomial gives the following possible combinations of zeros:

+	−	I
1	3	0
1	1	2

We form a synthetic division table:

	2	1	−4	−8	−3	
3	2	7	17	43	126	
−1	2	−1	−3	−5	2	
−3	2	−5	11	−41	120	
−$\frac{1}{2}$	2	0	−4	−6	0	−$\frac{1}{2}$ is a zero

$$\begin{aligned}
\text{So } P(x) &= (x-3)(x+\tfrac{1}{2})(2x^3 - 4x - 6) \\
&= (x-3)(x+\tfrac{1}{2})2(x^3 - 2x - 3)
\end{aligned}$$

We must search for a zero of $Q(x) = x^3 - 2x - 3$, between 1 and 2.

	1	0	−2	−3	
1	1	1	−1	-4	
2	1	2	2	1	zero is probably closer to 2 than to 1
1.9	1	1.9	1.61	0.059	
1.8	1	1.8	1.24	-0.768	zero is between 1.9 and 1.8, probably closer to 1.9 than to 1.8
1.89	1	1.89	1.5721	−0.0287	

The zero is between 1.89 and 1.9, to one decimal place 1.9.

The real zeros are $3, \frac{1}{2}$, and 1.9.

51. Let $x =$ the amount of increase.

$$\begin{aligned}
\text{Then old volume} &= 1 \times 2 \times 3 = 6 \\
\text{new volume} &= (x+1)(x+2)(x+3) = x^3 + 6x^2 + 11x + 6
\end{aligned}$$

Since (new volume) = 10 (old volume), we must solve

$$\begin{aligned}
x^3 + 6x^2 + 11x + 6 &= 10(6) \\
x^3 + 6x^2 + 11x + 6 &= 60 \\
x^3 + 6x^2 + 11x - 54 &= 0
\end{aligned}$$

The possible rational zeros are $\pm 1, \pm 2, \pm 3, \pm 9, \pm 18, \pm 27, \pm 54$.

Descartes' rule of signs indicates one positive zero, which is the only possibility of interest to us. We form a synthetic division table:

	1	6	11	−54	
1	1	7	18	−36	
2	1	8	27	0	2 is a zero

2 is the only positive zero. Hence the increase must equal 2 feet.

53.

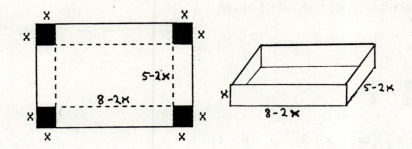

From the figure, it should be clear that

Volume $= x(5 - 2x)(8 - 2x) = 14$

Since $x, 5 - 2x$, and $8 - 2x$ must all be positive, the domain of x is $0 < x < \frac{5}{2}$ or $(0, 2.5)$. We solve $x(5 - 2x)(8 - 2x) = 14$, or $4x^3 - 26x^2 + 40x = 14$, for x in this domain.

$$4x^3 - 26x^2 + 40x = 14$$
$$4x^3 - 26x^2 + 40x - 14 = 0$$
$$2x^3 - 13x^2 + 20x - 7 = 0$$

Possible rational zeros: $\pm 1, \pm 7, \pm \frac{1}{2}, \pm \frac{7}{2}$.

Descartes' rule of signs gives the following possible combinations of zeros:

+	−	I
3	0	0
1	0	2

We form a synthetic division table

	2	−13	20	−7	
1	2	-11	9	2	
$\frac{1}{2}$	2	-12	14	0	$\frac{1}{2}$ is a zero

So

$$2x^3 - 13x^2 - 20x - 7 = (x - \tfrac{1}{2})(2x^2 - 12x + 14)$$
$$= (2x - 1)(x^2 - 6x + 7)$$

To find the remaining zeros, we solve $x^2 - 6x + 7 = 0$, by completing the square:

$$x^2 - 6x = 7$$
$$x^2 - 6x + 9 = 2$$
$$(x - 3)^2 = 2$$
$$x - 3 = \pm\sqrt{2}$$
$$x = 3 \pm \sqrt{2}$$

Hence the zeros are $\frac{1}{2}, 3 - \sqrt{2}, 3 + \sqrt{2}$, or 0.5, 1.59, 4.44 to two significant digits.

We discard $3 + \sqrt{2}$ or 4.44, since it is not in the interval $(0, 2.5)$

The square should be 0.5×0.5 inches or 1.59×1.59 inches.

Exercise 6-6

CHAPTER REVIEW

1.
$$
\begin{array}{r}
\;\;2 \quad\;\; 3 \quad\; 0 \;\; -1 \\
\;\; -4 \quad 2 \;\; -4 \\
\hline
-2\;\; \lfloor\;\; 2 \;\; -1 \quad 2 \;\; -5
\end{array}
$$

$$2x^3 + 3x^2 - 1 = (x + 2)(2x^2 - x + 2) - 5 \quad (6-1)$$

2.
$$
\begin{array}{r}
\;\; 1 \;\; -4 \quad\; 0 \quad\;\; 9 \quad 0 \;\; -8 \\
\;\;\;\; 3 \;\; -3 \;\; -9 \quad 0 \quad 0 \\
\hline
3\;\; \lfloor\;\; 1 \;\; -1 \;\; -3 \quad 0 \quad 0 \;\; -8
\end{array}
$$

$$P(3) = -8 \qquad (6-1, 6-2)$$

3. $2, -4, -1 \quad (6-2)$

4.   Since complex zeros come in conjugate pairs, $1 - i$ is a zero. $(6-3)$

5.   A. $P(x) = x^3 - x^2 - x + 3$ 2 variations in sign

   $$P(-x) = -x^3 - x^2 + x + 3 \quad \text{1 variation in sign}$$

   Possible Combinations of Zeros

+	−	I
2	1	0
0	1	2

   B. $P(x) = x^5 + x^3 + 4$ 0 variations in sign

   $$P(-x) = -x^5 - x^3 + 4 \quad \text{1 variation in sign}$$
   Possible Combinations of Zeros

+	−	I
0	1	4

   $$(6-4)$$

6.   We form a synthetic division table:

	1	−4	0	2	
−2	1	−6	12	−22	both are lower bounds, since both rows alternate
−1	1	−5	5	−3	in sign
3	1	−1	−3	−7	
4	1	0	0	2	4 is an upper bound, since this row is non-negative

$$(6-4)$$

7. We investigate $P(1)$ and $P(2)$ by forming a synthetic division table.

	2	−3	1	−5
1	2	−1	0	−5
2	2	1	3	1

Since $P(1)$ and $P(2)$ have opposite signs, there is at least one real zero between 1 and 2. $(6-4)$

8. The factors of 6 are $\pm 1, \pm 2, \pm 3, \pm 6$. $(6-5)$

9. Descartes' rule of signs gives the following possible combinations of zeros:

+	−	I
2	1	0
0	1	2

Using the possibilities found in problem 8, we form a synthetic division table.

	1	−4	1	6	
1	1	−3	−2	4	
2	1	−2	−3	0	2 is a zero

Thus $x^3 - 4x^2 + x + 6 = (x-2)(x^2 - 2x - 3) = (x-2)(x-3)(x+1)$.
The rational zeros are $2, 3, -1$. $(6-4, 6-5)$

10. We use synthetic division:

$$
\begin{array}{r|rrrrr}
 & 8 & -14 & -13 & -4 & 7 \\
 &   & 2 & -3 & -4 & -2 \\
\hline
\frac{1}{4} & 8 & -12 & -16 & -8 & 5
\end{array}
$$

Thus $P(x) = (x - \frac{1}{4})(8x^3 - 12x^2 - 16x - 8) + 5$
$\qquad P(\frac{1}{4}) = 5$ $(6-1, 6-2)$

11.
$$
\begin{array}{rrrrr}
& 4 & -8 & -3 & -3 \\
& & -2 & 5 & -1 \\
\hline
-\tfrac{1}{2}\,\big| & 4 & -10 & 2 & -4
\end{array}
$$

$P(-\tfrac{1}{2}) = -4 \quad (6-1, 6-2)$

12.   The quadratic formula tells us that $x^2 - 2x - 1 = 0$ if

$x = \dfrac{-b \pm \sqrt{b^2 - 4ac}}{2a} \quad a = 1, b - 2, c = -1$

$x = \dfrac{-(-2) \pm \sqrt{(-2)^2 - 4(1)(-1)}}{2(1)}$

$ = \dfrac{2 \pm \sqrt{8}}{2}$

$ = 1 \pm \sqrt{2}$

Since $1 \pm \sqrt{2}$ are zeros of $x^2 - 2x - 1$, its factors are $x - (1 + \sqrt{2})$ and $x - (1 - \sqrt{2})$, that is,

$x^2 - 2x - 1 = [x - (1 + \sqrt{2})][x - (1 - \sqrt{2})] \qquad (6-2)$

13. $x + 1$ will be a factor of $P(x)$ if $P(-1) = 0$.

Since $P(-1) = 9(-1)^{26} - 11(-1)^{17} + 8(-1)^{11} - 5(-1)^4 - 7 = 9 + 11 - 8 - 5 - 7 = 0$, the answer is yes, $x + 1$ is a factor. $(6-2)$

14.   A. $P(x) = 2x^4 - 3x^3 - 14x^2 + 2x + 4$ 2 variations in sign

$\ \ P(-x) = 2x^4 + 3x^3 - 14x^2 - 2x + 4$ 2 variations in sign

Possible Combinations of Zeros

+	−	I
2	2	0
2	0	2
0	2	0
0	0	4

B. We form a synthetic division table.

	2	−3	−14	2	4	
0	2	−3	−14	2	4	
1	2	−1	−15	−13	−11	zero between 0 and 1
2	2	1	−12	−22	−40	
3	2	3	−5	−13	−35	
4	2	5	6	26	108	zero between 3 and 4; 4 is an upper bound
−1	2	−5	−9	11	−7	zero between −1 and 0
−2	2	−7	0	2	0	−2 is a zero
−3	2	−9	13	−37	115	−3 is a lower bound

Lower Bound: -3 Upper Bound: 4

C. -2 is a zero; there are real zeros in the intervals $(-1,0), (0,1)$, and $(3,4)$. $(6-4)$

15. The possible rational zeros are $\pm 1, \pm 2, \pm 4, \pm 8, \pm \frac{1}{2}$.

Descartes' rule of signs gives the following possible combinations of zeros:

+	−	I
1	2	0
1	0	2

We form a synthetic division table:

	2	−3	−18	−8
1	2	−2	−20	−28
2	2	1	−16	−40
4	2	5	2	0

So $2x^3 - 3x^2 - 18 - 8 = (x-4)(2x^2 + 5x + 2)$
$$= (x-4)(2x+1)(x+2)$$

Zeros: $4, -\frac{1}{2}, -2$ $(6-4, 6-5)$

16. $(x-4)(2x+1)(x+2)$ $(6-2)$

17. The possible rational zeros are $\pm 1, \pm 5$. Descartes' rule of signs gives the following possible combinations of zeros:

+	−	I
2	1	0
0	1	2

We form a synthetic division table.

	1	−3	0	5
1	1	−2	−2	3
5	1	2	10	55
−1	1	−4	4	1
−5	1	−8	40	−195

There are no rational zeros, since all possibilities fail. $(6-4, 6-5)$

18. $P(x) = 2x^4 - x^3 + 2x - 1$

We can factor $P(x)$ by grouping into $(2x-1)(x^3+1) = (2x-1)(x+1)(x^2 - x + 1)$.

However, if we don't notice this, we find the possible rational zeros to be $\pm 1, \pm \frac{1}{2}$.

Descartes' rule of signs gives the following possible combinations of zeros:

+	−	I
3	1	0
1	1	2

We form a synthetic division table.

	2	−1	0	2	−1	
1	2	1	1	3	2	
−1	2	−3	3	−1	0	−1 is a zero

We now examine $2x^3 - 3x^2 + 3x - 1$. Only $-1, \frac{1}{2}$, and $-\frac{1}{2}$ remain as possible rational zeros; Descartes' rule of signs applied to this reduced polynomial gives
the following possible combinations of zeros:

+	−	I
3	0	0
1	0	2

We form a synthetic division table.

	2	−3	3	−1	
−1	2	−5	8	−9	Not a double zero
$\frac{1}{2}$	2	−2	2	0	$\frac{1}{2}$ is a zero

So $\ P(x) \ = \ (x+1)(x-\frac{1}{2})(2x^2 - 2x + 2)$
$\qquad\qquad\ = \ (x+1)(2x-1)(x^2 - x + 1)$

To find the remaining zeros, we solve $x^2 - x + 1 = 0$, by the quadratic formula.

$$x^2 - x + 1 \ = \ 0$$
$$x \ = \ \frac{-b \pm \sqrt{b^2 - 4ac}}{2a}$$
$$a = 1$$
$$b = -1$$
$$c = 1$$
$$x \ = \ \frac{-(-1) \pm \sqrt{(-1)^2 - 4(1)(1)}}{2(1)}$$
$$x \ = \ \frac{1 \pm \sqrt{-3}}{2}$$
$$x \ = \ \frac{1 \pm i\sqrt{3}}{2}$$

The four zeros are $-1, \frac{1}{2}$, and $\frac{1 \pm i\sqrt{3}}{2}$ $\quad (6-4, 6-5)$

19. $\quad (x+1)(x-\frac{1}{2})2(x - \frac{1+i\sqrt{3}}{2})(x - \frac{1-i\sqrt{3}}{2})$

$\quad = (x+1)(2x-1)(x - \frac{1+i\sqrt{3}}{2})(x - \frac{1-i\sqrt{3}}{2})$ $\quad (6-2)$

20. $2x^3 + 3x^2 \leq 11x + 6$

$2x^3 + 3x^2 - 11x - 6 \leq 0$

To factor $2x^3 + 3x^2 - 11x - 6$, we search for zeros of the polynomial.

Possible rational zeros are $\pm 1, \pm 2, \pm 3, \pm 6, \pm\frac{1}{2}$, and $\pm\frac{3}{2}$.

Descartes' rule of signs gives the following possible combinations of zeros:

+	−	I
1	2	0
1	0	2

We form a synthetic division table.

	2	3	−11	−6	
1	2	5	−6	−12	
2	2	7	3	0	2 is a zero

$2x^3 + 3x^2 - 11x - 6 = (x - 2)(2x^2 + 7x + 3) = (x - 2)(x + 3)(2x + 1)$

Hence we must examine the sign behavior of $(x - 2)(x + 3)(2x + 1)$.

We form a sign chart.

Zeros: $-3, -\frac{1}{2}, 2$

$2x^3 + 3x^2 - 11x - 6 = (x - 2)(x + 3)(2x + 1)$				
Test Number	−4	−1	0	3
Value of Polynomial for Test Number	−42	6	−6	42
Sign of Polynomial for Interval	−	+	−	+
Interval	$(-\infty, -3)$	$(-3, -\frac{1}{2})$	$(-\frac{1}{2}, 2)$	$(2, \infty)$

$2x^3 + 3x^2 - 11x - 6 \leq 0$ and $2x^3 + 3x^2 \leq 11x + 6$

within the intervals $(-\infty, -3)$ and $[-\frac{1}{2}, 2]$, or $x \leq -3$ or $-\frac{1}{2} \leq x \leq 2$. $(6 - 5, 3 - 7)$

21.

		1	0	3	2
			$1 + i$	$2i$	$1 + 5i$
$1 + i$		1	$1 + i$	$3 + 2i$	$3 + 5i$

$$(1+i)^2 \;=\; (1+i)(1+i) = 1 + 2i + i^2$$
$$\;=\; 1 + 2i - 1 = 2i$$
$$(1+i)(3+2i) \;=\; 3 + 5i + 2i^2 = 3 + 5i - 2 = 1 + 5i$$
$$P(x) \;=\; [x^2 + (1+i)x + (3+2i)][x - (1+i)] + 3 + 5i \quad (6-1)$$

22. $P(x) = (x - \frac{1}{2})^2(x + 3)(x - 1)^3$. The degree is 6. (6 − 2)

23. $P(x) = (x + 5)[x - (2 - 3i)][x - (2 + 3i)]$. The degree is 3. (6 − 2)

24. The possible rational zeros are $\pm 1, \pm 2, \pm 4, \pm \frac{1}{2}$.

Descartes' rule of signs gives the following possible combinations of zeros:

+	−	I
3	2	0
1	2	2
3	0	2
1	0	4

We form a synthetic division table:

	2	−5	−8	21	0	−4	
1	2	−3	−11	10	10	6	
2	2	−1	−10	1	2	0	−2 is a zero

We continue to examine $2x^4 - x^3 - 10x^2 + x + 2$.

The possible rational zeros are $-1, \pm 2$, and $\pm \frac{1}{2}$.

Descartes' rule of signs applied to this reduced polynomial gives the following possible combinations of zeros.

+	−	I
2	2	0
0	2	2
2	0	2
0	0	4

We form a synthetic division table:

	2	−1	−10	1	2	
2	2	3	−4	−7	−12	Not a double zero
−1	2	−3	−7	8	−6	
−2	2	−5	0	1	0	−2 is a zero

Hence $P(x) = (x - 2)(x + 2)(2x^3 - 5x^2 + 1)$. $2x^3 - 5x^2 + 1$ has been shown

(see Exercise 6-5, problem 15 for details)

to have zeros $\frac{1}{2}, 1 \pm \sqrt{2}$. Hence $P(x)$ has zeros $\frac{1}{2}, \pm 2, 1 \pm \sqrt{2}$.

25. $(x - 2)(x + 2)(x - \frac{1}{2})2[x - (1 - \sqrt{2})][x - (1 + \sqrt{2})] =$

$(x - 2)(x + 2)(2x - 1)[x - (1 - \sqrt{2})][x - (1 + \sqrt{2})]$ $(6 - 2)$

26. $\frac{4x^2 + 4x - 3}{2x^3 + 3x^2 - 11x - 6} \geq 0$

We need to form a sign chart for $\frac{P}{Q} = \frac{4x^2 + 4x - 3}{2x^3 + 3x^2 - 11x - 6}$. We first

locate the zeros of P and Q. $4x^2 + 4x - 3 = (2x - 1)(2x + 3)$, hence

P has zeros at $\frac{1}{2}$ and $-\frac{3}{2}$.

The zeros of Q were found in problem 20 to be $-3, -\frac{1}{2}$, and 2.

We now form the sign chart, with zeros $-3, -\frac{3}{2}, -\frac{1}{2}, \frac{1}{2}$, and 2.

Open dots at zeros of Q

Solid dots at zeros of P

$\frac{P}{Q} = \frac{(2x-1)(2x+3)}{(x-2)(2x+1)(x+3)}$						
Test Number	-4	-2	-1	0	1	3
Value of $\frac{P}{Q}$	$-\frac{15}{14}$	$\frac{5}{12}$	$-\frac{1}{2}$	$\frac{1}{2}$	$-\frac{5}{12}$	$\frac{15}{14}$
Sign of $\frac{P}{Q}$	$-$	$+$	$-$	$+$	$-$	$+$
Interval	$(-\infty, -3)$	$(-3, -\frac{3}{2})$	$(-\frac{3}{2}, -\frac{1}{2})$	$(-\frac{1}{2}, \frac{1}{2})$	$(\frac{1}{2}, 2)$	$(2, \infty)$

$\frac{4x^2 + 4x - 3}{2x^3 + 3x^2 - 11x - 6} \geq 0$ within the intervals $(-3, \frac{3}{2}] \cup (-\frac{1}{2}, \frac{1}{2}] \cup (2, \infty)$, or

$-3 < x \leq \frac{3}{2}$ or $-\frac{1}{2} < x \leq \frac{1}{2}$ or $x > 2$. $(6 - 5, 3 - 7)$

27. Since $P(1) = -2$ and $P(2) = 6$, we expect the root to be closer to 1

than to 2. We divide the interval into tenths and

form a synthetic division table:

	1	0	-1	0	-2	
1.1	1	1.1	0.21	0.231	-1.7459	
1.2	1	1.2	0.44	0.528	-1.3664	
1.3	1	1.3	0.69	0.897	-0.8339	
1.4	1	1.4	0.96	1.344	-0.1184	
1.5	1	1.5	1.25	1.875	0.8125	zero between 1.4 and 1.5

Since $\mid P(1.4)\mid < \mid P(1.5)\mid$ the zero probably lies closer to 1.4 than 1.5.

| 1.41 | 1 | 1.41 | 0.9881 | 1.3932 | -0.3556 |
| 1.42 | 1 | 1.42 | 1.0164 | 1.4433 | 0.4947 |

Thus the zero lies between 1.41 and 1.42, to one place accuracy 1.4.

(Check: by inspection $\sqrt{2}$ is a zero; to one decimal place $\sqrt{2} \approx 1.4$) $(6-5)$

28.   Let $P(x) = x^3 + 3x - 5$. The possible rational zeros are $\pm 1, \pm 5$.

Descartes' rule of signs gives the following possible combinations of zeros:

+	−	I
1	0	2

This eliminates -1 and -5, since there are no negative zeros.

We form a synthetic division table.

	1	0	3	-5	
1	1	1	4	-1	
2	1	2	7	9	zero between 1 and 2, probably closer to 1. This eliminates 5.

We divide the interval into tenths and proceed.

	1	0	3	-5	
1.1	1	1.1	4.21	-0.369	
1.2	1	1.2	4.44	0.328	a zero between 1.1 and 1.2, possibly closer to 1.2, but near the middle of the interval.

We divide the interval into hundredths and proceed, starting in the middle.

	1	0	3	-5
1.15	1	1.15	4.3225	-0.2913

There is a zero between 1.15 and 1.2, to one decimal place 1.2.

This is the only real root of the equation, as required. $(6-4, 6-5)$.

CHAPTER 7

Exercise 7-1
Key Ideas and Formulas

A linear system of equations

in two variables   or   three variables

$$ax + by = h \qquad a_1x + b_1y + c_1z = d_1$$
$$cx + dy = k \qquad a_2x + b_2y + c_2z = d_2$$
$$\qquad\qquad\qquad a_3x + b_3y + c_3z = d_3$$

must have

1. exactly one solution (consistent and independent) or

2. no solution (inconsistent) or

3. infinitely many solutions (consistent and dependent).

There are no other possibilities.

A linear system in two variables can be solved by graphing, by substitution, or by elimination using addition. The latter method is best adapted to solving a linear system in three variables. (See text for descriptions of each method.)

Equivalent Systems of equations are systems with the same solution set.

A system of linear equations is transformed into an equivalent system if:

1. Two equations are interchanged.

2. An equation is multiplied by a non-zero constant.

3. A constant multiple of another equation is added to a given equation.

1.

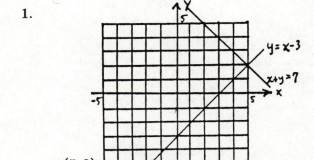

$(5, 2)$

3.

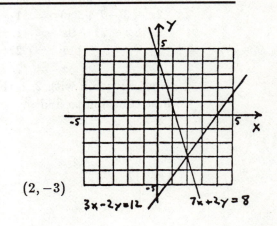

$(2, -3)$

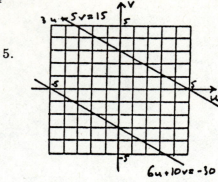

3u+5v=15

6u+10v=-30

5.

The lines are parallel.
No Solution

7.
$$x - y = 4$$
$$x + 3y = 12$$
Solve the first equation for x in terms of y.
$$x = 4 + y$$
Substitute into the second equation to eliminate x.
$$(4 + y) + 3y = 12$$
$$4y = 8$$
$$y = 2$$
Now replace y with 2 in the first equation to find x.
$$x - 2 = 4$$
$$x = 6$$
$$(6, 2)$$

9.
$$3x - y = 7$$
$$2x + 3y = 1$$
Solve the first equation for y in terms of x.
$$-y = 7 - 3x$$
$$y = -7 + 3x$$
Substitute into the second equation to eliminate y.
$$2x + 3(-7 + 3x) = 1$$
$$2x - 21 + 9x = 1$$
$$11x = 22$$
$$x = 2$$
Now replace x with 2 in the first equation to find y.
$$3 \cdot 2 - y = 7$$
$$6 - y = 7$$
$$y = -1$$
$$(2, -1)$$

11.
$$2x + 3y = 1$$
$$3x - y = 7$$
If we multiply the bottom equation by 3 and add, we can eliminate y.

$$
\begin{array}{rcr}
2x + 3y &=& 1 \\
9x - 3y &=& 21 \\
\hline
11x &=& 22 \\
x &=& 2
\end{array}
$$

Now substitute $x = 2$ back into the top equation and solve for y.
$$2(2) + 3y = 1$$
$$3y = -3$$
$$y = -1$$
$$(2, -1)$$

13.
$$4x + 3y = 26$$
$$3x - 11y = -7$$
If we multiply the top equation by 3, the bottom by -4, and add, we can eliminate x.

$$
\begin{array}{rcr}
12x + 9y &=& 78 \\
-12x + 44y &=& 28 \\
\hline
53y &=& 106 \\
y &=& 2
\end{array}
$$

Now substitute $y = 2$ back into the bottom equation and solve for x.
$$3x - 11(2) = -7$$
$$3x = 15$$
$$x = 5$$
$$(5, 2)$$

15.
$$3x - 6y = -9$$
$$-2x + 4y = 6$$

We choose elimination by addition, multiplying the top equation by 2 and the bottom by 3.

$$
\begin{array}{rrr}
6x & - \; 12y & = \; -18 \\
-6x & + \; 12y & = \;\;\; 18 \\
\hline
& 0 & = \;\;\;\; 0
\end{array}
$$

There are infinitely many solutions.

17.
$$7m + 12n = -1$$
$$5m - 3n = 7$$

We choose elimination by addition, multiplying the bottom equation by 4.

$$
\begin{array}{rrr}
7m & + \; 12n & = \; -1 \\
20m & - \; 12n & = \; 28 \\
\hline
27m & & = \; 27 \\
& m & = \;\;\; 1
\end{array}
$$

Substituting in the bottom equation, we have
$$5(1) - 3n = 7$$
$$-3n = 2$$
$$n = -\tfrac{2}{3}$$
$$(1, -\tfrac{2}{3})$$

19.
$$2x + 4y = -8$$
$$x + 2y = 4$$

We choose substitution, solving the bottom equation for x in terms of y, then substituting in the top equation.
$$x = 4 - 2y$$
$$2(4 - 2y) + 4y = -8$$
$$8 - 4y + 4y = -8$$
$$8 = -8$$

No solution

21.
$$y = 0.08x$$
$$y = 100 + 0.04x$$

We choose substitution, substituting y from the top equation into the bottom equation.
$$0.08x = 100 + 0.04x$$
$$0.04x = 100$$
$$x = 2500$$
$$y = 0.08(2500)$$
$$y = 200$$
$$(2500, 200)$$

23.
$$0.3u - 0.6v = 0.18$$
$$0.5u + 0.2v = 0.54$$

We choose elimination by addition, multiplying the bottom equation by 3.

$$
\begin{array}{rrr}
0.3u & - \; 0.6v & = \; 0.18 \\
1.5u & + \; 0.6v & = \; 1.62 \\
\hline
1.8u & & = \; 1.8 \\
& u & = \;\;\; 1 \\
0.5(1) & + \; 0.2v & = \; 0.54 \\
& 0.2v & = \; 0.04 \\
& v & = \; 0.2
\end{array}
$$

$$(1, 0.2)$$

25.
$$y + 2z = -5 \quad (1)$$
$$3y - z = 6 \quad (2)$$
$$2x - y + 4z = -9 \quad (3)$$

We eliminate z from equations (1) and (2)

$$
\begin{array}{rrrl}
y & + \; 2z & = \; -5 & (1) \\
6y & - \; 2z & = \; 12 & (2) \; [\text{equation } (2)] \\
\hline
7y & & = \; 7 & \\
& y & = \; 1 &
\end{array}
$$

Substituting $y = 1$ into equation (1), we have

$$1 + 2z = -5$$
$$2z = -6$$
$$z = -3$$

Substituting $y = 1$ and $z = -3$
into equation (3) , we have

$$2x - 1 + 4(-3) = -9$$
$$2x - 13 = -9$$
$$2x = 4$$
$$x = 2$$

$(2, 1, -3)$

27.
$$x - 3y + z = 4 \quad (1)$$
$$-x + 4y - 4z = 1 \quad (2)$$
$$2x - y + 5z = -3 \quad (3)$$

We eliminate x from equations (2) and (3).

$$
\begin{array}{rrrrrll}
x & - & 3y & + & z & = & 4 & (1) \\
-x & + & 4y & - & 4z & = & 1 & (2) \\
\hline
 & & y & - & 3z & = & 5 & (4) \\
-2x & + & 6y & - & 2z & = & -8 & -2[\text{equation}(1)] \\
2x & - & y & + & 5z & = & -3 & \text{Equation}(3) \\
\hline
 & & 5y & + & 3z & = & -11 & (5)
\end{array}
$$

We now solve the system

$$
\begin{array}{rrrrll}
y & - & 3z & = & 5 & (4) \\
5y & + & 3z & = & -11 & (5) \\
\hline
6y & & & = & -6 & (\text{adding}) \\
 & y & & = & -1 &
\end{array}
$$

Substituting $y = -1$ into equation (4), we have

$$-1 - 3z = 5$$
$$-3z = 6$$
$$z = -2$$

Substituting $y = -1$ and $z = -2$ into equation (1), we have

$$x - 3(-1) + (-2) = 4$$
$$x = 3$$

$(3, 1, -2)$

29.
$$3u - 2v + 3w = 11 \quad (1)$$
$$2u + 3v - 2w = -5 \quad (2)$$
$$u + 4v - w = -5 \quad (3)$$

We eliminate w from equations (1) and (2).

$$
\begin{array}{rrrrrrll}
3u & - & 2v & + & 3w & = & 11 & \text{Equation (1)}\\
3u & + & 12v & - & 3w & = & -15 & 3\,[\text{equation}(3)]\\
\hline
6u & + & 10v & & & = & -4 & (4)\\
2u & + & 3v & - & 2w & = & -5 & \text{Equation}(2)\\
-2u & - & 8v & + & 2w & = & 10 & -2\,[\text{equation}(3)]\\
\hline
 & - & 5v & & & = & 5 & (5)
\end{array}
$$

We now solve the system

$$6u + 10v = -4 \quad (4)$$
$$-5v = 5 \quad (5)$$
$$v = -1$$

Substituting $v = -1$ into equation(4), we have

$$6u + 10(-1) = -4$$
$$6u = 6$$
$$u = 1$$

Substituting $u = 1$ and $v = -1$ into equation (3), we have

$$1 + 4(-1) - w = -5$$
$$-w = -2$$
$$w = 2$$

$(1, -1, 2)$

31.
$$3x - 2y - 4z = -8 \quad (1)$$
$$4x + 3y - 5z = -5 \quad (2)$$
$$6x - 5y + 2z = -17 \quad (3)$$

We eliminate z from equations (1)and (2)

$$
\begin{array}{rrrrrrll}
3x & - & 2y & -4z & = & -8 & \text{Equation (1)}\\
12x & - & 10y & +4z & = & -34 & 2\,[\text{equation}(3)]\\
\hline
15x & - & 12y & & = & -42 & (4)\\
8x & + & 6y & -10z & = & -10 & 2\,[\text{equation (2)}]\\
30x & - & 25y & +10z & = & -85 & 5\,[\text{equation}(3)]\\
\hline
38x & - & 19y & & = & -95 & (5)
\end{array}
$$

We now solve the system

$$15x - 12y = -42 \quad (4)$$
$$38x - 19y = -95 \quad (5)$$

We eliminate y

$$15x - 12y = -42 \quad \text{Equation (4)}$$
$$\underline{-24x + 12y = 60} \quad -\tfrac{12}{19} \text{ [Equation (5)]}$$
$$-9x = 18$$
$$x = -2$$

Substituting $x = -2$ into equation (4), we have
$$15(-2) - 12y = -42$$
$$-12y = -12$$
$$y = 1$$
Substituting $x = -2$ and $y = 1$ into equation (1), we have
$$3(-2) - 2(1) - 4z = -8$$
$$-4z = 0$$
$$z = 0$$
$(-2, 1, 0)$

33.
$$-x + 2y - z = -4 \quad (1)$$
$$2x + 5y - 4z = -16 \quad (2)$$
$$x + y - z = -4 \quad (3)$$
We eliminate x from equations(1)and (2)

$$-x + 2y - z = -4 \quad (1)$$
$$\underline{x + y - z = -4} \quad (3)$$
$$3y - 2z = -8 \quad (4)$$

$$2x + 5y - 4z = -16 \quad \text{Equation(2)}$$
$$\underline{-2x - 2y + 2z = 8} \quad -2 \text{ [equation (3)]}$$
$$3y - 2z = -8 \quad (5)$$

We now solve the system
$$3y - 2z = -8 \quad (4)$$
$$3y - 2z = -8 \quad (5)$$
This system is clearly dependent. To determine whether the original system is inconsistent or dependent, we express y in terms of z from equation(5), then x in terms of z
from equation (3), and check.
$$3y = 2z - 8$$
$$y = \tfrac{2}{3}z - \tfrac{8}{3}$$
$$x + y - z = -4 \quad (3)$$
$$x + \tfrac{2}{3}z - \tfrac{8}{3} - z = -4$$
$$x - \tfrac{1}{3}z - \tfrac{8}{3} = -4$$
$$x = \tfrac{1}{3}z - \tfrac{4}{3}$$
Check in equations (1) and (2):

$$-\left(\tfrac{1}{3}z - \tfrac{4}{3}\right) + 2\left(\tfrac{2}{3}z - \tfrac{8}{3}\right) - z \overset{?}{=} +4 \text{ from}(1)$$

$$-\tfrac{1}{3}z + \tfrac{4}{3} + \tfrac{4}{3}z - \tfrac{16}{3} - z \overset{?}{=} -4$$

$$-4 \overset{\vee}{=} -4$$

$$2\left(\tfrac{1}{3}z - \tfrac{4}{3}\right) + 5\left(\tfrac{2}{3}z - \tfrac{8}{3}\right) - 4z \overset{?}{=} -16 \text{ from } (2)$$

$$\tfrac{2}{3}z - \tfrac{8}{3} + \tfrac{10}{3}z - \tfrac{40}{3} - 4z \overset{?}{=} -16$$

$$-16 \overset{\vee}{=} -16$$

The system is dependent, and has infinitely many solutions.

35.   Let x =   number of 20 − cent stamps
         y =   number of 15 − cent stamps

She bought 47 stamps, hence

$x + y = 47$ (1)

She spent $8.80, hence

$0.20x + 0.15y = 8.80$ (2)

We solve the system of equations (1), (2) by solving equation (1) for x in terms of y, then substituting into equation (2).

$$
\begin{aligned}
x + y &= 47 \\
x &= 47 - y \\
0.20(47 - y) + 0.15y &= 8.80 \\
9.4 - 0.20y + 0.15y &= 8.80 \\
-0.05y &= -0.6 \\
y &= 12 \\
x &= 47 - y \\
&= 35
\end{aligned}
$$

35 20-cent stamps, 12 15-cent stamps

37.   Let x =   airspeed of the plane
         y =   rate at which wind is blowing

Then

$x - y$ =   groundspeed flying from Atlanta to LosAngeles (headwind)
$x + y$ =   groundspeed flying from LosAngeles to Atlanta (tailwind)

Then, applying Distance = Rate × time, we have

$$
\begin{aligned}
2{,}100 &= 8.75(x - y) \\
2{,}100 &= 5(x + y)
\end{aligned}
$$

After simplification, we have

$$
\begin{aligned}
x - y &= 240 \quad (1) \\
x + y &= 420 \quad (2)
\end{aligned}
$$

Solve using elimination by addition:

We add equations (1) and (2) to obtain

$$
\begin{aligned}
2x &= 660 \\
x &= 330 \text{ mph} = \text{airspeed} \\
330 + y &= 420 \\
y &= 90 \text{ mph} = \text{windrate}
\end{aligned}
$$

39.   Let x =   amount of 50% solution
        y =   amount of 80% solution

100 milliliters are required, hence

$x + y = 100$ (1)

68% of the 100 milliliters must be acid, hence

$0.50x + 0.80y = 0.68(100)$ (2)

We solve the system of equations (1), (2) using elimination by addition.

$$
\begin{array}{rlll}
-0.50x & - & 0.50y & = & -0.50(100) & -0.50[\text{equation}(1)] \\
\underline{-0.50x} & + & \underline{0.80y} & = & \underline{0.68(100)} & \text{Equation}(2) \\
 & & 0.30y & = & -0.50(100) & + 0.68\,(100) \\
 & & 0.30y & = & 18 \\
 & & y & = & 60 \\
 & x + y & & = & 100 \\
 & x & & = & 40
\end{array}
$$

40 milliliters of 50% solution and 60 milliliters of 80% solution

41.   Let x =   number of grams of 12 − carat gold
        y =   number of grams of 18 − carat gold

10 grams are required, hence

$x + y = 10$ (1)

The gold must be $\frac{14}{24}$ pure, hence

$\frac{12}{24}x + \frac{18}{24}y = \frac{14}{24}(10)$ (2)

We solve the system of equations (1), (2) by solving equation (1) for x in terms of y and substituting into equation (2).

$$
\begin{array}{rcl}
x & = & 10 - y \\
\frac{12}{24}(10 - y) + \frac{18}{24}y & = & \frac{14}{24}(10) \\
12(10 - y) + 18y & = & 140 \\
120 - 12y + 18y & = & 140 \\
6y & = & 20 \\
y & = & \frac{10}{3} \text{ or } 3\frac{1}{3} \\
x & = & 10 - 3\frac{1}{3} = 6\frac{2}{3}
\end{array}
$$

$6\frac{2}{3}$ grams of 12 − carat gold

$3\frac{1}{3}$ grams of 18 − carat gold

43. $s = a + bt^2$

A. We are given: When $t = 1, s = 180$
                 When $t = 2, s = 132$

Substituting these values in the given equation, we have

$$
\begin{array}{rcll}
180 & = & a + b(1)^2 \\
132 & = & a + b(2)^2 & \text{or} \\
180 & = & a + b & (1) \\
132 & = & a + 4b & (2)
\end{array}
$$

We solve the system of equations (1), (2) by solving equation (1) for a in terms of

b and substituting into equation (2).

$$a = 180 - b$$
$$132 = 180 - b + 4b$$
$$132 = 180 + 3b$$
$$-48 = 3b$$
$$b = -16$$
$$180 = a - 16$$
$$a = 196$$

B. The height of the building is represented by s, the distance of the object above the ground, when $t = 0$. Since we now know

$$s = 196 - 16t^2$$

from part A, when $t = 0, s = 196$ feet is the height of the building.

C. The object falls until s, its distance above the ground, is zero.

Since

$$s = 196 - 16t^2$$

we substitute $s = 0$ and solve for t.

$$0 = 196 - 16t^2$$
$$16t^2 = 196$$
$$t^2 = \frac{196}{16}$$
$$t = \frac{14}{4} \qquad \text{(discarding the negative solution)}$$
$$t = 3.5 \qquad \text{seconds}$$

45.　　Let p = time of primary wave

s = time for secondary wave

We know

$s - p = 16$ (time difference)　(1)

To find a second equation, we have use Distance = rate $\times$ time

$5p$ = distance for primary wave

$3s$ = distance for secondary wave

These distance are equal, hence

$5p = 3s$　(2)

We solve the system of equations (1), (2) by solving equation (1) for s and substituting into equation (2)

$$s = p + 16$$
$$5p = 3(p + 16)$$
$$5p = 3p + 48$$
$$2p = 48$$
$$p = 24 \text{ seconds}$$
$$s = 24 + 16$$
$$s = 40 \text{ seconds}$$

The distance travelled $= 5p = 3s = 120$ miles

47.　　Let x = number of shirt A produced per week

Let y = number of shirt B produced per week

Let z = number of shirt C produced per week

We have

$$0.2x + 0.4y + 0.3z = 1160 \text{ Cutting department}$$
$$0.3x + 0.5y + 0.4z = 1560 \text{ Sewing department}$$
$$0.1x + 0.2y + 0.1z = 480 \text{ Packaging department}$$

Clearing of decimals for convenience:

$$2x + 4y + 3z = 11600 \quad (1)$$
$$3x + 5y + 4z = 15600 \quad (2)$$
$$x + 2y + z = 4800 \quad (3)$$

We eliminate x from equations (1) and (2)

$2x$	$+$	$4y$	$+$	$3z$	$=$	11600	Equation (1)
$-2x$	$-$	$4y$	$-$	$2z$	$=$	-9600	-2[equation (3)]
				z	$=$	2000	(4)
$3x$	$+$	$5y$	$+$	$4z$	$=$	15600	Equation(2)
$-3x$	$-$	$6y$	$-$	$3z$	$=$	-14400	-3[equation(3)]
		$-y$	$+$	z	$=$	1200 (5)	

Solving the system (4), (5) we have

$$z = 2000 (4)$$
$$-y + 2000 = 2000$$
$$-y = -800$$
$$y = 800$$

Substituting $y = 800, z = 2000$ into equation (3), we have

$$x + 2(800) + 2000 = 4800$$
$$x = 1200$$

1200 style A; 800 style B; 2000 style C

49.   Let x = amount of Mix A.

          y = amount of Mix B.

          z = amount of Mix C.

The amounts of protein must add up to 23 grams.

Protein in Mix A	$+$	Protein in Mix B	$+$	Protein in Mix C	$=$	Total Protein
$0.20x$	$+$	$0.10y$	$+$	$0.15z$	$=$	23

Common Error: The facts in this problem do not justify an equation like

$$0.20x + 0.02y + 0.15z = 23$$

The amounts of fat must add up to 6.2 grams.

Fat in Mix A	$+$	Fat in Mix B	$+$	Fat in Mix C	$=$	Total Fat
$0.02x$	$+$	$0.06y$	$+$	$0.05z$	$=$	6.2

The amounts of moisture must add up to 16 grams.

Moisture in Mix A	$+$	Moisture in Mix B	$+$	Moisture in Mix C	$=$	Total Moisture
$0.15x$	$+$	$0.10y$	$+$	$0.05z$	$=$	16

After simplication, we obtain

$$20x + 10y + 15z = 2300 \quad (1)$$
$$2x + 6y + 5z = 620 \quad (2)$$
$$15x + 10y + 5z = 1600 \quad (3)$$

We eliminate z from equations (1) and (2)

$20x$	$+$	$10y$	$+$	$15z$	$=$	2300	Equation(1)
$-45x$	$-$	$30y$	$-$	$15z$	$=$	-4800	$-3[\text{equation}(3)]$
$-25x$	$-$	$20y$			$=$	-2500	(4)
$-2x$	$-$	$6y$	$-$	$5z$	$=$	-620	$-1[\text{equation}(2)]$
$15x$	$+$	$10y$	$+$	$5z$	$=$	1600	Equation(3)
$13x$	$+$	$4y$			$=$	980	(5)

We solve the system (4), (5) using elimination by addition

$-25x$	$-$	$20y$	$=$	-2500	Equation(4)
$65x$	$+$	$20y$	$=$	4900	$5[\text{equation}(5)]$
$40x$			$=$	2400	
		x	$=$	60	

Substituting $x = 60$ into equation (5), we have

$$13(60) + 4y = 980$$
$$4y = 200$$
$$y = 50$$

Substituting $x = 60, y = 50$ into equation (2), we have

$$2(60) + 6(50) + 5z = 620$$
$$420 + 5z = 620$$
$$5z = 200$$
$$z = 40$$

60 grams Mix A, 50 grams Mix B, 40 grams Mix C

Exercise 7-2

Key Ideas and Formulas

Associated with each linear system:

$$a_1x_1 + b_1x_2 = k_1$$
$$a_2x_1 + b_2x_2 = k_2$$

is an augmented matrix:
$\begin{bmatrix} a_1 & b_1 & k_1 \\ a_2 & b_2 & k_2 \end{bmatrix}$

We solve a linear system by using row operations to transform it into a row equivalent matrix of one of these types:

$\begin{bmatrix} 1 & 0 & m \\ 0 & 1 & n \end{bmatrix}$ a unique solution (consistent and independent)

$\begin{bmatrix} 1 & m & n \\ 0 & 0 & 0 \end{bmatrix}$ infinitely many solutions (consistent and dependent)

$\begin{bmatrix} 1 & m & n \\ 0 & 0 & P \end{bmatrix}$ no solution (inconsistent)

An augmented matrix is transformed into a row-equivalent matrix if any of the following row operations is performed.

1. Two rows are interchanged. ($R_i \leftrightarrow R_j$ means "interchange row i and row j")

2. A row is multiplied by a non-zero constant ($kR_i \rightarrow R_i$ means "multiply row i by the constant k")

3. A constant multiple of another row is added to a given row ($R_i + kR_j \rightarrow R_i$ means "multiply row j by the constant k and add to R_i.")

---

1. $R_1 \leftrightarrow R_2$ means interchange Rows 1 and 2.

$\begin{bmatrix} 4 & 6 & -8 \\ 1 & -3 & 2 \end{bmatrix}$

3. $-4R_1 \rightarrow R_1$ means multiply Row 1 by -4.

$\begin{bmatrix} -4 & 12 & -8 \\ 4 & -6 & -8 \end{bmatrix}$

5. $2R_2 \rightarrow R_2$ means multiply Row 2 by 2.

$\begin{bmatrix} 1 & -3 & 2 \\ 8 & -12 & -16 \end{bmatrix}$

7. $R_2 + (-4)R_1 \rightarrow R_2$ means replace Row 2 by itself plus -4 times Row 1.

$\begin{bmatrix} 1 & -3 & 2 \\ 4 & -6 & -8 \end{bmatrix} \rightarrow \begin{bmatrix} 1 & -3 & 2 \\ 0 & 6 & -16 \end{bmatrix}$

$-4 \quad 12 \quad -8$

9. $R_2 + (-2)R_1 \rightarrow R_2$ means replace Row 2 by itself plus -2 times Row 1.

$\begin{bmatrix} 1 & -3 & 2 \\ 4 & -6 & -8 \end{bmatrix} \rightarrow \begin{bmatrix} 1 & -3 & 2 \\ 2 & 0 & -12 \end{bmatrix}$

$-2 \quad 6 \quad -4$

11. $R_2 + (-1)R_1 \rightarrow R_2$ means replace Row 2 by itself plus -1 times Row 1.

$$\begin{bmatrix} 1 & -3 & | & 2 \\ 4 & -6 & | & -8 \end{bmatrix} \rightarrow \begin{bmatrix} 1 & -3 & | & 2 \\ 3 & -3 & | & -10 \end{bmatrix}$$

$-1 \quad 3 \quad -2$

13. We write the augmented matrix:

$$\begin{bmatrix} 1 & 1 & | & 5 \\ 1 & -1 & | & 1 \end{bmatrix} \qquad R_2 + (-1)R_1 \rightarrow R_2$$

Need a 0 here

$-1 - 1 - 5$

$$\sim \begin{bmatrix} 1 & 1 & | & 5 \\ 0 & -2 & | & -4 \end{bmatrix} \qquad -\tfrac{1}{2}R_2 \rightarrow R_2$$

Need a 1 here
Need a 0 here

$$\sim \begin{bmatrix} 1 & 1 & | & 5 \\ 0 & 1 & | & 2 \end{bmatrix} \qquad R_1 + (-1)R_2 \rightarrow R_1$$

$0 \quad -1 \quad -2$

$$\sim \begin{bmatrix} 1 & 0 & | & 3 \\ 0 & 1 & | & 2 \end{bmatrix}$$

Therefore $x_1 = 3$ and $x_2 = 2$

15. We write the augmented matrix:

$$\begin{bmatrix} 1 & -2 & | & 1 \\ 2 & -1 & | & 5 \end{bmatrix} \qquad R_2 + (-2)R_1 \rightarrow R_2$$

Need a 0 here

$-2 \quad 4 - 2$

$$\sim \begin{bmatrix} 1 & -2 & | & 1 \\ 0 & 3 & | & 3 \end{bmatrix} \qquad \tfrac{1}{3}R_2 \rightarrow R_2$$

Need a 1 here
Need a 0 here

$$\sim \begin{bmatrix} 1 & -2 & | & 1 \\ 0 & 1 & | & 1 \end{bmatrix} \qquad R_1 + 2R_2 \rightarrow R_1$$

$0 \quad 2 \quad 2$

$$\sim \begin{bmatrix} 1 & 0 & | & 3 \\ 0 & 1 & | & 1 \end{bmatrix}$$

Therefore $x_1 = 3$ and $x_2 = 1$

17.
$$\begin{bmatrix} 1 & -4 & \vline & -2 \\ -2 & 1 & \vline & -3 \end{bmatrix} \quad R_2 + \underbrace{2R_1} \to R_2$$

Need a 0 here

$$2 \quad -8 \quad -4$$

$$\sim \begin{bmatrix} 1 & -4 & \vline & -2 \\ 0 & -7 & \vline & -7 \end{bmatrix} \quad -\tfrac{1}{7}R_2 \to R_2$$

Need a 1 here

Need a 0 here

$$\sim \begin{bmatrix} 1 & -4 & \vline & -2 \\ 0 & 1 & \vline & 1 \end{bmatrix} \quad R_1 + \underbrace{4R_2} \to R_1$$

$$0 \quad 4 \quad 4$$

$$\sim \begin{bmatrix} 1 & 0 & \vline & 2 \\ 0 & 1 & \vline & 1 \end{bmatrix}$$

Therefore $x_1 = 2$ and $x_2 = 1$

19.
$$\begin{bmatrix} 3 & -1 & \vline & 2 \\ 1 & 2 & \vline & 10 \end{bmatrix} \quad R_1 \leftrightarrow R_2$$

Need a 1 here

$$\sim \begin{bmatrix} 1 & 2 & \vline & 10 \\ 3 & -1 & \vline & 2 \end{bmatrix} \quad R_2 + \underbrace{(-3)R_1} \to R_2$$

Need a 0 here

$$-3 \quad -6 \quad -30$$

$$\sim \begin{bmatrix} 1 & 2 & \vline & 10 \\ 0 & -7 & \vline & -28 \end{bmatrix} \quad -\tfrac{1}{7}R_2 \to R_2$$

Need a 1 here

Need a 0 here

$$\sim \begin{bmatrix} 1 & 2 & \vline & 10 \\ 0 & 1 & \vline & 4 \end{bmatrix} \quad R_1 + \underbrace{(-2)R_2} \to R_1$$

$$0 \quad -2 \quad -8$$

$$\sim \begin{bmatrix} 1 & 0 & \vline & 2 \\ 0 & 1 & \vline & 4 \end{bmatrix}$$

Therefore $x_1 = 2$ and $x_2 = 4$

21.
$$\sim \begin{bmatrix} 1 & 2 & \vline & 4 \\ 2 & 4 & \vline & -8 \end{bmatrix} \quad R_2 + \underbrace{(-2)R_1} \to R_2$$

Need a 0 here

$$-2 \quad -4 \quad -8$$

$$\sim \begin{bmatrix} 1 & 2 & \vline & 4 \\ 0 & 0 & \vline & -16 \end{bmatrix}$$

This matrix corresponds to the system

$$x_1 + 2x_2 = 4$$
$$0x_1 + 0x_2 = -16$$

This system has no solution.

23.
$$\begin{bmatrix} 2 & 1 & \vline & 6 \\ 1 & -1 & \vline & -3 \end{bmatrix} \quad R_1 \leftrightarrow R_2$$

Need a 1 here

$$\sim \begin{bmatrix} 1 & -1 & \vline & -3 \\ 2 & 1 & \vline & 6 \end{bmatrix} \quad R_2 + \underbrace{(-2)R_1} \to R_2$$

Need a 0 here

$$-2 \quad 2 \quad 6$$

$$\sim \begin{bmatrix} 1 & -1 & \vline & -3 \\ 0 & 3 & \vline & 12 \end{bmatrix} \quad \tfrac{1}{3}R_2 \to R_2$$

Need a 1 here

Need a 0 here

$$\sim \begin{bmatrix} 1 & -1 & \vline & -3 \\ 0 & 1 & \vline & 4 \end{bmatrix} \quad R_1 + R_2 \to R_1$$

$$\sim \begin{bmatrix} 1 & 0 & \vline & 1 \\ 0 & 1 & \vline & 4 \end{bmatrix}$$

Therefore $x_1 = 1$ and $x_2 = 4$.

25. $\begin{bmatrix} 3 & -6 & \vline & -9 \\ -2 & 4 & \vline & 6 \end{bmatrix}$ $\quad \frac{1}{3}R_1 \to R_1$

Need a 1 here.

$\sim \begin{bmatrix} 1 & -2 & \vline & -3 \\ -2 & 4 & \vline & 6 \end{bmatrix}$ $\quad R_2 + 2R_1 \to R_2$

Need a 0 here
$2 \quad -4 \quad -6$

$\sim \begin{bmatrix} 1 & -2 & \vline & -3 \\ 0 & 0 & \vline & 0 \end{bmatrix}$

This matrix corresponds to the system

$$x_1 - 2x_2 = -3$$
$$0x_1 + 0x_2 = 0$$

Thus $x_1 = 2x_2 - 3$.
Hence there are infinitely many
solutions: for any real number s,
$x_2 = s, x_1 = 2s - 3$ is a solution.

27. $\begin{bmatrix} 4 & -2 & \vline & 2 \\ -6 & 3 & \vline & -3 \end{bmatrix}$ $\quad \frac{1}{4}R_1 \to R_2$

Need a 1 here

$\sim \begin{bmatrix} 1 & -\frac{1}{2} & \vline & \frac{1}{2} \\ -6 & 3 & \vline & -3 \end{bmatrix}$ $\quad R_2 + 6R_1 \to R_2$

Need a 0 here
$6 \quad -3 \quad 3$

$\sim \begin{bmatrix} 1 & -\frac{1}{2} & \vline & \frac{1}{2} \\ 0 & 0 & \vline & 0 \end{bmatrix}$

This matrix corresponds to the system

$$x_1 - \frac{1}{2}x_2 = \frac{1}{2}$$
$$0x_1 + 0x_2 = 0$$

Thus $x_1 = \frac{1}{2}x_2 + \frac{1}{2}$

Hence there are infinitely many solutions:
for any real number $s, x_2 = s, x_1 = \frac{1}{2}s + \frac{1}{2}$
is a solution.

29. $\begin{bmatrix} 3 & -1 & \vline & 7 \\ 2 & 3 & \vline & 1 \end{bmatrix}$ $\quad \frac{1}{3}R_1 \to R_1$

Need a 1 here

$\sim \begin{bmatrix} 1 & -\frac{1}{3} & \vline & \frac{7}{3} \\ 2 & 3 & \vline & 1 \end{bmatrix}$ $\quad R_2 + (-2)R_1 \to R_2$

Need a 0 here
$-2 \quad \frac{2}{3} \quad -\frac{14}{3}$

$\sim \begin{bmatrix} 1 & -\frac{1}{3} & \vline & \frac{7}{3} \\ 0 & \frac{11}{3} & \vline & -\frac{11}{3} \end{bmatrix}$ $\quad \frac{3}{11}R_2 \to R_2$

Need a 1 here
Need a 0 here

$\sim \begin{bmatrix} 1 & -\frac{1}{3} & \vline & \frac{7}{3} \\ 0 & 1 & \vline & -1 \end{bmatrix}$ $\quad R_1 + \frac{1}{3}R_2 \to R_1$

$0 \quad \frac{1}{3} \quad -\frac{1}{3}$

$\sim \begin{bmatrix} 1 & 0 & \vline & 2 \\ 0 & 1 & \vline & -1 \end{bmatrix}$

Therefore $x_1 = 2$ and $x_2 = -1$.

31. $\begin{bmatrix} 3 & 2 & | & 4 \\ 2 & -1 & | & 5 \end{bmatrix}$ $\frac{1}{3}R_1 \rightarrow R_1$

Need a 1 here

$\sim \begin{bmatrix} 1 & \frac{2}{3} & | & \frac{4}{3} \\ 2 & -1 & | & 5 \end{bmatrix}$ $R_2 + \underbrace{(-2)R_1}_{} \rightarrow R_2$

Need a 0 here

$-2 \quad -\frac{4}{3} \quad -\frac{8}{3}$

$\sim \begin{bmatrix} 1 & \frac{2}{3} & | & \frac{4}{3} \\ 0 & -\frac{7}{3} & | & \frac{7}{3} \end{bmatrix}$ $-\frac{3}{7}R_2 \rightarrow R_2$

Need a 1 here
Need a 0 here

$\sim \begin{bmatrix} 1 & \frac{2}{3} & | & \frac{4}{3} \\ 0 & 1 & | & -1 \end{bmatrix}$ $R_1 + \left(-\frac{2}{3}\right)R_2 \rightarrow R_1$

$0 \quad -\frac{2}{3} \quad \frac{2}{3}$

$\sim \begin{bmatrix} 1 & 0 & | & 2 \\ 0 & 1 & | & -1 \end{bmatrix}$

Therefore $x_1 = 2$ and $x_2 = -1$.

33. $\begin{bmatrix} 0.2 & -0.5 & | & 0.07 \\ 0.8 & -0.3 & | & 0.79 \end{bmatrix}$ $5R_1 \rightarrow R_1$

Need a 1 here

$\sim \begin{bmatrix} 1 & -2.5 & | & 0.35 \\ 0.8 & -0.3 & | & 0.79 \end{bmatrix}$ $R_2 + \underbrace{(-0.8)R_1}_{} \rightarrow R_2$

Need a 0 here

$-0.8 \quad 2 \quad -0.28$

$\sim \begin{bmatrix} 1 & -2.5 & | & 0.35 \\ 0 & 1.7 & | & 0.51 \end{bmatrix}$ $\frac{1}{1.7}R_2 \rightarrow R_2$

Need a 1 here
Need a 0 here

$\sim \begin{bmatrix} 1 & -2.5 & | & 0.35 \\ 0 & 1 & | & 0.3 \end{bmatrix}$ $R_1 + \underbrace{2.5R_2}_{} \rightarrow R_2$

$0 \quad 2.5 \quad 0.75$

$\sim \begin{bmatrix} 1 & 0 & | & 1.1 \\ 0 & 1 & | & 0.3 \end{bmatrix}$

Therefore $x_1 = 1.1$ and $x_2 = 0.3$

Exercise 7-3

Key Ideas and Formulas

The method used for solving large systems of linear equations is called Gauss-Jordan elimination. We use a step-by-step procedure to transform the augmented matrix into reduced form, from which the solution to the system can be read by inspection.

A matrix is in reduced form if:

1. Each row consisting entirely of 0's is below any row having at least one non-zero element.

2. The left-most non-zero element in each row is 1.

3. The column containing the left-most 1 of a given row has 0's above and below the 1.

4. The left-most 1 in any row is the right of the left-most 1 in the preceding row.

Gauss-Jordan Elimination:

Write the augmented matrix of the system. Then

1. Choose the left-most non-zero column and use appropriate row operations to get a 1 at the top.

2. Use multiples of the first row to get 0's in all places below the 1 obtained in step 1.

3. Delete (mentally) the top row and first column of the matrix, yielding a submatrix. Repeat steps 1 and 2 with this submatrix. Continue this process (steps 1-3) until it is not possible to go further.

4. Consider the whole matrix again. Begin with the bottom non-zero row and use appropriate multiples of it to get 0's above the left-most 1. Continue this process, moving up row by row, until the matrix is finally in reduced form. [Note: If at any point we obtain a row with all 0's to left of the vertical line and a non zero number to the right, we stop, having obtained a contradiction.]

---

1. Yes
3. No. Condition 1 is violated.
5. No. Condition 2 is violated.
7. Yes

9.
$$x_1 \qquad\qquad = -2$$
$$\qquad x_2 \qquad = 3$$
$$\qquad\qquad x_3 = 0$$

The system is already solved.

13.
$$x_1 \qquad\qquad = 0$$
$$\qquad x_2 \qquad = 0$$
$$\qquad\qquad 0 = 1$$
The system has no solution.

11.
$$x_1 \qquad - 2x_3 = 3$$
$$\qquad x_2 + x_3 = -5$$

Solution
$$x_3 = t$$
$$x_2 = -5 - x_3 = -5 - t$$
$$x_1 = 3 + 2x_3 = 3 + 2t$$
Thus $x_1 = 2t + 3, x_2 = -t - 5,$
$x_3 = t$ is the solution,
for t any real number.

15. $\begin{aligned} x_1 - 2x_2 \quad\quad\; - \;\; 3x_4 &= -5 \\ x_3 + 3x_4 &= 2 \end{aligned}$

Solution:

$x_4 = t$

$x_3 = 2 - 3x_4 = 2 - 3t$

$x_2 = s$

$x_1 = -5 + 2x_2 + 3x_4 = -5 + 2s + 3t$

Thus $x_1 = 2s + 3t - 5, x_2 = s,$

$x_3 = -3t + 2, x_4 = t$ is the solution,

for s and t any real numbers.

17. $\begin{bmatrix} 1 & 2 & -1 \\ 0 & 1 & 3 \end{bmatrix} \quad R_1 + (-2)R_2 \rightarrow R_1$

Need a 0 here

$\sim \begin{bmatrix} 1 & 0 & -7 \\ 0 & 1 & 3 \end{bmatrix}$

19. $\begin{bmatrix} 1 & 0 & -3 & 1 \\ 0 & 1 & 2 & 0 \\ 0 & 0 & 3 & -6 \end{bmatrix} \quad \frac{1}{3}R_3 \rightarrow R_3$

Need a 1 here

Need 0's here

$\sim \begin{bmatrix} 1 & 0 & -3 & 1 \\ 0 & 1 & 2 & 0 \\ 0 & 0 & 1 & -2 \end{bmatrix} \quad \begin{matrix} R_1 + 3R_3 \rightarrow R_1 \\ R_2 + (-2)R_3 \rightarrow R_2 \end{matrix}$

$\sim \begin{bmatrix} 1 & 0 & 0 & -5 \\ 0 & 1 & 0 & 4 \\ 0 & 0 & 1 & -2 \end{bmatrix}$

21. $\sim \begin{bmatrix} 1 & 2 & -2 & -1 \\ 0 & 3 & -6 & 1 \\ 0 & -1 & 2 & -\frac{1}{3} \end{bmatrix} \quad \frac{1}{3}R_2 \rightarrow R_2$

Need a 1 here

$\sim \begin{bmatrix} 1 & 2 & -2 & -1 \\ 0 & 1 & -2 & \frac{1}{3} \\ 0 & -1 & 2 & -\frac{1}{3} \end{bmatrix} \quad \begin{matrix} R_1 + (-2)R_2 \rightarrow R_1 \\ \\ R_2 + R_3 \rightarrow R_3 \end{matrix}$

Need 0's here

$\sim \begin{bmatrix} 1 & 0 & 2 & -\frac{5}{3} \\ 0 & 1 & -2 & \frac{1}{3} \\ 0 & 0 & 0 & 0 \end{bmatrix}$

23. $\begin{bmatrix} 2 & 4 & -10 & -2 \\ 3 & 9 & -21 & 0 \\ 1 & 5 & -12 & 1 \end{bmatrix} \quad R_1 \leftrightarrow R_3$

Need a 1 here

$$\sim \begin{bmatrix} 1 & 5 & -12 & | & 1 \\ 3 & 9 & -21 & | & 0 \\ 2 & 4 & -10 & | & -2 \end{bmatrix} \quad \begin{matrix} R_2 + (-3)R_1 \to R_2 \\ R_3 + (-2)R_1 \to R_3 \end{matrix}$$

Need 0's here

$$\sim \begin{bmatrix} 1 & 5 & -12 & | & 1 \\ 0 & -6 & 15 & | & -3 \\ 0 & -6 & 14 & | & -4 \end{bmatrix} \quad -\tfrac{1}{6}R_2 \to R_2$$

Need a 1 here

$$\sim \begin{bmatrix} 1 & 5 & -12 & | & 1 \\ 0 & 1 & -\tfrac{5}{2} & | & \tfrac{1}{2} \\ 0 & -6 & 14 & | & -4 \end{bmatrix} \quad \begin{matrix} R_1 + (-5)R_2 \to R_1 \\ \\ R_3 + 6R_2 \to R_3 \end{matrix}$$

Need 0's here

$$\sim \begin{bmatrix} 1 & 0 & \tfrac{1}{2} & | & -\tfrac{3}{2} \\ 0 & 1 & -\tfrac{5}{2} & | & \tfrac{1}{2} \\ 0 & 0 & -1 & | & -1 \end{bmatrix} \quad -R_3 \to R_3$$

Need a 1 here

$$\sim \begin{bmatrix} 1 & 0 & \tfrac{1}{2} & | & -\tfrac{3}{2} \\ 0 & 1 & -\tfrac{5}{2} & | & \tfrac{1}{2} \\ 0 & 0 & 1 & | & 1 \end{bmatrix} \quad \begin{matrix} R_1 + (-\tfrac{1}{2})R_3 \to R_1 \\ R_2 + \tfrac{5}{2}R_3 \to R_2 \end{matrix}$$

Need 0's here

$$\sim \begin{bmatrix} 1 & 0 & 0 & | & -2 \\ 0 & 1 & 0 & | & 3 \\ 0 & 0 & 1 & | & 1 \end{bmatrix}$$

Therefore $x_1 = -2$, $x_2 = 3$, and $x_3 = 1$.

25. $\quad \begin{bmatrix} 3 & 8 & -1 & | & -18 \\ 2 & 1 & 5 & | & 8 \\ 2 & 4 & 2 & | & -4 \end{bmatrix} \quad \begin{matrix} \tfrac{1}{2}R_3 \to R_3 \\ \\ R_3 \leftrightarrow R_1 \end{matrix}$

Need a 1 here

$$\sim \begin{bmatrix} 1 & 2 & 1 & | & -2 \\ 2 & 1 & 5 & | & 8 \\ 3 & 8 & -1 & | & -18 \end{bmatrix} \quad \begin{matrix} R_2 + (-2)R_1 \to R_2 \\ R_3 + (-3)R_1 \to R_3 \end{matrix}$$

Need 0's here

$$\sim \begin{bmatrix} 1 & 2 & 1 & | & -2 \\ 0 & -3 & 3 & | & 12 \\ 0 & 2 & -4 & | & -12 \end{bmatrix} \quad -\tfrac{1}{3}R_2 \to R_2$$

Need a 1 here

$$\sim \begin{bmatrix} 1 & 2 & 1 & -2 \\ 0 & 1 & -1 & -4 \\ 0 & 2 & -4 & -12 \end{bmatrix} \quad \begin{array}{l} R_1 + (-2)R_2 \to R_1 \\ \\ R_3 + (-2)R_2 \to R_3 \end{array}$$

Need 0's here

$$\sim \begin{bmatrix} 1 & 0 & 3 & 6 \\ 0 & 1 & -1 & -4 \\ 0 & 0 & -2 & -4 \end{bmatrix} \quad -\tfrac{1}{2}R_3 \to R_3$$

Need a 1 here

$$\sim \begin{bmatrix} 1 & 0 & 3 & 6 \\ 0 & 1 & -1 & -4 \\ 0 & 0 & 1 & 2 \end{bmatrix} \quad \begin{array}{l} R_1 + (-3)R_3 \to R_1 \\ R_2 + R_3 \to R_2 \end{array}$$

Need 0's here

$$\sim \begin{bmatrix} 1 & 0 & 0 & 0 \\ 0 & 1 & 0 & -2 \\ 0 & 0 & 1 & 2 \end{bmatrix}$$

Therefore $x_1 = 0, x_2 = -2$, and $x_3 = 2$.

27. $\begin{bmatrix} 2 & -1 & -3 & 8 \\ 1 & -2 & 0 & 7 \end{bmatrix} \quad R_1 \leftrightarrow R_2$

$$\sim \begin{bmatrix} 1 & -2 & 0 & 7 \\ 2 & -1 & -3 & 8 \end{bmatrix} \quad R_2 + (-2)R_1 \to R_2$$

$$\sim \begin{bmatrix} 1 & -2 & 0 & 7 \\ 0 & 3 & -3 & -6 \end{bmatrix} \quad \tfrac{1}{3}R_2 \to R_2$$

$$\sim \begin{bmatrix} 1 & -2 & 0 & 7 \\ 0 & 1 & -1 & -2 \end{bmatrix} \quad R_1 + 2R_2 \to R_1$$

$$\sim \begin{bmatrix} 1 & 0 & -2 & 3 \\ 0 & 1 & -1 & -2 \end{bmatrix}$$

Let $x_3 = t$. Then

$$\begin{aligned} x_2 - x_3 &= -2 \\ x_2 &= x_3 - 2 = t - 2 \\ x_1 - 2x_3 &= 3 \\ x_1 &= 2x_3 + 3 = 2t + 3 \end{aligned}$$

Solution: $x_1 = 2t + 3, x_2 = t - 2$,
$x_3 = t, t$ any real number

29. $\begin{bmatrix} 2 & 3 & -1 & 1 \\ 1 & -2 & 2 & -2 \end{bmatrix} \quad R_1 \leftrightarrow R_2$

$$\sim \begin{bmatrix} 1 & -2 & 2 & -2 \\ 2 & 3 & -1 & 1 \end{bmatrix} \quad R_2 + (-2)R_1 \to R_2$$

$$\sim \begin{bmatrix} 1 & -2 & 2 & -2 \\ 0 & 7 & -5 & 5 \end{bmatrix} \quad \tfrac{1}{7}R_2 \to R_2$$

$$\sim \begin{bmatrix} 1 & -2 & 2 & -2 \\ 0 & 1 & -\tfrac{5}{7} & \tfrac{5}{7} \end{bmatrix} \quad R_1 + 2R_2 \to R_1$$

$$\sim \begin{bmatrix} 1 & 0 & \frac{4}{7} & \Big| & -\frac{4}{7} \\ 0 & 1 & -\frac{5}{7} & \Big| & \frac{5}{7} \end{bmatrix}$$

Let $x_3 = t$. Then

$$x_2 \quad - \quad \frac{5}{7}x_3 = \frac{5}{7}$$
$$x_2 \quad = \quad \frac{5}{7}x_3 + \frac{5}{7}$$
$$= \quad \frac{5t+5}{7}$$
$$x_1 + \frac{4}{7}x_3 \quad = \quad -\frac{4}{7}$$
$$x_1 \quad = \quad \frac{-4t-4}{7}$$

Solution: $x_1 = \frac{-4t-4}{7}$,

$x_2 = \frac{5t+5}{7}, x_3 = t, t$ any real number

31. $\begin{bmatrix} 2 & 2 & \Big| & 2 \\ 1 & 2 & \Big| & 3 \\ 0 & -3 & \Big| & -6 \end{bmatrix}$ $R_1 \leftrightarrow R_2$

$$\sim \begin{bmatrix} 1 & 2 & \Big| & 3 \\ 2 & 2 & \Big| & 2 \\ 0 & -3 & \Big| & -6 \end{bmatrix} \quad R_2 + (-2)R_1 \to R_2$$

$$\sim \begin{bmatrix} 1 & 2 & \Big| & 3 \\ 0 & -2 & \Big| & -4 \\ 0 & -3 & \Big| & -6 \end{bmatrix} \quad -\frac{1}{2}R_2 \to R_2$$

$$\sim \begin{bmatrix} 1 & 2 & \Big| & 3 \\ 0 & 1 & \Big| & 2 \\ 0 & -3 & \Big| & -6 \end{bmatrix} \quad \begin{matrix} R_1 \; + \; (-2)R_2 \to R_1 \\ \\ R_3 \; + \; 3R_2 \to R_3 \end{matrix}$$

$$\sim \begin{bmatrix} 1 & 0 & \Big| & -1 \\ 0 & 1 & \Big| & 2 \\ 0 & 0 & \Big| & 0 \end{bmatrix}$$

Solution: $x_1 = -1, x_2 = 2$

33. $\begin{bmatrix} 2 & -1 & \Big| & 0 \\ 3 & 2 & \Big| & 7 \\ 1 & -1 & \Big| & -2 \end{bmatrix}$ $R_1 \leftrightarrow R_3$

$$\sim \begin{bmatrix} 1 & -1 & \Big| & -2 \\ 3 & -2 & \Big| & 7 \\ 2 & -1 & \Big| & 0 \end{bmatrix} \quad \begin{matrix} R_2 + (-3)R_1 \to R_2 \\ R_3 + (-2)R_1 \to R_3 \end{matrix}$$

$$\sim \begin{bmatrix} 1 & -1 & \Big| & -2 \\ 0 & 5 & \Big| & 13 \\ 0 & 1 & \Big| & 4 \end{bmatrix} \quad R_2 \leftrightarrow R_1$$

$$\sim \begin{bmatrix} 1 & -1 & \Big| & -2 \\ 0 & 1 & \Big| & 4 \\ 0 & 5 & \Big| & 13 \end{bmatrix} \quad R_3 + (-5)R_2 \to R_3$$

$$\sim \begin{bmatrix} 1 & -1 & -2 \\ 0 & 1 & 4 \\ 0 & 0 & -7 \end{bmatrix}$$

Since the last row corresponds to the equation $0x_2 + 0x_2 = -7$, there is no solution.

35. $\begin{bmatrix} 3 & -4 & -1 & 1 \\ 2 & -3 & 1 & 1 \\ 1 & -2 & 3 & 2 \end{bmatrix}$ $R_1 \leftrightarrow R_3$

$\sim \begin{bmatrix} 1 & -2 & 3 & 2 \\ 2 & -3 & 1 & 1 \\ 3 & -4 & -1 & 1 \end{bmatrix}$ $\begin{aligned} R_2 &+ (-2)R_1 \to R_2 \\ R_3 &+ (-3)R_1 \to R_3 \end{aligned}$

$\sim \begin{bmatrix} 1 & -2 & 3 & 2 \\ 0 & 1 & -5 & -3 \\ 0 & 0 & 0 & 1 \end{bmatrix}$

Since the last row corresponds to the equation $0x_1 + 0x_2 + 0x_3 = 1$,
there is no solution.

37. $\begin{bmatrix} -2 & 1 & 3 & -7 \\ 1 & -4 & 2 & 0 \\ 1 & -3 & 1 & 1 \end{bmatrix}$ $R_1 \leftrightarrow R_2$

$\sim \begin{bmatrix} 1 & -4 & 2 & 0 \\ -2 & 1 & 3 & -7 \\ 1 & -3 & 1 & 1 \end{bmatrix}$ $\begin{aligned} R_2 &+ 2R_1 \to R_2 \\ R_3 &+ (-1)R_1 \to R_3 \end{aligned}$

$\sim \begin{bmatrix} 1 & -4 & 2 & 0 \\ 0 & -7 & 7 & -7 \\ 0 & 1 & -1 & 1 \end{bmatrix}$ $R_2 \leftrightarrow R_3$

$\sim \begin{bmatrix} 1 & -4 & 2 & 0 \\ 0 & 1 & -1 & 1 \\ 0 & -7 & 7 & -7 \end{bmatrix}$ $\begin{aligned} R_1 &+ 4R_2 \to R_1 \\ R_3 &+ 7R_2 \to R_3 \end{aligned}$

$\sim \begin{bmatrix} 1 & 0 & -2 & 4 \\ 0 & 1 & -1 & 1 \\ 0 & 0 & 0 & 0 \end{bmatrix}$

Let $x_3 = t$. Then
$$\begin{aligned} x_2 - x_3 &= 1 \\ x_2 &= x_3 + 1 = t + 1 \\ x_1 - 2x_3 &= 4 \\ x_1 = 2x_3 + 4 &= 2t + 4 \end{aligned}$$
Solution: $x_1 = 2t + 4, x_2 = t + 1, x_3 = t$, t any real number.

39. $\begin{bmatrix} 2 & -2 & -4 & -2 \\ -3 & 3 & 6 & 3 \end{bmatrix}$ $\begin{aligned} \tfrac{1}{2}R_1 &\to R_1 \\ \tfrac{1}{3}R_2 &\to R_2 \end{aligned}$

$\sim \begin{bmatrix} 1 & 1 & -2 & -1 \\ -1 & -1 & 2 & 1 \end{bmatrix}$ $R_2 + R_1 \to R_2$

$$\sim \begin{bmatrix} 1 & -1 & -2 & | & -1 \\ 0 & 0 & 0 & | & 0 \end{bmatrix}$$

Let $x_3 = t, x_2 = s$. Then

$$x_1 - x_2 - 2x_3 = -1$$
$$x_1 = x_2 + 2x_3 - 1$$
$$= s + 2t - 1$$

Solution $x_1 = s + 2t - 1, x_2 = s, x_3 = t$ s and t any real numbers

41. $\begin{bmatrix} 2 & -3 & 3 & | & -15 \\ 3 & 2 & -5 & | & 19 \\ 5 & -4 & -2 & | & -2 \end{bmatrix}$ $\frac{1}{2}R_1 \rightarrow R_1$

$$\sim \begin{bmatrix} 1 & -\frac{3}{2} & \frac{3}{2} & | & -\frac{15}{2} \\ 3 & 2 & -5 & | & 19 \\ 5 & -4 & -2 & | & -2 \end{bmatrix} \quad \begin{matrix} R_2 + (-3)R_1 \rightarrow R_2 \\ R_3 + (-5)R_1 \rightarrow R_3 \end{matrix}$$

$$\sim \begin{bmatrix} 1 & -\frac{3}{2} & \frac{3}{2} & | & -\frac{15}{2} \\ 0 & \frac{13}{2} & -\frac{19}{2} & | & \frac{83}{2} \\ 0 & \frac{7}{2} & -\frac{19}{2} & | & \frac{71}{2} \end{bmatrix} \quad \frac{2}{13}R_2 \rightarrow R_2$$

$$\sim \begin{bmatrix} 1 & -\frac{3}{2} & \frac{3}{2} & | & -\frac{15}{2} \\ 0 & 1 & -\frac{19}{13} & | & \frac{83}{13} \\ 0 & \frac{7}{2} & -\frac{19}{2} & | & \frac{71}{2} \end{bmatrix} \quad \begin{matrix} R_1 + \frac{3}{2}R_2 \rightarrow R_1 \\ R_3 + (-\frac{7}{2})R_2 \rightarrow R_3 \end{matrix}$$

$$\sim \begin{bmatrix} 1 & 0 & -\frac{9}{13} & | & \frac{27}{13} \\ 0 & 1 & -\frac{19}{13} & | & \frac{83}{13} \\ 0 & 0 & -\frac{57}{13} & | & \frac{171}{13} \end{bmatrix} \quad -\frac{13}{57}R_3 \rightarrow R_3$$

$$\sim \begin{bmatrix} 1 & 0 & -\frac{9}{13} & | & \frac{27}{13} \\ 0 & 1 & -\frac{19}{13} & | & \frac{83}{13} \\ 0 & 0 & 1 & | & -3 \end{bmatrix} \quad \begin{matrix} R_1 + \frac{9}{13}R_3 \rightarrow R_1 \\ R_2 + \frac{19}{13}R_3 \rightarrow R_2 \end{matrix}$$

$$\sim \begin{bmatrix} 1 & 0 & 0 & | & 0 \\ 0 & 1 & 0 & | & 2 \\ 0 & 0 & 1 & | & -3 \end{bmatrix}$$

Solution: $x_1 = 0, x_2 = 2, x_3 = -3$

43. $\sim \begin{bmatrix} 5 & -3 & 2 & | & 13 \\ 2 & 4 & -3 & | & -9 \\ 4 & -2 & 5 & | & 13 \end{bmatrix} \quad R_1 + (-1)R_3 \rightarrow R_1$

$$\sim \begin{bmatrix} 1 & -1 & -3 & | & 0 \\ 2 & 4 & -3 & | & -9 \\ 4 & -2 & 5 & | & 13 \end{bmatrix} \quad \begin{matrix} R_2 + (-2)R_1 \rightarrow R_2 \\ R_3 + (-4)R_1 \rightarrow R_3 \end{matrix}$$

$$\sim \begin{bmatrix} 1 & -1 & -3 & | & 0 \\ 0 & 6 & 3 & | & -9 \\ 0 & 2 & 17 & | & 13 \end{bmatrix} \quad \frac{1}{6}R_2 \rightarrow R_2$$

$$\sim \begin{bmatrix} 1 & -1 & -3 & | & 0 \\ 0 & 1 & \frac{1}{2} & | & -\frac{3}{2} \\ 0 & 2 & 17 & | & 13 \end{bmatrix} \begin{matrix} R_1 + R_2 \to R_1 \\ \\ R_3 + (-2)R_2 \to R_3 \end{matrix}$$

$$\sim \begin{bmatrix} 1 & 0 & -\frac{5}{2} & | & -\frac{3}{2} \\ 0 & 1 & \frac{1}{2} & | & -\frac{3}{2} \\ 0 & 0 & 16 & | & 16 \end{bmatrix} \begin{matrix} \\ \\ \frac{1}{16}R_3 \to R_3 \end{matrix}$$

$$\sim \begin{bmatrix} 1 & 0 & -\frac{5}{2} & | & -\frac{3}{2} \\ 0 & 1 & \frac{1}{2} & | & -\frac{3}{2} \\ 0 & 0 & 1 & | & 1 \end{bmatrix} \begin{matrix} R_1 + \frac{5}{2}R_3 \to R_1 \\ R_2 + (-\frac{1}{2})R_3 \to R_2 \\ \end{matrix}$$

$$\sim \begin{bmatrix} 1 & 0 & 0 & | & 1 \\ 0 & 1 & 0 & | & -2 \\ 0 & 0 & 1 & | & 1 \end{bmatrix}$$

Solution: $x_1 = 1, x_2 = -2, x_3 = 1$

45. $\begin{bmatrix} 1 & 2 & -4 & -1 & | & 7 \\ 2 & 5 & -9 & -4 & | & 16 \\ 1 & 5 & -7 & -7 & | & 13 \end{bmatrix} \begin{matrix} \\ R_2 + (-2)R_1 \to R_2 \\ R_3 + (-1)R_1 \to R_3 \end{matrix}$

$$\sim \begin{bmatrix} 1 & 2 & -4 & -1 & | & 7 \\ 0 & 1 & -1 & -2 & | & 2 \\ 0 & 3 & -3 & -6 & | & 6 \end{bmatrix} \begin{matrix} R_1 + (-2)R_2 \to R_1 \\ \\ R_3 + (-3)R_2 \to R_3 \end{matrix}$$

$$\sim \begin{bmatrix} 1 & 0 & -2 & 3 & | & 3 \\ 0 & 1 & -1 & -2 & | & 2 \\ 0 & 0 & 0 & 0 & | & 0 \end{bmatrix}$$

Let $x_4 = t, x_3 = s$.
Then
$$\begin{aligned} x_2 - x_3 - 2x_4 &= 2 \\ x_2 &= s + 2t + 2 \\ x_1 - 2x_3 + 3x_4 &= 3 \\ x_1 &= 2x_3 - 3x_4 + 3 \\ &= 2s - 3t + 3 \end{aligned}$$

Solution:
$$\begin{aligned} x_1 &= 2s - 3t + 3, \\ x_2 &= s + 2t + 2, \\ x_3 &= s, x_4 = t, \end{aligned}$$
s and t any real numbers

47. Let x_1 = number of 15 − cent stamps
     x_2 = number of 20 − cent stamps
     x_3 = number of 35 − cent stamps

Then $x_1 + x_2 + x_3 = 45$ (total number of stamps)
      $15x_1 + 20x_2 + 35x_3 = 1400$ (total value of stamps)

We write the augmented matrix and solve by Gauss-Jordan elimination.

$$\begin{bmatrix} 1 & 1 & 1 & | & 45 \\ 15 & 20 & 35 & | & 1400 \end{bmatrix} \quad R_2 + (-15)R_1 \to R_2$$

$$\sim \begin{bmatrix} 1 & 1 & 1 & | & 45 \\ 0 & 5 & 20 & | & 725 \end{bmatrix} \quad \tfrac{1}{5}R_2 \rightarrow R_2$$

$$\sim \begin{bmatrix} 1 & 1 & 1 & | & 45 \\ 0 & 1 & 4 & | & 145 \end{bmatrix} \quad R_1 + (-1)R_2 \rightarrow R_1$$

$$\sim \begin{bmatrix} 1 & 0 & -3 & | & -100 \\ 0 & 1 & 4 & | & 145 \end{bmatrix}$$

This augmented matrix is in reduced form. It corresponds to the system:

$x_1 - 3x_3 = -100$

$x_2 + 4x_3 = 145$

Let $x_3 = t$. Then

$$\begin{aligned} x_2 &= -4x_3 + 145 \\ &= -4t + 145 \\ x_1 &= 3x_3 - 100 \\ &= 3t - 100 \end{aligned}$$

A solution is achieved, not for every real value of t, but for integer values of t that give rise to non-negative x_1, x_2, x_3.

$x_1 \geq 0$ means $3t - 100 \geq 0$ or $t \geq 33\tfrac{1}{3}$

$x_2 \geq 0$ means $-4t + 145 \leq 0$ or $t \geq 36\tfrac{1}{4}$

The only integer values of t that satisfy these conditions are $34, 35, 36$.

Thus we have the solutions:

$$\begin{aligned} x_1 &= (3t - 100)15 - \text{cent stamps} \\ x_2 &= (145 - 4t)20 - \text{cent stamps} \\ x_3 &= t\,35 - \text{cent stamps} \end{aligned}$$

where $t = 34, 35,$ or $36,$

49.   Let $x_1 = $ number of $500 - $ cc containers of 10% solution

$x_2 = $ number of $500 - $ cc containers of 20% solution

$x_3 = $ number of $1,000 - $ cc containers of 50% solution

Then $500x_1 + 500x_2 + 1,000x_3 = 12,000$(Total number of cc)

$.10(500x_1) + .20(500x_2) + .50(1,000x_3) = .30(12,000)$

(total amount of ingredient in solution)

After simplification, we have

$x_1 + x_2 + 2x_3 = 24$

$x_1 + 2x_2 + 10x_3 = 72$

We write the augmented matrix and solve by Gauss-Jordan elimination.

$$\begin{bmatrix} 1 & 1 & 2 & | & 24 \\ 1 & 2 & 10 & | & 72 \end{bmatrix} \quad R_2 + (-1)R_1 \rightarrow R_2$$

$$\sim \begin{bmatrix} 1 & 1 & 2 & | & 24 \\ 0 & 1 & 8 & | & 48 \end{bmatrix} \quad R_1 + (-1)R_2 \rightarrow R_1$$

$$\sim \begin{bmatrix} 1 & 0 & -6 & | & -24 \\ 0 & 1 & 8 & | & 48 \end{bmatrix}$$

This augmented matrix is in reduced form. It corresponds to the system:

$$x_1 - 6x_3 = -24$$
$$x_2 + 8x_3 = 48$$

Let $x_3 = t$. Then

$$x_2 = -8x_3 + 48$$
$$= -8t + 48$$
$$x_1 = 6x_3 - 24$$
$$= 6t - 24$$

Solution is achieved not for every real value of t, but for integer values of t that give rise to non-negative x_1, x_2, x_3.

$$x_1 \geq 0 \text{ means } 6t - 24 \geq 0 \text{ or } t \geq 4$$
$$x_2 \geq 0 \text{ means } -8t + 48 \geq 0 \text{ or } t \leq 6$$

Thus we have the solution:

$$x_1 = (6t - 24)500 - \text{cc containers of 10\% solution}$$
$$x_2 = (48 - 8t)500 - \text{cc containers of 20\% solution}$$
$$x_3 = t\, 1000 - \text{cc containers of 50\% solution}$$

where $t = 4, 5,$ or 6.

51. If the curve passes through a point, the coordinates of the point satisfy the equation of the curve. Hence,

$$3 = a + b(-2) + c(-2)^2$$
$$2 = a + b(-1) + c(-1)^2$$
$$6 = a + b(1) + c(1)^2$$

After simplification, we have

$$a - 2b + 4c = 3$$
$$a - b + c = 2$$
$$a + b + c = 6$$

We write the augmented matrix and solve by Gauss-Jordan elimination.

$$\begin{bmatrix} 1 & -2 & 4 & | & 3 \\ 1 & -1 & 1 & | & 2 \\ 1 & 1 & 1 & | & 6 \end{bmatrix} \quad \begin{matrix} \\ R_2 + (-1)R_1 \to R_2 \\ R_3 + (-1)R_1 \to R_3 \end{matrix}$$

$$\sim \begin{bmatrix} 1 & -2 & 4 & | & 3 \\ 0 & -1 & -3 & | & -1 \\ 0 & 3 & -3 & | & 3 \end{bmatrix} \quad \begin{matrix} R_1 + 2R_2 \to R_1 \\ \\ R_3 + (-3)R_2 \to R_3 \end{matrix}$$

$$\sim \begin{bmatrix} 1 & 0 & -2 & | & 1 \\ 0 & -1 & -3 & | & -1 \\ 0 & 0 & 6 & | & 6 \end{bmatrix} \quad \frac{1}{6}R_3 \to R_3$$

$$\sim \begin{bmatrix} 1 & 0 & -2 & | & 1 \\ 0 & 1 & -3 & | & -1 \\ 0 & 0 & 1 & | & 1 \end{bmatrix} \quad \begin{matrix} R_1 + 2R_3 \to R_1 \\ R_2 + 3R_3 \to R_2 \end{matrix}$$

$$\sim \begin{bmatrix} 1 & 0 & 0 & | & 3 \\ 0 & 1 & 0 & | & 2 \\ 0 & 0 & 1 & | & 1 \end{bmatrix}$$

Thus $a = 3, b = 2, c = 1$.

53. If the curve passes through a point, the coordinates of the point satisfy the

equation of the curve. Hence,

$$6^2 + 2^2 + a(6) + b(2) + c = 0$$
$$4^2 + 6^2 + a(4) + b(6) + c = 0$$
$$(-3)^2 + (-1)^2 + a(-3) + b(-1) + c = 0$$

After simplification, we have

$$6a + 2b + c = -40$$
$$4a + 6b + c = -52$$
$$-3a - b + c = -10$$

We write the augmented matrix and solve by Gauss-Jordan elimination.

$$\begin{bmatrix} 6 & 2 & 1 & | & -40 \\ 4 & 6 & 1 & | & -52 \\ -3 & -1 & 1 & | & -10 \end{bmatrix} \quad \frac{1}{6}R_1 \to R_1$$

$$\sim \begin{bmatrix} 1 & \frac{1}{3} & \frac{1}{6} & | & -\frac{20}{3} \\ 4 & 6 & 1 & | & -52 \\ -3 & -1 & 1 & | & -10 \end{bmatrix} \quad \begin{matrix} R_2 + (-4)R_1 \to R_2 \\ R_3 + 3R_1 \to R_3 \end{matrix}$$

$$\sim \begin{bmatrix} 1 & \frac{1}{3} & \frac{1}{6} & | & -\frac{20}{3} \\ 0 & \frac{14}{3} & \frac{1}{3} & | & -\frac{76}{3} \\ 0 & 0 & \frac{3}{2} & | & -30 \end{bmatrix} \quad \frac{2}{3}R_3 \to R_3$$

$$\sim \begin{bmatrix} 1 & \frac{1}{3} & \frac{1}{6} & | & -\frac{20}{3} \\ 0 & \frac{14}{3} & \frac{1}{3} & | & -\frac{76}{3} \\ 0 & 0 & 1 & | & -20 \end{bmatrix} \quad \begin{matrix} R_1 + (-\frac{1}{6})R_3 \to R_1 \\ R_2 + (-\frac{1}{3})R_3 \to R_2 \end{matrix}$$

$$\sim \begin{bmatrix} 1 & \frac{1}{3} & 0 & | & -\frac{10}{3} \\ 0 & \frac{14}{3} & 0 & | & -\frac{56}{3} \\ 0 & 0 & 1 & | & -20 \end{bmatrix} \quad \frac{3}{14}R_2 \to R_2$$

$$\sim \begin{bmatrix} 1 & \frac{1}{3} & 0 & | & -\frac{10}{3} \\ 0 & 1 & 0 & | & -4 \\ 0 & 0 & 1 & | & -20 \end{bmatrix} \quad R_1 + (-\frac{1}{3})R_2 \to R_1$$

$$\sim \begin{bmatrix} 1 & 0 & 0 & | & -2 \\ 0 & 1 & 0 & | & -4 \\ 0 & 0 & 1 & | & -20 \end{bmatrix}$$

Thus $a = -2, b = -4,$ and $c = -20$

55. Let x_1 = number of one − person boats
     x_2 = number of two − person boats
     x_3 = number of four − person boats
  We have

$$0.5x_1 + 1.0x_2 + 1.5x_3 = 380 \text{ cutting department}$$
$$0.6x_1 + 0.9x_2 + 1.2x_3 = 330 \text{ assembly department}$$
$$0.2x_1 + 0.3x_2 + 0.5x_3 = 120 \text{ packing department}$$

CommonError :
The facts in this problem do not
 justify the equation
$$0.5x_1 + 0.6x_2 + 0.2x_3 = 380$$

Clearing of decimals for convenience:

$$x_1 + 2x_2 + 3x_3 = 760$$
$$6x_1 + 9x_2 + 12x_3 = 3300$$
$$2x_1 + 3x_2 + 5x_3 = 1200$$

We write the augmented matrix and solve by Gauss-Jordan elimination:

$$\begin{bmatrix} 1 & 2 & 3 & 760 \\ 6 & 9 & 12 & 3300 \\ 2 & 3 & 5 & 1200 \end{bmatrix} \begin{array}{l} \\ R_2 + (-6)R_1 \to R_2 \\ R_3 + (-2)R_1 \to R_3 \end{array}$$

$$\sim \begin{bmatrix} 1 & 2 & 3 & 760 \\ 0 & -3 & -6 & -1260 \\ 0 & -1 & -1 & -320 \end{bmatrix} \begin{array}{l} \\ \frac{1}{3}R_2 \to R_2 \\ \\ \end{array}$$

$$\sim \begin{bmatrix} 1 & 2 & 3 & 760 \\ 0 & 1 & 2 & 420 \\ 0 & -1 & -1 & -320 \end{bmatrix} \begin{array}{l} R_1 + (-2)R_2 \to R_1 \\ \\ R_3 + R_2 \to R_3 \end{array}$$

$$\sim \begin{bmatrix} 1 & 0 & -1 & -80 \\ 0 & 1 & 2 & 420 \\ 0 & 0 & 1 & 100 \end{bmatrix} \begin{array}{l} R_1 + R_3 \to R_1 \\ R_2 + (-2)R_3 \to R_2 \\ \\ \end{array}$$

$$\sim \begin{bmatrix} 1 & 0 & 0 & 20 \\ 0 & 1 & 0 & 220 \\ 0 & 0 & 1 & 100 \end{bmatrix}$$

$$\begin{array}{rll} x_1 &=& 20 \text{ one } - \text{ person boats} \\ \text{Therefore} \quad x_2 &=& 220 \text{ two } - \text{ person boats} \\ x_3 &=& 100 \text{ four } - \text{ person boats} \end{array}$$

57.   This assumption discards the third equation. The system, cleared of decimals, reads

$$x_1 + 2x_2 + 3x_3 = 760$$
$$6x_1 + 9x_2 + 12x_3 = 3300$$

The augmented matrix becomes

$$\begin{bmatrix} 1 & 2 & 3 & 760 \\ 6 & 9 & 12 & 3300 \end{bmatrix}$$

We solve by Gauss-Jordan elimination. We start by introducing a 0 into the lower left corner:

$$\sim \begin{bmatrix} 1 & 2 & 3 & 760 \\ 0 & -3 & -6 & -1260 \end{bmatrix} \quad -\frac{1}{3}R_2 \to R_2$$

$$\sim \begin{bmatrix} 1 & 2 & 3 & 760 \\ 0 & 1 & 2 & 420 \end{bmatrix} \quad R_1 + (-2)R_2 \to R_1$$

$$\sim \begin{bmatrix} 1 & 0 & -1 & -80 \\ 0 & 1 & 2 & 420 \end{bmatrix}$$

This augmented matrix is in reduced form. It corresponds to the system:

$$x_1 - x_3 = -80$$
$$x_2 + 2x_3 = 420$$

Let $x_3 = t$. Then

$$x_2 = -2x_3 + 420$$
$$= -2t + 420$$
$$x_1 = x_3 - 80$$
$$= t - 80$$

A solution is achieved, not for every real value of t but for integer values of t that give rise to non-negative x_1, x_2, x_3.

$x_1 \geq 0$ means $t - 80 \geq 0$ or $t \geq 80$

$x_2 \geq 0$ means $-2t + 420 \geq 0$ or $210 \geq t$

Thus we have the solution

$$x_1 = (t - 80) \text{ one } - \text{person boats}$$
$$x_2 = (-2t + 420) \text{ two } - \text{person boats}$$
$$x_3 = t \text{ four } - \text{person boats}$$

$80 \leq t \leq 210, t$ an integer

59. In this case we have $x_3 = 0$ from the beginning. The three equations of problem 55, cleared of decimals, read:

$$x_1 + 2x_2 = 760$$
$$6x_1 + 9x_2 = 3300$$
$$2x_1 + 3x_2 = 1200$$

The augmented matrix becomes:

$$\begin{bmatrix} 1 & 2 & | & 760 \\ 6 & 9 & | & 3300 \\ 2 & 3 & | & 1200 \end{bmatrix}$$

Notice that the row operation

$R_2 + (-3)R_3 \rightarrow R_2$

transforms this into the equivalent augmented matrix:

$$\begin{bmatrix} 1 & 2 & | & 760 \\ 0 & 0 & | & -300 \\ 2 & 3 & | & 1200 \end{bmatrix}$$

Therefore, since the second row corresponds to the equation

$0x_1 + 0x_2 = -300$

there is no solution.

No production schedule will use all the work-hours in all departments.

61. Let $x_1 =$ number of ounces of food A.

$x_2 =$ number of ounces of food B.

$x_3 =$ number of ounces of food C.

Then

$$30x_1 + 10x_2 + 20c_3 = 340 \text{ (calcium)}$$
$$10x_1 + 10x_2 + 20x_3 = 180 \text{ (iron)}$$
$$10x_1 + 30x_2 + 20x_3 = 220 \text{ (vitamin} A)$$

Common Error:
The facts in this problem do not justify the equation
$30x_1 + 10x_2 + 10x_3 = 340$

or

$$3x_1 + x_2 + 2x_3 = 34$$
$$x_1 + x_2 + 2x_3 = 18$$
$$x_1 + 3x_2 + 2x_3 = 22$$

is the system to be solved. We form the augmented matrix and solve by Gauss-Jordan elimination.

$$\begin{bmatrix} 3 & 1 & 2 & | & 34 \\ 1 & 1 & 2 & | & 18 \\ 1 & 3 & 2 & | & 22 \end{bmatrix} \quad R_1 \leftrightarrow R_2$$

$$\sim \begin{bmatrix} 1 & 1 & 2 & | & 18 \\ 3 & 1 & 2 & | & 34 \\ 1 & 3 & 2 & | & 22 \end{bmatrix} \quad \begin{matrix} R_2 + (-3)R_1 \to R_2 \\ R_3 + (-1)R_1 \to R_3 \end{matrix}$$

$$\sim \begin{bmatrix} 1 & 1 & 2 & | & 18 \\ 0 & -2 & -4 & | & -20 \\ 0 & 2 & 0 & | & 4 \end{bmatrix} \quad -\tfrac{1}{2}R_2 \to R_2$$

$$\sim \begin{bmatrix} 1 & 1 & 2 & | & 18 \\ 0 & 1 & 2 & | & 10 \\ 0 & 2 & 0 & | & 4 \end{bmatrix} \quad \begin{matrix} R_1 + (-1)R_2 \to R_1 \\ R_3 + (-2)R_2 \to R_3 \end{matrix}$$

$$\sim \begin{bmatrix} 1 & 0 & 0 & | & 8 \\ 0 & 1 & 2 & | & 10 \\ 0 & 0 & -4 & | & -16 \end{bmatrix} \quad -\tfrac{1}{4}R_3 \to R_3$$

$$\sim \begin{bmatrix} 1 & 0 & 0 & | & 8 \\ 0 & 1 & 2 & | & 10 \\ 0 & 0 & 1 & | & 4 \end{bmatrix} \quad R_2 + (-2)R_3 \to R_2$$

$$\sim \begin{bmatrix} 1 & 0 & 0 & | & 8 \\ 0 & 1 & 0 & | & 2 \\ 0 & 0 & 1 & | & 4 \end{bmatrix}$$

Thus

$x_1 = $ 8 ounces food A

$x_2 = $ 2 ounces food B

$x_3 = $ 4 ounces food C

63. In this case we have $x_3 = 0$ from the beginning. The three equations of problem 61 become

$$30x_1 + 10x_2 = 340$$
$$10x_1 + 10x_2 = 180$$
$$10x_1 + 30x_2 = 220$$

or

$$3x_1 + x_2 = 34$$
$$x_1 + x_2 = 18$$
$$x_1 + 3x_2 = 22$$

The augmented matrix becomes

$$\begin{bmatrix} 3 & 1 & | & 34 \\ 1 & 1 & | & 18 \\ 1 & 3 & | & 22 \end{bmatrix}$$

We solve by Gauss-Jordan elimination,

starting by the row operation
$R_1 \leftrightarrow R_2$.

$$\begin{bmatrix} 1 & 1 & | & 18 \\ 3 & 1 & | & 34 \\ 1 & 3 & | & 22 \end{bmatrix} \quad \begin{matrix} \\ R_2 + (-3)R_1 \to R_2 \\ R_3 + (-1)R_1 \to R_3 \end{matrix}$$

$$\sim \begin{bmatrix} 1 & 1 & | & 18 \\ 0 & -2 & | & -20 \\ 0 & 2 & | & 4 \end{bmatrix} \quad \begin{matrix} \\ \\ R_3 + (-3)R_2 \to R_3 \end{matrix}$$

$$\sim \begin{bmatrix} 1 & 1 & | & 18 \\ 0 & -2 & | & -20 \\ 0 & 0 & | & -16 \end{bmatrix}$$

Since the third row corresponds to the equation
$0x_1 + 0x_2 = -16$, there is no solution.

65.   In this case we discard the third equation.
The system becomes
$$30x_1 + 10x_2 + 20x_3 = 340$$
$$10x_1 + 10x_2 + 20x_3 = 180$$
or
$$3x_1 + x_2 + 2x_3 = 34$$
$$x_1 + x_2 + 2x_3 = 18$$
The augmented matrix becomes
$$\begin{bmatrix} 3 & 1 & 2 & | & 34 \\ 1 & 1 & 2 & | & 18 \end{bmatrix}$$

We solve by Gauss-Jordan elimination, starting by the row
operation $R_1 \leftrightarrow R_2$.

$$\begin{bmatrix} 1 & 1 & 2 & | & 18 \\ 3 & 1 & 2 & | & 34 \end{bmatrix} \quad R_2 + (-3)R_1 \to R_2$$

$$\sim \begin{bmatrix} 1 & 1 & 2 & | & 18 \\ 0 & -2 & -4 & | & -20 \end{bmatrix} \quad -\tfrac{1}{2}R_2 \to R_2$$

$$\sim \begin{bmatrix} 1 & 1 & 2 & | & 18 \\ 0 & 1 & 2 & | & 10 \end{bmatrix} \quad R_1 + (-1)R_2 \to R_1$$

$$\sim \begin{bmatrix} 1 & 0 & 0 & | & 8 \\ 0 & 1 & 2 & | & 10 \end{bmatrix}$$

This augmented matrix is in reduced form.
It corresponds to the system
$$x_1 = 8$$
$$x_2 + 2x_3 = 10$$
$$\text{Let } x_3 = t$$
$$\text{Then } x_2 = -2x_3 + 10$$
$$= -2t + 10$$

A solution is achieved, not for every real value t, but for values of t that give rise to non-negative x_2, x_3.

$x_3 \geq 0$ means $t \geq 0$

$x_2 \geq 0$ means $-2t + 10 \geq 0, 5 \geq t$

Thus we have the solution

$x_1 = 8$ ounces food A

$x_2 = -2t + 10$ ounces food B

$x_3 = t$ ounces food C

$0 \leq t \leq 5$

67.   Let $x_1 = $ number of hours company A is to be scheduled

          $x_2 = $ number of hours company B is to be scheduled

In x_1 hours, company A can handle $30x_1$ telephone and $10x_1$ house contacts.

In x_2 hours, company B can handle $20x_2$ telephone and $20x_2$ house contacts.

We therefore have:

$30x_1 + 20x_2 = 600$ telephone contacts

$10x_1 + 20x_2 = 400$ telephone contacts

We form the augmented matrix and solve by Gauss-Jordan elimination.

$$\begin{bmatrix} 30 & 20 & | & 600 \\ 10 & 20 & | & 400 \end{bmatrix} \begin{matrix} \frac{1}{10}R_1 \to R_1 \\ \frac{1}{10}R_2 \to R_2 \end{matrix}$$

$$\sim \begin{bmatrix} 3 & 2 & | & 60 \\ 1 & 2 & | & 40 \end{bmatrix} R_1 \leftrightarrow R_2$$

$$\sim \begin{bmatrix} 1 & 2 & | & 40 \\ 3 & 2 & | & 60 \end{bmatrix} R_2 + (-3)R_1 \to R_2$$

$$\sim \begin{bmatrix} 1 & 2 & | & 40 \\ 0 & -4 & | & -60 \end{bmatrix} -\frac{1}{4}R_3 \to R_3$$

$$\sim \begin{bmatrix} 1 & 2 & | & 40 \\ 0 & 1 & | & 15 \end{bmatrix} R_1 + (-2)R_2 \to R_1$$

$$\sim \begin{bmatrix} 1 & 0 & | & 10 \\ 0 & 1 & | & 15 \end{bmatrix}$$

Therefore

$x_1 = 10$ hours company A

$x_2 = 15$ hours company B

Exercise 7-4

Key Ideas and Formulas

Nonlinear systems are systems that contain at least one nonlinear equation.

Nonlinear systems involving second degree terms can have at most four solutions, some of which may be imaginary.

Nonlinear systems can be solved by the substitution and, in some cases, the elimination methods of Section 7-1.

It is important to check the apparent solutions of any nonlinear system to insure that extraneous roots have not been introduced.

1.
$$x^2 + y^2 = 169$$
$$x = -12$$
$$(-12)^2 + y^2 = 169$$
$$y^2 = 25$$
$$y = \pm 5$$
Solutions: $(-12, 5), (-12, -5)$
Check: $-12 \overset{\vee}{=} -12$
$$(-12)^2 + (\pm 5)^2 \overset{\vee}{=} 169$$

3.
$$8x^2 - y^2 = 16$$
$$y = 2x$$
Substitute y from the second equation into the first equation.
$$8x^2 - (2x)^2 = 16$$
$$8x^2 - 4x^2 = 16$$
$$4x^2 = 16$$
$$x^2 = 4$$
$$x = \pm 2$$

For $x = 2$ For $x = -2$
$$y = 2(2) \qquad y = 2(-2)$$
$$y = 4 \qquad\quad y = -4$$
Solutions: $(2, 4), (-2, -4)$
Check:

For $(2, 4)$ For $(-2, -4)$
$$4 \overset{\vee}{=} 2 \cdot 2 \qquad\qquad -4 \overset{\vee}{=} 2(-2)$$
$$8(2)^2 - 4^2 \overset{\vee}{=} 16 \qquad 8(-2)^2 - (-4)^2 \overset{\vee}{=} 16$$

5.
$$3x^2 - 2y^2 = 25$$
$$x + y = 0$$
Solve for y in the first degree equation
$$y = -x$$
Substitute into the second degree equation.
$$3x^2 - 2(-x)^2 = 25$$
$$x^2 = 25$$
$$x = \pm 5$$
For $x = 5$ For $x = -5$
$$y = -5 \qquad y = 5$$
Solutions: $(5, -5), (-5, 5)$
Check:

$$\text{For } (5,-5) \qquad\qquad \text{For } (-5,5)$$
$$5+(-5) \overset{\vee}{=} 0 \qquad\qquad (-5)+5 \overset{\vee}{=} 0$$
$$3(5)^2-2(-5)^2 \overset{\vee}{=} 25 \quad 3(-5)^2-2(5)^2 \overset{\vee}{=} 25$$

From this point on we will not show the checking steps for lack of space. The student should perform these checking steps, however.

7.
$$y^2 = x$$
$$x - 2y = 2$$

Solve for x in the first degree equation.

$$x = 2y + 2$$

Substitute into the second degree equation.

$$y^2 = 2y + 2$$
$$y^2 - 2y - 2 = 0$$
$$y = \frac{-b \pm \sqrt{b^2 - 4ac}}{2a}$$
$$a = 1$$
$$b = -2$$
$$c = -2$$
$$y = \frac{-(-2) \pm \sqrt{(-2)^2 - 4(1)(-2)}}{2(1)}$$
$$y = \frac{2 \pm \sqrt{12}}{2}$$
$$y = 1 \pm \sqrt{3}$$

$$\text{For } y = 1 + \sqrt{3} \qquad\qquad \text{For } y = 1 - \sqrt{3}$$
$$x = 2(1 + \sqrt{3}) + 2 \qquad\qquad x = 2(1 + \sqrt{3}) + 2$$
$$x = 4 + 2\sqrt{3} \qquad\qquad x = 4 - 2\sqrt{3}$$

Solutions: $(4 + 2\sqrt{3}, 1 + 2\sqrt{3}), (4 - 2\sqrt{3}, 1 - 2\sqrt{3})$

9.
$$2x^2 + y^2 = 24$$
$$x^2 - y^2 = -12$$

Solve using elimination by addition. Adding, we obtain:

$$3x^2 = 12$$
$$x^2 = 4$$
$$x = \pm 2$$

$$\text{For } x = 2 \qquad\qquad \text{For } x = -2$$
$$4 - y^2 = -12 \qquad 4 - y^2 = -12$$
$$-y^2 = -16 \quad \text{Similarly}$$
$$y^2 = 16 \qquad\qquad y = \pm 4$$
$$y = \pm 4$$

Solutions: $(2, 4), (2, -4), (-2, 4), (-2, -4)$

11.
$$x^2 + y^2 = 1$$
$$16x^2 + y^2 = 25$$

Solve using elimination by addition. Multiply the top equation by -1 and add.

$$\begin{array}{rcr} -x^2 - y^2 &=& -10 \\ 16x^2 + y^2 &=& 25 \\ \hline 15x^2 &=& 15 \\ x^2 &=& 1 \\ x &=& \pm 1 \end{array}$$

For $x = 1$ For $x = -1$

$1 + y^2 = 10$ $1 + y^2 = 10$

$y^2 = 9$ $y = \pm 3$

$y = \pm 3$

Solutions: $(1,3),(1,-3),(-1,3),(-1,-3)$

13. $xy - 4 = 0$

     $x - y = 2$

Solve for x in the first degree equation.

$x = y + 2$

Substitute into the second degree equation

$(y + 2)y - 4 = 0$

$y^2 + 2y - 4 = 0$

$y = \frac{-b \pm \sqrt{b^2 - 4ac}}{2a}$

     $a = 1$

     $b = 2$

     $c = 4$

$y = \frac{-2 \pm \sqrt{(2)^2 - 4(1)(-4)}}{2(1)}$

$y = \frac{-2 \pm \sqrt{20}}{2}$

$y = -1 \pm \sqrt{5}$

For $y = -1 + \sqrt{5}$ For $y = -1 - \sqrt{5}$

   $y = -1 - \sqrt{5} + 2$ $x = -1 - \sqrt{5} + 2$

   $x = 1 + \sqrt{5}$ $x = 1 - \sqrt{5}$

Solutions: $(1 + \sqrt{5}, -1 + \sqrt{5}), (1 - \sqrt{5}, -1 - \sqrt{5})$

15. $x^2 + 2y^2 = 6$

      $xy = 2$

Solve for y in the second equation

$y = \frac{2}{x}$

Substitute into the first equation

$x^2 + 2(\frac{2}{x})^2 = 6$

$x^2 + \frac{8}{x^2} = 6 \quad x \neq 0$

$x^2 \cdot x^2 + x^2 \cdot \frac{8}{x^2} = 6x^2$

$x^4 + 8 = 6x^2$

$x^4 - 6x^2 + 8 = 0$

$(x^2 - 2)(x^2 - 4) = 0$

$(x - \sqrt{2})(x + \sqrt{2})(x - 2)(x + 2) = 0$

$x = \sqrt{2}, -\sqrt{2}, 2, -2$

For $x = \sqrt{2}$ For $x = -\sqrt{2}$ For $x = 2$ For $x = -2$

$y = \frac{2}{\sqrt{2}}$ $y = -\frac{2}{\sqrt{2}}$ $y = \frac{2}{2}$ $y = \frac{2}{-2}$

$y = \sqrt{2}$ $y = -\sqrt{2}$ $y = 1$ $y = -1$

Solutions: $(\sqrt{2}, \sqrt{2}), (-\sqrt{2}, -\sqrt{2}), (2, 1), (-2, -1)$

17. $2x^2 + 3y^2 = -4$

     $4x^2 + 2y^2 = 8$

Solve using elimination by addition. Multiply the second equation by $-\frac{1}{2}$ and add.

$$
\begin{array}{rcr}
2x^2 + 3y^2 &=& -4 \\
-2x^2 - y^2 &=& -4 \\
\hline
2y^2 &=& -8 \\
y^2 &=& -4 \\
\hline
y &=& \pm 2i
\end{array}
$$

For $y = 2i$ For $y = -2i$

$2x^2 + 3(2i)^2 = -4$ $2x^2 + 3(-2i)^2 = -4$

$2x^2 - 12 = -4$ $2x^2 - 12 = -4$

$2x^2 = 8$ Similarly,

$x^2 = 4$ $x = \pm 2$

$x = \pm 2$

Solutions: $(2, 2i), (-2, 2i), (2, -2i), (-2, -2i)$

---

19. $x^2 - y^2 = 2$

      $y^2 = x$

Substitute y^2 from the second equation into the first equation.

$x^2 - x = 2$

$x^2 - x - 2 = 0$

$(x - 2)(x + 1) = 0$

$x = 2, -1$

For $x = 2$ For $x = -1$

$y^2 = 2$ $y^2 = -1$

$y = \pm\sqrt{2}$ $y = \pm i$

Solutions; $(2, \sqrt{2}), (2, -\sqrt{2}),$

$(-1, i), (-1, -i)$

21. $x^2 + y^2 = 9$

      $x^2 = 9 - 2y$

Substitute x^2 from the second equation into the first equation.

$9 - 2y + y^2 = 9$

$y^2 - 2y = 0$

$y(y - 2) = 0$

$y = 0, 2$

For $y = 0$ For $y = 2$

$x^2 = 9 - 2(0)$ $x^2 = 9 - 2(2)$

$x^2 = 9$ $x^2 = 5$

$x = \pm 3$ $x = \pm\sqrt{5}$

Solutions: $(3, 0), (-3, 0), (\sqrt{5}, 2), (-\sqrt{5}, 2)$

---

23. $x^2 - y^2 = 3$

      $xy = 2$

Solve for y in the second equation.

$y = \frac{2}{x}$

Substitute into the first equation

$x^2 - \left(\frac{2}{x}\right)^2 = 3$

$x^2 - \frac{4}{x^2} = 3 \quad x \neq 0$

$x^4 - 4 = 3x^2$

$x^4 - 3x^2 - 4 = 0$

$(x^2 - 4)(x^2 + 1) = 0$

$x^2 - 4 = 0 \quad x^2 + 1 = 0$

$x^2 = 4 \quad\quad x^2 = = -1$

$x = \pm 2 \quad\quad x = \pm i$

For $x = 2$ For $x = -2$ For $x = i$ For $x = -i$

$y = \frac{2}{2}$ $y = \frac{2}{-2}$ $y = \frac{2}{i}$ $y = \frac{2}{-i}$

$y = 1$ $y = -1$ $y = -2i$ $y = 2i$

Solutions: $(2,1), (-2,-1), (i,-2i), (-i,2i)$

25.
$$2x + 5y + 7xy = 8$$
$$xy - 3 = 0$$
Solve for y in the second equation.
$$xy = 3$$
$$y = \frac{3}{x}$$
Substitute into the first equation.
$$2x + 5\left(\frac{3}{x}\right) + 7x\left(\frac{3}{x}\right) = 8$$
$$2x + \frac{15}{x} + 21 = 8 \quad x \neq 0$$
$$2x^2 + 15 + 21x = 8x$$
$$2x^2 + 13x + 15 = 0$$
$$(2x + 3)(x + 5) = 0$$
$$x = -\frac{3}{2}, -5$$

For $x = -\frac{3}{2}$ For $x = -5$
$$y = 3 \div \left(-\frac{3}{2}\right) \qquad y = \frac{3}{-5}$$
$$y = -2 \qquad\qquad y = -\frac{3}{5}$$
Solutions: $\left(-\frac{3}{2}, -2\right)\left(-5, -\frac{3}{5}\right)$

27.
$$x^2 - 2xy + y^2 = 1$$
$$x - 2y = 2$$
Solve for x in terms of y in the first-degree equation.
$$x = 2y + 2$$
Substitute into the second-degree equation.
$$(2y + 2)^2 - 2(2y + 2)y + y^2 = 1$$
$$4y^2 + 8y + 4 - 4y^2 - 4y + y^2 = 1$$
$$y^2 + 4y + 3 = 0$$
$$(y + 1)(y + 3) = 0$$
$$y = -1, -3$$

For $y = -1$ For $y = -3$
$$x = 2(-1) + 2 \qquad x = 2(-3) + 2$$
$$= 0 \qquad\qquad\qquad = -4$$
Solutions: $(0, -1), (-4, -3)$

29.
$$2x^2 - xy + y^2 = 8$$
$$x^2 - y^2 = 0$$
Factor the left side of the equation that has a zero constant term.
$$(x - y)(x + y) = 0 \qquad CommonError:$$
$$x = y \text{ or } x = -y \qquad$$
It is incorrect to replace
$x^2 - y^2 = 0$ or $x^2 = y^2$ by $x = y$.
This neglects the possibility $x = -y$.

Thus, the original system is equivalent to the two systems
$$2x^2 - xy + y^2 = 8 \qquad 2x^2 - xy + y^2 = 8$$
$$x = y \qquad\qquad\qquad x = -y$$
These systems are solved by substitution.

First system :   Second system :
$$2x^2 - xy + y^2 = 8 \qquad\qquad 2x^2 - xy + y^2 = 8$$
$$x = y \qquad\qquad\qquad\qquad x = -y$$
$$2y^2 - yy + y^2 = 8 \quad 2(-y)^2 - (-y)y + y^2 = 8$$
$$2y^2 = 8 \qquad\qquad 2y^2 + y^2 + y^2 = 8$$
$$y^2 = 4 \qquad\qquad\qquad 4y^2 = 8$$
$$y = \pm 2 \qquad\qquad\qquad y^2 = 2$$
$$y = \pm\sqrt{2}$$

For $y = 2$ For $y = -2$ For $y = \sqrt{2}$ For $y = -\sqrt{2}$
$$x = 2 \qquad x = -2 \qquad x = -\sqrt{2} \qquad x = \sqrt{2}$$
Solutions: $(2,2), (-2,-2), (-\sqrt{2}, \sqrt{2}), (\sqrt{2}, -\sqrt{2})$

31. $x^2 + xy - 3y^2 = 3$

$x^2 + 4xy + 3y^2 = 0$

Factor the left side of the equation that has a zero constant term.

$(x + y)(x + 3y) = 0$

$x = -y$ or $x = -3y$

Thus the original system is equivalent to the two systems

$x^2 + xy - 3y^2 = 3 \qquad x^2 + xy - 3y^2 = 3$

$x = -y \qquad\qquad x = -3y$

These systems are solved by substitution.

First system :   Second system :

$x^2 + xy - 3y^2 = 3 \qquad\qquad x^2 + xy - 3y^2 = 3$

$x = -y \qquad\qquad\qquad\qquad x = -3y$

$(-y)^2 + (-y)y - 3y^2 = 3 \qquad (-3y)^2 + (-3y)y - 3y^2 = 3$

$y^2 - y^2 - 3y^2 = 3 \qquad\qquad 9y^2 - 3y^2 - 3y^2 = 3$

$-3y^2 = 3 \qquad\qquad\qquad\qquad 3y^2 = 3$

$y^2 = -1 \qquad\qquad\qquad\qquad y^2 = 1$

$y = \pm i \qquad\qquad\qquad\qquad y = \pm 1$

For $y = i$ For $y = -i$ For $y = 1$ For $y = -1$

$x = -i \qquad\quad x = i \qquad\quad x = -3 \qquad\quad x = 3$

Solutions: $(-i, i), (i, -i), (-3, 1), (3, -1)$

33. Let x and y equal the two numbers. We have the system

$x + y = 3$

$xy = 1$

Solve the first degree equation for y in terms of x, then substitute into the second degree equation.

$y = 3 - x$

$x(3 - x) = 1$

$3x - x^2 = 1$

$-x^2 + 3x - 1 = 0$

$x^2 - 3x + 1 = 0$

$x = \frac{-b \pm \sqrt{b^2 - 4ac}}{2a}$

$a = 1$

$b = -3$

$c = 1$

$x = \frac{-(-3) \pm \sqrt{(-3)^2 - 4(1)(1)}}{2(1)}$

$x = \frac{3 \pm \sqrt{5}}{2}$

For $x = \frac{3+\sqrt{5}}{2}, y = 3 - x$ For $x = \frac{3+\sqrt{5}}{2}, y = 3 - x$

$= 3 - \frac{3+\sqrt{5}}{2}$ $= 3 - \frac{3-\sqrt{5}}{2}$

$= \frac{6-3-\sqrt{5}}{2}$ $= \frac{6-3+\sqrt{5}}{2}$

$= \frac{3-\sqrt{5}}{2}$ $= \frac{3+\sqrt{5}}{2}$

Thus the two numbers are $\frac{1}{2}(3 - \sqrt{5})$ and $\frac{1}{2}(3 + \sqrt{5})$.

35.   Sketch a figure.

Let x and y represent the lengths of the two legs.

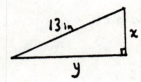

From the Pythagorean Theorem we have
$x^2 + y^2 = 13^2$
From the formula for the area of a triangle we have
$\frac{1}{2}xy = 30$
Thus the system of equations is

$$
\begin{aligned}
x^2 + y^2 &= 169 \\
\tfrac{1}{2}xy &= 30
\end{aligned}
$$

Solve the second equation for y in terms of x, then substitute into the first equation.

$$
\begin{aligned}
xy &= 60 \\
y &= \frac{60}{x} \\
x^2 + \left(\frac{60}{x}\right)^2 &= 169 \\
x^2 + \frac{3600}{x^2} &= 169 \quad x \neq 0 \\
x^4 + 3600 &= 169x^2 \\
x^4 - 169x^2 + 3600 &= 0 \\
(x^2 - 144)(x^2 - 25) &= 0 \\
(x - 12)(x + 12)(x - 5)(x + 5) &= 0 \\
x &= \pm 12, \pm 5.
\end{aligned}
$$

Discarding the negative solutions, we have
$x = 12$ or $x = 5$

$$
\begin{array}{ll}
\text{For } x = 12, y = \dfrac{60}{x} & \text{For } x = 5, y = \dfrac{60}{x} \\
\qquad\qquad\quad y = 5 & \qquad\qquad\quad y = 12
\end{array}
$$

The lengths of the legs are 5 inches and 12 inches.

37.   Let x = width of screen.

         y = height of screen.

From the Pythagorean Theorem we have
$x^2 + y^2 = (7.5)^2$
From the formula for the area of a rectangle we have
$xy = 27$
Thus the system of equations is

$$
\begin{aligned}
x^2 + y^2 &= 56.25 \\
xy &= 27
\end{aligned}
$$

Solve the second equation for y in terms of x, then substitute into the first equation.

$$y = \frac{27}{x}$$

$$x^2 + \left(\frac{27}{x}\right)^2 = 56.25 \quad x \neq 0$$

$$x^2 + \frac{729^2}{x} = 56.25 \quad x \neq 0$$

$$x^4 + 729 = 56.25x^2$$

$$x^4 - 56.25x^2 + 729 = 0 \quad \text{quadratic in } x^2$$

$$x^2 = \frac{-b \pm \sqrt{b^2 - 4ac}}{2a}$$

$$a = 1$$

$$b = -56.25$$

$$c = 729$$

$$x^2 = \frac{-(-56.25) \pm \sqrt{(-56.25)^2 - 4(1)(729)}}{2(1)}$$

$$x^2 = \frac{56.25 \pm 15.75}{2}$$

$$x^2 = 36, 20.25$$

$$x = 6, 4.5 \text{(discarding the negative solutions)}$$

For $x = 6, y = \frac{27}{6} = 4.5$

For $x = 4.5, y = \frac{27}{4.5} = 6$

The dimensions of the screen must be 6 inches by 4.5 inches.

39.

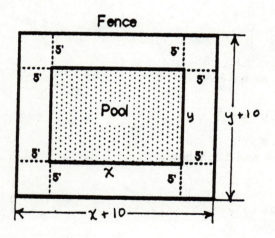

Fence

Redrawing and labelling the figure as above, we have

$$\text{Area of pool} = 572$$

$$xy = 572$$

$$\text{Area enclosed by fence} = 1,152$$

$$(x + 10)(y + 10) = 1,152$$

We solve this system by solving for y in terms of x in the first equation, then substituting into the second equation.

$$y = \frac{572}{x}$$
$$(x+10)\left(\frac{572}{x}+10\right) = 1{,}152$$
$$572 + 10x + \frac{5{,}720}{x} + 100 = 1{,}152$$
$$10x + \frac{5{,}720}{x} - 480 = 0 \qquad x \neq 0$$
$$10x^2 + 5{,}720 - 480x = 0$$
$$x^2 - 48x + 572 = 0$$
$$(x-26)(x-22) = 0$$
$$x = 26, 22$$
For $x = 26$, $y = \frac{572}{26}$ For; $x = 22$, $y = \frac{572}{22}$
$$y = 22 \qquad\qquad y = 26$$

The dimensions of the pool are 22 feet by 26 feet.

41. Let x = average speed of Boat B

Then $x + 5$ = average speed of Boat A

Let y = time of Boat B then $y - \frac{1}{2}$ = time of Boat A

Using Distance = rate × time, we have

$$75 = xy$$
$$75 = (x+5)\left(y-\tfrac{1}{2}\right)$$

Note: The *faster* boat, A, has the *shorter* time. It is a common error to confuse the signs here. Another common error: if rates are expressed in miles per hour, then $y - 30$ is not the correct time for boat A. Times must be expressed in hours.

Solve the first equation for y in terms of x, then substitute into the second equation.

$$y = \frac{75}{x}$$
$$75 = (x+5)\left(\frac{75}{x}-\tfrac{1}{2}\right)$$
$$75 = 75 - \tfrac{1}{2}x + \frac{375}{x} - \tfrac{5}{2}$$
$$0 = -\tfrac{1}{2}x + \frac{375}{x} - \tfrac{5}{2} \qquad x \neq 0$$
$$2x(0) = 2x\left(-\tfrac{1}{2}x\right) + 2x\left(\frac{375}{x}\right) - 2x\left(\tfrac{5}{2}\right)$$
$$0 = -x^2 + 750 - 5x$$
$$x^2 + 5x - 750 = 0$$
$$(x-25)(x+30) = 0$$
$$x = 25, -30$$

Discarding the negative solution, we have

$$x = 25 \text{ mph} = \text{average speed of Boat } B$$
$$x + 5 = 30 \text{ mph} = \text{average speed of Boat } A$$

Exercise 7-5

Key Ideas and Formulas

The graph of a linear inequality $Ax + By < C$ or $Ax + By > C$ (with $B \neq 0$) is either the upper half-plane or the lower half-plane (but not both) determined by the line $Ax + By = C$. If $B = 0$, then the graph of $Ax < C$ or $Ax > C$ is either the left half-plane or the right half-plane (but not both) determined by the line $Ax = C$.

Linear inequalities are graphed by the procedure given in the text (not repeated here for reasons of space.)

To graph systems of linear inequalities:

1. Graph each inequality, shading the solution half-plane lightly.

2. The multiply shaded region is the graph of the system, called the solution region or the feasible region.

3. Determine the corner points from the graph, and, as necessary, solving the systems of equations formed by pairs of equations of the appropriate lines.

A solution region of a system of linear inequalities is bounded if it can be enclosed in a circle; if it cannot be enclosed within a circle, then it is unbounded.

---

1.   Graph $2x - 3y = 6$ as a broken line, since equality is not included in the original statement. The origin is a suitable test point.
$2x - 3y < 6$
$2(0) - 3(0) = 0 < 6$
Hence $(0,0)$ is in the solution set. The graph is the half-plane containing $(0,0)$.

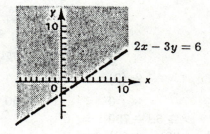

3.   Graph $3x + 2y = 18$ as a solid line, since equality is not included in the original statement. The origin is a suitable test point.
$3(0) + 2(0) = 0 \not\geq 18$
Hence $(0,0)$ is not in the solution set. The graph is the line $3x + 2y = 18$ and the half-plane not containing the origin.

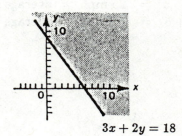

5.   Graph $y = \frac{2}{3}x + 5$ as a solid line, since
equality is not included in the original
statement. The origin is a suitable test
point.

$0 \overset{?}{\leq} (0) + 5$

$0 \leq 5$

Hence (0,0) is in the solution set. The
graph is the line $y = \frac{2}{3}x + 5$ and
the half-plane containing the origin.

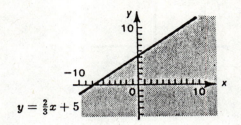

7.   Graph $y = 8$ as a broken line, since
equality is not included in the original
statement. Clearly the graph consists
of all points whose y–coordinates are less
than 8, that is, the lower half-plane.

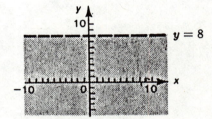

9.   This system is equivalent to the system

$y \geq -3$

$y < 2$

and its graph is the intersection of the graphs
of these inequalities.

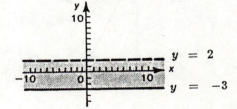

11. $x + 2y \leq 8$
$3x - 2y \geq 0$

Choose a suitable test point that lies on neither line, for example, (2,0).

$2 + 2(0) = 2 \leq 8$ Hence, the solution region is *below* the graph of $x + 2y = 8$
$3(2) - 2(0) = 6 \geq 0$ Hence, the solution region is *below* the graph of $3x - 2y = 0$

Thus the solution region is region IV in the diagram

13. $x + 2y \geq 8$
$3x - 2y \geq 0$

Choose a suitable test point that lies on neither line, for example, (2,0).

$2 + 2(0) = 2 \not\geq 8$ Hence, the solution region is *above* the graph of $x + 2y = 8$
$3(2) - 2(0) = 6 \geq 0$ Hence, the solution region is *below* the graph of $3x - 2y = 0$

Thus the solution region is region I in the diagram

15.

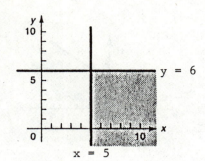

17.

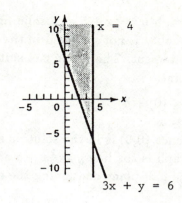

19.

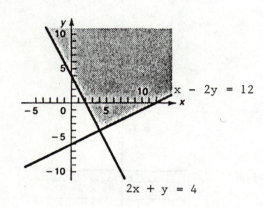

21.   Choose a suitable test point that lies on none of the lines, say (5,1).

$$5 + 3(1) = 8 \quad \leq \quad 18 \quad \text{Hence, the solution region is } below \text{ the graph of } x + 3y = 18$$
$$2(5) + 1 = 11 \quad \nleq \quad 16 \quad \text{Hence, the solution region is } above \text{ the graph of } 2x + y = 16$$
$$5 \geq 0$$
$$1 \geq 0$$

Thus, the solution region is region IV in the diagram

The corner points are the labelled points (6,4), (8,0), and (18,0).

23.   Choose a suitable test point that lies on none of the lines, say $(5,1)$.

$5 + 3(1) = 8 \not\geq 18$ Hence, the solution region is *above* the graph of $x + 3y = 18$

$2(5) + 1 = 11 \not\geq 16$ Hence, the solution region is *above* the graph of $2x + y = 16$

$$5 \geq 0$$
$$1 \geq 0$$

Thus the solution region is region I in the diagram.

The corner points are the labelled points $(0,16)$ $(6,4)$, and $(18,0)$.

25.   The solution region is bounded (contained in, for example, the circle $x^2 + y^2 = 16$).

The corner points are obvious from the graph: $(0,0)$, $(0,2)$, $(3,0)$.

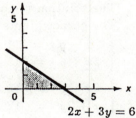

$2x + 3y = 6$

27.   The solution region is unbounded. The corner points are obvious from the graph: $(0,4)$ and $(5,0)$

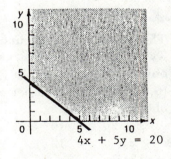

$4x + 5y = 20$

29.   The solution region is bounded.

Three corner points are obvious from the graph: $(0,4)$ and $(0,0)$ and $(0,4)$.

The fourth corner point is obtained by solving the system

$$x + 3y = 12$$
$$2x + y = 8 \text{ to obtain } \left(\tfrac{12}{5}, \tfrac{16}{5}\right).$$

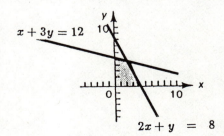

$x + 3y = 12$

$2x + y = 8$

31. The solution region is unbounded.
Two corner points are obvious from
the graph: (9,0) and (0,8).
The third corner point is obtained
by solving the system

$$4x + 3y = 24$$
$$2x + 3y = 18 \text{ to obtain } (3,4).$$

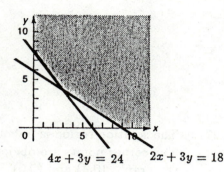

$$4x + 3y = 24 \qquad 2x + 3y = 18$$

33. The solution region is bounded.
Three of the corner points are
obvious from the graph:
(6,0),(0,0), and (0,5).
The other corner points are
obtained by solving:

$$2x + y = 12 \quad \text{and} \qquad x + y = 7$$
$$2x + y = 7 \qquad\qquad x + 2y = 10$$

to obtain   to obtain

(5,2)   (4,3)

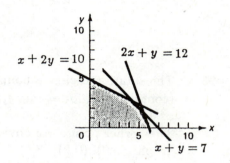

35. The solution region is unbounded.
Two of the corner points are
obvious from the graph: (16,0) and
(0,14). The other corner points are
obtained by solving:

$$x + 2y = 16 \quad \text{and} \qquad x + y = 12$$
$$x + y = 12 \qquad\qquad 2x + y = 14$$

to obtain   to obtain

(8,4)   (2,10)

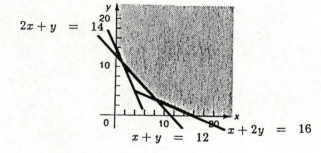

$$2x + y = 14 \qquad\qquad x + y = 12 \qquad x + 2y = 16$$

37. The solution region is unbounded.
The corner points are obtained by solving:

$$\begin{cases} x + y = 11 \\ 5x + y = 15 \end{cases} \text{to obtain } (1,10)$$

$$\begin{cases} 5x + y = 15 \\ x + 2y = 12 \end{cases} \text{to obtain } (2,5)$$

and

$$\begin{cases} x + y = 11 \\ x + 2y = 12 \end{cases} \text{to obtain } (10,1)$$

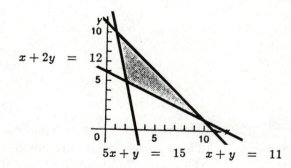

$$x + 2y = 12 \qquad 5x + y = 15 \qquad x + y = 11$$

39.   From the graph it should be clear
        that there is no point with x coordinate
        greater than 4 which satisfies both
        $3x + 2y \leq 24$ (arrows pointing, roughly, northeast)
        and $3x + y \leq 15$ (arrows pointing, roughly,
        southwest). The feasible region is
        empty.

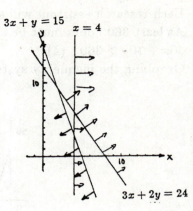

41.   The feasible region is bounded.
        The corner points are obtained by solving:
$$\begin{cases} x + y = 10 \\ 3x - 2y = 15 \end{cases} \text{ to obtain}(7,3)$$

$$\begin{cases} 3x - y = 15 \\ 3x + 5y = 15 \end{cases} \text{ to obtain}(5,0)$$

$$\begin{cases} 3x + 5y = 15 \\ -5x + 2y = 6 \end{cases} \text{ to obtain}(0,3)$$
        and
$$\begin{cases} -5x + 2y = 6 \\ x + y = 10 \end{cases} \text{ to obtain}(2,8)$$

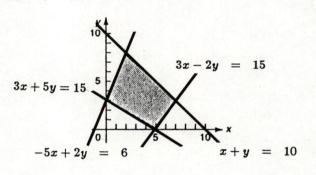

43.   Let x =   number of trick skis produced per day
            y =   number of slalom skis produced per day
        Clearly x and y must be non-negative.
            Hence $x \leq 0$ (1)
                  $y \leq 0$ (2)
        To fabricate x trick skis requires $6x$ hours.
        To fabricate y slalom skis requires $4y$ hours.
        108 hours are available for fabricating; hence
        $6x + 4y \leq 108$ (3)
        To finish x trick skis requires $1x$ hours.
        To finish y slalom skis requires $1y$ hours.
        24 hours are available for finishing, hence
        $x + y \leq 24$ (4)
        Graphing the inequality system (1),(2), (3), (4), we have:

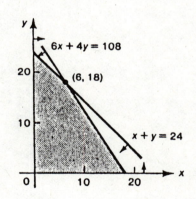

45.   Clearly x and y must be non negative
            Hence $x \geq 0$ (1)
                  $y \geq 0$ (2)
        Each sociologist will spend 10 hours collecting data: $10x$ hours.
        Each research assistant will spend 30 hours collecting data: $30y$ hours.
        At least 280 hours must be spent collecting data; hence
        $10x + 30y \geq 280$ (3)
        Each sociologist will spend 30 hours analyzing data: $30x$ hours.

Each research assistant will spend 10 hours analyzing data: $10y$ hours.

At least 360 hours must be spent analyzing data; hence

$$30x + 10y \geq 360 \quad (4)$$

Graphing the inequality system (1),(2),(3),(4), we have:

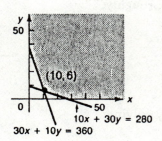

Exercise 7-6

Key Ideas and Formulas

A linear programming problem involves maximizing or minimizing an objective function subject to linear inequalities known as problem constraints.

Fundamental Theorem of Linear Programming:

Let S be the feasible region for a linear programming problem and let $z = ax + by$ be the objective function. If S is bounded, then z has both a maximum and a minimum value on S and each of these occurs at a corner point of S. If S is unbounded, then a maximum or minimum value of z on S may not exist. However, if either does exist, then it must occur at a corner point of S.

Solution of Linear Programming Problems:

1. Form a mathematical model for the problem:

    A. Introduce decision variables and write a linear objective function.

    B. Write problem constraints in the form of linear inequalities.

    C. Write nonnegative constraints.

2. Graph the feasible region and find the corner points.

3. Evaluate the objective function at each corner point to determine the optimal solution.

1.

Corner Point (x, y)	Objective Function $z = x + y$	
(0,12)	12	
(7,9)	16	Maximum value
(10,0)	10	
(0,0)	0	

The maximum value of z on S is 16 at (7,9).

3.

Corner Point (x, y)	Objective Function $z = 3x + 7y$	
(0,12)	84	Maximum value
(7,9)	84	Maximum value
(10,0)	30	
(0,0)	0	

Multiple optimal solution

The maximum value of z on S is 84 at both (0,12) and (7,9).

5.

Corner Point (x, y)	Objective Function $z = 7x + 4y$	
(0,12)	48	
(12,0)	84	
(4,3)	40	
(0,8)	32	Minimum value

The minimum value of z on S is 32 at (0,8).

7.

Corner Point (x, y)	Objective Function $3x + 8y$	
(0,12)	96	
(12,0)	36	Minimum value
		}Multiple Optimal Solution
(4,3)	36	Minimum value
(0,8)	64	

The minimum value of z on S is 36 at both (12,0) and (4,3).

9. The feasible region is graphed as follows:
The corner points (0,5), (5,0) and
(0,0) are obvious from the
graph. The corner point (4,3)
is obtained by solving the system
$$x + 2y = 10$$
$$3x + y = 15$$
We now evaluate the objective
function at each corner point.

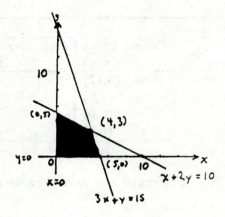

Corner Point (x, y)	Objective Function $z = 3x + 2y$	
(0,5)	10	
(0,0)	0	
(5,0)	15	
(4,3)	18	Maximum value

The maximum value of z on S is 18 at (4,3).

11. The feasible region is graphed as follows:
     The corner points $(4,0)$ and $(10,0)$
     are obvious from the graph.
     The corner point $(2,4)$ is
     obtained by solving the system

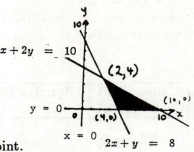

$$x + 2y = 10$$
$$2x + y = 8$$

We evaluate the objective function at each corner point.

Corner Point (x,y)	Objective Function $z = 3x + 4y$	
$(4,0)$	12	Minimum value
$(10,0)$	30	
$(2,4)$	22	

The minimum value of z on S is 12 at $(4,0)$.

13. The feasible region is graphed as follows:
     The corner points $(0,12)$, $(0,0)$ and $(12,0)$
     are obvious from the graph.
     The other corner points are
     obtained by solving:

$$x + 2y = 24 \qquad \text{and } x + y = 14$$
$$x + y = 14 \quad \text{to obtain } (4,10) \qquad 2x + y = 24 \quad \text{to obtain} (10,4)$$

We now evaluate the objective function at each corner point.

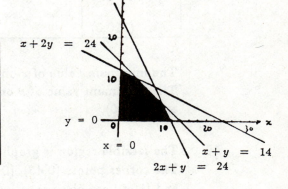

Corner Point (x,y)	Objective Function $z = 3x + 4y$	
$(0,12)$	48	
$(0,0)$	0	
$(12,0)$	36	
$(10,4)$	46	
$(4,10)$	52	Maximum value

The maximum value of z on S is 52 at $(4,10)$

15. The feasible region is graphed as follows:
     The corner points $(0,20)$ and $(20,0)$
     are obvious from the graph.
     The third corner
     point is obtained by solving

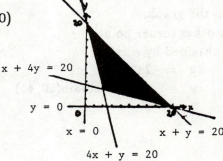

$$x + 4y = 20$$
$$4x + y = 20 \quad \text{to obtain}(4,4)$$
We now evaluate the objective function at each corner point.

Corner Point (x, y)	Objective Function $z = 5x + 6y$	
(0,20)	120	
(20,0)	100	
(4,4)	44	Minimum value

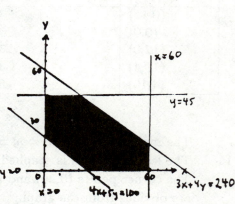

The minimum value of z on S is 44 at (4,4)

17. The feasible region is graphed as follows:
The corner points (60,0) and (120,0)
are obvious from the graph.
The other corner points are obtained by solving:
$$x + y = 60 \qquad\qquad \text{and}$$
$$x - 2y = 0 \quad \text{to obtain}(40,20) \quad x + 2y = 120$$
$$x - 2y = 0 \quad \text{to obtain } (60,30)$$

We now evaluate the objective function
at each corner point.

Corner Point (x, y)	Objective Function $25x + 50y$		
(60,0)	1,500	Minimum value	
(40,20)	2,000		
(60,30)	3,000	Maximum value	Multiple Optimal Solutions
(120,0)	3,000	Maximum value	

The minimum value of z on S is 1,500 at (60,0).
The maximum value of z on S is 3,000 at (60,30)
and (120,0).

19. The feasible region is graphed as follows:
The corner points (0,45), (0,20), (25,0),
and (60,0) are obvious
from the graph.
The other corner points
are obtained by solving:
$$3x + 4y = 240 \qquad\qquad \text{and}$$
$$y = 45 \quad \text{to obtain}(20,45) \quad 3x + 4y = 240$$
$$x = 60 \qquad \text{to obtain}(60,15)$$

We now evaluate the objective function at each corner point

Corner Point (x,y)	Objective Function $25x + 15y$	
(0,45)	675	
(0,20)	300	Minimum value
(25,0)	625	
(60,0)	1,500	Maximum value
(60,15)	1,725	
(20,45)	1,175	

The minimum value of z on S is 300 at (0,20).

The maximum value of z on S is 1,725 at (60,15).

21.   The feasible region is graphed as follows:
(heavily outlined for clarity)
Consider the objective function $x + y$.
It should be clear that it takes on
the value 2 along $x + y = 2$,
the value 4 along $x + y = 4$, the
value 7 along $x + y = 7$,
and so on. The maximum value
of the objective function,
then, on S, is 7,
which occurs at B. Graphically this
occurs when the line $x + y = c$ coincides
with the boundary of S. Thus
to answer questions $(A) - (E)$
we must determine values of a and b
such that the appropriate line $ax + by = c$
coincides with the boundary of S
only at the specified points.

A. The line $ax + by = c$ must have slope negative, but greater in absolute
value than that of line segment $AB, 2x + y = 10$. Therefore $a > 2b$.

B. The line $ax + by = c$ must have slope negative but between that of $x + 3y = 15$ and $2x + y = 10$.
Therefore $\frac{1}{3}b < a < 2b$.

C. The line $ax + by = c$ must have slope greater than that of line segment BC, $x + 3y = 15$.
Therefore $a < \frac{1}{3}b$ or $b > 3a$

D. The line $ax + by = c$ must be parallel to line segment AB, therefore
$a = 2b$.

E. The line $ax + by + c$ must be parallel to line segment BC, therefore $b = 3a$.

23.   Much of the mathematical model for this problem was formed in Section 7-5
problem 43. We let

x = the number of trick skis
y = the number of slalom skis

The problem constraints were

$$6x + 4y \leq 108$$
$$x + y \leq 24$$

The nonnegative constraints were

$$x \geq 0$$
$$y \geq 0$$

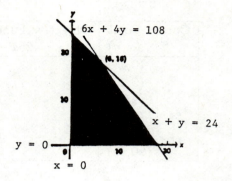

The feasible region was graphed there.

We note now: the linear objective function
$P = 40x + 30y$ represents the profit.

Three of the corner points are
obvious from the graph:$(0,24),(0,0)$, and $(18,0)$.

The fourth corner point is obtained by solving:

$$6x + 4y = 108$$
$$x + y = 24 \quad \text{to obtain } (6,18).$$

Summarizing: the mathematical model for this problem is:

Maximize $P = 40x + 30y$

  subject to

$$6x + 4y \leq 108$$
$$x + y \leq 24$$
$$x, y \geq 0$$

We now evaluate the objective function $40x + 30y$ at each corner point.

Corner Point (x,y)	Objective Function $40x + 30y$	
(0,0)	0	
(18,0)	720	
(6,18)	780	Maximum value
(0,24)	720	

The optimal value is 780 at the corner point $(6,18)$.

Thus, 6 trick skis and 18 slalom skis should be manufactured to obtain the maximum profit of $780.

25.   Let x = Number of model A trucks
          y = Number of model B trucks

We form the linear objective function

$C = 15,000x + 24,000y$

We wish to minimize C, the cost of buying x
trucks @ $15,000 and
y trucks @ $24,000, subject to the constraints.

$$x + y \leq 15 \text{ maximum number of trucks constraint}$$
$$2x + 3y \geq 36 \text{ capacity constraint}$$
$$x, y \geq 0 \text{ nonnegative constraints}$$

Solving the system of constraint inequalities

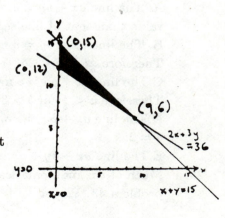

graphically, we obtain
the feasible region S shown in the diagram:

Next we evaluate the objective function at each corner point.

Corner Point (x, y)	Objective Function $C = 15,000x + 24,000y$	
(0,12)	288,000	
(0,15)	360,000	
(9,6)	279,000	Minimum value

The optimal value is $279,000 at the corner point (9,6).
Thus, the company should purchase 9 model A trucks
and 6 model B trucks to realize the minimum
cost of $279,000.

27.   A. Let
x = Number of tables
y = Number of chairs
We form the linear objective function
$P = 90x + 25y$
We wish to maximize P , the profit from
x tables @ $100 and y chairs @ $25,
subject to the constraints

$$8x + 2y \leq 400 \quad \text{assembly department constraint}$$
$$2x + y \leq 120 \quad \text{finishing department constraint}$$
$$x, y \geq 0 \quad \text{nonnegative constraints}$$

Solving the system of constraint inequalities graphically, we obtain
the feasible region S shown in the diagram.
Next we evaluate the objective function at each corner point.

Corner Point (x, y)	Objective Function $P = 90x + 25y$	
(0,0)	0	
(50,0)	4,500	
(40,40)	4,600	Maximum value
(0,120)	3,000	

The optimal value is 4,600 at the corner point (40,40).
Thus, the company should manufacture 40 tables and 40 chairs for a maximum
profit of $4,600.

B.   We are faced with the further condition that $y \geq 4x$.
We wish, then, to maximize $P = 90x + 25y$

under the constraints

$$8x + 2y \leq 400$$
$$2x + y \leq 120$$
$$y \geq 4x$$
$$x, y \geq 0$$

The feasible region is now S' as graphed.

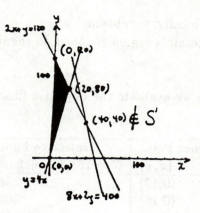

Note that the new condition has the effect of excluding (40,40) from the feasible region. We now evaluate the objective function at the new corner points.

Corner Point (x, y)	Objective Function $P = 90x + 25y$	
(0,120)	3,000	
(0,0)	0	
(20,80)	3,800	Maximum value

The optimal value is now 3,800 at the corner point (20,80).

Thus the company should manufacture 20 tables and 80 chairs for a maximum profit of $3,800.

29.   A. Let x =   Number of 40 − megabyte disk drives
              y =   Number of 80 − megabyte disk drives

We form the linear objective function

$C = 600x + 1500y$

We wish to maximize C , the cost of x 40-megabyte disk drives @ $600 and y 80-megabyte disk drives @ $1,500, subject to the constraints

$$x + y \leq 24 \qquad \text{capacity constraint}$$
$$40x + 80y \geq 1200$$
$$\text{or } x + 2y \geq 30 \qquad \text{storage requirement constraint}$$
$$x, y \geq 0 \qquad \text{nonnegative constraints}$$

Solving the system of constraint inequalities graphically, we obtain the feasible region S shown below.

Next we evaluate the objective function at each corner point.

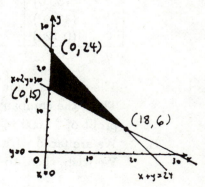

Corner Point (x, y)	Objective Function $C = 600x + 1500y$	
(0,15)	22,500	
(0,24)	36,000	
(18,6)	19,800	Minimum value

The optimal value is 19,800 at the corner point (18,6).

Thus the company should purchase 18 40-megabyte disk drives and 6 80-megabyte disk drives for a minimum cost of $19,800.

B.   We are faced with the further condition that $y \geq 2x$.

We wish, then, to maximize $C = 600x + 1500y$ under the constraints

$$
\begin{aligned}
x + y &\leq 24 \\
x + 2y &\geq 30 \\
y &\geq 2x \\
x, y &\geq 0
\end{aligned}
$$

The feasible region is now S'
as graphed.

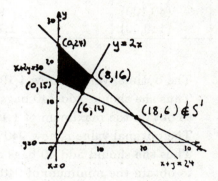

We now evaluate the objective function at the
new corner points.

Corner Point (x,y)	Objective Function $C = 600x + 1,500y$	
(0,15)	22,500	
(0,24)	36,000	
(8,16)	28,800	
(6,12)	21,600	Minimum value

The optimal value is now 21,600 at the corner point (6,12).

Thus the company should purchase 6 40-megabyte disk drives and 12 80-megabyte disk drives for a minimum cost of $21,600.

31.   A. Let x =   Number of bags of Brand A
          y =   Number of bags of Brand B

We form the objective function

$N = 6x + 7y$

N represents the amount of nitrogen in x bags @ 6 pounds per bag
and y @7 pounds per bag.

We wish to optimize N subject to the constraints

$$
\begin{aligned}
2x + 4y &\geq 480 \quad \text{phosphoric acid constraint} \\
6x + 3y &\geq 540 \quad \text{potash constraint} \\
3x + 4y &\leq 620 \quad \text{chlorine constraint} \\
x, y &\geq 0 \quad \text{nonnegative constraint}
\end{aligned}
$$

Solving the system of constraint inequalities graphically, we obtain
the feasible region S shown in the diagram.

Next we evaluate the objective function at corner point.

Corner Point (x, y)	Objective Function $N = 6x + 7y$	
(20,140)	1,100	
(40,100)	940	Minimum value
(140,50)	1,190	Maximum value

A.   The optimal value is now 1,190 at the corner point (140,50).
      Thus she should add 140 bags of Brand A, 50 bags of Brand B
      to obtain the maximum of 1,190 pounds of nitrogen.

B.   The optimal value is now 940 at the corner point (40,100).
      Thus she should add 40 bags of Brand A, 100 bags of Brand B
      to obtain the minimum of 940 pounds of nitrogen.

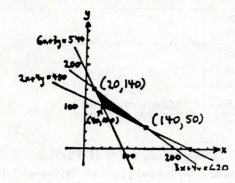

Exercise 7-7

CHAPTER REVIEW

1.
$$2x + y = 7$$
$$3x - 2y = 0$$
We choose elimination by addition.
We multiply the top equation by 2 and add.

$$
\begin{array}{rcrcr}
4x & + & 2y & = & 14 \\
3x & - & 2y & = & 0 \\
\hline
7x & & & = & 14 \\
& & x & = & 2
\end{array}
$$

Substituting $x = 2$ in the top equation, we have
$$2(2) + y = 7$$
$$y = 3$$
Solution: $(2,3)$ $(7-1)$

2.
$$
\begin{array}{rcrcr}
3x & - & 6y & = & 5 \\
-2x & + & 4y & = & 1
\end{array}
$$

We choose elimination by addition.
We multiply the top equation by 2,
the bottom by 3,
and add.

$$
\begin{array}{rcrcr}
6x & - & 12y & = & 10 \\
-6x & + & 12y & = & 3 \\
\hline
& & 0 & = & 13
\end{array}
$$

No solution $(7-1)$

3.
$$4x - 3y = -8$$
$$-2x + \tfrac{3}{2}y = 4$$
We choose elimination by addition.
We multiply the bottom equation by 2 and add.

$$
\begin{array}{rcrcr}
4x & - & 3y & = & -8 \\
-4x & + & 3y & = & 8 \\
\hline
& & 0 & = & 0
\end{array}
$$

There are infinitely many solutions.
For any real number $x, 4x - 3y = -8$, hence,

$$
\begin{array}{rcl}
-3y & = & -4x - 8 \\
y & = & \frac{4x+8}{3}
\end{array}
$$

Thus $(x, \frac{4x+8}{3})$ is a solution for any real number x. $(7-1)$

4.
$$
\begin{array}{rcrcrcrl}
x & + & 2y & + & z & = & 3 & (1) \\
2x & + & 3y & + & 4z & = & 3 & (2) \\
x & + & 2y & + & 3z & = & 1 & (3)
\end{array}
$$

$$
\begin{array}{rcrcrcrl}
2x & + & 3y & + & 4z & = & 3 & \text{Equation (2)} \\
-2x & - & 4y & - & 2z & = & -6 & -2\,[\text{equation}(1)] \\
\hline
& - & y & + & 2z & = & -3 & (4)\ (\text{adding}) \\
x & + & 2y & + & 3z & = & 1 & \text{Equation (3)} \\
-x & - & 2y & - & z & = & -3 & -1\,[\text{equation }(1)] \\
\hline
& & & & 2z & = & -2 & (5)\ (\text{adding})
\end{array}
$$

We solve the system $(4),(5)$ by substituting $2z$ from equation (5) into equation (4).

$$-y + 2z = -3$$
$$2z = -2$$
$$-y - 2 = -3$$
$$-y = -1$$
$$y = 1$$
$$z = -1$$

Substituting $y = 1, z = -1$ into equation (1) we have

$$x + 2(1) - 1 = 3$$
$$x = 2$$

Solution: $(2, 1, -1)$ $(7-1)$

5. $$x^2 + y^2 = 2$$
$$2x - y = 3$$

We solve the first-degree equation for y in terms of x, then substitute into the second-degree equation,

$$2x - y = 3$$
$$-y = 3 - 2x$$
$$y = 2x - 3$$
$$x^2 + (2x - 3)^2 = 2$$
$$x^2 + 4x^2 - 12x + 9 = 2$$
$$5x^2 - 12x + 7 = 0$$
$$(5x - 7)(x - 1) = 0$$
$$x = \tfrac{7}{5}, 1$$

For $x = \tfrac{7}{5}$ For $x = 1$
$$y = 2(\tfrac{7}{5}) - 3 \qquad y = 2(1) - 3$$
$$= -\tfrac{1}{5} \qquad\qquad = -1$$

Solutions: $(1, -1), (\tfrac{7}{5}, -\tfrac{1}{5})$
$(7-4)$

The checking steps are omitted for lack of space.

6. $$3x^2 - y^2 = -6$$
$$2x^2 + 3y^2 = 29$$

We use elimination by addition. We multiply the top equation by 3 and add.

$$9x^2 - 3y^2 = -18$$
$$2x^2 + 3y^2 = 29$$
$$\overline{11x^2 = 11}$$
$$x^2 = 1$$
$$x = \pm 1$$

For $x = 1$ For $x = -1$
$$3(1)^2 - y^2 = -6 \quad 3(-1)^2 - y^2 = -6$$
$$-y^2 = -9 \qquad\qquad -y^2 = -9$$
$$y^2 = 9 \qquad\qquad\quad y = \pm 3$$
$$y = \pm 3$$

Solutions: $(1,3), (1,-3), (-1,3), (-1,-3)$
$(7-4)$

7.

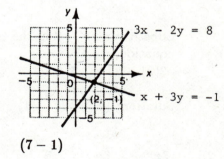

$(7-1)$

8. Graph $3x - 4y = 24$ as a solid line since equality is included in the original statement. $(0,0)$ is a suitable test point
$3(0) - 4(0) = 0 \not\geq 24$
Therefore $(0,0)$ is not in the solution set. The line $3x - 4y = 24$ and the half-plane not containing $(0,0)$ form the graph.

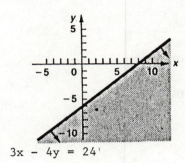

$3x - 4y = 24$

$(7 - 5)$

9.

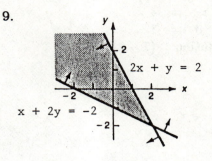

$(7 - 5)$

10. $R_1 \leftrightarrow R_2$ means interchange Rows 1 and 2
$$\begin{bmatrix} 3 & -6 & | & 12 \\ 1 & -4 & | & 5 \end{bmatrix}$$
$(7 - 2)$

11. $\frac{1}{3}R_2 \rightarrow R_2$ means multiply Row 2 by $\frac{1}{3}$
$$\begin{bmatrix} 1 & -4 & | & 5 \\ 1 & -2 & | & 4 \end{bmatrix}$$
$(7 - 2)$

12. $R_2 + (-3)R_1 \rightarrow R_2$ means replace Row 2 by itself plus -3 times Row 1
$$\begin{bmatrix} 1 & -4 & | & 5 \\ 1 & 6 & | & -3 \end{bmatrix}$$
$(7 - 2)$

13. $x_1 = 4$
$x_2 = -7$
The solution is $(4, -7)$
$(7 - 2)$

14. $x_1 - x_2 = 4$
$$0 = 1$$
No solution $(7-2)$

15. $x_1 - x_2 = 4$
$$0 = 0$$
Solution:
$$x_2 = t$$
$$x_1 = x_2 + 4 = t + 4$$
Thus $x_1 = t + 4$, $x_2 = t$ is the solution,
for t any real number.
$(7-2)$

16.

Corner Point (x,y)	Objective Function $z = 5x + 3y$	
(0,10)	30	
(0,6)	18	Minimum value
(4,2)	26	
(6,4)	42	Maximum value

The maximum value of z on S is 42 at (6,4)
The minimum value of z on S is 18 at (0,6)
$(7-6)$

17. $\begin{bmatrix} 3 & 2 & | & 3 \\ 1 & 3 & | & 8 \end{bmatrix}$ $R_1 \leftrightarrow R_2$

$\sim \begin{bmatrix} 1 & 3 & | & 8 \\ 3 & 2 & | & 3 \end{bmatrix}$ $R_2 + (-3)R_1 \rightarrow R_2$

$-3 - 9 - 24 \leftarrow$

$\sim \begin{bmatrix} 1 & 3 & | & 8 \\ 0 & -7 & | & -21 \end{bmatrix}$ $-\frac{1}{7}R_2 \rightarrow R_2$

$\sim \begin{bmatrix} 1 & 3 & | & 8 \\ 0 & 1 & | & 3 \end{bmatrix}$ $R_1 + (-3)R_2 \rightarrow R_1$

$0 - 3 - 9 \leftarrow$

$\sim \begin{bmatrix} 1 & 0 & | & -1 \\ 0 & 1 & | & 3 \end{bmatrix}$

Solution: $x_1 = -1, x_2 = 3$

$(7-2)$

18. $\begin{bmatrix} 1 & 1 & 0 & | & 1 \\ 1 & 0 & -1 & | & -2 \\ 0 & 1 & 2 & | & 4 \end{bmatrix}$ $R_2 + (-1)R_1 \rightarrow R_2$

$$\sim \begin{bmatrix} 1 & 1 & 0 & | & 1 \\ 0 & -1 & -1 & | & -3 \\ 0 & 1 & 2 & | & 4 \end{bmatrix} \quad (-1)R_2 \;\rightarrow\; R_2$$

$$\sim \begin{bmatrix} 1 & 1 & 0 & | & 1 \\ 0 & 1 & 1 & | & 3 \\ 0 & 1 & 2 & | & 4 \end{bmatrix} \quad \begin{matrix} R_1 + (-1)R_2 \;\rightarrow\; R_1 \\ R_3 + (-1)R_2 \;\rightarrow\; R_3 \end{matrix}$$

$$\sim \begin{bmatrix} 1 & 0 & -1 & | & -2 \\ 0 & 1 & 1 & | & 3 \\ 0 & 0 & 1 & | & 1 \end{bmatrix} \quad \begin{matrix} R_1 + R_3 \;\rightarrow\; R_1 \\ R_2 + (-1)R_3 \;\rightarrow\; R_2 \end{matrix}$$

$$\sim \begin{bmatrix} 1 & 0 & 0 & | & -1 \\ 0 & 1 & 0 & | & 2 \\ 0 & 0 & 1 & | & 1 \end{bmatrix}$$

Solution: $x_1 = -1, x_2 = 2, x_3 = 1$

(7 − 3)

19. $\begin{bmatrix} 1 & 2 & 3 & | & 1 \\ 2 & 3 & 4 & | & 3 \\ 1 & 2 & 1 & | & 3 \end{bmatrix} \quad \begin{matrix} R_2 + (-2)R_1 \;\rightarrow\; R_2 \\ R_3 + (-1)R_1 \;\rightarrow\; R_3 \end{matrix}$

$$\sim \begin{bmatrix} 1 & 2 & 3 & | & 1 \\ 0 & -1 & -2 & | & 1 \\ 0 & 0 & -2 & | & 2 \end{bmatrix} \quad \begin{matrix} (-1)R_2 \;\rightarrow\; R_2 \\ -\frac{1}{2}R_3 \;\rightarrow\; R_3 \end{matrix}$$

$$\sim \begin{bmatrix} 1 & 2 & 3 & | & 1 \\ 0 & 1 & 2 & | & -1 \\ 0 & 0 & 1 & | & -1 \end{bmatrix} \quad R_1 + (-2)R_2 \;\rightarrow\; R_1$$

$$\sim \begin{bmatrix} 1 & 0 & -1 & | & -1 \\ 0 & 1 & 2 & | & -1 \\ 0 & 0 & 1 & | & -1 \end{bmatrix} \quad \begin{matrix} R_1 + R_3 \;\rightarrow\; R_1 \\ R_2 + (-2)R_3 \;\rightarrow\; R_2 \end{matrix}$$

$$\sim \begin{bmatrix} 1 & 0 & 0 & | & -2 \\ 0 & 1 & 0 & | & 1 \\ 0 & 0 & 1 & | & -1 \end{bmatrix}$$

Solution: $x_1 = -2, x_2 = 1, x_3 = -1$

(7 − 3)

20. $\begin{bmatrix} 1 & 2 & -1 & | & 2 \\ 2 & 3 & 1 & | & -3 \\ 3 & 5 & 0 & | & -1 \end{bmatrix} \quad \begin{matrix} R_2 + (-2)R_1 \;\rightarrow\; R_2 \\ R_3 + (-3)R_1 \;\rightarrow\; R_3 \end{matrix}$

$$\sim \begin{bmatrix} 1 & 2 & -1 & | & 2 \\ 0 & -1 & 3 & | & -7 \\ 0 & -1 & 3 & | & -7 \end{bmatrix} \quad R_3 + (-1)R_2 \;\rightarrow\; R_3$$

$$\sim \begin{bmatrix} 1 & 2 & -1 & 2 \\ 0 & -1 & 3 & -7 \\ 0 & 0 & 0 & 0 \end{bmatrix} \quad R_1 + R_2 \ \rightarrow \ R_1$$

$$\sim \begin{bmatrix} 1 & 0 & 5 & -12 \\ 0 & -1 & 3 & -7 \\ 0 & 0 & 0 & 0 \end{bmatrix} \quad (-1)R_2 \ \rightarrow \ R_2$$

$$\sim \begin{bmatrix} 1 & 0 & 5 & -12 \\ 0 & 1 & -3 & 7 \\ 0 & 0 & 0 & 0 \end{bmatrix}$$

This corresponds to the system

$$\begin{aligned} x_1 \quad + \quad 5x_3 &= -12 \\ x_2 - \quad 3x_3 &= 7 \end{aligned}$$

$$\begin{aligned} \text{Let } x_3 &= t \\ \text{Then } x_2 &= 3x_3 + 7 \\ &= 3t + 7 \\ x_1 &= -5x_3 - 12 \\ &= -5t - 12 \end{aligned}$$

Hence $x_1 = -5t - 12, x_2 = 3t + 7, x_3 = t$

is a solution for every real number t. There are infinitely many solutions.

(7 – 3)

21.
$$\begin{bmatrix} 1 & -2 & 1 \\ 2 & -1 & 0 \\ 1 & -3 & -2 \end{bmatrix} \quad \begin{array}{l} R_2 + (-2)R_1 \ \rightarrow \ R_2 \\ R_3 + (-1)R_2 \ \rightarrow \ R_3 \end{array}$$

$$\sim \begin{bmatrix} 1 & -2 & 1 \\ 0 & 3 & -2 \\ 0 & -1 & -3 \end{bmatrix} \quad R_2 + 3R_3 \ \rightarrow \ R_2$$

$$\sim \begin{bmatrix} 1 & -2 & 1 \\ 0 & 0 & 11 \\ 0 & -1 & -3 \end{bmatrix}$$

The second row corresponds to the equation
$0x_1 + 0x_2 = -11$,
hence there is no solution.

(7 – 3)

22.
$$\begin{bmatrix} 1 & 2 & -1 & 2 \\ 3 & -1 & 2 & -3 \end{bmatrix} \quad R_2 + (-3)R_1 \rightarrow R_2$$

$$\sim \begin{bmatrix} 1 & 2 & -1 & 2 \\ 0 & -7 & 5 & -9 \end{bmatrix} \quad -\tfrac{1}{7}R_2 \ \rightarrow \ R_2$$

$$\sim \begin{bmatrix} 1 & 2 & -1 & \Big| & 2 \\ 0 & 1 & -\frac{5}{7} & \Big| & \frac{9}{7} \end{bmatrix} \quad R_1 + (-2)R_2 \;\rightarrow\; R_1$$

$$\sim \begin{bmatrix} 1 & 0 & \frac{3}{7} & \Big| & -\frac{4}{7} \\ 0 & 1 & -\frac{5}{7} & \Big| & \frac{9}{7} \end{bmatrix}$$

This corresponds to the system

$$x_1 \qquad + \tfrac{3}{7}x_3 \;=\; -\tfrac{4}{7}$$
$$x_2 \;-\; \tfrac{5}{7}x_3 \;=\; \tfrac{9}{7}$$

$$\text{Let } x_3 \;=\; t$$
$$\text{Then } x_2 \;=\; \tfrac{5}{7}x_3 + \tfrac{9}{7}$$
$$\qquad\;=\; \tfrac{5}{7}t + \tfrac{9}{7}$$
$$x_1 \;=\; -\tfrac{3}{7}x_3 - \tfrac{4}{7}$$
$$\qquad\;=\; -\tfrac{3}{7}t - \tfrac{4}{7}$$

Hence $x_1 = -\tfrac{3}{7}t - \tfrac{4}{7}, x_2 = \tfrac{5}{7}t + \tfrac{9}{7}, x_3 = t$ is a solution for every real number t.
There are infinitely many solutions.
(7 – 3)
The checking steps are omitted in problems 23-25 for lack of space.

23. $x^2 - y^2 \;=\; 2$
$\qquad\;\; y^2 \;=\; x$

Substitute y^2 in the second equation into the first equation.

$$x^2 - x \;=\; 2$$
$$x^2 - x - 2 \;=\; 0$$
$$(x - 2)(x + 1) \;=\; 0$$
$$x \;=\; 2, -1$$

For $x = 2$ $\qquad$ For $x = -1$
$\;\;\; y^2 = 2$ $\qquad\qquad\;\; y^2 = -1$
$\;\;\;\; y = \pm\sqrt{2}$ $\qquad\qquad\; y = \pm i$

Solutions: $(2, \sqrt{2}), (2, -\sqrt{2}), (-1, i), (-1, i)$
(7 – 4)

24. $x^2 + 2xy + y^2 \;=\; 1$
$\qquad\qquad\;\;\; xy \;=\; -2$

Solve for y in the second equation.

$$y = \tfrac{-2}{x}$$

Substitute into the first equation

$$x^2 + 2x\left(-\tfrac{2}{x}\right) + \left(\tfrac{2}{x}\right)^2 \;=\; 1$$
$$x^2 - 4 + \tfrac{4}{x^2} \;=\; 1 \quad x \neq 0$$
$$x^4 - 4x^2 + 4 \;=\; x^2$$
$$x^4 - 5x^2 + 4 \;=\; 0$$
$$(x^2 - 4)(x^2 - 1) \;=\; 0$$
$$(x - 2)(x + 2)(x - 1)(x + 1) \;=\; 0$$
$$x = 2, -2, 1, -1$$

For $x = 2$ For $x = -2$ For $x = 1$ For $x = 1$
$y = \frac{-2}{2}$ $y = \frac{-2}{-2}$ $y = \frac{-2}{1}$ $y = \frac{-2}{-1}$
$y = -1$ $y = 1$ $y = -2$ $y = 2$

Solutions: (2,-1) (-2,1) (1,-2), (-1,2)
(7 − 4)

25. $\quad 2x^2 + xy + y^2 = 8$
$\quad\quad\quad x^2 - y^2 = 0$

We factor the left side of the equation that has a zero constant term.
$(x - y)(x + y) = 0$
$x = y$ or $x = -y$
Thus, the original system is equivalent to the two systems
$2x^2 + xy + y^2 + \;=\; 8 \quad 2x^2 + xy + y^2 = 8$
$\quad\quad\quad\quad x \;=\; y \quad\quad\quad\quad\quad x = -y$

These systems are solved by substitution

First System :   Second System :
$2x^2 + xy + y^2 = 8$ $\quad\quad$ $2x^2 + xy + y^2 = 8$
$\quad\quad\quad x = y$ $\quad\quad\quad\quad\quad\quad$ $x = -y$
$2y^2 + yy + y^2 = 8$ $\quad$ $2(-y)^2 + (-y)y + y^2 = 8$
$\quad\quad 4y^2 = 8$ $\quad\quad\quad\quad\quad\quad 2y^2 = 8$
$\quad\quad\; y^2 = 2$ $\quad\quad\quad\quad\quad\quad\; y^2 = 4$
$\quad\quad\;\; y = \pm\sqrt{2}$ $\quad\quad\quad\quad\quad\;\; y = \pm\sqrt{2}$

For $y = \sqrt{2}$ For $y = -\sqrt{2}$ For $y = 2$ For $y = -2$
$\;\; x = \sqrt{2}$ $\;\; x = -\sqrt{2}$ $\;\; x = -2$ $\;\; x = 2$

Solutions: $(\sqrt{2}, \sqrt{2}), (-\sqrt{2}, -\sqrt{2}), (2, -2), (-2, 2)$
(7 − 4)

26. The solution region is bounded (contained
in, for example, the circle $x^2 + y^2 = 36$)
Three corner points are obvious from
the graph: (0,4),(0,0), and (4,0).
The fourth corner point is obtained
by solving the system
$\quad 2x + y = 8$
$\quad 2x + 3y = 12$ to obtain $(3, 2)$.
(7 − 5)

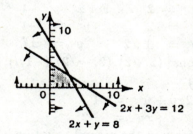

27. The solution region is unbounded.
Two corner points are obvious from
the graph: $(0,8)$, and $(12,0)$.
The third corner point is obtained
by solving the system
$$2x + y = 8$$
$$x + 3y = 12$$
to obtain $\left(\frac{12}{5}, \frac{16}{5}\right)$.
$(7 - 5)$

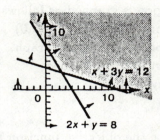

28. The solution region is bounded.
The corner point $(20,0)$ is obvious from
the graph. The other corner
points are found by solving the system
$$\begin{cases} x + y = 20 \\ x - y = 0 \end{cases}$$ to obtain $(10, 10)$
and
$$\begin{cases} x - y = 0 \\ x + 4y = 20 \end{cases}$$ to obtain $(4, 4)$.
$(7 - 5)$

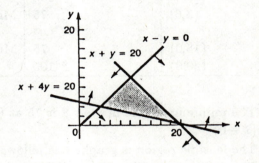

29. The feasible region is graphed as follows:
The corner points $(0,4),(0,0)$,
and $(5,0)$ are obvious from the graph.
The corner point $(4,2)$ is
obtained by solving the system
$$x + 2y = 8$$
$$2x + y = 10$$

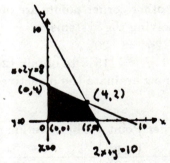

We now evaluate the objective function at each corner point.

Corner Point (x, y)	Objective Function $z = 7x + 9y$	
$(0,4)$	36	
$(0,0)$	0	
$(5,0)$	35	
$(4,2)$	46	Maximum value

The maximum value of z on S is 46 at $(4,2)$.
$(7-6)$

30.   The feasible region is graphed as follows:
The corner points $(0,20), (0,15)$, $(15,0)$
and $(20,0)$ are obvious from the graph.
The corner point $(3,6)$ is obtained
by solving the system:
$$3x + y = 15$$
$$x + 2y = 15$$
We now evaluate the objective function at
each corner point.

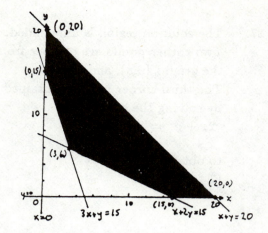

Corner Point (x,y)	Objective Function $z = 5x + 10y$		
$(0,20)$	200		
$(0,15)$	150		
$(3,6)$	75	Minimum value	
$(15,0)$	75	Minimum value	Multiple Optimal Solutions
$(20,0)$	100		

The minimum value of z on S is 75 at $(3,6)$ and $(15,0)$
$(7-6)$

31.   The feasible region is graphed as follows:
The corner points $(0,10)$ and $(0,7)$
are obvious from the graph.
The other corner points are obtained
by solving the systems
$$x + 2y = 20 \qquad \text{and } 3x + y = 15$$
$$3x + y = 15 \quad \text{to obtain } (2,9) \qquad x + y = 7 \quad \text{to obtain } (4,3)$$
We now evaluate the objective function at each corner point.

Corner Point (x,y)	Objective Function $5x + 8y$	
$(0,10)$	80	
$(0,7)$	56	
$(4,3)$	44	Minimum value
$(2,9)$	82	Maximum value

The minimum value of z on S is 44 at $(4,3)$.
The maximum value of z on S is 82 at $(2,9)$.

32.
$$\begin{bmatrix} 1 & 1 & 1 & | & 7000 \\ 0.04 & 0.05 & 0.06 & | & 360 \\ 0.04 & 0.05 & -0.06 & | & 120 \end{bmatrix} \begin{array}{l} \\ R_2 + (-0.04)R_1 \rightarrow R_2 \\ R_3 + (-0.04)R_1 \rightarrow R_3 \end{array}$$

$$\sim \begin{bmatrix} 1 & 1 & 1 & 7000 \\ 0 & 0.01 & 0.02 & 80 \\ 0 & 0.01 & -0.1 & -160 \end{bmatrix} \quad 100R_2 \rightarrow R_2$$

$$\sim \begin{bmatrix} 1 & 1 & 1 & 7000 \\ 0 & 1 & 2 & 8000 \\ 0 & 0.01 & -0.1 & -160 \end{bmatrix} \begin{array}{l} R_1 + (-1)R_2 \rightarrow R_1 \\ \\ R_3 + (-0.01)R_2 \rightarrow R_3 \end{array}$$

$$\sim \begin{bmatrix} 1 & 0 & -1 & -1000 \\ 0 & 1 & 2 & 8000 \\ 0 & 0 & -0.12 & -240 \end{bmatrix} \quad -\frac{25}{3}R_3 \rightarrow R_3$$

$$\sim \begin{bmatrix} 1 & 0 & -1 & -1000 \\ 0 & 1 & 2 & 8000 \\ 0 & 0 & 1 & -2000 \end{bmatrix} \begin{array}{l} R_1 + R_3 \rightarrow R_1 \\ R_2 + (-2)R_3 \rightarrow R_2 \end{array}$$

$$\sim \begin{bmatrix} 1 & 0 & 0 & 1000 \\ 0 & 1 & 0 & 4000 \\ 0 & 0 & 1 & 2000 \end{bmatrix}$$

Solution: $x_1 = 1000, x_2 = 4000, x_3 = 2000$

(7 − 3)

33. $x^2 - xy + y^2 = 4$
$x^2 + xy - 2y^2 = 0$

Factor the left side of the equation that has a zero constant term.

$(x + 2y)(x - y) = 0$

$x = -2y$ or $x = y$

Thus , the original system is equivalent to the two systems

$x^2 - xy + y^2 = 4 \qquad x^2 - xy + y^2 = 4$
$x = -2y \qquad\qquad x = y$

These systems are solved by substitution:

First system : $\qquad$ Second system :
$x^2 - xy + y^2 = 4 \qquad x^2 - xy + y^2 = 4$
$x = -2y \qquad\qquad x = 4$
$(-2y)^2 - (-2y)y + y^2 = 4 \qquad y^2 - yy + y^2 = 4$
$4y^2 + 2y^2 + y^2 = 4 \qquad\qquad y^2 = 4$
$7y^2 = 4 \qquad\qquad y = \pm 2$
$y = \pm\sqrt{\frac{4}{7}}$
$y = \pm\frac{2\sqrt{7}}{7}$

For $y = \frac{2\sqrt{7}}{7}$ For $y = -\frac{2\sqrt{7}}{7}$ For $y = -2$ For $y = 2$
$x = -2(\frac{2\sqrt{7}}{7})$ $x = -2(-\frac{2\sqrt{7}}{7})$ $x = -2$ $x = 2$
$= -\frac{4\sqrt{7}}{7}$ $= \frac{4\sqrt{7}}{7}$

Solutions:
$(\frac{4\sqrt{7}}{7}, -\frac{2\sqrt{7}}{7}), (-\frac{4\sqrt{7}}{7}, \frac{2\sqrt{7}}{7}), (2,2), (-2,-2).$
The checking steps are omitted for lack of space

(7 − 4)

34.   For convenience we rewrite the constraint conditions:

$$1.2x + 0.6y \leq 960 \quad \text{becomes } 2x + y \leq 1,600$$
$$0.04x + 0.03y \leq 36 \quad \text{becomes } 4x + 3y \leq 3,600$$
$$0.2x + .03y \leq 270 \quad \text{becomes } 2x + 3y \leq 2,700$$
$$x, y \geq 0 \qquad \text{are unaltered.}$$

The feasible region is graphed in the diagram.

The corner points $(0,900)$, $(0,0)$, and $(800,0)$ are obvious from the graph.

The other corner points are obtained by solving the system

$$2x + y = 1,600 \qquad\qquad \text{and } 4x + 3y = 3,600$$
$$4x + 3y = 3,600 \quad \text{to obtain}(600, 400) \qquad 2x + 3y = 2,700 \quad \text{to obtain}(450, 600)$$

We now evaluate the objective function at each corner point.

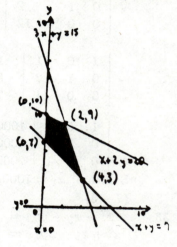

Corner Point (x,y)	Objective Function $30x + 20y$	
$(0,900)$	18,000	
$(0,0)$	0	
$(800,0)$	24,000	
$(600,400)$	26,000	Maximum value
$(450,600)$	25,500	

The maximum value of z on S is 26,000 at $(600,400)$.

$(7-6)$

35.   Let

$x = $ number of $\frac{1}{2}$ pound packages

$y = $ number of $\frac{1}{3}$ pound packages

There are 120 packages. Hence

$x + y = 120 \quad (1)$

Since x $\frac{1}{2}$- pound packages weigh $\frac{1}{2}x$ pounds and y $\frac{1}{3}$- pound packages weigh $\frac{1}{3}y$ pounds, we have $\frac{1}{2}x + \frac{1}{3}y = 48 \quad (2)$

We solve the system (1), (2) using elimination by addition. We multiply the bottom equation by -3 and add.

$$x + y = 120$$
$$-\tfrac{3}{2}x - y = -144$$
$$-\tfrac{1}{2}x = -24$$
$$x = 48$$

Substituting into equation (1), we have

$$48 + y = 120$$
$$y = 72$$

48 $\frac{1}{2}$-pound packages and 72 $\frac{1}{3}$-pound packages

$(7-1)$

36.   Using the formulas for perimeter and area of a rectangle, we have

$$2a + 2b = 28$$
$$ab = 48$$

Solving the first-degree equation for b and substituting into the second-degree equation, we have

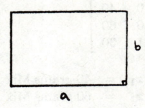

$$2b = 28 - 2a$$
$$b = 14 - a$$
$$a(14 - a) = 48$$
$$-a^2 + 14a = 48$$
$$a^2 - 14a + 48 = 0$$
$$(a - 6)(a - 8) = 0$$
$$a = 6, 8$$

For $a = 6$ For $a = 8$
$$b = 14 - 6 \qquad b = 14 - 6$$
$$= 8 \qquad\qquad = 8$$

Dimensions: 6 meters by 8 meters

(7 – 4)

37. Let $x_1 =$ number of grams of mix A
   $x_2 =$ number of grass of mix B
   $x_3 =$ number of grass of mix C

We have

$$0.30x_1 + 0.20x_2 + 0.10x_3 = 27 \text{(protein)}$$
$$0.03x_1 + 0.05x_2 + 0.04x_3 = 5.4 \text{(fat)}$$
$$0.10x_1 + 0.20x_2 + 0.10x_3 = 19 \text{(moisture)}$$

Clearing of decimals for convenience, we have

$$3x_1 + 2x_2 + x_3 = 270$$
$$3x_1 + 5x_2 + 4x_3 = 540$$
$$x_1 + 2x_2 + x_3 = 190$$

Form the augmented matrix and solve by Gauss-Jordan elimination.

$$\begin{bmatrix} 3 & 2 & 1 & | & 270 \\ 3 & 5 & 4 & | & 540 \\ 1 & 2 & 1 & | & 190 \end{bmatrix} \quad R_3 \leftrightarrow R_1$$

$$\sim \begin{bmatrix} 1 & 2 & 1 & | & 190 \\ 3 & 5 & 4 & | & 540 \\ 3 & 2 & 1 & | & 270 \end{bmatrix} \quad \begin{array}{l} R_2 + (-3)R_1 \rightarrow R_2 \\ R_3 + (-3)R_1 \rightarrow R_3 \end{array}$$

$$\sim \begin{bmatrix} 1 & 2 & 1 & | & 190 \\ 0 & -1 & 1 & | & -30 \\ 0 & -4 & -2 & | & -300 \end{bmatrix} \quad (-1)R_2 \rightarrow R_2$$

$$\sim \begin{bmatrix} 1 & 2 & 1 & | & 190 \\ 0 & 1 & -1 & | & 30 \\ 0 & -4 & -2 & | & -300 \end{bmatrix} \quad \begin{array}{l} R_1 + (-2)R_2 \rightarrow R_1 \\ R_3 + 4R_2 \rightarrow R_3 \end{array}$$

$$\sim \begin{bmatrix} 1 & 0 & 3 & | & 130 \\ 0 & 1 & -1 & | & 30 \\ 0 & 0 & -6 & | & -180 \end{bmatrix} \quad -\tfrac{1}{6}R_3 \rightarrow R_3$$

$$\sim \begin{bmatrix} 1 & 0 & 3 & | & 130 \\ 0 & 1 & -1 & | & 30 \\ 0 & 0 & 1 & | & 30 \end{bmatrix} \quad \begin{array}{rcl} R_2 + (-3)R_3 & \to & R_1 \\ R_2 + R_3 & \to & R_2 \end{array}$$

$$\sim \begin{bmatrix} 1 & 0 & 0 & | & 40 \\ 0 & 1 & 0 & | & 60 \\ 0 & 0 & 1 & | & 30 \end{bmatrix}$$

Therefore, $x_1 = 40$ grams Mix A

$\qquad\qquad x_2 = 60$ grams Mix B

$\qquad\qquad x_3 = 30$ grams Mix C

$(7-1, 7-3)$

38.   A. Let $x_1 = $ Number of nickels

$\qquad\qquad x_2 = $ Number of dimes

Then $x_1 + x_2 = 30$(total number of coins)

$\qquad 5x_1 + 10x_2 = 190$(total value of coins)

We form the augmented matrix and solve by Gauss-Jordan elimination

$$\begin{bmatrix} 1 & 1 & | & 30 \\ 4 & 10 & | & 190 \end{bmatrix} \quad R_2 + (-5)R_1 \to R_2$$

$$\sim \begin{bmatrix} 1 & 1 & | & 30 \\ 0 & 5 & | & 40 \end{bmatrix} \quad \tfrac{1}{5}R_2 \to R_2$$

$$\sim \begin{bmatrix} 1 & 1 & | & 30 \\ 0 & 1 & | & 8 \end{bmatrix} \quad R_1 + (-1)R_2 \to R_1$$

$$\sim \begin{bmatrix} 1 & 0 & | & 22 \\ 0 & 1 & | & 8 \end{bmatrix}$$

The augmented matrix is in reduced form. It corresponds to the system

$x_1 = 22$ nickels

$x_2 = 8$ dimes

B. Let $x_1 = $ Number of nickels

$\qquad\ x_2 = $ Number of dimes

$\qquad\ x_3 = $ Number of quarters

Then

$$x_1 + x_2 + x_3 = 30\text{(total number of coins)}$$
$$5x_1 + 10x_2 + 25x_3 = 190\text{(total value of coins)}$$

We form the augmented matrix and solve by Gauss-Jordan elimination

$$\begin{bmatrix} 1 & 1 & 1 & | & 30 \\ 5 & 10 & 25 & | & 190 \end{bmatrix} \quad R_2 + (-5)R_1 \to R_2$$

$$\sim \begin{bmatrix} 1 & 1 & 1 & | & 30 \\ 0 & 5 & 20 & | & 40 \end{bmatrix} \quad \tfrac{1}{5}R_2 \to R_1$$

$$\sim \begin{bmatrix} 1 & 1 & 1 & | & 30 \\ 0 & 1 & 4 & | & 8 \end{bmatrix} \quad R_1 + (-1)R_2 \;\rightarrow\; R_1$$

$$\sim \begin{bmatrix} 1 & 0 & -3 & | & 22 \\ 0 & 1 & 4 & | & 8 \end{bmatrix}$$

The augmented matrix is in reduced form. It corresponds to the system.

$$
\begin{aligned}
x_1 \quad\;\; - \;\; 3x_3 &= 22 \\
x_2 + \;\; 4x_3 &= 8
\end{aligned}
$$

Let $x_3 = t$. Then

$$
\begin{aligned}
x_2 &= -4x_3 + 8 \\
&= -4t + 8 \\
x_1 &= 3x_3 + 22 \\
&= 3t + 22
\end{aligned}
$$

A solution is achieved, not for every real value of t, but for integer values of t that give rise to non-negative x_1, x_2, x_3.

$$
\begin{aligned}
x_1 \geq 0 \quad &\text{means} \quad 3t + 22 \geq 0 \quad \text{or} \quad t \geq -7\tfrac{1}{3} \\
x_2 \geq 0 \quad &\text{means} \quad -4t + 8 \geq 0 \quad \text{or} \quad t \leq 2 \\
x_3 \geq 0 \quad &\text{means} \quad t \geq 0
\end{aligned}
$$

The only integer values of t that satisfy these conditions are 0, 1, 2. Thus we have the solutions

$$
\begin{aligned}
x_1 &= 3t + 22 \;\; \text{nickels} \\
x_2 &= 8 - 4t \;\; \text{dimes} \\
x_3 &= t \;\; \text{quarters}
\end{aligned}
$$

where $t = 0, 1$, or 2

39. Let x = Number of regular sails

  y = Number of competition sails

We form the linear objective function

$P = 60x + 100y$

We wish to maximize P, the profit from x regular sails @ \$60 and y competition sails @ \$100, subject to the constraints

$$
\begin{aligned}
x + 2y &\leq 140 \quad &\text{cutting department constraint} \\
3x + 4y &\leq 360 \quad &\text{sewing department constraint} \\
x, y &\geq 0 \quad &\text{nonnegative constraints}
\end{aligned}
$$

Solving the system of constraint inequalities graphically, we obtain the feasible region S shown below.

Next we evaluate the objective function at each corner point.

Corner Point (x, y)	Objective Function $P = 60x + 100y$	
(0, 0)	0	
(0, 70)	7,000	
(80, 30)	7,800	Maximum value
(120, 0)	7,200	

The optimal value is 7,800 at the corner point (80, 30). Thus, the company should manufacture 80 regular sails and 30 competition sails for a maximum profit of $7,800.   (7 – 6)

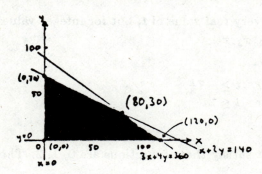

CHAPTER 8

Exercise 8-1

Key Ideas and Formulas

A matrix is a rectangular array of numbers (enclosed within brackets). The numbers are called elements.

A matrix with m rows and n columns has dimension $m \times n$.

If $m = n$, the matrix is a square matrix.
If $m = 1$, the matrix has 1 row and is called a row matrix.
If $n = 1$, the matrix has 1 column and is called a column matrix.

Two matrices are equal if they have the same dimension and their corresponding elements are equal.

The sum of two matrices A and B, denoted $A + B$, is a matrix with elements that are the sums of the corresponding elements of A and B.

$A + B = B + A$ (Commutative property of addition).
$(A + B) + C = A + (B + C)$ (Associative property of addition).

The negative of a matrix M, denoted $-M$, is a matrix with elements that are the negative of the elements of M.

A matrix with elements that are all 0's is called a zero matrix, denoted 0.

$M + (-M) = 0.$

If A and B are matrices of the same dimension, we define subtraction as follows: $A - B = A + (-B)$.

The product of a number k and a matrix M, denoted kM, is a matrix formed by multiplying each element of M by k.

---

1. $2 \times 2, 1 \times 4$

3. 2

5. $\begin{bmatrix} 0 & 0 \\ 0 & 0 \end{bmatrix}$

7. C, D

9. A, B

11. $A + B = \begin{bmatrix} 2 & -1 \\ -3 & 0 \end{bmatrix} + \begin{bmatrix} -3 & 1 \\ 2 & -3 \end{bmatrix} = \begin{bmatrix} -1 & 0 \\ 5 & -3 \end{bmatrix}$

13. E and F do not have the same dimension. $E + F$ is not defined.

15. $\quad -C = -\begin{bmatrix} 2 \\ -3 \\ 0 \end{bmatrix} = \begin{bmatrix} -2 \\ 3 \\ 0 \end{bmatrix}$

17. $\quad D - C = \begin{bmatrix} 1 \\ 3 \\ 5 \end{bmatrix} - \begin{bmatrix} 2 \\ -3 \\ 0 \end{bmatrix} = \begin{bmatrix} -1 \\ 6 \\ 5 \end{bmatrix}$

19. $\quad 5B = 5\begin{bmatrix} -3 & 1 \\ 2 & -3 \end{bmatrix} = \begin{bmatrix} -15 & 5 \\ 10 & -15 \end{bmatrix}$

21. $\quad \begin{bmatrix} 4 & 0 & -2 & 1 \\ 3 & -2 & -3 & 5 \\ 0 & 2 & -1 & 4 \end{bmatrix} + \begin{bmatrix} -3 & -1 & 6 & 0 \\ -5 & -4 & 7 & -3 \\ 2 & 5 & 1 & -6 \end{bmatrix} = \begin{bmatrix} 1 & -1 & 4 & 1 \\ -2 & -6 & 4 & 2 \\ 2 & 7 & 0 & -2 \end{bmatrix}$

23. $\quad \begin{bmatrix} 1 & -2 & 4 \\ -3 & 4 & -6 \end{bmatrix} - \begin{bmatrix} -1 & 3 & 4 \\ 2 & 1 & -7 \end{bmatrix}$

$\quad = \begin{bmatrix} 1 - (-1) & (-2) - 3 & 4 - 4 \\ (-3) - 2 & 4 - 1 & (-6) - (-7) \end{bmatrix} = \begin{bmatrix} 2 & -5 & 0 \\ -5 & 3 & 1 \end{bmatrix}$

25. $\quad 1{,}000 \begin{bmatrix} 0.25 & 0.36 \\ 0.04 & 0.35 \end{bmatrix} = \begin{bmatrix} 250 & 360 \\ 40 & 350 \end{bmatrix}$

27. $\quad 2\begin{bmatrix} 1 & 2 \\ -1 & 5 \end{bmatrix} + 3\begin{bmatrix} 4 & 1 & -2 \\ 3 & 2 & -1 \end{bmatrix} = \begin{bmatrix} 2 & 4 \\ -2 & 10 \end{bmatrix} + \begin{bmatrix} 12 & 3 & -6 \\ 9 & 6 & -3 \end{bmatrix}$ is not defined.

The two matrices to be added do not have the same dimension.

29. $\quad 0.05\begin{bmatrix} 400 & 600 \\ 200 & 700 \end{bmatrix} + 0.04\begin{bmatrix} 350 & 200 \\ 450 & 800 \end{bmatrix} = \begin{bmatrix} 20 & 30 \\ 10 & 35 \end{bmatrix} + \begin{bmatrix} 14 & 8 \\ 18 & 32 \end{bmatrix} = \begin{bmatrix} 34 & 38 \\ 28 & 67 \end{bmatrix}$

31. $\quad \begin{bmatrix} a & b \\ c & d \end{bmatrix} + \begin{bmatrix} 2 & -3 \\ 0 & 1 \end{bmatrix} = \begin{bmatrix} a+2 & b-3 \\ c & d+1 \end{bmatrix} = \begin{bmatrix} 1 & -2 \\ 3 & -4 \end{bmatrix}$

if and only if corresponding elements are equal.

$$a + 2 = 1 \quad b - 3 = -2 \quad c = 3 \quad d + 1 = -4$$
$$a = -1 \quad\quad b = 1 \quad\quad c = 3 \quad\quad d = -5$$

33. $\quad \begin{bmatrix} 3x & 5 \\ -1 & 4x \end{bmatrix} + \begin{bmatrix} 2y & -3 \\ -6 & -y \end{bmatrix} = \begin{bmatrix} 3x+2y & 2 \\ -7 & 4x-y \end{bmatrix} = \begin{bmatrix} 7 & 2 \\ -7 & 2 \end{bmatrix}$

if and only if corresponding elements are equal.

$$3x + 2y \ = \ 7 \qquad\qquad 2 \ \overset{\checkmark}{=} \ 2 \ \Big\} \text{ Two conditions are already met.}$$
$$-7 \ \overset{\checkmark}{=} \ -7 \quad 4x - y \ = \ 2$$

To find x and y, we solve the system:

$$3x + 2y \ = \ 7$$
$$4x - y \ = \ 2 \text{ to obtain } x = 1, y = 2.$$

35. $\frac{1}{2}(A + B) = \frac{1}{2}\left(\begin{bmatrix} 30 & 25 \\ 60 & 80 \end{bmatrix} + \begin{bmatrix} 36 & 27 \\ 54 & 74 \end{bmatrix} \right)$

Guitar Banjo

$= \frac{1}{2}\begin{bmatrix} 66 & 52 \\ 114 & 154 \end{bmatrix} = \begin{bmatrix} 33 & 26 \\ 57 & 77 \end{bmatrix} \begin{matrix} \text{Materials} \\ \text{Labor} \end{matrix}$

37. $A + B = \begin{bmatrix} 70 & 122 & 20 \\ 30 & 118 & 80 \end{bmatrix} + \begin{bmatrix} 65 & 160 & 30 \\ 25 & 140 & 75 \end{bmatrix}$

Under 5 ft $\quad$ $5 - 5\frac{1}{2}$ ft $\quad$ Over $5\frac{1}{2}$ ft.

$= \begin{bmatrix} 135 & 282 & 50 \\ 55 & 258 & 155 \end{bmatrix} \begin{matrix} \text{Passive} \\ \text{Aggressive} \end{matrix}$

Decimal fraction of total sample in each category:

$\frac{1}{935}(A + B) = \frac{1}{935}\begin{bmatrix} 135 & 282 & 50 \\ 55 & 258 & 155 \end{bmatrix}$

Under 5 ft $\quad$ $5 - 5\frac{1}{2}$ ft $\quad$ Over $5\frac{1}{2}$ ft.

$= \begin{bmatrix} 0.14 & 0.30 & 0.05 \\ 0.06 & 0.27 & 0.17 \end{bmatrix} \begin{matrix} \text{Passive} \\ \text{Aggressive} \end{matrix}$

Exercise 8-2

Key Ideas and Formulas

The dot product of a $1 \times n$ row matrix and an $n \times 1$ column matrix is a real number given by

$$[a_1 \; a_2 \; \cdots \; a_n] \cdot \begin{bmatrix} b_1 \\ b_2 \\ \vdots \\ b_n \end{bmatrix} = a_1 b_1 + a_2 b_2 + \cdots + a_n b_n$$

The product of two matrices A and B is defined only on the assumption that the number of columns in A is equal to the number of rows in B.

$$\begin{array}{ccc} A & B & = & AB \\ m \times p & p \times n & & m \times n \end{array}$$

number of columns in $A = p =$ number of rows in B

If A is an $m \times p$ matrix and B is a $p \times n$ matrix, then AB is an $m \times n$ matrix whose element in the ith row and jth column is the dot product of the ith row matrix of A and the jth column matrix of B.

Matrix multiplication is not commutative. Depending on A and B, either AB or BA could be not defined, or defined but of different dimensions, or of the same dimension but AB still unequal to BA.

Also AB could be zero without $A = 0$ or $B = 0$.

Assuming all products and sums are defined for matrices A, B, C, with k any real number,

$A(BC) = (AB)C$	(associative property)
$A(B + C) = AB + AC$	(left distributive property)
$(B + C)A = BA + CA$	(right distributive property)
If $A = B$ then $CA = CB$	(left multiplication property)
If $A = B$ then $AC = BC$	(right multiplication property)
$k(AB) = (kA)B = A(kB)$	*Common Error:* $k(AB) \neq (kA)(kB)$.

1. $[2 \; 4] \cdot \begin{bmatrix} 3 \\ 1 \end{bmatrix} = 2 \cdot 3 + 4 \cdot 1 = 10$

3. $[-3 \; 2] \cdot \begin{bmatrix} -1 \\ -2 \end{bmatrix} = (-3)(-1) + 2(-2) = -1$

5. $[2 \; 5] \cdot \begin{bmatrix} 1 & -1 \\ 2 & 3 \end{bmatrix} = [2 \cdot 1 + 5 \cdot 2 \;\; 2(-1) + 5 \cdot 3] = [12 \; 13]$

7. $\begin{bmatrix} 3 & 4 \\ -1 & -2 \end{bmatrix} \begin{bmatrix} -1 \\ 2 \end{bmatrix} = \begin{bmatrix} 3(-1) + 4(2) \\ (-1)(-1) + (-2)2 \end{bmatrix} = \begin{bmatrix} 5 \\ -3 \end{bmatrix}$

9. $\begin{bmatrix} 2 & -3 \\ 1 & 2 \end{bmatrix} \begin{bmatrix} 1 & -1 \\ 0 & -2 \end{bmatrix} = \begin{bmatrix} 2 \cdot 1 + (-3)0 & 2(-1) + (-3)(-2) \\ 1 \cdot 1 + 2 \cdot 0 & 1(-1) + 2(-2) \end{bmatrix} = \begin{bmatrix} 2 & 4 \\ 1 & -5 \end{bmatrix}$

11. $\begin{bmatrix} 1 & -1 \\ 0 & -2 \end{bmatrix} \begin{bmatrix} 2 & -3 \\ 1 & 2 \end{bmatrix} = \begin{bmatrix} 1 \cdot 2 + (-1)1 & 1(-3) + (-1)2 \\ 0 \cdot 2 + (-2)1 & 0(-3) + (-2)2 \end{bmatrix} = \begin{bmatrix} 1 & -5 \\ -2 & -4 \end{bmatrix}$

13. $\begin{bmatrix} 4 & -2 \end{bmatrix} \begin{bmatrix} -5 \\ -3 \end{bmatrix} = [4(-5) + (-2)(-3)] = [-14]$

15. $\begin{bmatrix} -5 \\ -3 \end{bmatrix} \begin{bmatrix} 4 & -2 \end{bmatrix} = \begin{bmatrix} (-5)4 & (-5)(-2) \\ (-3)4 & (-3)(-2) \end{bmatrix} = \begin{bmatrix} -20 & 10 \\ -12 & 6 \end{bmatrix}$

17. $\begin{bmatrix} -1 & -2 & 2 \end{bmatrix} \cdot \begin{bmatrix} 2 \\ -1 \\ 3 \end{bmatrix} = (-1)2 + (-2)(-1) + 2 \cdot 3 = 6$

19. $\begin{bmatrix} -1 & -3 & 0 & 5 \end{bmatrix} \cdot \begin{bmatrix} 4 \\ -3 \\ -1 \\ 2 \end{bmatrix} = (-1)4 + (-3)(-3) + 0(-1) + 5 \cdot 2 = 15$

21. $\begin{bmatrix} 2 & -1 & 1 \\ 1 & 3 & -2 \end{bmatrix} \begin{bmatrix} 1 & 3 \\ 0 & -1 \\ -2 & 2 \end{bmatrix} = \begin{bmatrix} 2 \cdot 1 + (-1)0 + 1(-2) & 2 \cdot 3 + (-1)(-1) + 1 \cdot 2 \\ 1 \cdot 1 + 3 \cdot 0 + (-2)(-2) & 1 \cdot 3 + 3(-1) + (-2)2 \end{bmatrix}$

$= \begin{bmatrix} 0 & 9 \\ 5 & -4 \end{bmatrix}$

23. $\begin{bmatrix} 1 & 3 \\ 0 & -1 \\ -2 & 2 \end{bmatrix} \begin{bmatrix} 2 & -1 & 1 \\ 1 & 3 & -2 \end{bmatrix} = \begin{bmatrix} 1 \cdot 2 + 3 \cdot 1 & 1(-1) + 3 \cdot 3 & 1 \cdot 1 + 3(-2) \\ 0 \cdot 2 + (-1)1 & 0(-1) + (-1)3 & 0 \cdot 1 + (-1)(-2) \\ (-2)2 + 2 \cdot 1 & (-2)(-1) + 2 \cdot 3 & -2 \cdot 1 + 2(-2) \end{bmatrix}$

$= \begin{bmatrix} 5 & 8 & -5 \\ -1 & -3 & 2 \\ -2 & 8 & -6 \end{bmatrix}$

25. $\begin{bmatrix} 3 & -2 & -4 \end{bmatrix} \begin{bmatrix} 1 \\ 2 \\ -3 \end{bmatrix} = [3 \cdot 1 + (-2)2 + (-4)(-3)] = [11]$

27. $\begin{bmatrix} 1 \\ 2 \\ -3 \end{bmatrix} \begin{bmatrix} 3 & -2 & -4 \end{bmatrix} = \begin{bmatrix} 1 \cdot 3 & 1(-2) & 1(-4) \\ 2 \cdot 3 & 2(-2) & 2(-4) \\ (-3)3 & (-3)(-2) & (-3)(-4) \end{bmatrix} = \begin{bmatrix} 3 & -2 & -4 \\ 6 & -4 & -8 \\ -9 & 6 & 12 \end{bmatrix}$

29. The first matrix printed has 2 columns. The second matrix has 1 row. Therefore, their product is not defined.

31.
$$\begin{bmatrix} 1 & 2 & -1 \\ 3 & -1 & 4 \\ 2 & -4 & 5 \end{bmatrix} \begin{bmatrix} 4 \\ 5 \\ 7 \end{bmatrix} = \begin{bmatrix} 1 \cdot 4 + 2 \cdot 5 + (-1)7 \\ 3 \cdot 4 + (-1)5 + 4 \cdot 7 \\ 2 \cdot 4 + (-4)5 + 5 \cdot 7 \end{bmatrix} = \begin{bmatrix} 7 \\ 35 \\ 23 \end{bmatrix}$$

33.
$$\begin{bmatrix} 1 & -1 & 2 & 0 \\ 3 & -4 & -2 & 1 \\ 1 & 0 & -1 & 2 \end{bmatrix} \begin{bmatrix} 1 & -2 & 1 \\ 3 & 0 & -4 \\ -1 & 2 & -2 \\ 3 & -3 & 4 \end{bmatrix}$$

$$= \begin{bmatrix} 1 \cdot 1 + (-1)3 + 2(-1) + 0 \cdot 3 & 1(-2) + (-1)0 + 2 \cdot 2 + 0 \cdot (-3) \\ 3 \cdot 1 + (-4)3 + (-2)(-1) + 1 \cdot 3 & 3(-2) + (-4)0 + (-2)2 + 1(-3) \\ 1 \cdot 1 + 0 \cdot 3 + (-1)(-1) + 2 \cdot 3 & 1(-2) + 0 \cdot 0 + (-1)2 + 2(-3) \end{bmatrix}$$

$$\begin{matrix} 1 \cdot 1 + (-1)(-4) + 2(-2) + 0 \cdot 4 \\ 3 \cdot 1 + (-4)(-4) + (-2)(-2) + 1 \cdot 4 \\ 1 \cdot 1 + 0(-4) + (-1)(-2) + 2 \cdot 4 \end{matrix} = \begin{bmatrix} -4 & 2 & 1 \\ -4 & -13 & 27 \\ 8 & -10 & 11 \end{bmatrix}$$

35. $A = \begin{bmatrix} 6 & 9 \\ -4 & -6 \end{bmatrix}$ $A^2 = AA = \begin{bmatrix} 6 & 9 \\ -4 & -6 \end{bmatrix} \begin{bmatrix} 6 & 9 \\ -4 & -6 \end{bmatrix}$

$$= \begin{bmatrix} 6 \cdot 6 + 9(-4) & 6 \cdot 9 + 9(-6) \\ (-4)6 + (-6)(-4) & (-4)9 + (-6)(-6) \end{bmatrix} = \begin{bmatrix} 0 & 0 \\ 0 & 0 \end{bmatrix}$$

Common Error: $A^2 \neq \begin{bmatrix} 6^2 & 9^2 \\ (-4)^2 & (-6)^2 \end{bmatrix}$

37. $A = \begin{bmatrix} \frac{1}{3} & \frac{1}{3} \\ \frac{2}{3} & \frac{2}{3} \end{bmatrix}$ $A^2 = AA = \begin{bmatrix} \frac{1}{3} & \frac{1}{3} \\ \frac{2}{3} & \frac{2}{3} \end{bmatrix} \begin{bmatrix} \frac{1}{3} & \frac{1}{3} \\ \frac{2}{3} & \frac{2}{3} \end{bmatrix} = \begin{bmatrix} \frac{1}{3} \cdot \frac{1}{3} + \frac{1}{3} \cdot \frac{2}{3} & \frac{1}{3} \cdot \frac{1}{3} + \frac{1}{3} \cdot \frac{2}{3} \\ \frac{2}{3} \cdot \frac{1}{3} + \frac{2}{3} \cdot \frac{2}{3} & \frac{2}{3} \cdot \frac{1}{3} + \frac{2}{3} \cdot \frac{2}{3} \end{bmatrix}$

$$= \begin{bmatrix} \frac{1}{3} & \frac{1}{3} \\ \frac{2}{3} & \frac{2}{3} \end{bmatrix}$$

39. $AB = \begin{bmatrix} 1 & 2 \\ 0 & 1 \end{bmatrix} \begin{bmatrix} 1 & 1 \\ 2 & 3 \end{bmatrix} = \begin{bmatrix} 1 \cdot 1 + 2 \cdot 2 & 1 \cdot 1 + 2 \cdot 3 \\ 0 \cdot 1 + 1 \cdot 2 & 0 \cdot 1 + 1 \cdot 3 \end{bmatrix} = \begin{bmatrix} 5 & 7 \\ 2 & 3 \end{bmatrix}$

$BA = \begin{bmatrix} 1 & 1 \\ 2 & 3 \end{bmatrix} \begin{bmatrix} 1 & 2 \\ 0 & 1 \end{bmatrix} = \begin{bmatrix} 1 \cdot 1 + 1 \cdot 0 & 1 \cdot 2 + 1 \cdot 1 \\ 2 \cdot 1 + 3 \cdot 0 & 2 \cdot 2 + 3 \cdot 1 \end{bmatrix} = \begin{bmatrix} 1 & 3 \\ 2 & 7 \end{bmatrix}$

Since corresponding elements are not equal, $AB \neq BA$.

41. $A(B+C) = \begin{bmatrix} 1 & 2 \\ 0 & 1 \end{bmatrix} (\begin{bmatrix} 1 & 1 \\ 2 & 3 \end{bmatrix} + \begin{bmatrix} -3 & 1 \\ -1 & 2 \end{bmatrix}) = \begin{bmatrix} 1 & 2 \\ 0 & 1 \end{bmatrix} \begin{bmatrix} -2 & 2 \\ 1 & 5 \end{bmatrix}$

$$= \begin{bmatrix} 1(-2)+2\cdot1 & 1\cdot2+2\cdot5 \\ 0(-2)+1\cdot1 & 0\cdot2+1\cdot5 \end{bmatrix} = \begin{bmatrix} 0 & 12 \\ 1 & 5 \end{bmatrix}$$

$$AB + AC = \begin{bmatrix} 1 & 2 \\ 0 & 1 \end{bmatrix}\begin{bmatrix} 1 & 1 \\ 2 & 3 \end{bmatrix} + \begin{bmatrix} 1 & 2 \\ 0 & 1 \end{bmatrix}\begin{bmatrix} -3 & 1 \\ -1 & 2 \end{bmatrix}$$

$$= \begin{bmatrix} 1\cdot1+2\cdot2 & 1\cdot1+2\cdot3 \\ 0\cdot1+1\cdot2 & 0\cdot1+1\cdot3 \end{bmatrix} + \begin{bmatrix} 1(-3)+2(-1) & 1\cdot1+2\cdot2 \\ 0(-3)+1(-1) & 0\cdot1+1\cdot2 \end{bmatrix} = \begin{bmatrix} 5 & 7 \\ 2 & 3 \end{bmatrix} + \begin{bmatrix} -5 & 5 \\ -1 & 2 \end{bmatrix}$$

$$= \begin{bmatrix} 0 & 12 \\ 1 & 5 \end{bmatrix}$$

$$A(B+C) = AB + AC$$

43. $A^2 - B^2 = AA - BB = \begin{bmatrix} 1 & 2 \\ 0 & 1 \end{bmatrix}\begin{bmatrix} 1 & 2 \\ 0 & 1 \end{bmatrix} - \begin{bmatrix} 1 & 1 \\ 2 & 3 \end{bmatrix}\begin{bmatrix} 1 & 1 \\ 2 & 3 \end{bmatrix}$

$$= \begin{bmatrix} 1\cdot1+2\cdot0 & 1\cdot2+2\cdot1 \\ 0\cdot1+1\cdot0 & 0\cdot2+1\cdot1 \end{bmatrix} - \begin{bmatrix} 1\cdot1+1\cdot2 & 1\cdot1+1\cdot3 \\ 2\cdot1+3\cdot2 & 2\cdot1+3\cdot3 \end{bmatrix} = \begin{bmatrix} 1 & 4 \\ 0 & 1 \end{bmatrix} - \begin{bmatrix} 3 & 4 \\ 8 & 11 \end{bmatrix}$$

$$= \begin{bmatrix} -2 & 0 \\ -8 & -10 \end{bmatrix}$$

$$A - B = \begin{bmatrix} 1 & 2 \\ 0 & 1 \end{bmatrix} - \begin{bmatrix} 1 & 1 \\ 2 & 3 \end{bmatrix} = \begin{bmatrix} 0 & 1 \\ -2 & -2 \end{bmatrix}$$

$$A + B = \begin{bmatrix} 1 & 2 \\ 0 & 1 \end{bmatrix} + \begin{bmatrix} 1 & 1 \\ 2 & 3 \end{bmatrix} = \begin{bmatrix} 2 & 3 \\ 2 & 4 \end{bmatrix}$$

Hence

$$(A - B)(A + B) = \begin{bmatrix} 0 & 1 \\ -2 & -2 \end{bmatrix}\begin{bmatrix} 2 & 3 \\ 2 & 4 \end{bmatrix} = \begin{bmatrix} 0\cdot2+1\cdot2 & 0\cdot3+1\cdot4 \\ (-2)2+(-2)2 & (-2)3+(-2)4 \end{bmatrix} = \begin{bmatrix} 2 & 4 \\ -8 & -14 \end{bmatrix}$$

Since corresponding elements are not equal, $A^2 - B^2 \neq (A - B)(A + B)$

45.
A. $[0.6\ 0.6\ 0.2] \cdot \begin{bmatrix} 8 \\ 10 \\ 5 \end{bmatrix} = (0.6)8 + (0.6)10 + (0.2)5 = 11.80$ dollars per boat

B. $[1.5\ 1.2\ 0.4] \cdot \begin{bmatrix} 9 \\ 12 \\ 6 \end{bmatrix} = (1.5)9 + (1.2)12 + (0.4)6 = 30.30$ dollars per boat

C. Since M is a 3×3 matrix and N is a 3×2 matrix, MN is a 3×2 matrix.

D. $MN = \begin{bmatrix} 0.6 & 0.6 & 0.2 \\ 1.0 & 0.9 & 0.3 \\ 1.5 & 1.2 & 0.4 \end{bmatrix}\begin{bmatrix} 8 & 9 \\ 10 & 12 \\ 5 & 6 \end{bmatrix}$

$$\left[\begin{array}{cc} [0.6\ 0.6\ 0.2]\cdot\begin{bmatrix}8\\10\\5\end{bmatrix} & [0.6\ 0.6\ 0.2]\cdot\begin{bmatrix}9\\12\\6\end{bmatrix} \\[2em] [1.0\ 0.9\ 0.3]\cdot\begin{bmatrix}8\\10\\5\end{bmatrix} & [1.0\ 0.9\ 0.3]\cdot\begin{bmatrix}9\\12\\6\end{bmatrix} \\[2em] [1.5\ 1.2\ 0.4]\cdot\begin{bmatrix}8\\10\\5\end{bmatrix} & [1.5\ 1.2\ 0.4]\cdot\begin{bmatrix}9\\12\\6\end{bmatrix} \end{array}\right]$$

$$=\begin{bmatrix} (0.6)8+(0.6)10+(0.2)5 & (0.6)9+(0.6)12+(0.2)6 \\ (1.0)8+(0.9)10+(0.3)5 & (1.0)9+(0.9)12+(0.3)6 \\ (1.5)8+(1.2)10+(0.4)5 & (1.5)9+(1.2)12+(0.4)6 \end{bmatrix}$$

	Plant I	Plant II	
$=$	$11.80	$13.80	One − person
	$18.50	$21.60	Two − person
	$26.00	$30.30	Four − person

This matrix gives the labor costs for each type of boat at each plant.

47. A. $[1{,}000\ 500\ 5{,}000]\cdot\begin{bmatrix}\$0.40\\\$0.75\\\$0.25\end{bmatrix}$ $\begin{aligned}&=\ 1{,}000(\$0.40)+500(\$0.75)+5{,}000(\$0.25)\\&=\ \$2{,}025\end{aligned}$

   B. $[2{,}000\ 800\ 8{,}000]\cdot\begin{bmatrix}\$0.40\\\$0.75\\\$0.25\end{bmatrix}$ $\begin{aligned}&=\ 2{,}000(\$0.40)+800(\$0.75)+8{,}000(\$0.25)\\&=\ \$3{,}400\end{aligned}$

   C. $NM=\begin{bmatrix}1{,}000 & 500 & 5{,}000\\2{,}000 & 800 & 8{,}000\end{bmatrix}\begin{bmatrix}\$0.40\\\$0.75\\\$0.25\end{bmatrix}$

   $$=\begin{bmatrix}[1{,}000\ 500\ 5{,}000]\cdot\begin{bmatrix}\$0.40\\\$0.75\\\$0.25\end{bmatrix}\\[2em][2{,}000\ 800\ 8{,}000]\cdot\begin{bmatrix}\$0.40\\\$0.75\\\$0.25\end{bmatrix}\end{bmatrix}$$

   $$=\begin{bmatrix}\$2{,}025\\\$3{,}400\end{bmatrix}\ \begin{array}{l}\text{Berkeley}\\\text{Oakland}\end{array}$$

   $=$ Cost of all contacts in each town.

   D. $[1\ 1]\begin{bmatrix}1{,}000 & 500 & 5{,}000\\2{,}000 & 800 & 8{,}000\end{bmatrix}$

   $$=\begin{bmatrix}[1\ 1]\cdot\begin{bmatrix}1{,}000\\2{,}000\end{bmatrix} & [1\ 1]\cdot\begin{bmatrix}500\\800\end{bmatrix} & [1\ 1]\cdot\begin{bmatrix}5{,}000\\8{,}000\end{bmatrix}\end{bmatrix}$$

   [Telephone House Letter] $= [3{,}000\ 1{,}300\ 13{,}000]$

   This matrix represents the number of each of the three types of contact that was made.

Exercise 8-3

Key Ideas and Formulas

The identity element for multiplication for square matrices of order n (dimension $n \times n$), denoted I, is the square matrix of order n with 1's along the principal diagonal and 0's elsewhere. For A a square matrix of order n.

$$IA = AI = A$$

If M is a square matrix of order n and if there exists a square matrix M^{-1} such that $MM^{-1} = M^{-1}M = I$, M^{-1} is called the (multiplicative) inverse of M.

To find M^{-1} if it exists, form the augmented matrix $[M|I]$ and use row operations to transform into $[I|B]$ if possible. Then $B = M^{-1}$. If, however, all 0's are obtained at some stage in one or more rows to the left of the vertical line, then M^{-1} will not exist.

A system of n equations in a variables can be written as

$$AX = B$$

where A is the coefficient matrix of the system,

$$X = \begin{bmatrix} x_1 \\ x_2 \\ \vdots \\ x_n \end{bmatrix}$$

and B is

the column matrix of constant terms. Then if A^{-1} exists, the solution of the system can be written as $X = A^{-1}B$.

---

1. $\begin{bmatrix} 2 & -3 \\ 4 & 5 \end{bmatrix}$

3. $\begin{bmatrix} -2 & 1 & 3 \\ 2 & 4 & -2 \\ 5 & 1 & 0 \end{bmatrix}$

5. $\begin{bmatrix} 3 & -4 \\ -2 & 3 \end{bmatrix}\begin{bmatrix} 3 & 4 \\ 2 & 3 \end{bmatrix} = \begin{bmatrix} 3\cdot 3 + (-4)2 & 3\cdot 4 + (-4)3 \\ (-2)3 + 3\cdot 2 & (-2)4 + 3\cdot 3 \end{bmatrix} = \begin{bmatrix} 1 & 0 \\ 0 & 1 \end{bmatrix}$

7. $\begin{bmatrix} 1 & -1 & 1 \\ 0 & 2 & -1 \\ 2 & 3 & 0 \end{bmatrix} \begin{bmatrix} 3 & 3 & -1 \\ -2 & -2 & 1 \\ -4 & -5 & 2 \end{bmatrix}$

$= \begin{bmatrix} 1\cdot 3+(-1)(-2)+1(-4) & 1\cdot 3+(-1)(-2)+1(-5) & 1\cdot(-1)+(-1)\cdot 1+1\cdot 2 \\ 0\cdot 3+2(-2)+(-1)(-4) & 0\cdot 3+2(-2)+(-1)(-5) & 0(-1)+2\cdot 1+(-1)\cdot 2 \\ 2\cdot 3+3(-2)+0(-4) & 2\cdot 3+3(-2)+0(-5) & 2(-1)+3\cdot 1+0\cdot 2 \end{bmatrix}$

$= \begin{bmatrix} 1 & 0 & 0 \\ 0 & 1 & 0 \\ 0 & 0 & 1 \end{bmatrix}$

9. Since $\begin{bmatrix} 3 & -2 \\ 1 & 4 \end{bmatrix} \begin{bmatrix} -2 \\ 1 \end{bmatrix}$

$= \begin{bmatrix} 3(-2)+(-2)1 \\ 1(-2)+4\cdot 1 \end{bmatrix} = \begin{bmatrix} -8 \\ 2 \end{bmatrix}, \begin{bmatrix} x_1 \\ x_2 \end{bmatrix} = \begin{bmatrix} -8 \\ 2 \end{bmatrix}$ if and only if $x_1 = -8$ and $x_2 = 2$.

11. Since

$\begin{bmatrix} -2 & 3 \\ 2 & -1 \end{bmatrix} \begin{bmatrix} 3 \\ 2 \end{bmatrix} = \begin{bmatrix} (-2)3+3\cdot 2 \\ 2\cdot 3+(-1)2 \end{bmatrix} = \begin{bmatrix} 0 \\ 4 \end{bmatrix}$

$\begin{bmatrix} x_1 \\ x_2 \end{bmatrix} = \begin{bmatrix} 0 \\ 4 \end{bmatrix}$ if and only if $x_1 = 0$ and $x_2 = 4$.

13. $\begin{bmatrix} 1 & 2 & | & 1 & 0 \\ 1 & 3 & | & 0 & 1 \end{bmatrix} R_2+(-1)R_1 \to R_2$

$\sim \begin{bmatrix} 1 & 2 & | & 1 & 0 \\ 0 & 1 & | & -1 & 1 \end{bmatrix} R_1+(-2)R_2 \to R_1$

$\sim \begin{bmatrix} 1 & 0 & | & 3 & -2 \\ 0 & 1 & | & -1 & 1 \end{bmatrix}$

Hence $M^{-1} = \begin{bmatrix} 3 & 2 \\ -1 & 1 \end{bmatrix}$

Check: $M^{-1}M = \begin{bmatrix} 3 & -2 \\ -1 & 1 \end{bmatrix} \begin{bmatrix} 1 & 2 \\ 1 & 3 \end{bmatrix} = \begin{bmatrix} 3\cdot 1+(-2)1 & 3\cdot 2+(-2)3 \\ (-1)1+1\cdot 1 & (-1)2+1\cdot 3 \end{bmatrix} = \begin{bmatrix} 1 & 0 \\ 0 & 1 \end{bmatrix}$

15. $\begin{bmatrix} 1 & 3 & | & 1 & 0 \\ 2 & 7 & | & 0 & 1 \end{bmatrix} R_2+(-2)R_2 \to R_2$

$\sim \begin{bmatrix} 1 & 3 & | & 1 & 0 \\ 0 & 1 & | & -2 & 1 \end{bmatrix} R_1+(-3)R_2 \to R_1$

$\begin{bmatrix} 1 & 0 & | & 7 & -3 \\ 0 & 1 & | & -2 & 1 \end{bmatrix}$

Hence $M^{-1} = \begin{bmatrix} 7 & -3 \\ -2 & 1 \end{bmatrix}$

Check: $M^{-1}M = \begin{bmatrix} 7 & -3 \\ -2 & 1 \end{bmatrix} \begin{bmatrix} 1 & 3 \\ 2 & 7 \end{bmatrix} = \begin{bmatrix} 7\cdot 1 + (-3)2 & 7\cdot 3 + (-3)7 \\ (-2)1 + 1\cdot 2 & (-2)3 + 1\cdot 7 \end{bmatrix} = \begin{bmatrix} 1 & 0 \\ 0 & 1 \end{bmatrix}$

17. $\begin{bmatrix} 1 & -3 & 0 & | & 1 & 0 & 0 \\ 0 & 3 & 1 & | & 0 & 1 & 0 \\ 2 & -1 & 2 & | & 0 & 0 & 1 \end{bmatrix}$ $R_3 + (-2)R_1 \to R_3$

$\sim \begin{bmatrix} 1 & -3 & 0 & | & 1 & 0 & 0 \\ 0 & 3 & 1 & | & 0 & 1 & 0 \\ 0 & 5 & 2 & | & -2 & 0 & 1 \end{bmatrix}$ $\frac{1}{3}R_2 \to R_2$

$\sim \begin{bmatrix} 1 & -3 & 1 & | & 1 & 0 & 0 \\ 0 & 1 & \frac{1}{3} & | & 1 & \frac{1}{3} & 0 \\ 0 & 5 & 2 & | & -2 & 0 & 1 \end{bmatrix}$ $\begin{array}{l} R_1 + 3R_2 \to R_1 \\[4pt] R_3 + (-5)R_2 \to R_3 \end{array}$

$\sim \begin{bmatrix} 1 & 0 & 1 & | & 1 & 1 & 0 \\ 0 & 1 & \frac{1}{3} & | & 0 & \frac{1}{3} & 0 \\ 0 & 0 & \frac{1}{3} & | & -2 & -\frac{5}{3} & 1 \end{bmatrix}$ $3R_3 \to R_3$

$\sim \begin{bmatrix} 1 & 0 & 1 & | & 1 & 1 & 0 \\ 0 & 1 & \frac{1}{3} & | & 0 & \frac{1}{3} & 0 \\ 0 & 0 & 1 & | & -6 & -5 & 3 \end{bmatrix}$ $\begin{array}{l} R_1 + (-1)R_3 \to R_1 \\[4pt] R_2 + (-\frac{1}{3})R_3 \to R_2 \end{array}$

$\sim \begin{bmatrix} 1 & 0 & 0 & | & 7 & 6 & -3 \\ 0 & 1 & 0 & | & 2 & 2 & -1 \\ 0 & 0 & 1 & | & -6 & -5 & 3 \end{bmatrix}$

Hence $M^{-1} = \begin{bmatrix} 7 & 6 & -3 \\ 2 & 2 & -1 \\ -6 & -5 & 3 \end{bmatrix}$

Check: $M^{-1}M =$
$\begin{bmatrix} 7 & 6 & -3 \\ 2 & 2 & -1 \\ -6 & -5 & 3 \end{bmatrix} \begin{bmatrix} 1 & -3 & 0 \\ 0 & 3 & 1 \\ 2 & -1 & 2 \end{bmatrix}$

$= \begin{bmatrix} 7\cdot 1 + 6\cdot 0 + (-3)2 & 7(-3) + 6\cdot 3 + (-3)(-1) & 7\cdot 0 + 6\cdot 1 + (-3)2 \\ 2\cdot 1 + 2\cdot 0 + (-1)2 & 2(-3) + 2\cdot 3 + (-1)(-1) & 2\cdot 0 + 2\cdot 1 + (-1)2 \\ -6\cdot 1 + (-5)0 + 3\cdot 2 & (-6)(-3) + (-5)3 + 3(-1) & -6\cdot 0 + (-5)\cdot 1 + 3\cdot 2 \end{bmatrix} = \begin{bmatrix} 1 & 0 & 0 \\ 0 & 1 & 0 \\ 0 & 0 & 1 \end{bmatrix}$

19. $\begin{bmatrix} 1 & 1 & 0 & | & 1 & 0 & 0 \\ 0 & 3 & -1 & | & 0 & 1 & 0 \\ 1 & 0 & 1 & | & 0 & 0 & 1 \end{bmatrix}$ $R_3 + (-1)R_1 \to R_3$

$\sim \begin{bmatrix} 1 & 1 & 0 & | & 1 & 0 & 0 \\ 0 & 3 & -1 & | & 0 & 1 & 0 \\ 0 & -1 & 1 & | & -1 & 0 & 1 \end{bmatrix}$ $R_2 \leftrightarrow R_3$

$$\sim \begin{bmatrix} 1 & 1 & 0 & 1 & 0 & 0 \\ 0 & -1 & 1 & -1 & 0 & 1 \\ 0 & 3 & -1 & 0 & 1 & 0 \end{bmatrix} \quad \begin{matrix} R_1 + R_2 & \to & R_1 \\[4pt] R_3 + 3R_2 & \to & R_3 \end{matrix}$$

$$\sim \begin{bmatrix} 1 & 0 & 1 & 0 & 0 & 1 \\ 0 & -1 & 1 & -1 & 0 & 1 \\ 0 & 0 & 2 & -3 & 1 & 3 \end{bmatrix} \quad \begin{matrix} (-1)R_2 & \to & R_2 \\[4pt] \frac{1}{2}R_3 & \to & R_3 \end{matrix}$$

$$\sim \begin{bmatrix} 1 & 0 & 1 & 0 & 0 & 1 \\ 0 & 1 & -1 & 1 & 0 & -1 \\ 0 & 0 & 1 & -\frac{3}{2} & \frac{1}{2} & \frac{3}{2} \end{bmatrix} \quad \begin{matrix} R_1 + (-1)R_3 \to R_1 \\[4pt] R_2 + R_3 \to R_2 \end{matrix}$$

$$\sim \begin{bmatrix} 1 & 0 & 0 & \frac{3}{2} & -\frac{1}{2} & -\frac{1}{2} \\ 0 & 1 & 0 & -\frac{1}{2} & \frac{1}{2} & \frac{1}{2} \\ 0 & 0 & 1 & -\frac{3}{2} & \frac{1}{2} & \frac{3}{2} \end{bmatrix}$$

Hence

$$M^{-1} = \begin{bmatrix} \frac{3}{2} & -\frac{1}{2} & -\frac{1}{2} \\ -\frac{1}{2} & \frac{1}{2} & \frac{1}{2} \\ -\frac{3}{2} & \frac{1}{2} & \frac{3}{2} \end{bmatrix} = \frac{1}{2} \begin{bmatrix} 3 & -1 & -1 \\ -1 & 1 & 1 \\ -3 & 1 & 3 \end{bmatrix}$$

Check: $M^{-1}M = \frac{1}{2} \begin{bmatrix} 3 & -1 & -1 \\ -1 & 1 & 1 \\ -3 & 1 & 3 \end{bmatrix} \begin{bmatrix} 1 & 1 & 0 \\ 0 & 3 & -1 \\ 1 & 0 & 1 \end{bmatrix}$

$$= \frac{1}{2} \begin{bmatrix} 3 \cdot 1 + (-1)0 + (-1)1 & 3 \cdot 1 + (-1)3 + (-1)0 & 3 \cdot 0 + (-1)(-1) + (-1)1 \\ (-1)1 + 1 \cdot 0 + 1 \cdot 1 & -1 \cdot 1 + 1 \cdot 3 + 1 \cdot 0 & -1 \cdot 0 + 1(-1) + 1 \cdot 1 \\ (-3)1 + 1 \cdot 0 + 3 \cdot 1 & -3 \cdot 1 + 1 \cdot 3 + 3 \cdot 0 & -3 \cdot 0 + 1(-1) + 3 \cdot 1 \end{bmatrix}$$

$$= \frac{1}{2} \begin{bmatrix} 2 & 0 & 0 \\ 0 & 2 & 0 \\ 0 & 0 & 2 \end{bmatrix} = \begin{bmatrix} 1 & 0 & 0 \\ 0 & 1 & 0 \\ 0 & 0 & 1 \end{bmatrix}$$

21. $\begin{bmatrix} 1 & 2 \\ 1 & 3 \end{bmatrix} \begin{bmatrix} x_1 \\ x_2 \end{bmatrix} = \begin{bmatrix} k_1 \\ k_2 \end{bmatrix} \quad MX = K$ has solution $X = M^{-1}K$.

We find $M^{-1}K$ for each given K. From problem 13, $M^{-1} = \begin{bmatrix} 3 & -2 \\ -1 & 1 \end{bmatrix}$

A. $K = \begin{bmatrix} 1 \\ 3 \end{bmatrix} \quad M^{-1}K = \begin{bmatrix} 3 & -2 \\ -1 & 1 \end{bmatrix} \begin{bmatrix} 1 \\ 3 \end{bmatrix} = \begin{bmatrix} -3 \\ 2 \end{bmatrix} = \begin{bmatrix} x_1 \\ x_2 \end{bmatrix} \quad x_1 = -3, x_2 = 2$

B. $K = \begin{bmatrix} 3 \\ 5 \end{bmatrix} \quad M^{-1}K = \begin{bmatrix} 3 & -2 \\ -1 & 1 \end{bmatrix} \begin{bmatrix} 3 \\ 5 \end{bmatrix}$

$$= \begin{bmatrix} -1 \\ 2 \end{bmatrix} = \begin{bmatrix} x_1 \\ x_2 \end{bmatrix} \quad x_1 = -1, x_2 = 2$$

C. $K = \begin{bmatrix} -2 \\ 1 \end{bmatrix}$

$M^{-1}K = \begin{bmatrix} 3 & -2 \\ -1 & 1 \end{bmatrix} \begin{bmatrix} -2 \\ 1 \end{bmatrix}$

$= \begin{bmatrix} -5 \\ 3 \end{bmatrix} = \begin{bmatrix} x_1 \\ x_2 \end{bmatrix}$ $x_1 = -5, x_2 = 3$

23. $\begin{bmatrix} 1 & 3 \\ 2 & 7 \end{bmatrix} \begin{bmatrix} x_1 \\ x_2 \end{bmatrix} = \begin{bmatrix} k_1 \\ k_2 \end{bmatrix}$ $MX = K$ has solution $X = M^{-1}K$.

We find $M^{-1}K$ for each given K. From problem 15, $M^{-1} = \begin{bmatrix} 7 & -3 \\ -2 & 1 \end{bmatrix}$

A. $K = \begin{bmatrix} 2 \\ -1 \end{bmatrix}$ $M^{-1}K = \begin{bmatrix} 7 & -3 \\ -2 & 1 \end{bmatrix} \begin{bmatrix} 2 \\ -1 \end{bmatrix} = \begin{bmatrix} 17 \\ -5 \end{bmatrix} = \begin{bmatrix} x_1 \\ x_2 \end{bmatrix}$ $x_1 = 17, x_2 = -5$

B. $K = \begin{bmatrix} 1 \\ 0 \end{bmatrix}$ $M^{-1}K = \begin{bmatrix} 7 & -3 \\ -2 & 1 \end{bmatrix} \begin{bmatrix} 1 \\ 0 \end{bmatrix} = \begin{bmatrix} 7 \\ -2 \end{bmatrix} = \begin{bmatrix} x_1 \\ x_2 \end{bmatrix}$ $x_1 = -7, x_2 = -2$

C. $K = \begin{bmatrix} 3 \\ -1 \end{bmatrix}$ $M^{-1}K = \begin{bmatrix} 7 & -3 \\ -2 & 1 \end{bmatrix} \begin{bmatrix} 3 \\ -1 \end{bmatrix} = \begin{bmatrix} 24 \\ -7 \end{bmatrix} = \begin{bmatrix} x_1 \\ x_2 \end{bmatrix}$ $x_1 = 24, x_2 = -7$

25. $\begin{bmatrix} 1 & -3 & 0 \\ 0 & 3 & 1 \\ 2 & -1 & 2 \end{bmatrix} \begin{bmatrix} x_1 \\ x_2 \\ x_3 \end{bmatrix} = \begin{bmatrix} k_1 \\ k_2 \\ k_3 \end{bmatrix}$

$MX = K$ has solution $X = M^{-1}K$.

We find $M^{-1}K$ for each given K. From problem 17, $M^{-1} = \begin{bmatrix} 7 & 6 & -3 \\ 2 & 2 & -1 \\ -6 & -5 & 3 \end{bmatrix}$

A. $K = \begin{bmatrix} 1 \\ 0 \\ 2 \end{bmatrix}$ $M^{-1}K = \begin{bmatrix} 7 & 6 & -3 \\ -2 & 2 & -1 \\ 6 & -5 & 3 \end{bmatrix} \begin{bmatrix} 1 \\ 0 \\ 2 \end{bmatrix} = \begin{bmatrix} 1 \\ 0 \\ 0 \end{bmatrix} = \begin{bmatrix} x_1 \\ x_2 \\ x_3 \end{bmatrix}$ $\begin{aligned} x_1 &= 1 \\ x_2 &= 0 \\ x_3 &= 0 \end{aligned}$

B. $K = \begin{bmatrix} -1 \\ 1 \\ 0 \end{bmatrix}$ $M^{-1}K = \begin{bmatrix} 7 & 6 & -3 \\ -2 & 2 & -1 \\ 6 & -5 & 3 \end{bmatrix} \begin{bmatrix} -1 \\ 1 \\ 0 \end{bmatrix} = \begin{bmatrix} -1 \\ 0 \\ 1 \end{bmatrix} = \begin{bmatrix} x_1 \\ x_2 \\ x_3 \end{bmatrix}$ $\begin{aligned} x_1 &= -1 \\ x_2 &= 0 \\ x_3 &= -1 \end{aligned}$

C. $K = \begin{bmatrix} 2 \\ -2 \\ 1 \end{bmatrix}$ $M^{-1}K = \begin{bmatrix} 7 & 6 & -3 \\ -2 & 2 & -1 \\ 6 & -5 & 3 \end{bmatrix} \begin{bmatrix} 2 \\ -2 \\ 1 \end{bmatrix} = \begin{bmatrix} -1 \\ -1 \\ 1 \end{bmatrix} = \begin{bmatrix} x_1 \\ x_2 \\ x_3 \end{bmatrix}$ $\begin{aligned} x_1 &= -1 \\ x_2 &= -1 \\ x_3 &= 1 \end{aligned}$

27. $\begin{bmatrix} 1 & 1 & 0 \\ 0 & 3 & -1 \\ 1 & 0 & 1 \end{bmatrix} = \begin{bmatrix} x_1 \\ x_2 \\ x_3 \end{bmatrix} = \begin{bmatrix} k_1 \\ k_2 \\ k_3 \end{bmatrix}$ $MX = K$ has solution $X = M^{-1}K$.

We find $M^{-1}K$ for each given K. From problem 19, $M^{-1} = \frac{1}{2} \begin{bmatrix} 3 & -1 & -1 \\ -1 & 1 & 1 \\ -3 & 1 & 3 \end{bmatrix}$

A. $K = \begin{bmatrix} 2 \\ 0 \\ 4 \end{bmatrix}$ $M^{-1}K = \frac{1}{2} \begin{bmatrix} 3 & -1 & -1 \\ -1 & 1 & 1 \\ -3 & 1 & 3 \end{bmatrix} \begin{bmatrix} 2 \\ 0 \\ 4 \end{bmatrix} = \frac{1}{2} \begin{bmatrix} 2 \\ 2 \\ 6 \end{bmatrix} = \begin{bmatrix} 1 \\ 1 \\ 3 \end{bmatrix} = \begin{bmatrix} x_1 \\ x_2 \\ x_3 \end{bmatrix}$

$x_1 = 1, x_2 = 1, x_3 = 3$

B. $K = \begin{bmatrix} 0 \\ 4 \\ -2 \end{bmatrix}$ $M^{-1}K = \frac{1}{2} \begin{bmatrix} 3 & -1 & -1 \\ -1 & 1 & 1 \\ -3 & 1 & 3 \end{bmatrix} \begin{bmatrix} 0 \\ 4 \\ -2 \end{bmatrix} = \frac{1}{2} \begin{bmatrix} -2 \\ 2 \\ -2 \end{bmatrix} = \begin{bmatrix} -1 \\ 1 \\ -1 \end{bmatrix} = \begin{bmatrix} x_1 \\ x_2 \\ x_3 \end{bmatrix}$

$x_1 = -1, x_2 = 1, x_3 = -1$

C. $K = \begin{bmatrix} 4 \\ 2 \\ 0 \end{bmatrix}$ $M^{-1}K = \frac{1}{2} \begin{bmatrix} 3 & -1 & -1 \\ -1 & 1 & 1 \\ -3 & 1 & 3 \end{bmatrix} \begin{bmatrix} 4 \\ 2 \\ 0 \end{bmatrix} = \frac{1}{2} \begin{bmatrix} 10 \\ -2 \\ -10 \end{bmatrix} = \begin{bmatrix} 5 \\ -1 \\ -5 \end{bmatrix} = \begin{bmatrix} x_1 \\ x_2 \\ x_3 \end{bmatrix}$

$x_1 = 5, x_2 = -1, x_3 = -5$

29. If the inverse existed we would find it by row operations on the following matrix:

$$= \begin{bmatrix} 3 & 9 & 1 & 0 \\ 2 & 6 & 0 & 1 \end{bmatrix}$$

But consider what happens if we perform $R_3 + (-\frac{2}{3})R_2 \to R_3$:

$$= \begin{bmatrix} 3 & 9 & 1 & 0 \\ 0 & 0 & -\frac{2}{3} & 1 \end{bmatrix}$$

Since a row of zeros results to the left of the vertical line, no inverse exists.

31. $\begin{bmatrix} 3 & 5 & 1 & 0 \\ 2 & 4 & 0 & 1 \end{bmatrix}$ $R_1 \to R_2$

$\sim \begin{bmatrix} 2 & 4 & 0 & 1 \\ 3 & 5 & 1 & 0 \end{bmatrix}$ $\frac{1}{2}R_1 \to R_1$

$\sim \begin{bmatrix} 1 & 2 & 0 & \frac{1}{2} \\ 3 & 5 & 1 & 0 \end{bmatrix}$ $R_2 + (-3)R_1 \to R_2$

$\sim \begin{bmatrix} 1 & 2 & 0 & \frac{1}{2} \\ 0 & -1 & 1 & -\frac{3}{2} \end{bmatrix}$ $R_1 + 2R_2 \to R_1$

$\sim \begin{bmatrix} 1 & 0 & 2 & -\frac{5}{2} \\ 0 & -1 & 1 & -\frac{3}{2} \end{bmatrix}$ $(-1)R_2 \to R_2$

$\sim \begin{bmatrix} 1 & 0 & 2 & -\frac{5}{2} \\ 0 & 1 & -1 & \frac{3}{2} \end{bmatrix}$

The inverse is
$$\begin{bmatrix} 2 & -\frac{5}{2} \\ -1 & \frac{3}{2} \end{bmatrix}$$

The check is omitted for lack of space in this and subsequent problems.

33. $\begin{bmatrix} 2 & 2 & -1 & 1 & 0 & 0 \\ 0 & 3 & -1 & 0 & 1 & 0 \\ -1 & -2 & 1 & 0 & 0 & 1 \end{bmatrix} R_1 \leftrightarrow R_3$

$\sim \begin{bmatrix} -1 & -2 & 1 & 0 & 0 & 1 \\ 0 & 3 & -1 & 0 & 1 & 0 \\ 2 & 2 & -1 & 1 & 0 & 0 \end{bmatrix} R_3 + 2R_1 \rightarrow R_3$

$\sim \begin{bmatrix} -1 & -2 & 1 & 0 & 0 & 1 \\ 0 & 3 & -1 & 0 & 1 & 0 \\ 0 & -2 & 1 & 1 & 0 & 2 \end{bmatrix} \begin{matrix} R_1 + (-1)R_3 \rightarrow R_1 \\ R_2 + R_3 \rightarrow R_2 \end{matrix}$

$\sim \begin{bmatrix} -1 & 0 & 0 & -1 & 0 & -1 \\ 0 & 1 & 0 & 1 & 1 & 2 \\ 0 & -2 & 1 & 1 & 0 & 2 \end{bmatrix} R_3 + 2R_2 \rightarrow R_3$

$\sim \begin{bmatrix} -1 & 0 & 0 & -1 & 0 & -1 \\ 0 & 1 & 0 & 1 & 1 & 2 \\ 0 & 0 & 1 & 3 & 2 & 6 \end{bmatrix} (-1)R_1 \rightarrow R_1$

$\sim \begin{bmatrix} 1 & 0 & 0 & 1 & 0 & 1 \\ 0 & 1 & 0 & 1 & 1 & 2 \\ 0 & 0 & 1 & 3 & 2 & 6 \end{bmatrix}$

The inverse is $\begin{bmatrix} 1 & 0 & 1 \\ 1 & 1 & 2 \\ 3 & 2 & 6 \end{bmatrix}$

35. If the inverse existed we would find it by row operations on the following matrix:

$\begin{bmatrix} 2 & 1 & 1 & 1 & 0 & 0 \\ 1 & 1 & 0 & 0 & 1 & 0 \\ -1 & -1 & 0 & 0 & 0 & 1 \end{bmatrix}$

But consider what happens if we perform $R_3 + R_2 \rightarrow R_2$:

$\begin{bmatrix} 2 & 1 & 1 & 1 & 0 & 0 \\ 0 & 0 & 0 & 0 & 1 & 1 \\ -1 & -1 & 0 & 0 & 0 & 1 \end{bmatrix}$

Since a row of zeros results to the left of the vertical line, no inverse exists.

37. $\begin{bmatrix} 1 & 5 & 10 & 1 & 0 & 0 \\ 0 & 1 & 4 & 0 & 1 & 0 \\ 1 & 6 & 15 & 0 & 0 & 1 \end{bmatrix} R_3 + (-1)R_1 \rightarrow R_3$

$\sim \begin{bmatrix} 1 & 5 & 10 & 1 & 0 & 0 \\ 0 & 1 & 4 & 0 & 1 & 0 \\ 0 & 1 & 5 & -1 & 0 & 1 \end{bmatrix} \begin{matrix} R_1 + (-5)R_2 \rightarrow R_1 \\ R_3 + (-1)R_2 \rightarrow R_3 \end{matrix}$

$$\sim \begin{bmatrix} 1 & 0 & -10 \\ 0 & 1 & 4 \\ 0 & 0 & 1 \end{bmatrix} \left| \begin{matrix} 1 & -5 & 0 \\ 0 & 1 & 0 \\ -1 & -1 & 1 \end{matrix} \right| \quad \begin{matrix} R_1 + 10R_3 \to R_1 \\ R_2 + (-4)R_3 \to R_2 \end{matrix}$$

$$\sim \begin{bmatrix} 1 & 0 & 0 \\ 0 & 1 & 0 \\ 0 & 0 & 1 \end{bmatrix} \left| \begin{matrix} -9 & -15 & 10 \\ 4 & 5 & -4 \\ -1 & -1 & 1 \end{matrix} \right|$$

The inverse is

$$\begin{bmatrix} -9 & -15 & 10 \\ 4 & 5 & -4 \\ -1 & -1 & 1 \end{bmatrix}$$

39. We calculate A^{-1} by row operations on

$$\begin{bmatrix} 3 & 4 \\ 2 & 3 \end{bmatrix} \left| \begin{matrix} 1 & 0 \\ 0 & 1 \end{matrix} \right| \quad R_1 + (-1)R_2 \to R_1$$

$$\sim \begin{bmatrix} 1 & 1 \\ 2 & 3 \end{bmatrix} \left| \begin{matrix} 1 & -1 \\ 0 & 1 \end{matrix} \right| \quad R_2 + (-2)R_1 \to R_2$$

$$\sim \begin{bmatrix} 1 & 1 \\ 0 & 1 \end{bmatrix} \left| \begin{matrix} 1 & -1 \\ -2 & 3 \end{matrix} \right| \quad R_1 + (-1)R_2 \to R_1$$

$$\sim \begin{bmatrix} 1 & 0 \\ 0 & 1 \end{bmatrix} \left| \begin{matrix} 3 & -4 \\ -2 & 3 \end{matrix} \right|$$

Hence $A^{-1} = \begin{bmatrix} 3 & -4 \\ -2 & 3 \end{bmatrix}$

We calculate $(A^{-1})^{-1}$ by row operations on

$$\begin{bmatrix} 3 & -4 \\ -2 & 3 \end{bmatrix} \left| \begin{matrix} 1 & 0 \\ 0 & 1 \end{matrix} \right| \quad R_1 + R_2 \to R_1$$

$$\sim \begin{bmatrix} 1 & -1 \\ -2 & 3 \end{bmatrix} \left| \begin{matrix} 1 & 1 \\ 0 & 1 \end{matrix} \right| \quad R_2 + 2R_1 \to R_2$$

$$\sim \begin{bmatrix} 1 & -1 \\ 0 & 1 \end{bmatrix} \left| \begin{matrix} 1 & 1 \\ 2 & 3 \end{matrix} \right| \quad R_1 + R_2 \to R_1$$

$$\sim \begin{bmatrix} 1 & 0 \\ 0 & 1 \end{bmatrix} \left| \begin{matrix} 3 & 4 \\ 2 & 3 \end{matrix} \right|$$

Hence $(A^{-1})^{-1} = \begin{bmatrix} 3 & 4 \\ 2 & 3 \end{bmatrix} = A$

41. The system to be solved, for an arbitrary return, is derived as follows.
Let
x_1 = number of $4 tickets sold
x_2 = number of $8 tickets sold

Then $x_1 + x_2 = 10,000$ number of seats

$\quad\quad 4x_1 + 8x_2 = k_2$ return required

We solve the system by writing it as a matrix equation.

$$\sim \begin{matrix} A \\ \begin{bmatrix} 1 & 1 \\ 4 & 8 \end{bmatrix} \end{matrix} \begin{matrix} X \\ \begin{bmatrix} x_1 \\ x_2 \end{bmatrix} \end{matrix} = \begin{matrix} B \\ \begin{bmatrix} 10,000 \\ k_2 \end{bmatrix} \end{matrix}$$

If A^{-1} exists, then $X = A^{-1}B$.

To find A^{-1}, we perform row operations on

$$\begin{bmatrix} 1 & 1 & \bigm| & 1 & 0 \\ 4 & 8 & \bigm| & 0 & 1 \end{bmatrix} R_2 + (-4)R_1 \rightarrow R_2$$

$$\sim \begin{bmatrix} 1 & 1 & \bigm| & 1 & 0 \\ 0 & 4 & \bigm| & -4 & 1 \end{bmatrix} 0.25R_2 \rightarrow R_2$$

$$\sim \begin{bmatrix} 1 & 1 & \bigm| & 1 & 0 \\ 0 & 1 & \bigm| & -1 & 0.25 \end{bmatrix} R_1 + (-1)R_2 \rightarrow R_1$$

$$\sim \begin{bmatrix} 1 & 0 & \bigm| & 2 & -0.25 \\ 0 & 1 & \bigm| & -1 & 0.25 \end{bmatrix}$$

Hence $A^{-1} = \begin{bmatrix} 2 & -0.25 \\ -1 & 0.25 \end{bmatrix}$

Check: $A^{-1}A = \begin{bmatrix} 2 & 0.25 \\ -1 & 0.25 \end{bmatrix} \begin{bmatrix} 1 & 1 \\ 4 & 8 \end{bmatrix} = \begin{bmatrix} 1 & 0 \\ 0 & 1 \end{bmatrix}$

We can now solve the system as

$$\begin{matrix} X \\ \begin{bmatrix} x_1 \\ x_2 \end{bmatrix} \end{matrix} = \begin{matrix} A^{-1} \\ \begin{bmatrix} 2 & -0.25 \\ -1 & 0.25 \end{bmatrix} \end{matrix} \begin{matrix} B \\ \begin{bmatrix} 10,000 \\ k_2 \end{bmatrix} \end{matrix}$$

If $k_2 = 56,000$ (Concert 1),

$$\begin{bmatrix} x_1 \\ x_2 \end{bmatrix} = \begin{bmatrix} 2 & -0.25 \\ -1 & 0.25 \end{bmatrix} \begin{bmatrix} 10,000 \\ 56,000 \end{bmatrix} = \begin{bmatrix} 6,000 \\ 4,000 \end{bmatrix}$$

Concert 1: 6,000 \$4 tickets and 4,000 \$8 tickets

If $k_2 = 60,000$ (Concert 2)

$$\begin{bmatrix} x_1 \\ x_2 \end{bmatrix} = \begin{bmatrix} 2 & -0.25 \\ -1 & 0.25 \end{bmatrix} \begin{bmatrix} 10,000 \\ 60,000 \end{bmatrix} = \begin{bmatrix} 5,000 \\ 5,000 \end{bmatrix}$$

Concert 2: 5,000 \$4 tickets and 5,000 \$8 tickets

If $k_2 = 68,000$ (Concert 3)

$$\begin{bmatrix} x_1 \\ x_2 \end{bmatrix} = \begin{bmatrix} 2 & -0.25 \\ -1 & 0.25 \end{bmatrix} \begin{bmatrix} 10,000 \\ 68,000 \end{bmatrix} \begin{bmatrix} 3,000 \\ 7,000 \end{bmatrix}$$

Concert 3: 3,000 $4 tickets and 7,000 $8 tickets

43.   We solve the system, for arbitrary V_1 and V_2, by writing it as a matrix equation.

$$
\overset{A}{\begin{bmatrix} 1 & -1 & 1 \\ 1 & 1 & 0 \\ 0 & 1 & 2 \end{bmatrix}} \overset{J}{\begin{bmatrix} I_1 \\ I_2 \\ I_3 \end{bmatrix}} = \overset{B}{\begin{bmatrix} 0 \\ V_1 \\ V_2 \end{bmatrix}}
$$

If A^{-1} exists, then $J = A^{-1}B$.

To find A^{-1} , we perform row operations on

$$
\begin{bmatrix} 1 & -1 & 1 & | & 1 & 0 & 0 \\ 1 & 1 & 0 & | & 0 & 1 & 0 \\ 0 & 1 & 2 & | & 0 & 0 & 1 \end{bmatrix} \quad R_2 + (-1)R_1 \to R_2
$$

$$
\sim \begin{bmatrix} 1 & -1 & 1 & | & 1 & 0 & 0 \\ 0 & 2 & -1 & | & -1 & 1 & 0 \\ 0 & 1 & 2 & | & 0 & 0 & 1 \end{bmatrix} \quad R_2 \leftrightarrow R_3
$$

$$
\sim \begin{bmatrix} 1 & -1 & 1 & | & 1 & 0 & 0 \\ 0 & 1 & 2 & | & 0 & 0 & 1 \\ 0 & 2 & -1 & | & -1 & 1 & 0 \end{bmatrix} \quad \begin{matrix} R_1 + R_2 \to R_1 \\ \\ R_3 + (-2)R_2 \to R_3 \end{matrix}
$$

$$
\sim \begin{bmatrix} 1 & 0 & 3 & | & 1 & 0 & 1 \\ 0 & 1 & 2 & | & 0 & 0 & 1 \\ 0 & 0 & -5 & | & -1 & 1 & -2 \end{bmatrix} \quad -\tfrac{1}{5}R_3 \to R_3
$$

$$
\sim \begin{bmatrix} 1 & 0 & 3 & | & 1 & 0 & 1 \\ 0 & 1 & 2 & | & 0 & 0 & 1 \\ 0 & 0 & 1 & | & \tfrac{1}{5} & -\tfrac{1}{5} & \tfrac{2}{5} \end{bmatrix} \quad \begin{matrix} R_1 + (-3)R_1 \to R_1 \\ R_2 + (-2)R_3 \to R_2 \end{matrix}
$$

$$
\sim \begin{bmatrix} 1 & 0 & 0 & | & \tfrac{2}{5} & \tfrac{3}{5} & -\tfrac{1}{5} \\ 0 & 1 & 0 & | & -\tfrac{2}{5} & \tfrac{2}{5} & \tfrac{1}{5} \\ 0 & 0 & 1 & | & \tfrac{1}{5} & -\tfrac{1}{5} & \tfrac{2}{5} \end{bmatrix}
$$

Hence $A^{-1} = \tfrac{1}{5} \begin{bmatrix} 2 & 3 & -1 \\ -2 & 2 & 1 \\ 1 & -1 & 2 \end{bmatrix}$

Check: $A^{-1}A = \tfrac{1}{5} \begin{bmatrix} 2 & 3 & -1 \\ -2 & 2 & 1 \\ 1 & -1 & 2 \end{bmatrix} \begin{bmatrix} 1 & -1 & 1 \\ 1 & 1 & 0 \\ 0 & 1 & 2 \end{bmatrix} = \begin{bmatrix} 1 & 0 & 0 \\ 0 & 1 & 0 \\ 0 & 0 & 1 \end{bmatrix}$

We can now solve the system as

$$
\overset{J}{\begin{bmatrix} I_1 \\ I_2 \\ I_3 \end{bmatrix}} = \tfrac{1}{5} \overset{A^{-1}}{\begin{bmatrix} 2 & 3 & -1 \\ -2 & 2 & 1 \\ 1 & -1 & 2 \end{bmatrix}} \overset{B}{\begin{bmatrix} 0 \\ V_1 \\ V_2 \end{bmatrix}}
$$

A. $V_1 = 10$ $V_2 = 10$

$$\begin{bmatrix} I_1 \\ I_2 \\ I_3 \end{bmatrix} = \tfrac{1}{5} \begin{bmatrix} 2 & 3 & -1 \\ -2 & 2 & 1 \\ 1 & -1 & 2 \end{bmatrix} \begin{bmatrix} 0 \\ 10 \\ 10 \end{bmatrix} = \begin{bmatrix} 4 \\ 6 \\ 2 \end{bmatrix}$$

$I_1 = 4, \quad I_2 = 6, \quad I_3 = 2$ (amperes)

B. $V_1 = 10$ $V_2 = 15$

$$\begin{bmatrix} I_1 \\ I_2 \\ I_3 \end{bmatrix} = \tfrac{1}{5} \begin{bmatrix} 2 & 3 & -1 \\ -2 & 2 & 1 \\ 1 & -1 & 2 \end{bmatrix} \begin{bmatrix} 0 \\ 10 \\ 15 \end{bmatrix} = \begin{bmatrix} 3 \\ 7 \\ 4 \end{bmatrix}$$

$I_1 = 3, \quad I_2 = 7, \quad I_3 = 4$ (amperes)

C. $V_1 = 15$ $V_2 = 10$

$$\begin{bmatrix} I_1 \\ I_2 \\ I_3 \end{bmatrix} = \tfrac{1}{5} \begin{bmatrix} 2 & 3 & -1 \\ -2 & 2 & 1 \\ 1 & -1 & 2 \end{bmatrix} \begin{bmatrix} 0 \\ 15 \\ 10 \end{bmatrix} = \begin{bmatrix} 7 \\ 8 \\ 1 \end{bmatrix}$$

$I_1 = 7, \quad I_2 = 8, \quad I_3 = 1$ (amperes)

45.   If the graph of $f(x) = ax^2 + bx + c$ passes through a point, its coordinates must satisfy the equation. Hence

$$\begin{aligned} k_1 &= a(1)^2 + b(1) + c \\ k_2 &= a(2)^2 + b(2) + c \\ k_3 &= a(3)^2 + b(3) + c \end{aligned}$$

After simplification, we obtain:

$$\begin{aligned} a + b + c &= k_1 \\ 4a + 2b + c &= k_2 \\ 9a + 3b + c &= k_3 \end{aligned}$$

We solve this system, for arbitrary k_1, k_2, k_3, by writing it as a matrix equation.

$$\overset{A}{\begin{bmatrix} 1 & 1 & 1 \\ 4 & 2 & 1 \\ 9 & 3 & 1 \end{bmatrix}} \overset{X}{\begin{bmatrix} a \\ b \\ c \end{bmatrix}} = \overset{B}{\begin{bmatrix} k_1 \\ k_2 \\ k_3 \end{bmatrix}}$$

If A^{-1} exists, then $X = A^{-1}B$.

To find A^{-1} we perform row operations on

$$\begin{bmatrix} 1 & 1 & 1 & 1 & 0 & 0 \\ 4 & 2 & 1 & 0 & 1 & 0 \\ 9 & 3 & 1 & 0 & 0 & 1 \end{bmatrix} \begin{matrix} \\ R_2 + (-4)R_1 \to R_2 \\ R_3 + (-9)R_1 \to R_3 \end{matrix}$$

$$\sim \begin{bmatrix} 1 & 1 & 1 & 1 & 0 & 0 \\ 0 & -2 & -3 & -4 & 1 & 0 \\ 0 & -6 & 8 & -9 & 0 & 1 \end{bmatrix} \tfrac{1}{2}R_2 \to R_2$$

$$\sim \begin{bmatrix} 1 & 1 & 1 & 1 & 0 & 0 \\ 0 & 1 & \tfrac{3}{2} & 2 & -\tfrac{1}{2} & 0 \\ 0 & -6 & -8 & -9 & 0 & 1 \end{bmatrix} \begin{matrix} R_1 + (-1)R_2 \to R_1 \\ \\ R_3 + 6R_2 \to R_3 \end{matrix}$$

$$\sim \begin{bmatrix} 1 & 0 & -\tfrac{1}{2} & -1 & \tfrac{1}{2} & 0 \\ 0 & 1 & \tfrac{3}{2} & 2 & -\tfrac{1}{2} & 0 \\ 0 & 0 & 1 & 3 & -3 & 1 \end{bmatrix} \begin{matrix} R_1 + \tfrac{1}{2}R_3 \to R_1 \\ R_2 + (-\tfrac{3}{2})R_3 \to R_2 \end{matrix}$$

$$\sim \begin{bmatrix} 1 & 0 & 0 & \tfrac{1}{2} & -1 & \tfrac{1}{2} \\ 0 & 1 & 0 & -\tfrac{5}{2} & 4 & -\tfrac{3}{2} \\ 0 & 0 & 1 & 3 & -3 & 1 \end{bmatrix}$$

Hence

$$A^{-1} = \tfrac{1}{2} \begin{bmatrix} 1 & -2 & 1 \\ -5 & 8 & -3 \\ 6 & -6 & 2 \end{bmatrix}$$

Check:

$$A^{-1}A = \tfrac{1}{2} \begin{bmatrix} 1 & -2 & 1 \\ -5 & 8 & -3 \\ 6 & -6 & 2 \end{bmatrix} \begin{bmatrix} 1 & 1 & 1 \\ 4 & 2 & 1 \\ 9 & 3 & 1 \end{bmatrix} = \begin{bmatrix} 1 & 0 & 0 \\ 0 & 1 & 0 \\ 0 & 0 & 1 \end{bmatrix}$$

We can now solve the system as

$$\begin{matrix} X & & A^{-1} & & B \end{matrix}$$
$$\begin{bmatrix} a \\ b \\ c \end{bmatrix} = \tfrac{1}{2} \begin{bmatrix} 1 & -2 & 1 \\ -5 & 8 & -3 \\ 6 & -6 & 2 \end{bmatrix} \begin{bmatrix} k_1 \\ k_2 \\ k_3 \end{bmatrix}$$

A. $\begin{bmatrix} a \\ b \\ c \end{bmatrix} = \tfrac{1}{2} \begin{bmatrix} 1 & -2 & 1 \\ -5 & 8 & -3 \\ 6 & -6 & 2 \end{bmatrix} \begin{bmatrix} -2 \\ 1 \\ 6 \end{bmatrix} = \begin{bmatrix} 1 \\ 0 \\ -3 \end{bmatrix}$

$a = 1, \quad b = 0, \quad c = -3$

B. $\begin{bmatrix} a \\ b \\ c \end{bmatrix} = \tfrac{1}{2} \begin{bmatrix} 1 & -2 & 1 \\ -5 & 8 & -3 \\ 6 & -6 & 2 \end{bmatrix} \begin{bmatrix} 4 \\ 3 \\ -2 \end{bmatrix} = \begin{bmatrix} -2 \\ 5 \\ 1 \end{bmatrix}$

$a = -2, \quad b = 5, \quad c = 1$

C. $\begin{bmatrix} a \\ b \\ c \end{bmatrix} = \frac{1}{2} \begin{bmatrix} 1 & -2 & 1 \\ -5 & 8 & -3 \\ 6 & -6 & 2 \end{bmatrix} \begin{bmatrix} 8 \\ -5 \\ 4 \end{bmatrix} = \begin{bmatrix} 11 \\ -46 \\ 43 \end{bmatrix}$

$a = 11, b = -46, c = 43$

47. The system to be solved, for an arbitrary diet, is derived as follows.

Let

x_1 = amount of mix A

x_2 = amount of mix B

Then $0.20x_1 + 0.10x_2 = k_1$ (k_1 = amount of protein)

$ 0.20x_1 + 0.06x_2 = k_2$ (k_2 = amount of fat)

We solve the system by writing it as a matrix equation.

$\begin{array}{ccc} A & X & B \end{array}$

$\begin{bmatrix} 0.20 & 0.10 \\ 0.02 & 0.06 \end{bmatrix} \begin{bmatrix} x_1 \\ x_2 \end{bmatrix} = \begin{bmatrix} k_1 \\ k_2 \end{bmatrix}$

If A^{-1} exists, then $X = A^{-1}B$.

To find A^{-1}, we perform row operations on

$\left[\begin{array}{cc|cc} 0.20 & 0.10 & 1 & 0 \\ 0.02 & 0.06 & 0 & 1 \end{array} \right] \begin{array}{c} 5R_1 \rightarrow R_2 \\ 50R_2 \rightarrow R_2 \end{array}$

$\sim \left[\begin{array}{cc|cc} 1 & 0.5 & 5 & 0 \\ 1 & 3 & 0 & 50 \end{array} \right] R_2 + (-1)R_1 \rightarrow R_2$

$\sim \left[\begin{array}{cc|cc} 1 & 0.5 & 5 & 0 \\ 0 & 2.5 & -5 & 50 \end{array} \right] 0.4R_2 \rightarrow R_2$

$\sim \left[\begin{array}{cc|cc} 1 & 0.5 & 5 & 0 \\ 0 & 1 & -2 & 20 \end{array} \right] R_1 + (-0.5)R_2 \rightarrow R_1$

$\sim \left[\begin{array}{cc|cc} 1 & 0 & 6 & -10 \\ 0 & 1 & -2 & 20 \end{array} \right]$

Hence $A^{-1} = \begin{bmatrix} 6 & -10 \\ -2 & 20 \end{bmatrix}$.

Check: $A^{-1}A = \begin{bmatrix} 6 & -10 \\ -2 & 20 \end{bmatrix} \begin{bmatrix} 0.20 & 0.10 \\ 0.02 & 0.06 \end{bmatrix} = \begin{bmatrix} 1 & 0 \\ 0 & 1 \end{bmatrix}$

We can now solve the system as

$\begin{array}{ccc} X & A^{-1} & B \end{array}$

$\begin{bmatrix} x_1 \\ x_2 \end{bmatrix} = \begin{bmatrix} 6 & -10 \\ -2 & 20 \end{bmatrix} \begin{bmatrix} k_1 \\ k_2 \end{bmatrix}$

For Diet 1, $k_1 = 20$ and $k_2 = 6$

$\begin{bmatrix} x_1 \\ x_2 \end{bmatrix} = \begin{bmatrix} 6 & -10 \\ -2 & 20 \end{bmatrix} \begin{bmatrix} 20 \\ 6 \end{bmatrix} = \begin{bmatrix} 60 \\ 80 \end{bmatrix}$

Diet 1: 60 ounces Mix A and 80 ounces Mix B

For Diet 2, $k_1 = 10$ and $k_2 = 4$

$$\begin{bmatrix} x_1 \\ x_2 \end{bmatrix} = \begin{bmatrix} 6 & -10 \\ -2 & 20 \end{bmatrix} \begin{bmatrix} 10 \\ 4 \end{bmatrix} = \begin{bmatrix} 20 \\ 60 \end{bmatrix}$$

Diet 2: 20 ounces Mix A and 60 ounces Mix B

For Diet 3: $k_1 = 10$ and $k_2 = 6$

$$\begin{bmatrix} x_1 \\ x_2 \end{bmatrix} = \begin{bmatrix} 6 & -10 \\ -2 & 20 \end{bmatrix} \begin{bmatrix} 10 \\ 6 \end{bmatrix} = \begin{bmatrix} 0 \\ 100 \end{bmatrix}$$

Diet 3: 0 ounces Mix A and 100 ounces Mix B

Exercise 8-4

Key Ideas and Formulas

The determinant of a square matrix A is a number, denoted det A, or by writing the array of elements in A using vertical lines instead of squares brackets.

Second-order determinant:

$$\begin{vmatrix} a_{11} & a_{12} \\ a_{21} & a_{22} \end{vmatrix} = a_{11}a_{22} - a_{21}a_{12}$$

Third-order determinant:

$$\begin{vmatrix} a_{11} & a_{12} & a_{13} \\ a_{21} & a_{22} & a_{23} \\ a_{31} & a_{32} & a_{33} \end{vmatrix}$$

The minor of an element a_{ij} is the determinant array obtained by deleting the row and column that contain a_{ij}, that is, row i and column j.

The cofactor of $a_{ij} = (-1)^{i+j}$ (Minor of a_{ij}).

The value of a determinant of order 3 (or $n > 3$) is the sum of the three (or n) products obtained by multiplying each element of any one row (or each element of any one column) by its cofactor.

---

1. $\begin{vmatrix} 2 & 2 \\ -3 & 1 \end{vmatrix} = 2 \cdot 1 - (-3)2 = 8$

3. $\begin{vmatrix} 6 & -2 \\ -1 & -3 \end{vmatrix} = 6(-3) - (-1)(-2) = -20$

5. $\begin{vmatrix} 1.8 & -1.6 \\ -1.9 & 1.2 \end{vmatrix} = (1.8)(1.2) - (-1.9)(-1.6)$
   $= -0.88$

7. $\begin{vmatrix} a_{11} & a_{12} & a_{13} \\ a_{21} & a_{22} & a_{23} \\ a_{31} & a_{32} & a_{33} \end{vmatrix} = \begin{vmatrix} a_{22} & a_{23} \\ a_{32} & a_{33} \end{vmatrix}$

9. $\begin{vmatrix} a_{11} & a_{12} & a_{13} \\ a_{21} & a_{22} & a_{23} \\ a_{31} & a_{32} & a_{33} \end{vmatrix} = \begin{vmatrix} a_{11} & a_{12} \\ a_{31} & a_{32} \end{vmatrix}$

11. $(-1)^{1+1} \begin{vmatrix} a_{22} & a_{23} \\ a_{32} & a_{33} \end{vmatrix}$

13. $(-1)^{2+3} \begin{vmatrix} a_{11} & a_{12} \\ a_{31} & a_{32} \end{vmatrix}$

15. $\begin{vmatrix} 2 & 3 & 0 \\ 5 & 1 & -2 \\ 7 & -4 & 8 \end{vmatrix} = \begin{vmatrix} 1 & -2 \\ -4 & 8 \end{vmatrix}$

17. $\begin{vmatrix} -2 & 3 & 0 \\ 5 & 1 & -2 \\ 7 & -4 & 8 \end{vmatrix} = \begin{vmatrix} -2 & 0 \\ 5 & -2 \end{vmatrix}$

19. $(-1)^{1+1} \begin{vmatrix} 1 & -2 \\ -4 & 8 \end{vmatrix} = (-1)^2[1 \cdot 8 - (-2)(-4)] = 0$

21. $(-1)^{3+2}\begin{vmatrix} -2 & 0 \\ 5 & -2 \end{vmatrix} = (-1)^5[(-2)(-2) - 0(5)] = -4$

23. We expand by row 1

$\begin{vmatrix} 1 & 0 & 0 \\ -2 & 4 & 3 \\ 5 & -2 & 1 \end{vmatrix} = a_{11}(\text{cofactor of } a_{11}) + a_{12}(\text{cofactor of } a_{12}) + a_{13}(\text{cofactor of } a_{13})$

$= 1(-1)^{1+1}\begin{vmatrix} 4 & 3 \\ -2 & 1 \end{vmatrix} + 0(\diagup) + (\diagdown)$

It is unnecessary to evaluate these since they are multiplied by 0.

$= (-1)^2[4 \cdot 1 - (-2)3]$

$= 10$

25. We expand by column 1

$\begin{vmatrix} 0 & 1 & 5 \\ 3 & -7 & 6 \\ 0 & -2 & -3 \end{vmatrix} = a_{11}(\text{cofactor of } a_{11}) + a_{21}(\text{cofactor of } a_{21}) + a_{31}(\text{cofactor of } a_{31})$

$= 0(\diagup) + 3(-1)^{2+1}\begin{vmatrix} 1 & 5 \\ -2 & -3 \end{vmatrix} + 0(\diagdown)$

It is unnecessary to evaluate these since they are multiplied by 0.

$= 3(-1)^3[1(-3) - (-2)5]$

$= -21$

Common Error: Neglecting the sign of the cofactor.

The cofactor is often called the "signed" minor.

27. We expand by column 2

$\begin{vmatrix} -1 & 2 & -3 \\ -2 & 0 & -6 \\ 4 & -3 & 2 \end{vmatrix} = a_{12}(\text{cofactor of } a_{12}) + a_{22}(\text{cofactor of } a_{22}) + a_{32}(\text{cofactor of } a_{32})$

$= 2(-1)^{1+2}\begin{vmatrix} -2 & -6 \\ 4 & 2 \end{vmatrix} + 0(\quad) + (-3)(-1)^{3+2}\begin{vmatrix} -1 & -3 \\ -2 & -6 \end{vmatrix}$

$= 2(-1)^3[(-2)2 - 4(-6)] + (-3)(-1)^5[(-1)(-6) - (-2)(-3)]$

$= (-2)(20) + 3(0) = -40$

29. $(-1)^{1+1}\begin{vmatrix} a_{11} & a_{12} & a_{13} & a_{14} \\ a_{21} & a_{22} & a_{23} & a_{24} \\ a_{31} & a_{32} & a_{33} & a_{34} \\ a_{41} & a_{42} & a_{43} & a_{44} \end{vmatrix} = (-1)^{1+1}\begin{vmatrix} a_{22} & a_{23} & a_{24} \\ a_{32} & a_{33} & a_{34} \\ a_{42} & a_{43} & a_{44} \end{vmatrix}$

31. $(-1)^{4+3}\begin{vmatrix} a_{11} & a_{12} & a_{13} & a_{14} \\ a_{21} & a_{22} & a_{23} & a_{24} \\ a_{31} & a_{32} & a_{33} & a_{34} \\ a_{41} & a_{42} & a_{43} & a_{44} \end{vmatrix} = (-1)^{4+3}\begin{vmatrix} a_{11} & a_{12} & a_{14} \\ a_{21} & a_{22} & a_{24} \\ a_{31} & a_{32} & a_{34} \end{vmatrix}$

33. We expand by the second column

$$\begin{vmatrix} 3 & -2 & -8 \\ -2 & 0 & -3 \\ 1 & 0 & -4 \end{vmatrix} = a_{12}(\text{cofactor of } a_{12}) + a_{22}(\text{cofactor of } a_{22}) + a_{32}(\text{cofactor of } a_{32})$$

$$= (-2)(-1)^{1+2}\begin{vmatrix} -2 & -3 \\ 1 & -4 \end{vmatrix} + 0 + 0$$

$$= (-2)(-1)^3[(-2)(-4) - 1(-3)]$$
$$= 2(11)$$
$$= 22$$

35. We expand by the first row

$$\begin{vmatrix} 1 & 4 & 1 \\ 1 & 1 & -2 \\ 2 & 1 & -1 \end{vmatrix} = a_{11}(\text{cofactor of } a_{11}) + a_{12}(\text{cofactor of } a_{12}) + a_{13}(\text{cofactor of } a_{13})$$

$$= 1(-1)^{1+1}\begin{vmatrix} 1 & -2 \\ 1 & -1 \end{vmatrix} + 4(-1)^{1+2}\begin{vmatrix} 1 & -2 \\ 2 & -1 \end{vmatrix} + 1(-1)^{1+3}\begin{vmatrix} 1 & 1 \\ 2 & 1 \end{vmatrix}$$

$$= (-1)^2[1(-1) - 1(-2)] + 4(-1)^3[1(-1) - 2(-2)] + (-1)^4[1 \cdot 1 - 2 \cdot 1]$$

$$= 1 + (-12) + (-1)$$
$$= -12$$

37. We expand by the first row.

$$\begin{vmatrix} 1 & 4 & 3 \\ 2 & 1 & 6 \\ 3 & -2 & 9 \end{vmatrix} = a_{11}(\text{cofactor of } a_{11}) + a_{12}(\text{cofactor of} a_{12}) + a_{13}(\text{cofactor of } a_{13})$$

$$= 1(-1)^{1+1}\begin{vmatrix} 1 & 6 \\ -2 & 9 \end{vmatrix} + 4(-1)^{1+2}\begin{vmatrix} 2 & 6 \\ 3 & 9 \end{vmatrix} + 3(-1)^{1+3}\begin{vmatrix} 2 & 1 \\ 3 & -2 \end{vmatrix}$$

$$= (-1)^2[1 \cdot 9 - (-2)6] + 4(-1)^3[2 \cdot 9 - 3 \cdot 6] + 3(-1)^4[2(-2) - 1 \cdot 3]$$

$$= 21 + 0 - 21$$
$$= 0$$

39. We expand by the second row. Clearly the only non-zero term will be a_{22} (cofactor of a_{22}), which is

$$3(-1)^{2+2}\begin{vmatrix} 2 & 1 & 7 \\ 3 & 2 & 5 \\ 0 & 0 & 2 \end{vmatrix}$$

The order 3 determinant is expanded by the third row. Again there is only one non-zero term, a_{33} (cofactor of a_{33}). So the original determinant is reduced to

$$3(-1)^{2+2}2(-1)^{3+3}\begin{vmatrix} 2 & 1 \\ 3 & 2 \end{vmatrix} = 6(-1)^{10}(2 \cdot 2 - 3 \cdot 1) = 6$$

41.
$$\begin{vmatrix} -2 & 0 & 0 & 0 & 0 \\ 9 & -1 & 0 & 0 & 0 \\ 2 & 1 & 3 & 0 & 0 \\ -1 & 4 & 2 & 2 & 0 \\ 7 & -2 & 3 & 5 & 5 \end{vmatrix} = (-2)(-1)^{1+1} \begin{vmatrix} -1 & 0 & 0 & 0 \\ 1 & 3 & 0 & 0 \\ 4 & 2 & 2 & 0 \\ -2 & 3 & 5 & 5 \end{vmatrix} + 0 \text{ terms}$$

$$= -2 \begin{vmatrix} -1 & 0 & 0 & 0 \\ 1 & 3 & 0 & 0 \\ 4 & 2 & 2 & 0 \\ -2 & 3 & 5 & 5 \end{vmatrix} = (-2)\left[(-1)(-1)^{1+1} \begin{vmatrix} 3 & 0 & 0 \\ 2 & 2 & 0 \\ 3 & 5 & 5 \end{vmatrix} + 0 \text{ terms} \right]$$

$$= (-2)(-1) \begin{vmatrix} 3 & 0 & 0 \\ 2 & 2 & 0 \\ 3 & 5 & 5 \end{vmatrix} = (-2)(-1)[3(-1)^{1+1} \begin{vmatrix} 2 & 0 \\ 5 & 5 \end{vmatrix} + 0 \text{ terms}].$$

$$= (-2)(-1)3 \begin{vmatrix} 2 & 0 \\ 5 & 5 \end{vmatrix}$$

$$= (-2)(-1)(3)[2 \cdot 5 - 5 \cdot 0]$$

$$= (-2)(-1)(3)(2)(5)$$

$$= -60$$

43.
$$\begin{vmatrix} 2 & 6 & -1 & 2 & 6 \\ 5 & 3 & -7 & 5 & 3 \\ -4 & -2 & 1 & -4 & -2 \end{vmatrix}$$

$$2 \cdot 3 \cdot 1 + 6(-7)(-4) + (-1)(5)(-2) - (-4)(3)(-1) - (-2)(-7)2 - 6 \cdot 5 \cdot 1$$
$$= 6 + 168 + 10 - 12 - 28 - 30$$
$$= 114$$

45.
$$\begin{vmatrix} a & a \\ ka & kb \end{vmatrix} = akb - kab = 0$$

47.
$$\begin{vmatrix} a & b \\ c & d \end{vmatrix} = ad - cb = ad - bc = \begin{vmatrix} a & c \\ b & d \end{vmatrix}$$

49. Expanding by the first column

$$\begin{vmatrix} a_{11} & a_{12} & a_{13} \\ a_{21} & a_{22} & a_{23} \\ a_{31} & a_{32} & a_{33} \end{vmatrix} = a_{11}(-1)^{1+1} \begin{vmatrix} a_{22} & a_{23} \\ a_{32} & a_{33} \end{vmatrix} + a_{21}(-1)^{2+1} \begin{vmatrix} a_{12} & a_{13} \\ a_{32} & a_{33} \end{vmatrix}$$

$$+ a_{31}(-1)^{3+1} \begin{vmatrix} a_{12} & a_{13} \\ a_{22} & a_{23} \end{vmatrix}$$

$$= a_{11} \begin{vmatrix} a_{22} & a_{23} \\ a_{32} & a_{33} \end{vmatrix} - a_{21} \begin{vmatrix} a_{12} & a_{13} \\ a_{32} & a_{33} \end{vmatrix} + a_{31} \begin{vmatrix} a_{12} & a_{13} \\ a_{22} & a_{23} \end{vmatrix}$$

$$= a_{11}(a_{22}a_{33} - a_{32}a_{23}) - a_{21}(a_{12}a_{33} - a_{32}a_{13})$$
$$+ a_{31}(a_{12}a_{23} - a_{22}a_{13})$$

$$\qquad\quad (1) \qquad\quad (2) \qquad\quad (3) \qquad\quad (4)$$
$$= a_{11}a_{22}a_{33} - a_{11}a_{32}a_{23} - a_{21}a_{12}a_{33} + a_{21}a_{32}a_{13}$$

$$\overset{\text{⑤}}{+a_{31}a_{12}a_{23}} \overset{\text{⑥}}{- a_{31}a_{22}a_{13}}$$

Expanding by the third row

$$\begin{vmatrix} a_{11} & a_{12} & a_{13} \\ a_{21} & a_{22} & a_{23} \\ a_{31} & a_{32} & a_{33} \end{vmatrix} = a_{31}(-1)^{3+1} \begin{vmatrix} a_{12} & a_{13} \\ a_{22} & a_{23} \end{vmatrix} + a_{32}(-1)^{3+2} \begin{vmatrix} a_{11} & a_{13} \\ a_{21} & a_{23} \end{vmatrix} + a_{33}(-1)^{3+3} \begin{vmatrix} a_{11} & a_{12} \\ a_{21} & a_{22} \end{vmatrix}$$

$$= a_{31} \begin{vmatrix} a_{12} & a_{13} \\ a_{22} & a_{23} \end{vmatrix} - a_{32} \begin{vmatrix} a_{11} & a_{13} \\ a_{21} & a_{23} \end{vmatrix} + a_{33} \begin{vmatrix} a_{11} & a_{12} \\ a_{21} & a_{22} \end{vmatrix}$$

$$= a_{31}(a_{12}a_{23} - a_{13}a_{22}) - a_{32}(a_{11}a_{23} - a_{13}a_{21}) + a_{33}(a_{11}a_{22} - a_{12}a_{21})$$

$$\overset{\text{⑤}}{= a_{31}a_{12}a_{23}} \overset{\text{⑥}}{- a_{31}a_{13}a_{22}} \overset{\text{②}}{- a_{32}a_{11}a_{23}} \overset{\text{④}}{+ a_{32}a_{13}a_{21}} \overset{\text{①}}{+ a_{33}a_{11}a_{22}} \overset{\text{③}}{- a_{33}a_{12}a_{21}}$$

Comparing the two expressions, with the aid of the numbers over the terms, shows that the expressions are the same.

51. $A = \begin{bmatrix} 2 & 3 \\ 1 & -2 \end{bmatrix} B = \begin{bmatrix} -1 & 3 \\ 2 & 1 \end{bmatrix}$

We calculate

$$AB = \begin{bmatrix} 2 & 3 \\ 1 & -2 \end{bmatrix} \begin{bmatrix} -1 & 3 \\ 2 & 1 \end{bmatrix} = \begin{bmatrix} 2(-1) + 3 \cdot 2 & 2 \cdot 3 + 3 \cdot 1 \\ 1(-1) + (-2)2 & 1 \cdot 3 + (-2) \cdot 1 \end{bmatrix} = \begin{bmatrix} 4 & 9 \\ -5 & 1 \end{bmatrix}$$

$$\det (AB) = \begin{vmatrix} 4 & 9 \\ -5 & 1 \end{vmatrix} = 4 \cdot 1 - (-5)9 = 49$$

$$\det A = \begin{vmatrix} 2 & 3 \\ 1 & -2 \end{vmatrix} = 2(-2) - 1 \cdot 3 = -7$$

$$\det B = \begin{vmatrix} -1 & 3 \\ 2 & 1 \end{vmatrix} = (-1)1 - 2 \cdot 3 = -7$$

Therefore
$$\det (AB) = 49 = (-7)(-7) = \det A \cdot \det B$$

53. The matrix $xI - A$ is calculated as

$$x \begin{bmatrix} 1 & 0 \\ 0 & 1 \end{bmatrix} - \begin{bmatrix} 5 & -4 \\ 2 & -1 \end{bmatrix} = \begin{bmatrix} x & 0 \\ 0 & x \end{bmatrix} - \begin{bmatrix} 5 & -4 \\ 2 & -1 \end{bmatrix} = \begin{bmatrix} x-5 & 4 \\ -2 & x+1 \end{bmatrix}$$

The characteristic polynomial is the determinant of this matrix:

$$\begin{vmatrix} x-5 & 4 \\ -2 & x+1 \end{vmatrix} = (x-5)(x+1) - (4)(-2) = x^2 - 4x - 5 + 8 = x^2 - 4x + 3$$

The zeros of this polynomial are the solutions of $x^2 - 4x + 3 = 0$

$$x^2 - 4x + 3 = 0$$
$$(x-1)(x-3) = 0$$
$$x = 1, 3$$

The eigenvalues of this matrix are 1 and 3.

55. The matrix $xI - A$ is calculated as

$$x \begin{bmatrix} 1 & 0 & 0 \\ 0 & 1 & 0 \\ 0 & 0 & 1 \end{bmatrix} - \begin{bmatrix} 4 & -4 & 0 \\ 2 & -2 & 0 \\ 4 & -8 & -4 \end{bmatrix} = \begin{bmatrix} x & 0 & 0 \\ 0 & x & 0 \\ 0 & 0 & x \end{bmatrix} - \begin{bmatrix} 4 & -4 & 0 \\ 2 & -2 & 0 \\ 4 & -8 & -4 \end{bmatrix}$$

$$= \begin{bmatrix} x-4 & 4 & 0 \\ -2 & x+2 & 0 \\ -4 & 8 & x+4 \end{bmatrix}$$

The characteristic polynomial is the determinant of this matrix.

$$\begin{vmatrix} x-4 & 4 & 0 \\ -2 & x+2 & 0 \\ -4 & 8 & x+4 \end{vmatrix} = (x+4)(-1)^{3+3} \begin{vmatrix} x-4 & 4 \\ -2 & x+2 \end{vmatrix} \text{ expanding by the third column}$$

$$= (x+4)(1)[(x-4)(x+2) - 4(-2)]$$

$$= (x+4)(x^2 - 2x - 8 + 8)$$

$$= (x+4)(x^2 - 2x)$$

$$= x^3 + 2x^2 - 8x$$

The zeros of this polynomial are the solutions of

$$
\begin{aligned}
x^3 + 2x^2 - 8x &= 0 \\
x(x^2 + 2x - 8) &= 0 \\
x(x+4)(x-2) &= 0 \\
x &= 0, -4, 2
\end{aligned}
$$

The eigenvalues of this matrix are $0, -4,$ and 2.

Exercise 8-5

Key Ideas and Formulas

If each element of any row (or column) of a determinant is multiplied by a constant k, the new determinant is k times the original. (Theorem 4) Caution: This operation and its result is very different from the operation of multiplying a matrix by a number k.

If every element in a row (or column) is 0, the value of the determinant is 0 (Theorem 5)

If two rows (or two columns) of a determinant are interchanged, the new determinant is the negative of the old. (Theorem 6)

If the corresponding elements are equal in two rows (or columns) the value of the determinant is 0. (Theorem 7)

If a multiple of any row (or column) of a determinant is added to any other row (or column) the value of the determinant is unchanged. (Theorem 8)

---

1.   Theorem 4
3.   Theorem 4
5.   Theorem 5
7.   Theorem 6
9.   Theorem 8
11. $C_2 + 3C_1 \rightarrow C_2$ has been used. Hence $x = 3 + 3(-1) = 0$
13. $C_3 + 3C_1 \rightarrow C_3$ has been used. Hence $x = 2 + 3(-1) = 5$

15.
$$\begin{vmatrix} -1 & 0 & 3 \\ 2 & 5 & 4 \\ 1 & 5 & 2 \end{vmatrix} = \begin{vmatrix} -1 & 0 & 3 \\ 1 & 0 & 2 \\ 1 & 5 & 2 \end{vmatrix} R_2 + (-1)R_3 \rightarrow R_2$$
$$= 5(-1)^{3+2} \begin{vmatrix} -1 & 3 \\ 1 & 2 \end{vmatrix} \begin{aligned} &= -5[(-1)2 - 1(3)] \\ &= 25 \end{aligned}$$

17.
$$\begin{vmatrix} 3 & 5 & 0 \\ 1 & 1 & -2 \\ 2 & 1 & -1 \end{vmatrix} = \begin{vmatrix} 3 & 5 & 0 \\ -3 & -1 & 0 \\ 2 & 1 & -1 \end{vmatrix} R_2 + (-2)R_3 \rightarrow R_2$$
$$= (-1)(-1)^{3+3} \begin{vmatrix} 3 & 5 \\ -3 & -1 \end{vmatrix} \begin{aligned} &= (-1)[3(-1) - (-3)5] \\ &= -12 \end{aligned}$$

19.   Theorem 4
21.   Theorem 5
23.   Theorem 8
25. $C_1 + 2C_3 \rightarrow C_1$ has been used. Hence $x = 3 + 2 \cdot 1 = 5$
      $C_2 + C_3 \rightarrow C_2$ has been used. Hence $y = 2 + (-2) = 0$
27. $R_1 + (-4)R_2 \rightarrow R_1$ has been used. Hence $x = 9 + (-4)3 = -3$
      $R_3 + 2R_2 \rightarrow R_3$ has been used. Hence $y = 4 + 2 \cdot 3 = 10$

29.   We will generate zeros in the first column by row operations.

$$\begin{vmatrix} 1 & 5 & 3 \\ 4 & 2 & 1 \\ 3 & 1 & 2 \end{vmatrix} = \begin{vmatrix} 1 & 5 & 3 \\ 0 & -18 & -11 \\ 0 & -14 & -7 \end{vmatrix} \quad \begin{matrix} R_2 + (-4)R_1 & \to & R_2 \\ R_3 + (-3)R_1 & \to & R_3 \end{matrix}$$

$$= 1(-1)^{1+1} \begin{vmatrix} -18 & -11 \\ -14 & -7 \end{vmatrix} + 0 + 0$$

$$= (-1)^2[(-18)(-7) - (-14)(-11)] = -28$$

31.   We will generate zeros in the second row by column operations.

$$\begin{vmatrix} 5 & 2 & -3 \\ -2 & 4 & 4 \\ 1 & -1 & 3 \end{vmatrix} = \begin{vmatrix} 5 & 12 & 7 \\ -2 & 0 & 0 \\ 1 & 1 & 5 \end{vmatrix} \quad \begin{matrix} C_2 + 2C_1 & \to & C_2 \\ C_3 + 2C_1 & \to & C_3 \end{matrix}$$

$$= (-2)(-1)^{2+1} \begin{vmatrix} 12 & 7 \\ 1 & 5 \end{vmatrix} + 0 + 0$$

$$= (-2)(-1)^3[12 \cdot 5 - 1 \cdot 7]$$

$$= 106$$

33.   The column operation $C_1 + (-3)C_3 \to C_1$ transforms this determinant into

$$\begin{vmatrix} 0 & -4 & 1 \\ 0 & -1 & 2 \\ 0 & 2 & 3 \end{vmatrix}$$

By Theorem 5, the value of this determinant is 0

35.   We start by generating one more zero in the first row.

$$\begin{vmatrix} 0 & 1 & 0 & 1 \\ 1 & -2 & 4 & 3 \\ 2 & 1 & 5 & 4 \\ 1 & 2 & 1 & 2 \end{vmatrix} = \begin{vmatrix} 0 & 0 & 0 & 1 \\ 1 & -5 & 4 & 3 \\ 2 & -3 & 5 & 4 \\ 1 & 0 & 1 & 2 \end{vmatrix} \quad C_2 + (-1)C_4 \to C_2$$

$$= 1(-1)^{1+4} \begin{vmatrix} 1 & -5 & 4 \\ 2 & -3 & 5 \\ 1 & 0 & 1 \end{vmatrix} + 0 + 0 + 0$$

$$= (-1) \begin{vmatrix} 1 & -5 & 4 \\ 2 & -3 & 5 \\ 1 & 0 & 1 \end{vmatrix}$$

$$= \begin{vmatrix} 1 & 5 & 4 \\ 2 & 3 & 5 \\ 1 & 0 & 1 \end{vmatrix} \quad \text{by Theorem 4}$$

We now generate one more zero in the third row.

$$\begin{vmatrix} 1 & 5 & 4 \\ 2 & 3 & 5 \\ 1 & 0 & 1 \end{vmatrix} = \begin{vmatrix} 1 & 5 & 3 \\ 2 & 3 & 3 \\ 1 & 0 & 0 \end{vmatrix} \quad C_3 + (-1)C_1 \to C_3$$

$$= 1(-1)^{3+1} \begin{vmatrix} 5 & 3 \\ 3 & 3 \end{vmatrix} = (-1)^4[5 \cdot 3 - 3 \cdot 3] = 6$$

37.   We start by generating zeros in the third row.

$$\begin{vmatrix} 3 & 2 & 3 & 1 \\ 3 & -2 & 8 & 5 \\ 2 & 1 & 3 & 1 \\ 4 & 5 & 4 & -3 \end{vmatrix} = \begin{vmatrix} 1 & 1 & 0 & 1 \\ -7 & -7 & -7 & 5 \\ 0 & 0 & 0 & 1 \\ 10 & 8 & 13 & -3 \end{vmatrix} \quad \begin{array}{l} C_1 + (-2)C_4 \rightarrow C_1 \\ C_2 + (-1)C_4 \rightarrow C_2 \\ C_3 + (-3)C_4 \rightarrow C_3 \end{array}$$

$$= 1(-1)^{3+4} \begin{vmatrix} 1 & 1 & 0 \\ -7 & -7 & -7 \\ 10 & 8 & 13 \end{vmatrix} + 0 + 0 + 0$$

$$= (-1) \begin{vmatrix} 1 & 1 & 0 \\ -7 & -7 & -7 \\ 10 & 8 & 13 \end{vmatrix}$$

$$= \begin{vmatrix} 1 & 1 & 0 \\ 7 & 7 & 7 \\ 10 & 8 & 13 \end{vmatrix} \text{ by Theorem 4}$$

We now generate one more zero in the first row.

$$\begin{vmatrix} 1 & 1 & 0 \\ 7 & 7 & 7 \\ 10 & 8 & 13 \end{vmatrix} = \begin{vmatrix} 1 & 0 & 0 \\ 7 & 0 & 7 \\ 10 & -2 & 13 \end{vmatrix} \quad C_2 + (-1)C_1 \rightarrow C_2$$

$$= (-2)(-1)^{3+2} \begin{vmatrix} 1 & 0 \\ 7 & 7 \end{vmatrix} + 0 + 0$$

$$= (-2)(-1)^5 [1 \cdot 7 - 0 \cdot 7]$$
$$= 14$$

39. Expand, for example, by the first column.

$$\begin{vmatrix} a & b & a \\ d & e & d \\ g & h & g \end{vmatrix} = a(-1)^{1+1} \begin{vmatrix} e & d \\ h & g \end{vmatrix} + d(-1)^{2+1} \begin{vmatrix} b & a \\ h & g \end{vmatrix} + g(-1)^{3+1} \begin{vmatrix} b & a \\ e & d \end{vmatrix}$$

$$= a \begin{vmatrix} e & d \\ h & g \end{vmatrix} - d \begin{vmatrix} b & a \\ h & g \end{vmatrix} + g \begin{vmatrix} b & a \\ e & d \end{vmatrix}$$
$$= a(eg - hd) - d(bg - ha) + g(bd - ea)$$
$$= aeg - adh - bdg + adh + bdg - aeg$$
$$= 0$$

41. We expand the left side by the first column, the right side by the second column.

$$\begin{vmatrix} a_1 & b_1 & c_1 \\ a_2 & b_2 & c_2 \\ a_3 & b_3 & c_3 \end{vmatrix} = a_1(-1)^{1+1} \begin{vmatrix} b_2 & c_2 \\ b_3 & c_3 \end{vmatrix} + a_2(-1)^{2+1} \begin{vmatrix} b_1 & c_1 \\ b_3 & c_3 \end{vmatrix}$$

$$+ a_3(-1)^{3+1} \begin{vmatrix} b_1 & c_1 \\ b_2 & c_2 \end{vmatrix}$$
$$= a_1 \begin{vmatrix} b_2 & c_2 \\ b_3 & c_3 \end{vmatrix} - a_2 \begin{vmatrix} b_1 & c_1 \\ b_3 & c_3 \end{vmatrix} + a_3 \begin{vmatrix} b_1 & c_1 \\ b_2 & c_2 \end{vmatrix}$$

$$- \begin{vmatrix} b_1 & a_1 & c_1 \\ b_2 & a_2 & c_2 \\ b_3 & a_3 & c_3 \end{vmatrix} = - \left[a_1(-1)^{1+2} \begin{vmatrix} b_2 & c_2 \\ b_3 & c_3 \end{vmatrix} + a_2(-1)^{2+2} \begin{vmatrix} b_1 & c_1 \\ b_3 & c_3 \end{vmatrix} \right.$$

$$+ a_3(-1)^{3+2} \begin{vmatrix} b_1 & c_1 \\ b_2 & c_2 \end{vmatrix} \Big]$$

$$= - \left[-a_1 \begin{vmatrix} b_2 & c_2 \\ b_3 & c_3 \end{vmatrix} + a_2 \begin{vmatrix} b_1 & c_1 \\ b_3 & c_3 \end{vmatrix} - a_3 \begin{vmatrix} b_1 & c_1 \\ b_2 & c_2 \end{vmatrix} \right]$$

$$= a_1 \begin{vmatrix} b_2 & c_2 \\ b_3 & c_3 \end{vmatrix} - a_2 \begin{vmatrix} b_1 & c_1 \\ b_3 & c_3 \end{vmatrix} + a_3 \begin{vmatrix} b_1 & c_1 \\ b_2 & c_2 \end{vmatrix}$$

Hence the two original expressions are equal.

43. The statements: (2,5) satisfies the equation

$$\begin{vmatrix} x & y & 1 \\ 2 & 5 & 1 \\ -3 & 4 & 1 \end{vmatrix} = 0$$

and $(-3,4)$ satisfies the equation $\begin{vmatrix} x & y & 1 \\ 2 & 5 & 1 \\ -3 & 4 & 1 \end{vmatrix} = 0$

are equivalent to the statements $\begin{vmatrix} 2 & 5 & 1 \\ 2 & 5 & 1 \\ -3 & 4 & 1 \end{vmatrix} = 0$ and $\begin{vmatrix} -3 & 4 & 1 \\ 2 & 5 & 1 \\ -3 & 4 & 1 \end{vmatrix} = 0.$

The latter statements are true by Theorem 7.

45. The statement $\begin{vmatrix} x & y & 1 \\ x_1 & y_2 & 1 \\ x_2 & y_2 & 1 \end{vmatrix} = 0$ is the equation of a line because, expanding

the first row, we have

$$x(-1)^{-1} \begin{vmatrix} y_1 & 1 \\ y_2 & 1 \end{vmatrix} + y(-1)^{1+2} \begin{vmatrix} x_1 & 1 \\ x_2 & 1 \end{vmatrix} + 1(-1)^{1+3} \begin{vmatrix} x_1 & y_1 \\ x_2 & y_2 \end{vmatrix} = 0$$

This is in the standard form for the equation of a line
$Ax + By + C = 0$
To show that the line passes through (x_1, y_1) and (x_2, y_2), we note merely that (x_1, y_1) and (x_2, y_2) satisfy the equation, because

$$\begin{vmatrix} x_1 & y_1 & 1 \\ x_1 & y_1 & 1 \\ x_2 & y_2 & 1 \end{vmatrix} = 0 \text{ and } \begin{vmatrix} x_2 & y_2 & 1 \\ x_1 & y_1 & 1 \\ x_2 & y_2 & 1 \end{vmatrix} = 0$$

are true by Theorem 7.

47. Using the result stated in problem 46, we have

$$\begin{vmatrix} x_1 & y_1 & 1 \\ x_2 & y_2 & 1 \\ x_3 & y_3 & 1 \end{vmatrix} = 2 \text{ (area of triangle formed by the three points)}.$$

If the determinant is 0, then the area of the triangle formed by the three points is zero. The only way this can happen is if the three points are on the same line; that is, the points are collinear.

Exercise 8-6
Key Ideas and Formulas

Cramer's Rule for Two Equations and Two Variables

Given the system

$$a_{11}x + a_{12}y = k_1$$
$$a_{21}x + a_{22}y = k_2$$

with

$$D = \begin{vmatrix} a_{11} & a_{12} \\ a_{21} & a_{21} \end{vmatrix} \neq 0$$

then

$$x = \frac{\begin{vmatrix} k_1 & a_{12} \\ k_2 & a_{22} \end{vmatrix}}{D} \text{ and } y = \frac{\begin{vmatrix} a_{11} & k_1 \\ a_{21} & k_2 \end{vmatrix}}{D}$$

Cramer's Rule for Three Equations and Three Variables

Given the system

$$a_{11}x + a_{12}y + a_{13}z = k_1$$
$$a_{21}x + a_{22}y + a_{23}z = k_2$$
$$a_{31}x + a_{32}y + a_{33}z = k_3$$

with

$$D = \begin{vmatrix} a_{11} & a_{12} & a_{13} \\ a_{21} & a_{22} & a_{23} \\ a_{31} & a_{32} & a_{33} \end{vmatrix} \neq 0$$

then

$$x = \frac{\begin{vmatrix} k_1 & a_{12} & a_{13} \\ k_2 & a_{22} & a_{23} \\ k_3 & a_{32} & a_{33} \end{vmatrix}}{D} \quad y = \frac{\begin{vmatrix} a_{11} & k_1 & a_{13} \\ a_{21} & k_2 & a_{23} \\ a_{31} & k_3 & a_{33} \end{vmatrix}}{D} \quad z = \frac{\begin{vmatrix} a_{11} & a_{12} & k_1 \\ a_{21} & a_{22} & k_2 \\ a_{31} & a_{32} & k_3 \end{vmatrix}}{D}$$

The determinant D is called the coefficient determinant. If $D \neq 0$, the system has exactly one solution, given by Cramer's rule. If $D = 0$, the system is either inconsistent or dependent. Cramer's rule does not apply. We use methods of Chapter 7 to find solutions (if any) of the system.

---

1. $\quad D = \begin{vmatrix} 1 & 2 \\ 1 & 3 \end{vmatrix} = 1 \quad x = \dfrac{\begin{vmatrix} 1 & 2 \\ -1 & 3 \end{vmatrix}}{D} = \dfrac{5}{1} = 5 \quad y = \dfrac{\begin{vmatrix} 1 & 1 \\ 1 & -1 \end{vmatrix}}{D} = \dfrac{-2}{1} = -2 \quad x = 5, y = -2$

3. $\quad D = \begin{vmatrix} 2 & 1 \\ 5 & 3 \end{vmatrix} = 1 \quad x = \dfrac{\begin{vmatrix} 1 & 1 \\ 2 & 3 \end{vmatrix}}{D} = \dfrac{1}{1} = 1 \quad y = \dfrac{\begin{vmatrix} 2 & 1 \\ 5 & 2 \end{vmatrix}}{D} = \dfrac{-1}{1} = -1 \quad x = 1, y = -1$

5. $\quad D = \begin{vmatrix} 2 & -1 \\ -1 & 3 \end{vmatrix} = 5 \quad x = \dfrac{\begin{vmatrix} -3 & -1 \\ 4 & 3 \end{vmatrix}}{D} = \dfrac{-5}{5} = 1 \quad y = \dfrac{\begin{vmatrix} 2 & -3 \\ -1 & 4 \end{vmatrix}}{D} = \dfrac{5}{5} = 1$

7. $\quad D = \begin{vmatrix} 1 & 1 & 0 \\ 0 & 2 & 1 \\ -1 & 0 & 1 \end{vmatrix} = 1 \quad x = \dfrac{\begin{vmatrix} 0 & 1 & 0 \\ -5 & 2 & 1 \\ -3 & 0 & 1 \end{vmatrix}}{D} = \dfrac{2}{1} = 2 \quad y = \dfrac{\begin{vmatrix} 1 & 0 & 1 \\ 0 & -5 & 1 \\ -1 & -3 & 1 \end{vmatrix}}{D} = \dfrac{-2}{1} = -2$

$$z = \dfrac{\begin{vmatrix} 1 & 1 & 0 \\ 0 & 2 & -5 \\ -1 & 0 & -3 \end{vmatrix}}{D} = \dfrac{-1}{1} = -1$$

9. $\quad D = \begin{vmatrix} 1 & 1 & 0 \\ 0 & 2 & 1 \\ -1 & 0 & 1 \end{vmatrix} = 1 \quad x = \dfrac{\begin{vmatrix} 1 & 1 & 0 \\ 0 & 2 & 0 \\ 0 & 0 & 1 \end{vmatrix}}{D} = \dfrac{2}{1} = 2 \quad y = \dfrac{\begin{vmatrix} 1 & 1 & 0 \\ 0 & 0 & 1 \\ -1 & 0 & 1 \end{vmatrix}}{D} = \dfrac{-1}{1} = -1$

$$z = \dfrac{\begin{vmatrix} 1 & 1 & 1 \\ 0 & 2 & 0 \\ -1 & 0 & 0 \end{vmatrix}}{D} = \dfrac{2}{1} = 2$$

11. $\quad D = \begin{vmatrix} 0 & 1 & 1 \\ 1 & 0 & 2 \\ 1 & -1 & 0 \end{vmatrix} = 1 \quad x = \dfrac{\begin{vmatrix} -4 & 1 & 1 \\ 0 & 0 & 2 \\ 5 & -1 & 0 \end{vmatrix}}{D} = \dfrac{2}{1} = 2 \quad y = \dfrac{\begin{vmatrix} 0 & -4 & 1 \\ 1 & 0 & 2 \\ 1 & 5 & 0 \end{vmatrix}}{D} = \dfrac{-3}{1} = -3$

$$z = \dfrac{\begin{vmatrix} 0 & 1 & -4 \\ 1 & 0 & 0 \\ 1 & -1 & 5 \end{vmatrix}}{D} = \dfrac{-1}{1} = -1$$

13. $\quad D = \begin{vmatrix} 0 & 2 & -1 \\ 1 & -1 & -1 \\ 1 & -1 & 2 \end{vmatrix} = -6 \quad x = \dfrac{\begin{vmatrix} -4 & 2 & -1 \\ 0 & -1 & -1 \\ 6 & -1 & 2 \end{vmatrix}}{D} = \dfrac{-6}{-6} = 1 \quad y = \dfrac{\begin{vmatrix} 0 & -4 & 1 \\ 1 & 0 & -1 \\ 1 & 6 & 2 \end{vmatrix}}{D} = \dfrac{6}{-6} = -1$

$$z = \dfrac{\begin{vmatrix} 0 & 2 & -4 \\ 1 & -1 & 0 \\ 1 & -1 & 6 \end{vmatrix}}{-6} = \dfrac{6}{-6} = -1$$

15. $\quad x = \dfrac{\begin{vmatrix} -3 & -3 & 1 \\ -11 & 3 & 2 \\ 3 & -1 & -1 \end{vmatrix}}{\begin{vmatrix} 2 & -3 & 1 \\ -4 & 3 & 2 \\ 1 & -1 & -1 \end{vmatrix}} = \dfrac{20}{5} = 4$

17. $y = \dfrac{\begin{vmatrix} 12 & 5 & 11 \\ 15 & -13 & -9 \\ 5 & 0 & 2 \end{vmatrix}}{\begin{vmatrix} 12 & -14 & 11 \\ 15 & 7 & -9 \\ 5 & -3 & 2 \end{vmatrix}} = \dfrac{28}{14} = 2$

19. $z = \dfrac{\begin{vmatrix} 3 & -4 & 18 \\ -9 & 8 & -13 \\ 5 & -7 & 33 \end{vmatrix}}{\begin{vmatrix} 3 & -4 & 5 \\ -9 & 8 & 7 \\ 5 & -7 & 10 \end{vmatrix}} = \dfrac{5}{2}$

21. $D = \begin{vmatrix} 1 & -4 & 9 \\ 4 & -1 & 6 \\ 1 & -1 & 3 \end{vmatrix}$

$= \begin{vmatrix} -2 & -1 & 0 \\ 2 & 1 & 0 \\ 1 & -1 & 3 \end{vmatrix} \quad \begin{array}{l} R_1 + (-3)R_3 \;\rightarrow\; R_1 \\ R_2 + (-2)R_3 \;\rightarrow\; R_2 \end{array}$

$= - \begin{vmatrix} 2 & 1 & 0 \\ 2 & 1 & 0 \\ 1 & -1 & 3 \end{vmatrix} \quad \text{by Theorem 4}$

$= 0$ by Theorem 7

Since $D = 0$, the system either has no solution or infinitely

many. Since $x = 0, y = 0, z = 0$ is a solution, the second case must hold.

23. We start with

$$a_{11}x + a_{12}y = k_1$$
$$a_{21}x + a_{22}y = k_2$$

We wish to eliminate x. We multiply the top equation by $-a_{21}$

and the bottom equation by a_{11}, then add.

$$
\begin{array}{rcl}
-a_{11}a_{21}x - a_{21}a_{12}y &=& -k_1 a_{21} \\
a_{11}a_{21}x - a_{11}a_{22}y &=& k_2 a_{11} \\
\hline
a_{11}a_{22}y - a_{21}a_{12}y &=& k_2 a_{11} - k_1 a_{21} \\
(a_{11}a_{22} - a_{21}a_{12})y &=& a_{11}k_2 - a_{21}k_1
\end{array}
$$

with $0x +$ in front of the third line.

$y = \dfrac{a_{11}k_2 - a_{21}k_1}{a_{11}a_{22} - a_{21}a_{12}}$

$= \dfrac{\begin{vmatrix} a_{11} & k_1 \\ a_{21} & k_2 \end{vmatrix}}{\begin{vmatrix} a_{11} & a_{12} \\ a_{21} & a_{22} \end{vmatrix}}$

Exercise 8-7

CHAPTER REVIEW

1. $A + B = \begin{bmatrix} 1 & 2 \\ 1 & 3 \end{bmatrix} + \begin{bmatrix} 2 & 1 \\ 1 & 1 \end{bmatrix} = \begin{bmatrix} 1+2 & 2+1 \\ 3+1 & 1+1 \end{bmatrix} = \begin{bmatrix} 3 & 3 \\ 4 & 2 \end{bmatrix}$
$(8-1)$

2. $B + D$ is not defined. $(8-1)$

3. $A - 2B = \begin{bmatrix} 1 & 2 \\ 3 & 1 \end{bmatrix} - 2\begin{bmatrix} 2 & 1 \\ 1 & 1 \end{bmatrix} = \begin{bmatrix} 1 & 2 \\ 3 & 1 \end{bmatrix} - \begin{bmatrix} 4 & 2 \\ 2 & 2 \end{bmatrix} = \begin{bmatrix} -3 & 0 \\ 1 & -1 \end{bmatrix}$

4. $AB = \begin{bmatrix} 1 & 2 \\ 3 & 1 \end{bmatrix}\begin{bmatrix} 2 & 1 \\ 1 & 1 \end{bmatrix} = \begin{bmatrix} 1\cdot2+2\cdot1 & 1\cdot1+2\cdot1 \\ 3\cdot2+1\cdot1 & 3\cdot1+1\cdot1 \end{bmatrix} = \begin{bmatrix} 4 & 3 \\ 7 & 4 \end{bmatrix}$
$(8-1, 8-2)$

5. AC is not defined. $(8-2)$

6. $AD = \begin{bmatrix} 1 & 2 \\ 3 & 1 \end{bmatrix}\begin{bmatrix} 1 \\ 2 \end{bmatrix} = \begin{bmatrix} 1\cdot1+2\cdot2 \\ 3\cdot1+1\cdot2 \end{bmatrix} = \begin{bmatrix} 5 \\ 5 \end{bmatrix}$
$(8-2)$

7. $DC = \begin{bmatrix} 1 \\ 2 \end{bmatrix}[2 \quad 3] = \begin{bmatrix} 1\cdot2 & 1\cdot3 \\ 2\cdot2 & 2\cdot3 \end{bmatrix} = \begin{bmatrix} 2 & 3 \\ 4 & 6 \end{bmatrix}$
$(8-2)$

8. $C \cdot D = [2 \quad 3] \cdot \begin{bmatrix} 1 \\ 2 \end{bmatrix} = 2\cdot1 + 3\cdot2 = 8$
$(8-2)$

9. $C + D$ is not defined. $(8-1)$

10. $\begin{bmatrix} 3 & 2 & | & 1 & 0 \\ 4 & 3 & | & 0 & 1 \end{bmatrix} R_2 + (-1)R_1 \to R_2$

$\sim \begin{bmatrix} 3 & 2 & | & 1 & 0 \\ 1 & 1 & | & -1 & 1 \end{bmatrix} R_1 \leftrightarrow R_2$

$\sim \begin{bmatrix} 1 & 1 & | & -1 & 1 \\ 3 & 2 & | & 1 & 0 \end{bmatrix} R_2 + (-3)R_1 \to R_2$

$\sim \begin{bmatrix} 1 & 1 & | & -1 & 1 \\ 0 & -1 & | & 4 & -3 \end{bmatrix} R_1 + R_2 \to R_1$

$\sim \begin{bmatrix} 1 & 0 & | & 3 & -2 \\ 0 & -1 & | & 4 & -3 \end{bmatrix} (-1)R_2 \to R_2$

$\sim \begin{bmatrix} 1 & 0 & | & 3 & -2 \\ 0 & 1 & | & -4 & 3 \end{bmatrix}$

Hence $A^{-1} = \begin{bmatrix} 3 & -2 \\ -4 & 3 \end{bmatrix}$

$$A^{-1}A = \begin{bmatrix} 3 & -2 \\ -4 & 3 \end{bmatrix} \begin{bmatrix} 3 & 2 \\ 4 & 3 \end{bmatrix} = \begin{bmatrix} 3\cdot3 + (-2)4 & 3\cdot2 + (-2)3 \\ (-4)3 + 3\cdot4 & (-4)2 + 3\cdot3 \end{bmatrix} = \begin{bmatrix} 1 & 0 \\ 0 & 1 \end{bmatrix} = I \quad (8-3)$$

11. As a matrix equation the system becomes

$$\begin{array}{ccc} A & X & B \end{array}$$

$$\begin{bmatrix} 3 & 2 \\ 4 & 3 \end{bmatrix} \begin{bmatrix} x_1 \\ x_2 \end{bmatrix} = \begin{bmatrix} k_1 \\ k_2 \end{bmatrix}$$

The solution of $AX = B$ is $X = A^{-1}B$.

Using the result of problem 10, we have

$$X = \begin{bmatrix} 3 & -2 \\ -4 & 3 \end{bmatrix} \begin{bmatrix} k_1 \\ k_2 \end{bmatrix}$$

A. $\begin{bmatrix} x_1 \\ x_2 \end{bmatrix} = \begin{bmatrix} 3 & -2 \\ -4 & 3 \end{bmatrix} \begin{bmatrix} 3 \\ 5 \end{bmatrix} = \begin{bmatrix} 3\cdot3 + (-2)5 \\ (-4)3 + 3\cdot5 \end{bmatrix} = \begin{bmatrix} -1 \\ 3 \end{bmatrix} \quad x_1 = -1, x_2 = 3$

B. $\begin{bmatrix} x_1 \\ x_2 \end{bmatrix} = \begin{bmatrix} 3 & -2 \\ -4 & 3 \end{bmatrix} \begin{bmatrix} 7 \\ 10 \end{bmatrix} = \begin{bmatrix} 3\cdot7 + (-2)10 \\ (-4)7 + 3\cdot10 \end{bmatrix} = \begin{bmatrix} 1 \\ 2 \end{bmatrix} \quad x_1 = 1, x_2 = 2$

C. $\begin{bmatrix} x_1 \\ x_2 \end{bmatrix} = \begin{bmatrix} 3 & -2 \\ -4 & 3 \end{bmatrix} \begin{bmatrix} 4 \\ 2 \end{bmatrix} = \begin{bmatrix} 3\cdot4 + (-2)2 \\ (-4)4 + 3\cdot2 \end{bmatrix} = \begin{bmatrix} 8 \\ -10 \end{bmatrix} \quad x_1 = 8, x_2 = -10 \quad (8-3)$

12. $\begin{vmatrix} 2 & -3 \\ -5 & -1 \end{vmatrix} = 2(-1) - (-5)(-3) = -17 \quad (8-4)$

13. $\begin{vmatrix} 2 & 3 & -4 \\ 0 & 5 & 0 \\ 1 & -4 & -2 \end{vmatrix} = 0 + 5(-1)^{2+2} \begin{vmatrix} 2 & -4 \\ 1 & -2 \end{vmatrix} + 0$

   $= 5(-1)^4[2(-2) - 1(-4)]$

   $= 0 \quad (8-4)$

14. $D = \begin{vmatrix} 3 & -2 \\ 1 & 3 \end{vmatrix} = 11$

   $x = \dfrac{\begin{vmatrix} 8 & -2 \\ -1 & 3 \end{vmatrix}}{D} = \dfrac{22}{11} = 2 \quad y = \dfrac{\begin{vmatrix} 3 & 8 \\ 1 & -1 \end{vmatrix}}{D} = \dfrac{-11}{11} = -1 \quad (8-6)$

15. $A + D$ is not defined $(8-1)$.

16. $DA = \begin{bmatrix} 3 & -2 & 1 \\ -1 & 1 & 2 \end{bmatrix} \begin{bmatrix} 2 & -2 \\ 1 & 0 \\ 3 & 2 \end{bmatrix} = \begin{bmatrix} 3\cdot2 + (-2)1 + 1\cdot3 & 3(-2) + (-2)0 + 1\cdot2 \\ (-1)2 + 1\cdot1 + 2\cdot3 & (-1)(-2) + 1\cdot0 + 2\cdot2 \end{bmatrix}$

   $= \begin{bmatrix} 7 & -4 \\ 5 & 6 \end{bmatrix}$

$$E + DA = \begin{bmatrix} 3 & -4 \\ -1 & 0 \end{bmatrix} + \begin{bmatrix} 7 & -4 \\ 5 & 6 \end{bmatrix} = \begin{bmatrix} 10 & -8 \\ 4 & 6 \end{bmatrix} \quad (8-1, 8-2)$$

17.   From problem 16, $DA = \begin{bmatrix} 7 & -4 \\ 5 & 6 \end{bmatrix}$

$$DA - 3E = \begin{bmatrix} 7 & -4 \\ 5 & 6 \end{bmatrix} - 3\begin{bmatrix} 3 & -4 \\ -1 & 0 \end{bmatrix} = \begin{bmatrix} 7 & -4 \\ 5 & 6 \end{bmatrix} - \begin{bmatrix} 9 & -12 \\ -3 & 0 \end{bmatrix} = \begin{bmatrix} -2 & 8 \\ 8 & 6 \end{bmatrix} \quad (8-1)$$

18. $C \cdot B = [2 \quad 1 \quad 3] \cdot \begin{bmatrix} -1 \\ 2 \\ 3 \end{bmatrix} = 2(-1) + 1 \cdot 2 + 3 \cdot 3 = 9 \quad (8-2)$

19. $CB = [C \cdot B] = [9] \quad (8-2)$

20. $AD = \begin{bmatrix} 2 & -2 \\ 1 & 0 \\ 3 & 2 \end{bmatrix} \begin{bmatrix} 3 & -2 & 1 \\ -1 & 1 & 2 \end{bmatrix} =$

$$= \begin{bmatrix} 2 \cdot 3 + (-2)(-1) & 2(-2) + (-2)1 & 2 \cdot 1 + (-2)2 \\ 1 \cdot 3 + 0(-1) & 1(-2) + 0 \cdot 1 & 1 \cdot 1 + 0 \cdot 2 \\ 3 \cdot 3 + 2(-1) & 3(-2) + 2 \cdot 1 & 3 \cdot 1 + 2 \cdot 2 \end{bmatrix}$$

$$= \begin{bmatrix} 8 & -6 & -2 \\ 3 & -2 & 1 \\ 7 & -4 & 7 \end{bmatrix}$$

$$BC = \begin{bmatrix} -1 \\ 2 \\ 3 \end{bmatrix} [2 \quad 1 \quad 3] = \begin{bmatrix} -1 \cdot 2 & -1 \cdot 1 & -1 \cdot 3 \\ 2 \cdot 2 & 2 \cdot 1 & 2 \cdot 3 \\ 3 \cdot 2 & 3 \cdot 1 & 3 \cdot 3 \end{bmatrix}$$

$$= \begin{bmatrix} -2 & -1 & -3 \\ 4 & 2 & 6 \\ 6 & 3 & 9 \end{bmatrix}$$

$$AD - BC = \begin{bmatrix} 8 & -6 & -2 \\ 3 & -2 & 1 \\ 7 & -4 & 7 \end{bmatrix} - \begin{bmatrix} -2 & -1 & -3 \\ 4 & 2 & 6 \\ 6 & 3 & 9 \end{bmatrix} = \begin{bmatrix} 10 & -5 & 1 \\ -1 & -4 & -5 \\ 1 & -7 & -2 \end{bmatrix} \quad (8-1, 8-2)$$

21. $\begin{bmatrix} 1 & 2 & 3 & | & 1 & 0 & 0 \\ 2 & 3 & 4 & | & 0 & 1 & 0 \\ 1 & 2 & 1 & | & 0 & 0 & 1 \end{bmatrix}$ $\begin{matrix} R_2 + (-2)R_1 & \to & R_2 \\ R_3 + (-1)R_1 & \to & R_3 \end{matrix}$

$\sim \begin{bmatrix} 1 & 2 & 3 & | & 1 & 0 & 0 \\ 0 & -1 & -2 & | & -2 & 1 & 0 \\ 0 & 0 & -2 & | & -1 & 0 & 1 \end{bmatrix}$ $R_1 + 2R_2 \to R_1$

$\sim \begin{bmatrix} 1 & 0 & -1 & | & -3 & 2 & 0 \\ 0 & -1 & -2 & | & -2 & 1 & 0 \\ 0 & 0 & -2 & | & -1 & 0 & 1 \end{bmatrix}$ $\begin{matrix} R_1 + (-\frac{1}{2})R_3 & \to & R_1 \\ R_2 + (-1)R_3 & \to & R_2 \end{matrix}$

$$\sim \begin{bmatrix} 1 & 0 & 0 & \bigm| & -\frac{5}{2} & 2 & -\frac{1}{2} \\ 0 & -1 & 0 & \bigm| & -1 & 1 & -1 \\ 0 & 0 & -2 & \bigm| & -1 & 0 & 1 \end{bmatrix} \quad \begin{array}{l} (-1)R_2 \rightarrow R_2 \\ (-\frac{1}{2})R_3 \rightarrow R_3 \end{array}$$

$$\sim \begin{bmatrix} 1 & 0 & 0 & \bigm| & -\frac{5}{2} & 2 & -\frac{1}{2} \\ 0 & 1 & 0 & \bigm| & 1 & -1 & 1 \\ 0 & 0 & 1 & \bigm| & \frac{1}{2} & 0 & -\frac{1}{2} \end{bmatrix}$$

Hence

$$A^{-1} = \begin{bmatrix} -\frac{5}{2} & 2 & -\frac{1}{2} \\ 1 & -1 & 1 \\ \frac{1}{2} & 0 & -\frac{1}{2} \end{bmatrix} \text{ or } \frac{1}{2}\begin{bmatrix} -5 & 4 & -1 \\ 2 & -2 & 2 \\ 1 & 0 & -1 \end{bmatrix}$$

$$A^{-1}A = \begin{bmatrix} -\frac{5}{2} & 2 & -\frac{1}{2} \\ 1 & -1 & 1 \\ \frac{1}{2} & 0 & -\frac{1}{2} \end{bmatrix} \begin{bmatrix} 1 & 2 & 3 \\ 2 & 3 & 4 \\ 1 & 2 & 1 \end{bmatrix}$$

$$\begin{bmatrix} (-\frac{5}{2})1 + 2\cdot 2 + (-\frac{1}{2})1 & (-\frac{5}{2})2 + 2\cdot 3 + (-\frac{1}{2})2 & (-\frac{5}{2})3 + 2\cdot 4 + (-\frac{1}{2})1 \\ 1\cdot 1 + (-1)2 + 1\cdot 1 & 1\cdot 2 + (-1)3 + 1\cdot 2 & 1\cdot 3 + (-1)4 + 1\cdot 1 \\ (\frac{1}{2})1 + 0\cdot 2 + (-\frac{1}{2})1 & (\frac{1}{2})2 + 0\cdot 3 + (-\frac{1}{2})2 & (\frac{1}{2})3 + 0\cdot 4 + (-\frac{1}{2})1 \end{bmatrix}$$

$$= \begin{bmatrix} 1 & 0 & 0 \\ 0 & 1 & 0 \\ 0 & 0 & 1 \end{bmatrix} = I \quad (8-3)$$

22.
$$\begin{array}{ccc} A & X & B \end{array}$$

$$\begin{bmatrix} 1 & 2 & 3 \\ 2 & 3 & 4 \\ 1 & 2 & 1 \end{bmatrix} \begin{bmatrix} x_1 \\ x_2 \\ x_3 \end{bmatrix} = \begin{bmatrix} k_1 \\ k_2 \\ k_3 \end{bmatrix}$$

The solution to $AX = B$ is $X = A^{-1}B$.

Applying the A^{-1} found in problem 21, we have

A. $B = \begin{bmatrix} 1 \\ 2 \\ 3 \end{bmatrix} \quad X = \begin{bmatrix} x_1 \\ x_2 \\ x_3 \end{bmatrix} = \begin{bmatrix} -\frac{5}{2} & 2 & -\frac{1}{2} \\ 1 & -1 & 1 \\ \frac{1}{2} & 0 & -\frac{1}{2} \end{bmatrix} \begin{bmatrix} 1 \\ 3 \\ 3 \end{bmatrix} = \begin{bmatrix} 2 \\ 1 \\ -1 \end{bmatrix}$

$$x_1 = 2, x_2 = 1, x_3 = -1$$

B. $B = \begin{bmatrix} 0 \\ 0 \\ -2 \end{bmatrix} \quad X = \begin{bmatrix} x_1 \\ x_2 \\ x_3 \end{bmatrix} = \begin{bmatrix} -\frac{5}{2} & 2 & -\frac{1}{2} \\ 1 & -1 & 1 \\ \frac{1}{2} & 0 & -\frac{1}{2} \end{bmatrix} \begin{bmatrix} 0 \\ 0 \\ -2 \end{bmatrix} = \begin{bmatrix} 1 \\ -2 \\ 1 \end{bmatrix}$

$$x_1 = 1, x_2 = -2, x_3 = -1$$

C. $B = \begin{bmatrix} -3 \\ -4 \\ 1 \end{bmatrix} \quad X = \begin{bmatrix} x_1 \\ x_2 \\ x_3 \end{bmatrix} = \begin{bmatrix} -\frac{5}{2} & 2 & -\frac{1}{2} \\ 1 & -1 & 1 \\ \frac{1}{2} & 0 & -\frac{1}{2} \end{bmatrix} \begin{bmatrix} -3 \\ -4 \\ 1 \end{bmatrix} = \begin{bmatrix} -1 \\ 2 \\ -2 \end{bmatrix}$

$$x_1 = -1, x_2 = 2, x_3 = -2 \quad (8-3)$$

23. $\begin{vmatrix} -\frac{1}{4} & \frac{3}{2} \\ \frac{1}{2} & \frac{2}{3} \end{vmatrix} = (-\frac{1}{4})(\frac{2}{3}) - (\frac{1}{2})(\frac{3}{2}) = -\frac{1}{6} - \frac{3}{4} = -\frac{11}{12} \quad (8-4)$

24. $\begin{vmatrix} 2 & -1 & 1 \\ -3 & 5 & 2 \\ 1 & -2 & 4 \end{vmatrix} = \begin{vmatrix} 0 & 0 & 1 \\ -7 & 7 & 2 \\ -7 & 2 & 4 \end{vmatrix}$ $\begin{matrix} C_1 + (-2)C_3 & \to & C_1 \\ C_2 + C_3 & \to & C_2 \end{matrix}$

$$= 0 + 0 + 1(-1)^{1+3}\begin{vmatrix} -7 & 7 \\ -7 & 2 \end{vmatrix}$$

$$= (-1)^4[(-7)2 - (-7)7]$$

$$= 35 \qquad (8-5)$$

25. $y = \dfrac{\begin{vmatrix} 1 & -6 & 1 \\ 0 & 4 & -1 \\ 2 & 2 & 1 \end{vmatrix}}{\begin{vmatrix} 1 & -2 & 1 \\ 0 & 1 & -1 \\ 2 & 2 & 1 \end{vmatrix}} = \dfrac{\begin{vmatrix} 1 & -6 & 1 \\ 0 & 4 & -1 \\ 0 & 14 & -1 \end{vmatrix}}{\begin{vmatrix} 1 & -2 & 1 \\ 0 & 1 & 0 \\ 2 & 2 & 3 \end{vmatrix}} = \dfrac{1(-1)^{1+1}\begin{vmatrix} 4 & -1 \\ 14 & -1 \end{vmatrix}}{1(-1)^{2+2}\begin{vmatrix} 1 & -1 \\ 2 & 3 \end{vmatrix}} = \dfrac{(-1)^2[4(-1)-14(-1)]}{(-1)^4[1\cdot3-2(-1)]}$

$$= \tfrac{10}{5} = 2 \quad (8-6)$$

26. $\begin{bmatrix} 4 & 5 & 6 & | & 1 & 0 & 0 \\ 4 & 5 & -6 & | & 0 & 1 & 0 \\ 1 & 1 & 1 & | & 0 & 0 & 1 \end{bmatrix}$ $R_2 + (-1)R_1 \to R_2$

$\sim \begin{bmatrix} 4 & 5 & 6 & | & 1 & 0 & 0 \\ 0 & 0 & -12 & | & -1 & 1 & 0 \\ 1 & 1 & 1 & | & 0 & 0 & 1 \end{bmatrix}$ $R_1 \leftrightarrow R_3$

$\sim \begin{bmatrix} 1 & 1 & 1 & | & 0 & 0 & 1 \\ 0 & 0 & -12 & | & -1 & 1 & 0 \\ 4 & 5 & 6 & | & 1 & 0 & 0 \end{bmatrix}$ $R_3 + (-4)R_1 \to R_3$

$\sim \begin{bmatrix} 1 & 1 & 1 & | & 0 & 0 & 1 \\ 0 & 0 & -12 & | & -1 & 1 & 0 \\ 0 & 1 & 2 & | & 1 & 0 & -4 \end{bmatrix}$ $R_2 \leftrightarrow R_3$

$\sim \begin{bmatrix} 1 & 1 & 1 & | & 0 & 0 & 1 \\ 0 & 1 & 2 & | & 1 & 0 & -4 \\ 0 & 0 & -12 & | & -1 & 1 & 0 \end{bmatrix}$ $R_1 + (-1)R_2 \to R_1$

$\sim \begin{bmatrix} 1 & 0 & -1 & | & -1 & 0 & 5 \\ 0 & 1 & 2 & | & 1 & 0 & -4 \\ 0 & 0 & -12 & | & -1 & 1 & 0 \end{bmatrix}$ $-\tfrac{1}{12}R_3 \to R_3$

$\sim \begin{bmatrix} 1 & 0 & -1 & | & -1 & 0 & 5 \\ 0 & 1 & 2 & | & 1 & 0 & -4 \\ 0 & 0 & 1 & | & \tfrac{1}{12} & -\tfrac{1}{12} & 0 \end{bmatrix}$ $\begin{matrix} R_1 + R_3 & \to & R_1 \\ R_2 + (-2)R_3 & \to & R_2 \end{matrix}$

$$\sim \begin{bmatrix} 1 & 0 & 0 \\ 0 & 1 & 0 \\ 0 & 0 & 1 \end{bmatrix} \begin{array}{|ccc} -\frac{11}{12} & -\frac{1}{12} & 5 \\ \frac{10}{12} & \frac{2}{12} & -4 \\ \frac{1}{12} & -\frac{1}{12} & 0 \end{array}$$

Hence

$$A^{-1} = \begin{bmatrix} -\frac{11}{12} & -\frac{1}{12} & 5 \\ \frac{10}{12} & \frac{2}{12} & -4 \\ \frac{1}{12} & -\frac{1}{-12} & 0 \end{bmatrix} \text{ or } \frac{1}{12} \begin{bmatrix} -11 & -1 & 60 \\ 10 & 2 & -48 \\ 1 & -1 & 0 \end{bmatrix}$$

$$A^{-1}A = \frac{1}{12} \begin{bmatrix} -11 & -1 & 60 \\ 10 & 2 & -48 \\ 1 & -1 & 0 \end{bmatrix} \begin{bmatrix} 4 & 5 & 6 \\ 4 & 5 & -6 \\ 1 & 1 & 1 \end{bmatrix}$$

$$= \frac{1}{12} \begin{bmatrix} (-11)4+(-1)4+60\cdot 1 & (-11)5+(-1)5+60\cdot 1 & (-11)6+(-1)(-6)+60\cdot 1 \\ 10\cdot 4+2\cdot 4+(-48)1 & 10\cdot 5+2\cdot 5+(-48)1 & 10\cdot 6+2(-6)+(-48)1 \\ 1\cdot 4+(-1)4+0\cdot 1 & 1\cdot 5+(-1)5+0\cdot 1 & 1\cdot 6+(-1)(-6)+0\cdot 1 \end{bmatrix}$$

$$= \frac{1}{12} \begin{bmatrix} 12 & 0 & 0 \\ 0 & 12 & 0 \\ 0 & 0 & 12 \end{bmatrix} = \begin{bmatrix} 1 & 0 & 0 \\ 0 & 1 & 0 \\ 0 & 0 & 1 \end{bmatrix} = I \quad (8-3)$$

27. Multiplying the first two equations by 100, the system becomes

$$4x_1 + 5x_2 + 6x_3 = 36,000$$
$$4x_1 + 5x_2 - 6x_3 = 12,000$$
$$x_1 + x_2 + x_3 = 7,000$$

As a matrix equation, we have

$$\begin{array}{ccc} A & X & B \end{array}$$
$$\begin{bmatrix} 4 & 5 & 6 \\ 4 & 5 & -6 \\ 1 & 1 & 1 \end{bmatrix} \begin{bmatrix} x_1 \\ x_2 \\ x_3 \end{bmatrix} = \begin{bmatrix} 36,000 \\ 12,000 \\ 7,000 \end{bmatrix}$$

The solution to $AX = B$ is $X = A^{-1}B$. Using A^{-1} from problem 26, we have

$$X = \begin{bmatrix} x_1 \\ x_2 \\ x_3 \end{bmatrix} = \frac{1}{12} \begin{bmatrix} -11 & -1 & 60 \\ 10 & 2 & -48 \\ 1 & -1 & 0 \end{bmatrix} \begin{bmatrix} 36,000 \\ 12,000 \\ 7,000 \end{bmatrix}$$

$$= \frac{1}{12} \begin{bmatrix} (-11)(36,000)+(-1)(12,000)+(60)(7,000) \\ (10)(36,000)+(2)(12,000)+(-48)(7,000) \\ 1(36,000)+(-1)(12,000)+(0)(7,000) \end{bmatrix} = \frac{1}{12} \begin{bmatrix} 12,000 \\ 48,000 \\ 24,000 \end{bmatrix} = \begin{bmatrix} 1,000 \\ 4,000 \\ 2,000 \end{bmatrix}$$

Hence, $x_1 = 1,000, x_2 = 4,000, x_3 = 2,000$

28. $$\begin{vmatrix} -1 & 4 & 1 & 1 \\ 5 & -1 & 2 & -1 \\ 2 & -1 & 0 & 3 \\ -3 & 3 & 0 & 3 \end{vmatrix} = \begin{vmatrix} -1 & 4 & 1 & 1 \\ 7 & -9 & 0 & 3 \\ 2 & -1 & 0 & 3 \\ -3 & 3 & 0 & 3 \end{vmatrix} \quad R_2 + (-2)R_1 \to R_2$$

$$= 1(-1)^{1+3} \begin{vmatrix} 7 & -9 & -3 \\ 2 & -1 & 3 \\ -3 & 3 & 3 \end{vmatrix} + 0 + 0 + 0$$

$$= \begin{vmatrix} 7 & -9 & -3 \\ 2 & -1 & 3 \\ -3 & 3 & 3 \end{vmatrix}$$

$$= \begin{vmatrix} 7 & -9 & -3 \\ 9 & -10 & 0 \\ 4 & -6 & 0 \end{vmatrix} \quad \begin{matrix} R_2 + R_1 & \to & R_2 \\ R_3 + R_1 & \to & R_3 \end{matrix}$$

$$= (-3)(-1)^{1+3} \begin{vmatrix} 9 & -10 \\ 4 & -6 \end{vmatrix} + 0 + 0$$

$$= (-3)(-1)^4 [9(-6) - 4(-10)]$$

$$= (-3)(-14)$$

$$= 42 \quad (8-4)$$

29. $\begin{vmatrix} u + kv & v \\ w + kx & x \end{vmatrix} = (u + kv)x - (w + kx)v = ux + kvx - wv - kvx = ux - wv$

$$= \begin{vmatrix} u & v \\ w & x \end{vmatrix}$$

30. Let
x_1 = number of tons at Big Bend
x_2 = number of tons at Saw Pit
Then
$0.05\ x_1 + 0.03\ x_2$ = number of tons of nickel at both mines = k_1
$0.07\ x_1 + 0.04\ x_2$ = number of tons of copper at both mines = k_2
We solve
$0.05\ x_1 + 0.03\ x_2 = k_1$
$0.07\ x_1 + 0.04\ x_4 = k_2$,
for arbitrary k_1 and k_2, by writing the system as a matrix equation.

$$\overset{A}{\begin{bmatrix} 0.05 & 0.03 \\ 0.07 & 0.04 \end{bmatrix}} \overset{X}{\begin{bmatrix} x_1 \\ x_2 \end{bmatrix}} = \overset{B}{\begin{bmatrix} k_1 \\ k_2 \end{bmatrix}}$$

If A^{-1} exists, then $X = A^{-1}B$
To find A^{-1}, we perform row operations on

$$\begin{bmatrix} 0.05 & 0.03 & | & 1 & 0 \\ 0.07 & 0.04 & | & 0 & 1 \end{bmatrix} \quad 20R_1 \to R_1$$

$$\sim \begin{bmatrix} 1 & 0.6 & | & 20 & 0 \\ 0.07 & 0.04 & | & 0 & 1 \end{bmatrix} \quad R_2 + (-0.07)R_1 \to R_2$$

$$\sim \begin{bmatrix} 1 & 0.6 & | & 20 & 0 \\ 0 & -0.002 & | & -1.4 & 1 \end{bmatrix} \quad -500R_2 \to R_2$$

$$\sim \begin{bmatrix} 1 & 0.6 & 20 & 0 \\ 0 & 1 & 700 & -500 \end{bmatrix} R_2 + (-0.6)R_1 \rightarrow R_2$$

$$\sim \begin{bmatrix} 1 & 0 & -400 & 300 \\ 0 & 1 & 700 & -500 \end{bmatrix}$$

Hence

$$A^{-1} = \begin{bmatrix} -400 & 300 \\ 700 & -500 \end{bmatrix}$$

Check: $A^{-1}A = \begin{bmatrix} -400 & 300 \\ 700 & -500 \end{bmatrix} \begin{bmatrix} 0.05 & 0.03 \\ 0.07 & -0.04 \end{bmatrix} = \begin{bmatrix} 1 & 0 \\ 0 & 1 \end{bmatrix}$

We can now solve the system as:

$$\begin{matrix} X & A^{-1} & B \end{matrix}$$
$$\begin{bmatrix} x_1 \\ x_2 \end{bmatrix} = \begin{bmatrix} -400 & 300 \\ -=700 & -500 \end{bmatrix} \begin{bmatrix} k_1 \\ k_2 \end{bmatrix}$$

A. If $k_1 = 3.6, k_2 = 5$,

$$\begin{bmatrix} x_1 \\ x_2 \end{bmatrix} = \begin{bmatrix} -400 & 300 \\ 700 & -500 \end{bmatrix} \begin{bmatrix} 3.6 \\ 5 \end{bmatrix} = \begin{bmatrix} 60 \\ 20 \end{bmatrix}$$

60 tons of ore must be produced at Big Bend, 20 tons of ore at Saw Pit.

B. If $k_1 = 3, k_2 = 4.1$

$$\begin{bmatrix} x_1 \\ x_2 \end{bmatrix} = \begin{bmatrix} -400 & 300 \\ 700 & -500 \end{bmatrix} \begin{bmatrix} 3 \\ 4.1 \end{bmatrix} = \begin{bmatrix} 30 \\ 50 \end{bmatrix}$$

30 tons of ore must be produced at Big Bend, 50 tons of ore at Saw Pit.

C. If $k_1 = 3.2, k_2 = 4.4$

$$\begin{bmatrix} x_1 \\ x_2 \end{bmatrix} = \begin{bmatrix} -400 & 300 \\ 700 & -500 \end{bmatrix} \begin{bmatrix} 3.2 \\ 4.4 \end{bmatrix} = \begin{bmatrix} 40 \\ 40 \end{bmatrix}$$

40 tons of ore must be produced at Big Bend, 40 tons of ore at Saw Pit.

31.   A. The labor cost of producing one printer stand at the South Carolina plant
is the dot product of the stand row of L with the South Carolina column of H.

$$[0.9 \quad 1.8 \quad 0.6] \cdot \begin{bmatrix} 10.00 \\ 8.50 \\ 4.50 \end{bmatrix} = 27 \text{ dollars}$$

B. $LH = \begin{bmatrix} 1.7 & 2.4 & 0.8 \\ 0.9 & 1.8 & 0.6 \end{bmatrix} \begin{bmatrix} 11.50 & 10.00 \\ 9.50 & 8.50 \\ 5.00 & 4.50 \end{bmatrix}$

$$= \begin{bmatrix} (1.7)(11.50) + (2.4)(9.50) + (0.8)(5.00) & (1.7)(10.00) + (2.4)(8.50) + (0.8)(4.50) \\ (0.9)(11.50) + (1.8)(9.50) + (0.6)(5.00) & (0.9)(11.50) + (1.8)(8.50) + (0.6)(4.50) \end{bmatrix}$$

N. C.   S. C.
$$\begin{bmatrix} \$46.35 & \$41.00 \\ \$30.45 & \$27.00 \end{bmatrix} \begin{matrix} \text{Desk} \\ \text{Stands} \end{matrix}$$

This matrix represents the total labor costs for each item at each plant. $(8-2)$

32.   A. The average monthly production for the months of January and February is represented by the matrix $\frac{1}{2}(J + F)$

$$\frac{1}{2}(J + F) = \frac{1}{2}\left(\begin{bmatrix} 1,500 & 1,650 \\ 850 & 700 \end{bmatrix} + \begin{bmatrix} 1,700 & 1,810 \\ 930 & 740 \end{bmatrix}\right) = \frac{1}{2}\begin{bmatrix} 3,200 & 3,460 \\ 1,780 & 1,440 \end{bmatrix}$$

$$= \begin{matrix} & \text{N.C.} & \text{S.C.} \\ & \begin{bmatrix} 1,600 & 1,730 \\ 890 & 720 \end{bmatrix} & \begin{matrix} \text{Desks} \\ \text{Stands} \end{matrix} \end{matrix}$$

B. The increase in production from January to February is represented by the matrix $F - J$.

$$F - J = \begin{bmatrix} 1,700 & 1,810 \\ 930 & 740 \end{bmatrix} - \begin{bmatrix} 1,500 & 1,650 \\ 850 & 700 \end{bmatrix}$$

$$= \begin{matrix} \text{N.C.} & \text{S.C.} \\ \begin{bmatrix} 200 & 160 \\ 80 & 40 \end{bmatrix} & \begin{matrix} \text{Desks} \\ \text{Stands} \end{matrix} \end{matrix}$$

C. $J\begin{bmatrix} 1 \\ 1 \end{bmatrix} = \begin{bmatrix} 1,500 & 1,650 \\ 850 & 700 \end{bmatrix}\begin{bmatrix} 1 \\ 1 \end{bmatrix} = \begin{bmatrix} 3,150 \\ 1,550 \end{bmatrix} \begin{matrix} \text{Desks} \\ \text{Stands} \end{matrix}$

This matrix represents the total production of each item in January. $(8 - 1, 8 - 2)$

CHAPTER 9

Exercise 9-1

Key Ideas and Formulas

A sequence is a function with domain the natural numbers.

Notation: for $f(n)$ we write a_n. The elements of the range, $f(n)$, are called terms of the sequence, a_n. a_1 is the first term, a_2 is the second term, a_n is the nth term.

Finite sequence: domain is finite set of successive natural numbers.
Infinite sequence: domain is infinite set of successive natural numbers.

Some sequences are specified by a recursion formula, a formula that defines each term in terms of one or more preceding terms.

The sum of the terms of a sequence is called a series. If $a_1, a_2, \cdots, a_n$ are the terms of the sequence, the series is written

$$\sum_{k=1}^{n} a_k = a_1 + a_2 + \cdots + a_n \quad k \text{ is called the summing index.}$$

1. $a_1 = 1 - 2 = -1 \quad a_2 = 2 - 2 = -0 \quad a_3 = 3 - 2 = -1 \quad a_4 = 4 - 2 = 2$
3. $a_1 = \frac{1-1}{1+1} = 0 \quad a_2 = \frac{2-1}{2+1} = \frac{1}{3} \quad a_3 = \frac{3-1}{3+1} = \frac{2}{4} = \frac{1}{2} \quad a_4 = \frac{4-1}{4+1} = \frac{3}{5}$
5. $a_1 = (-2)^{1+1} = (-2)^2 = 4 \quad a_2 = (-2)^{2+1} = (-2)^3 = -8$
   $a_3 = (-2)^{3+1} = (-2)^4 = 16 \quad a_4 = (-2)^{4+1} = (-2)^5 = -32$
7. $a_8 = 8 - 2 = 6$
9. $a_{100} = \frac{100-1}{100+1} = \frac{99}{101}$
11. $S_5 = 1 + 2 + 3 + 4 + 5$
13. $S_3 = \frac{1}{10^1} + \frac{1}{10^2} + \frac{1}{10^3} = \frac{1}{10} + \frac{1}{100} + \frac{1}{1000}$
15. $S_4 = (-1)^1 + (-1)^2 + (-1)^3 + (-1)^4 = (-1) + 1 + (-1) + 1 = -1 + 1 - 1 + 1$
17. $a_1 = (-1)^{1+1}1^2 = 1$
    $a_2 = (-1)^{2+1}2^2 = -4$
    $a_3 = (-1)^{3+1}3^2 = 9$
    $a_4 = (-1)^{4+1}4^2 = -16$
    $a_5 = (-1)^{5+1}5^2 = 25$
19. $a_1 = \frac{1}{3}[1 - \frac{1}{10^1}] = \frac{1}{3} \cdot \frac{9}{10} = \frac{3}{10} = 0.3$
    $a_2 = \frac{1}{3}[1 - \frac{1}{10^2}] = \frac{1}{3} \cdot \frac{99}{100} = \frac{33}{100} = 0.33$
    $a_3 = \frac{1}{3}[1 - \frac{1}{10^3}] = \frac{1}{3} \cdot \frac{999}{1,000} = \frac{333}{1,000} = 0.333$
    $a_4 = \frac{1}{3}[1 - \frac{1}{10^4}] = \frac{1}{3} \cdot \frac{9,999}{10,000} = \frac{3,333}{10,000} = 0.3333$
    $a_5 = \frac{1}{3}[1 - \frac{1}{10^5}] = \frac{1}{3} \cdot \frac{99,999}{100,000} = \frac{33,333}{100,000} = 0.33333$

21. $\quad a_1 \;=\; (-\tfrac{1}{2})^{1-1} = (-\tfrac{1}{2})^0 = 1$
$\quad\;\; a_2 \;=\; (-\tfrac{1}{2})^{2-1} = -\tfrac{1}{2}$
$\quad\;\; a_3 \;=\; (-\tfrac{1}{2})^{3-1} = \tfrac{1}{4}$
$\quad\;\; a_4 \;=\; (-\tfrac{1}{2})^{4-1} = -\tfrac{1}{8}$
$\quad\;\; a_5 \;=\; (-\tfrac{1}{2})^{5-1} = \tfrac{1}{16}$

23. $\quad a_1 \;=\; 7$
$\quad\;\; a_2 \;=\; a_{2-1} - 4 = a_1 - 4 = 7 - 4 = 3$
$\quad\;\; a_3 \;=\; a_{3-1} - 4 = a_2 - 4 = 3 - 4 = -1$
$\quad\;\; a_4 \;=\; a_{4-1} - 4 = a_3 - 4 = -1 - 4 = -5$
$\quad\;\; a_5 \;=\; a_{5-1} - 4 = a_4 - 4 = -5 - 4 = -9$

Common Error:
$a_2 \neq a_2 - 1 - 4$. The 1 is in the subscript: a_{2-1}; that is, a_1

25. $\quad a_1 \;=\; 4$
$\quad\;\; a_2 \;=\; \tfrac{1}{4}a_{2-1} = \tfrac{1}{4}a_1 = \tfrac{1}{4} \cdot 4 = 1$
$\quad\;\; a_3 \;=\; \tfrac{1}{4}a_{3-1} = \tfrac{1}{4}a_2 = \tfrac{1}{4} \cdot 1 = \tfrac{1}{4}$
$\quad\;\; a_4 \;=\; \tfrac{1}{4}a_{4-1} = \tfrac{1}{4}a_3 = \tfrac{1}{4} \cdot \tfrac{1}{4} = \tfrac{1}{16}$
$\quad\;\; a_5 \;=\; \tfrac{1}{4}a_{5-1} = \tfrac{1}{4}a_4 = \tfrac{1}{4} \cdot \tfrac{1}{16} = \tfrac{1}{64}$

27. $\quad a_n : 4, 5, 6, 7, \cdots$
$\quad\;\; n = 1, 2, 3, 4, \cdots$
Comparing a_n with n, we see that
$a_n = n + 3$

29. $\quad a_n : 3, 6, 9, 12, \cdots$
$\quad\;\; n = 1, 2, 3, 4, \cdots$
Comparing a_n with n, we see that
$a_n = 3n$

31. $\quad a_n : \tfrac{1}{2}, \tfrac{2}{3}, \tfrac{3}{4}, \tfrac{4}{5}, \cdots$
$\quad\;\; n = 1, 2, 3, 4, \cdots$
Comparing a_n with n, we see that
$a_n = \dfrac{n}{n+1}$

33. $\quad a_n : 1, -1, 1, -1, \cdots$
$\quad\;\; n = 1, 2, 3, 4, \cdots$
Comparing a_n with n we see that a_n involves (-1) to successively
even and odd powers, hence to a power that depends on n. We could write
$a_n = (-1)^{n-1}$ or $a_n = (-1)^{n+1}$
or other choices. $a_n = (-1)^{n+1}$ is one of many correct answers.

35. $\quad a_n : -2, 4, -8, 16 \cdots$
$\quad\;\; n = 1, 2, 3, 4, \cdots$
Comparing a_n with n, we see that a_n involves -1 and 2 to successively
higher powers, hence to powers that depend on n. We write
$a_n = (-1)^n (2)^n$ or $a_n = (-2)^n$

37. $\quad a_n : x, \tfrac{x^2}{2}, \tfrac{x^3}{3}, \tfrac{x^4}{4}, \cdots$
$\quad\;\;$ or $\tfrac{x^1}{1}, \tfrac{x^2}{2}, \tfrac{x^3}{3}, \tfrac{x^4}{4}, \cdots$
Comparing a_n with n, we see that $a_n = \dfrac{x^n}{n}$

39. $S_4 = \frac{(-2)^{1+1}}{1} + \frac{(-2)^{2+1}}{2} + \frac{(-2)^{3+1}}{3} + \frac{(-2)^{4+1}}{4}$

$\quad = \frac{4}{1} - \frac{8}{2} + \frac{16}{3} - \frac{32}{4}$

41. $S_3 = \frac{1}{1}x^{1+1} + \frac{1}{2}x^{2+1} + + \frac{1}{3}x^{3+1}$

$\quad = x^2 + \frac{x^3}{2} + \frac{x^4}{3}$

43. $S_5 = \frac{(-1)^{1+1}}{1}x^1 + \frac{(-1)^{2+1}}{2}x^2 + \frac{(-1)^{3+1}}{3}x^3 + \frac{(-1)^{4+1}}{4}x^4 + \frac{(-1)^{5+1}}{5}x^5$

$\quad = x - \frac{x^2}{2} + \frac{x^3}{3} - \frac{x^4}{4} + \frac{x^5}{5}$

45. $S_4 = 1^2 + 2^2 + 3^2 + 4^2$

$k = 1, 2, 3, 4, \cdots$

Clearly, $a_k = k^2, k = 1, 2, 3, 4$

$S_4 = \sum_{k=1}^{4} k^2$

47. $S_5 = \frac{1}{2^1} + \frac{1}{2^2} + \frac{1}{2^3} + \frac{1}{2^4} + \frac{1}{2^5}$

$k = 1, 2, 3, 4, \cdots$

Clearly, $a_k = \frac{1}{2^k}, k = 1, 2, 3, 4, 5$

$S_5 = \sum_{k=1}^{5} \frac{1}{2^k}$

49. $S_n = 1 + \frac{1}{2^2} + \frac{1}{3^2} + \cdots + \frac{1}{n^2}$

$k = 1, 2, 3, \cdots, n$

Clearly, $a_k = \frac{1}{k^2}, k = 1, 2, 3, \cdots, n$

$S_n = \sum_{k=1}^{n} \frac{1}{k^2}$

51. $S_n = 1 - 4 + 9 + \cdots + (-1)^{n+1}n^2$

$k = 1, 2, 3, \cdots, n$

Clearly, $a_k = (-1)^{k+1}k^2, k = 1, 2, 3, \cdots, n$

$S_n = \sum_{k=1}^{n} (-1)^{k+1}k^2$

53. $\sum_{k=1}^{n} ca_k = ca_1 + ca_2 + ca_3 + \cdots + ca_n = c(a_1 + a_2 + a_3 + \cdots a_n) = c\sum_{k=1}^{n} a_k$

55. A. $a_1 = 3$

$a_2 = \frac{a_{2-1}^2 + 2}{2a_{2-1}} = \frac{a_1^2 + 2}{2a_1} = \frac{3^2 + 2}{2 \cdot 3} \approx 1.83$

$a_3 = \frac{a_{3-1}^2 + 2}{2a_{3-1}} = \frac{a_2^2 + 2}{2a_2} = \frac{(1.83)^2 + 2}{2(1.83)} \approx 1.46$

$a_4 = \frac{a_{4-1}^2 + 2}{2a_{4-1}} = \frac{a_2^3 + 2}{2a_3} = \frac{(1.46)^2 + 2}{2(1.46)} \approx 1.415$

B. Calculator $\sqrt{2} = 1.4142135 \cdots$

C. $a_1 = 1$

$a_2 = \frac{a_1^2 + 2}{2a_1} = \frac{1^2 + 2}{2 \cdot 1} = 1.5$

$a_3 = \frac{a_2^2 + 2}{2a_2} = \frac{(1.5)^2 + 2}{2(1.5)} \approx 1.417$

$a_4 = \frac{a_3^2 + 2}{2a_3} = \frac{(1.417)^2 + 2}{2(1.417)} \approx 1.414$

57. $e^{0.2} = 1 + \frac{0.2}{1!} + \frac{(0.2)^2}{2!} + \frac{(0.2)^3}{3!} + \frac{(0.2)^4}{4!}$

$\quad = 1 + 0.2 + \frac{(0.2)^2}{1 \cdot 2} + \frac{(0.2)^3}{1 \cdot 2 \cdot 3} + \frac{(0.2)^4}{1 \cdot 2 \cdot 3 \cdot 4}$

$\quad = 1.2 + 0.2[\frac{0.2}{2} + \frac{(0.2)^2}{6} + \frac{(0.2)^3}{24}]$

$\quad = 1.2 + 0.2\{0.2[\frac{1}{2} + \frac{0.2}{6} + \frac{(0.2)^2}{24}]\}$

$\quad = 1.2 + 0.2\{0.2[\frac{1}{2} + 0.2(\frac{1}{6} + \frac{0.2}{24})]\}$In this form this is more easily entered into the calculator.)

$\quad = 1.2214000$

$e^{0.2} = 1.2214028$ (calculator $--$direct evaluation)

Exercise 9-2

Key Ideas and Formulas

A set of integers is inductive if, whenever k is in the set, then $k + 1$ is in the set.

Principle of Mathematical Induction.

If p is a positive integer and S is a set of integers such that

1. $p \in S$
2. S is inductive

then S contains all integers greater than or equal to p.

Proof by Mathematical Induction: To prove that a statement P_n holds for all integers $n \geq p$. (p is often, but not always, equal to 1.)

1. Prove P_p is true.

2. Write out P_k and P_{k+1}

3. Starting with the assumption that P_k holds, derive the fact that P_{k+1} then holds. This shows that the set S of integers, for which P holds, is inductive.

---

1. $(3 + 5)^1 = 3^1 + 5^1$ True   3. $1^2 = 3 \cdot 1 - 2$ True
       $(3 + 5)^2 = 3^2 + 5^2$ False $2^2 = 3 \cdot 2 - 2$ True
       Fails at $n = 2$ $3^2 = 3 \cdot 3 - 2$ False
                                                 Fails at $n = 3$

5. $P_1 : 2 \ = \ 2 \cdot 1^2 \quad 2 \ = \ 2$
        $P_2 : 2 + 6 \ = \ 2 \cdot 2^2 \quad 8 \ = \ 8$
   $P_3 : 2 + 6 + 10 \ = \ 2 \cdot 3^2 \quad 18 \ = \ 18$

7. $P_1 : a^5 a^1 \ = \ a^{5+1} \quad a^6 \ = \ 6$
       $P_2 : a^5 a^2 \ = \ a^{5+2} \quad a^5 a^2 \ = \ a^5(a^1 a) = (a^5 a)a = a^6 a = a^7 = a^{5+2}$
       $P_3 : a^5 a^3 \ = \ a^{5+3} \quad a^5 a^3 \ = \ a^5(a^2 a) = a^5(a^1 a)a = [(a^5 a)a]a = a^8 = a^{5+3}$

9. $P_1 : 9^1 - 1 \ = \ 8$ is divisible by 4
       $P_2 : 9^2 - 1 \ = \ 80$ is divisible by 4
       $P_3 : 9^3 - 1 \ = \ 728$ is divisible by 4

11. $P_k : 2 + 6 + 10 + \cdots + (4k - 2) = 2k^2$
       $P_{k+1} : 2 + 6 + 10 + \cdots + (4k - 2) + [4(k + 1) - 2] = 2(k + 1)^2$
       or $2 + 6 + 10 + \cdots + (4k - 2) + (4k + 2) = 2(k + 1)^2$

13. $$P_k : a^5 a^k = a^{5+k}$$
    $$P_{k+1} : a^5 a^{k+1} = a^{5+k+1}$$

15. $$P_k : 9^k - 1 = 4r, r \in N$$
    $$P_{k+1} : 9^{k+1} - 1 = 4s, s \in N$$

17. Prove: $2 + 6 + 10 + \cdots + (4n - 2) = 2n^2, n \in N$

Write: $P_n : 2 + 6 + 10 + \cdots + (4n - 2) = 2n^2$

$S = \{n \in N \mid P_n \text{ is true}\}$

To show $S = N$, we show

Part 1: $1 \in S$

$\quad\quad P_1 : 2 = 2 \cdot 1^2$ is true

Part 2: S is inductive.

$\quad\quad P_k : 2 + 6 + 10 + \cdots + (4k - 2) = 2k^2$

$\quad\quad P_{k+1} : 2 + 6 + 10 + \cdots + (4k - 2) + (4k + 2) = 2(k + 1)^2$

We start with P_k:

$\quad\quad 2 + 6 + 10 + \cdots + (4k - 2) = 2k^2$

$\rightarrow$ Adding $(4k + 2)$ to both sides, we get

$$\begin{aligned} 2 + 6 + 10 + \cdots + (4k - 2) + (4k + 2) &= 2k^2 + 4k + 2 \\ &= 2(k^2 + 2k + 1) \\ &= 2(k + 1)^2 \end{aligned}$$

Thus, $P_k \Rightarrow P_{k+1}, k \in S \Rightarrow k + 1 \in S$ and S is inductive. Conclusion: $S = N$.

Common Errors:

P_1 is not the nonsensical statement

$2 + 6 + 10 + \cdots + (4 \cdot 1 - 2) = 2 \cdot 1^2$

achieved by "substitution" of 1 for n. The left side of P_n has n terms; the

left side of P_1 must have 1 term. Also, $P_{k+1} \neq P_k + 1$

Nor is P_{k+1} simply its $k + 1st$ term. $4k + 2 \neq 2(k + 1)^2$

19. Prove: $a^5 a^n = a^{5+n}, n \in N$

Write: $P_n : a^5 a^n = a^{5+n}, S = \{n \in N \mid P_n \text{ is true }\}$

To show $S = N$, we show

Part 1: $1 \in S$

$\quad\quad P_1 : a^5 a^1 = a^{5+1}$ is true by the redefinition of a^n

Part 2: S is inductive.

$\quad\quad P_k : a^5 a^k = a^{5+k}$

$\quad\quad P_{k+1} : a^5 a^{k+1} = a^{5+k+1}$

We start with P_k:

$\quad\quad a^5 a^k = a^{5+k}$

Multiply both sides by a:

$$\begin{aligned} a^5 a^k a &= a^{5+k} a \\ a^5 a^{k+1} a &= a^{5+k+1} a \text{ by the redefinition of } a^n \end{aligned}$$

Thus $P_k \Rightarrow P_{k+1}, k \in S \Rightarrow k + 1 \in S$ and S is inductive. Conclusion: $S = N$.

21. Prove: $9^n - 1$ is divisible by 4.

Write: $P_n : 9^n - 1$ is divisible by 4, $S = \{n \in N \mid P_n \text{ is true}\}$

To show $S = N$, we show

Part 1: $1 \in S$

$\quad$ $P_1 : 9^1 - 1 = 8 = 4 \cdot 2$ is divisible by 4. Clearly, true.

Part 2: S is inductive.

$\quad$ $P_k : 9^k - 1$ is divisible by 4.

$\quad$ $P_{k+1} : 9^{k+1} - 1$ is divisible by 4.

We start with P_k :

$\quad$ $9^k - 1 = 4r$ for some integer r

$\quad$ Now, $\quad$
$$
\begin{aligned}
9^{k+1} - 1 &= 9^{k+1} - 9^k + 9^k - 1 \\
&= 9^k(9 - 1) + 9^k - 1 \\
&= 9^k \cdot 8 + 9^k - 1 \\
&= 4 \cdot 9^k \cdot 2 + 4r \\
&= 4(9^k \cdot 2 + r)
\end{aligned}
$$

Therefore,

$9^{k+1} - 1 = 4s$ for some integer $s(= 2 \cdot 9^k + r)$

$9^{k+1} - 1$ is divisible by 4.

Thus, $P_k \Rightarrow P_{k+1}, k \in S \Rightarrow k + 1 \in S$ and S is inductive.

Conclusion: $S = N$.

23. $\quad$ Prove: $2 + 2^2 + 2^3 + \cdots + 2^n = 2^{n+1} - 2$

$\quad$ Write: $P_n : 2 + 2^2 + 2^3 + \cdots + 2^n = 2^{n+1} - 2, S = \{ n \in N \mid P_n$ is true $\}$

$\quad$ To show $S = N$ we show:

$\quad$ *Part 1:* $1 \in S$

$$
\begin{aligned}
P_1 : 2 &= 2^{1+1} - 2 \\
&= 2^2 - 2 \\
&= 4 - 2 \\
&= 2 \quad P_1 \text{ is true}
\end{aligned}
$$

$\quad$ *Part 2:* S is inductive.

$\qquad$ $P_k : 2 + 2^2 + 2^3 + \cdots + 2^k = 2^{k+1} - 2$

$\qquad$ $P_{k+1} : 2 + 2^2 + 2^3 + \cdots + 2^k + 2^{k+1} = 2^{k+2} - 2$

$\quad$ We start with P_k :

$\qquad$ $2 + 2^2 + 2^3 + \cdots + 2^k = 2^{k+1} - 2$

$\quad$ Adding 2^{k+1} to both sides,

$$
\begin{aligned}
2 + 2^2 + 2^3 + \cdots + 2^k + 2^{k+1} &= 2^{k+1} - 2 + 2^{k+1} \\
&= 2^{k+1} + 2^{k+1} - 2 \\
&= 2 \cdot 2^{k+1} - 2 \\
&= 2^1 \cdot 2^{k+1} - 2 \\
&= 2^{k+2} - 2
\end{aligned}
$$

$\quad$ Thus, $P_k \Rightarrow P_{k+1}, k \in S \Rightarrow k + 1 \in S$ and S is inductive. Conclusion: $S = N$.

25. $\quad$ Prove: $1^2 + 3^2 + 5^2 + \cdots + (2n - 1)^2 = \frac{1}{3}(4n^3 - n)$

$\quad$ Write: $P_n : 1^2 + 3^2 + 5^2 + \cdots + (2n - 1)^2 = \frac{1}{3}(4n^3 - n)$

$\quad$ $S = \{ n \in N \mid P_n$ is true $\}$

$\quad$ To show $S = N$, we show

$\quad$ *Part 1:* $1 \in S$

$$P_1 : 1^2 = \tfrac{1}{3}(4 \cdot 1^3 - 1)$$
$$= \tfrac{1}{3}(4 - 1)$$
$$= 1 \text{ True}$$

Part 2: S is inductive.

$$P_k : 1^2 + 3^2 + 5^2 + \cdots + (2k-1)^2 = \tfrac{1}{3}(4k^3 - k)$$
$$P_{k+1} : 1^2 + 3^2 + 5^2 + \cdots + (2k-1)^2 + (2k+1)^2 = \tfrac{1}{3}[4(k+1)^3 - (k+1)]$$

We start with P_k :

$$1^2 + 3^2 + 5^2 + \cdots + (2k-1)^2 = \tfrac{1}{3}(4k^3 - k)$$

Adding $(2k+1)^2$ to both sides:

$$1^2 + 3^2 + 5^2 + \cdots + (2k-1)^2 + (2k+1)^2 = \tfrac{1}{3}(4k^3 - k) + (2k+1)^2$$
$$= \tfrac{1}{3}[4k^3 - k + 3(2k+1)^2]$$
$$= \tfrac{1}{3}[4k^3 - k + 3(4k^2 + 4k + 1)]$$
$$= \tfrac{1}{3}[4k^3 - k + 12k^2 + 12k + 3]$$
$$= \tfrac{1}{3}[4k^3 + 12k^2 + 12k - k + 3]$$
$$= \tfrac{1}{3}[4k^3 + 12k^2 + 12k + 4 - k - 1]$$
$$= \tfrac{1}{3}[4(k^3 + 3k^2 + 3k + 1) - (k+1)]$$
$$= \tfrac{1}{3}[4(k+1)^3 - (k+1)]$$

Thus, $P_k \Rightarrow P_{k+1}, k \in S \Rightarrow k+1 \in S$ and S is inductive. Conclusion: $S = N$.

27. Prove: $1^2 + 2^2 + 3^2 + \cdots + n^2 = \dfrac{n(n+1)(2n+1)}{6}$

Write: $P_n : 1^2 + 2^2 + 3^2 + \cdots + n^2 = \dfrac{n(n+1)(2n+1)}{6}$

$S = \{n \in N \mid P_n \text{ is true }\}$

To show $S = N$, we show:

Part 1: $1 \in S$

$$P_1 : 1^2 = \dfrac{1(1+1)(2 \cdot 1 + 1)}{6}$$
$$= \dfrac{1 \cdot 2 \cdot 3}{6}$$
$$= 1 \quad P_1 \text{ is true}$$

Part 2: S is inductive.

$$P_k : 1^2 + 2^2 + 3^2 + \cdots + k^2 = \dfrac{k(k+1)(2k+1)}{6}$$
$$P_{k+1} : 1^2 + 2^2 + 3^2 + \cdots + k^2 + (k+1)^2 = \dfrac{(k+1)(k+2)(2k+3)}{6}$$

We start with P_k :

$$1^2 + 2^2 + 3^2 + \cdots + k^2 = \dfrac{k(k+1)(2k+1)}{6}$$

Adding $(k+1)^2$ to both sides:

$$1^2 + 2^2 + 3^2 + \cdots + k^2 + (k+1)^2 \; = \; \frac{k(k+1)(2k+1)}{6} + (k+1)^2$$

$$= \; \frac{k(k+1)(2k+1)}{6} + \frac{6(k+1)^2}{6}$$

$$= \; \frac{k(k+1)(2k+1) + 6(k+1)^2}{6}$$

$$= \; \frac{(k+1)[k(2k+1) + 6(k+1)]}{6}$$

$$= \; \frac{(k+1)[2k^2 + 7k + 6]}{6}$$

$$= \; \frac{(k+1)(k+2)(2k+3)}{6}$$

Thus, $P_k \Rightarrow P_{k+1}, k \in S \Rightarrow k+1 \in S$ and S is inductive.

29.   Prove: $\frac{a^n}{a^3} = a^{n-3}$, $n > 3$

Write: $P_n : \frac{a^n}{a^3} = a^{n-3}$, $S = \{n \in N \mid P_n \text{ is true}\}$

To show $S = \{n \in N \mid n > 3\}$, we show:

Part 1: $4 \in S$

$\quad\quad P_4 : \frac{a^4}{a^3} = a^{4-3}$

$\quad\quad$ But, $\frac{a^4}{a^3} = \frac{a^3 a}{a^3} = a = a^1 = a^{4-3}$ So, P_4 is true.

Part 2: S is inductive.

$\quad\quad P_k : \frac{a^k}{a^3} = a^{k-3}$

$\quad\quad P_{k+1} : \frac{a^{k+1}}{a^3} = a^{k-2}$

We start with P_k :

$\quad\quad \frac{a^k}{a^3} = a^{k-3}$

Multiply both sides by a :

$\quad\quad \frac{a^k}{a^3} a = a^{k-3} a$

$\quad\quad \frac{a^{k+1}}{a^3} = a^{k-3+1}$ by the redefinition of a^n

$\quad\quad \frac{a^{k+1}}{a^3} = a^{k-2}$

Thus, $P_k \Rightarrow P_{k+1}, k \in S \Rightarrow k+1 = S$ and S is inductive.

Conclusion: $S = \{n \in N \mid n > 3\}$.

31.   Prove: $a^m a^n = a^{m+n}, m, n \in N$. Let m be an arbitrary element of N.

Write: $P_n : a^m a^n = a^{m+n}$, $S_m = \{n \in N \mid P_n \text{ is true }\}$

To show $S_m = N$, we show:

Part 1: $1 \in S_m$

$\quad\quad P_1 : a^m a = a^{m+1}$ This is true by the redefinition of a^n.

Part 2: S_m is inductive.

$\quad\quad P_k : a^m a^k = a^{m+k}$

$\quad\quad P_{k+1} : a^m a^{k+1} = a^{m+k+1}$

We start with P_k :

$\quad\quad a^m a^k = a^{m+k}$

Multiplying both sides by a:
$$a^m a^k a = a^{m+k} a$$
$$a^m a^{k+1} = a^{m+k+1} \text{ by the redefinition of } a^n$$

Thus, $P_k \Rightarrow P_{k+1}, k \in S_m \Rightarrow k+1 \in S_m$ and S_m is inductive.

Conclusion: $S_m = N$. But, m was arbitrary. Therefore, $a^m a^n = a^{m+n}$ for any $m, n \in N$.

33.   Prove: $x^n - 1 = (x-1)Q_n(x)$ for some polynomial $Q_n(x)$

Write: $P_n : x^n - 1 = (x-1)Q_n(x), S = \{ n \in N \mid P_n \text{ is true} \}$

To show $S = N$, we show:

Part 1: $1 \in S$

$\qquad P_1 : x^1 - 1 = (x-1)Q_1(x)$. This is true since we choose $Q_1(x) = 1$.

Part 2: S is inductive.

$\qquad P_k : x^k - 1 = (x-1)Q_k(x)$ for some $Q_k(x)$

$\qquad P_{k+1} : x^{k+1} - 1 = (x-1)Q_{k+1}(x)$ for some $Q_{k+1}(x)$

We start with P_k:

$$x^k - 1 = (x-1)Q_k(x) \text{ for some } Q_k(x)$$

$$\begin{aligned} \text{Now, } x^{k+1} - 1 &= x^{k+1} - x^k + x^k - 1 \\ &= x^k(x-1) + x^k - 1 \\ &= x^k(x-1) + (x-1)Q_k(x) \\ &= (x-1)(x^k + Q_k(x)) \\ &= (x-1)Q_{k+1}(x), \text{where} Q_{k+1}(x) = x^k + Q_k(x) \end{aligned}$$

Thus, $P_k \Rightarrow P_{k+1}, k \in S \Rightarrow k+1 \in S$ and S is inductive.

Conclusion: $S = N$.

35.   Prove: $x^{2n} - 1$ is divisible by $x - 1, x \neq 1$

Write: $P_n : x^{2n} - 1 = (x-1)Q_n(x), S = \{ n \in N \mid P_n \text{ is true} \}$

To show $S = N$, we show:

Part 1: $1 \in S$

$\qquad P_1 : x^2 - 1 = (x-1)Q_1(x)$ This is true since we choose

$\qquad Q_1(x) = x + 1$.

Part 2: S is inductive.

$\qquad P_k : x^{2k} - 1 = (x-1)Q_k(x)$ for some $Q_k(x)$

$\qquad P_{k+1} : x^{2k+2} - 1 = (x-1)Q_{k+1}(x)$ for some $Q_{k+1}(x)$

$$\begin{aligned} \text{Now, } x^{2k+2} - 1 &= x^{2k+2} - x^{2+k} x^{2k} - 1 \\ &= x^{2k}(x^2 - 1) + x^{2k} - 1 \\ &= x^{2k}(x+1)(x-1) + (x-1)Q_k(x) \\ &= (x-1)[x^{2k}(x+1) + Q_k(x)] \\ &= (x-1)Q_{k+1}(x), \text{where } Q_{k+1}(x) = x^{2k}(x+1) + Q_k(x) \end{aligned}$$

Thus, $P_k \Rightarrow P_{k+1}, k \in S \Rightarrow k+1 \in S$ and S is inductive.

Conclusion: $S = N$.

37.   Prove: $1^3 + 2^3 + 3^3 + \cdots + n^3 = (1 + 2 + 3 + \cdots + n)^2$

Write: $P_n : 1^3 + 2^3 + 3^3 + \cdots n^3 = (1 + 2 + 3 + \cdots + n)^2$

To show $S = N$, we show: $S = \{ n \in N \mid P_n \text{ is true} \}$

Part 1: $1 \in S$

$\qquad P_1 : 1^3 = 1^2$ True

Part 2: S is inductive.

$\qquad P_k : 1^3 + 2^3 + 3^3 + \cdots + k^3 = (1 + 2 + 3 + \cdots + k)^2$

$$P_{k+1} : 1^3 + 2^3 + 3^3 + \cdots + k^3 + (k+1)^3 = (1 + 2 + 3 + \cdots + k + k + 1)^2$$

We take it as proved in text problem 5 that

$$1 + 2 + 3 + \cdots + n = \frac{n(n+1)}{2} \text{ for } n \in N$$

Thus, $1 + 2 + 3 + \cdots + k = \frac{k(k+1)}{2}$

$$1 + 2 + 3 + \cdots + k + (k+1) = \frac{(k+1)(k+2)}{2} \text{ are known}$$

We start with P_k :

$$1^3 + 2^3 + 3^3 + \cdots + k^3 = (1 + 2 + 3 + \cdots + k)^2$$

Adding $(k+1)^3$ to both sides:

$$
\begin{aligned}
1^3 + 2^3 + 3^3 + \cdots + k^3 + (k+1)^3 &= (1 + 2 + 3 + \cdots + k)^2 + (k+1)^3 \\
&= [\tfrac{k(k+1)}{2}]^2 + (k+1)^3 \\
&= (k+1)^2[(\tfrac{k}{2})^2 + (k+1)] \\
&= (k+1)^2[\tfrac{k^2}{4} + \tfrac{4(k+1)}{4}] \\
&= (k+1)^2[\tfrac{k^2+4k+4}{4}] \\
&= \tfrac{(k+1)^2(k+2)^2}{4} \\
&= [\tfrac{(k+1)(k+2)}{2}]^2 \\
&= [1 + 2 + 3 + \cdots + k + (k+1)]^2
\end{aligned}
$$

Thus, $P_k \Rightarrow P_{k+1}, k \in S \Rightarrow k + 1 \in S$ and S is inductive.

Conclusion: $S = N$.

39.   We note:

$$
\begin{aligned}
2 &= 2 &&= 1 \cdot 2 & n &= 1 \\
2 + 4 &= 6 &&= 2 \cdot 3 & n &= 2 \\
2 + 4 + 6 &= 12 &&= 3 \cdot 4 & n &= 3 \\
2 + 4 + 6 + 8 &= 20 &&= 4 \cdot 5 & n &= 4
\end{aligned}
$$

Hypothesis: $P_n : 2 + 4 + 6 + \cdots + 2n = n(n+1), S = \{ n \in N \mid P_n \text{ is true} \}$

To show $S = N$, we show:

Part 1: $1 \in S$

$P_1 : 2 = 1 \cdot 2$ True

Part 2: S is inductive.

$$P_k : 2 + 4 + 6 + \cdots + 2k = k(k+1)$$
$$P_{k+1} : 2 + 4 + 6 + \cdots + 2k + (2k+2) = (k+1)(k+2)$$

We start with P_k :

$$2 + 4 + 6 + \cdots + 2k = k(k+1)$$

Adding $2k + 2$ to both sides:

$$
\begin{aligned}
2 + 4 + 6 + \cdots + 2k + (2k+2) &= k(k+1) + 2k + 2 \\
&= (k+1)k + (k+1)2 \\
&= (k+1)(k+2)
\end{aligned}
$$

Thus, $P_k \Rightarrow P_{k+1}, k \in S \Rightarrow k + 1 \in S$ and S is inductive.

Conclusion $S = N$.

41. $n = 1$; no line is determined.

$n = 2$: one line is determined.

$n = 3$: three lines are determined.

$n = 4$: six lines are determined.

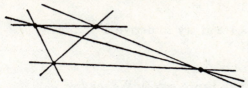

$n = 5$: ten lines are determined.

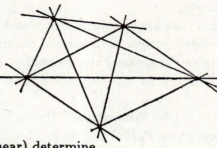

Hypothesis: P_n : n points (no three collinear) determine
$1 + 2 + 3 + \cdots + (n - 1) = \frac{n(n-1)}{2}$ lines, $n \geq 2$.
$S = \{n \in N \mid P_n$ is true $\}$
To show $S = \{n \in N \mid n \geq 2\}$ we show:
Part 1: $2 \in S$

P_2 : two points determine one line:
$$1 = \frac{2(2-1)}{2} \text{ is true}$$
$$= \frac{2 \cdot 1}{2}$$
$$= 1$$

Part 2: S is inductive.

P_k : k points determine $1 + 2 + 3 + \cdots + (k - 1) = \frac{k(k-1)}{2}$ lines

P_{k+1} : $k + 1$ points determine $1 + 2 + 3 + \cdots + (k - 1) + k = \frac{(k+1)k}{2}$ lines

We start with P_k :
k points determine $1 + 2 + 3 + \cdots + (k - 1) = \frac{k(k-1)}{2}$ lines

As already remarked, the $k + 1st$ point will determine a total of k new
lines, one with each of the previously existing k points. These k new
lines will be added to the previously existing lines. Hence, $k + 1$ points determine
$$1 + 2 + 3 + \cdots + (k - 1) + k = \frac{k(k-1)}{2} + k \text{ lines}$$
$$= k[\frac{k-1}{2} + 1] \text{lines}$$
$$= k[\frac{k-1+2}{2}] \text{lines}$$
$$= \frac{k(k+1)}{2} \text{lines}$$
Thus, $P_k \Rightarrow P_{k+1}, k \in S \Rightarrow k + 1 \in S$ and S is inductive.
Conclusion: $S = \{n \in N \mid n \geq 2\}$

43.   Prove: $a > 1 \Rightarrow a^n > 1, n \in N$
Write: $P_n : a > 1 \Rightarrow a^n > 1, S = \{n \in N \mid P_n$ is true $\}$
To show $S = N$, we show:
Part 1: $1 \in S$

$P_1 : a > 1 \Rightarrow a > 1$. This is automatically true.
Part 2: S is inductive.

$$P_k : a > 1 \Rightarrow a^k > 1$$
$$P_{k+1} : a > 1 \Rightarrow a^{k+1} > 1$$

We start with P_k. Further, assume $a > 1$ and try to derive $a^{k+1} > 1$.

If this succeeds, we have proved P_{k+1}.

Assume $a > 1$.

From P_k we know that $a^k > 1$, also $1 > 0$, hence $a > 0$. We may therefore multiply both sides of the inequality $a^k > 1$ by a without changing the sense of the inequality. Hence,

$$a^k a > 1a$$
$$a^{k+1} > a$$

But, $a > 1$ by assumption .. Hence, $a^{k+1} > 1$. We have derived this from P_k and $a > 1$.

Hence, $P_k \Rightarrow P_{k+1}, k \in S \Rightarrow k + 1 \in S$ and S is inductive.

Conclusion: $S = N$.

45.   Prove: $n^2 > 2n, n \geq 3$

Write: $P_n : n^2 > 2n, n \geq 3$; $S = \{ n \in N \mid P_n \text{ is true} \}$

To show $S = \{ n \in N \mid n \geq 3 \}$, we show:

Part 1: $3 \in S$

$$3^2 > 2 \cdot 3$$
$$9 > 6 : \quad \text{true}$$

Part 2: S is inductive

$$P_k : k^2 > 2k$$
$$P_{k+1} : (k+1)^2 > 2(k+1)$$

We start with P_k :

$$k^2 > 2k$$

Adding $2k + 1$ to both sides: $k^2 + 2k + 1 > 2k + 2k + 1$

$$(k+1)^2 > 2k + 2 + 2k - 1$$

Now, $2k - 1 > 0, k \in N$, hence $2k + 2 + 2k - 1 > 2k + 2$.

Therefore, $(k+1)^2 \quad > \quad 2k + 2$

$$(k+1)^2 \quad > \quad 2(k+1)$$

Thus, $P_k \Rightarrow P_{k+1}, k \in S \Rightarrow k + 1 \in S$ and S is inductive.

Thus, $S = \{ n \in N \mid n \geq 3 \}$.

47. $$3^4 + 4^4 + 5^4 + 6^4 \ = \ 2,258$$
$$7^4 \ = \ 2,401$$
$$3^4 + 4^4 + 5^4 + 6^4 \neq 7^4$$

49.   Prove: $a_n = b_n, n \in N$

Write: $P_n : a_n = b_n, S = \{ n \in N \mid P_n \text{ is true} \}$

To show $S = N$, we show:

Part 1: $1 \in S$

$$a_1 \ = \ 1, b_1 = 2 \cdot 1 - 1 = 1$$
$$a_1 \ = \ b_1$$

Part 2: S is inductive

$$P_k : a_k = b_k$$
$$P_{k+1} : a_{k+1} = b_{k+1}$$

We start with P_k :
$$a_k = b_k$$
Now, $a_{k+1} = a_{k+1-1} + 2 = a_k + 2 = b_k + 2 \quad = \quad 2k - 1 + 2 = 2k + 1$
$$= \quad 2(k+1) - 1 = b_{k+1}$$
Therefore, $a_{k+1} = b_{k+1}$

Thus, $P_k \Rightarrow P_{k+1}, k \in S \Rightarrow k + 1 \in S$ and S is inductive.

Thus, $S = N, a_n = b_n$ for all $n \in N$. Conclusion: $\{a_n\} = \{b_n\}$.

51.   Prove: $a_n = b_n, n \in N$

Write: $P_n : a_n = b_n, S = \{n \in N \mid P_n \text{ is true }\}$

To show $S = N$, we show:

Part 1: $1 \in S$
$$a_1 \quad = \quad 2, b_1 = 2^{2 \cdot 1 - 1} = 2^1 = 2$$
$$a_1 \quad = \quad b_1$$
Part 2: S is inductive.
$$P_k : a_k = b_k$$
$$P_{k+1} : a_{k+1} = b_{k+1}$$
We start with P_k :
$$a_k = b_k$$
Now, $a_{k+1} = 2^2 a_{k+1-1} = 2^2 a_k = 2^2 b_k = 2^2 2^{2k-1} = 2^{2(k+1)-1} = b_{k+1}$

Therefore, $a_{k+1} = b_{k+1}$.

Thus, $P_k \Rightarrow P_{k+1}, k \in S \Rightarrow k + 1 \in S$ and S is inductive.

Thus, $S = N, a_n = b_n$ for all $n \in N$. Conclusion: $\{a_n\} = \{b_n\}$.

Exercise 9-3

Key Ideas and Formulas

A sequence $a_1, a_2, a_3 \cdots, a_n, \cdots$ is called an arithmetic sequence if there exists a constant d, called the *common difference*, such that $a_n - a_{n-1} = d$, that is, $a_n = a_{n-1} + d$ for every $n > 1$ Then $a_n = a_1 + (n-1)d$ for every $n > 1$

Sum of an arithmetic series: Let $S_n = \sum_{k=1}^{n} a_n$. Then

$$S_n = \frac{n}{2}[2a_1 + (n-1)d] = \frac{n}{2}(a_1 + a_n)$$

---

1. A. $4 - 2 = 2, 8 - 4 = 4$. Since these are different, (A) cannot be the first three terms of an arithmetic sequence.

    B. $6.5 - 7 = -0.5, 6 - 6.5 = -0.5$. (B) is an arithmetic sequence with $d = -0.5; a_4 = 6 + (-0.5) = 5.5; a_5 = 5.5 + (-0.5) = 5$

    C. $-16 - (-11) = -5, (-21) - (-16) = 5$. (C) is an arithmetic sequence with $d = -5; a_4 = -21 + (-5) = -26; a_5 = -26 + (-5) = -31$

    D. $\frac{1}{6} - \frac{1}{2} = -\frac{1}{3}, \frac{1}{18} - \frac{1}{6} = -\frac{1}{9}$.

    Since these are different, (D) cannot be the first three terms of an arithmetic sequence.

3. $\begin{aligned} a_2 &= a_1 + d = -5 + 4 = -1 \\ a_3 &= a_2 + d = -1 + 4 = 3 \\ a_4 &= a_3 + d = -3 + 4 = 7 \\ a_n &= a_1 + (n-1)d \end{aligned}$

5. $\begin{aligned} a_{15} &= a_1 + 14d = -3 + 14 \cdot 5 = 67 \\ S_n &= \frac{n}{2}[2a_1 + (n-1)d] \\ S_{11} &= \frac{11}{2}[2(-3) + (11-1)5] \\ &= \frac{11}{2}(44) \\ &= 242 \end{aligned}$

7. $\begin{aligned} a_2 - a_1 &= 5 - 1 = d = 4 \\ S_n &= \frac{n}{2}[2a_1 + (n-1)d] \\ S_{21} &= \frac{21}{2}[2 \cdot 1 + (21-1)4] \\ &= \frac{21}{2}(82) \\ &= 861 \end{aligned}$

9. $\begin{aligned} a_2 - a_1 &= 5 - 7 = -2 = d \\ a_n &= a_1 + (n-1)d \\ a_{15} &= 7 + (15-1)(-2) \\ &= -21 \end{aligned}$

11. $\begin{aligned} a_n &= a_1 + (n-1)d \\ a_{20} &= a_1 + 19d \\ 117 &= 3 + 19d \end{aligned}$

So, $d = 6.$ Therefore

$$
\begin{aligned}
a_{101} &= a_1 + (100)d \\
&= 3 + 100(6) \\
&= 603
\end{aligned}
$$

13. $$
\begin{aligned}
S_n &= \frac{n}{2}(a_1 + a_n) \\
S_{40} &= \frac{40}{2}(-12 + 22) \\
&= 200
\end{aligned}
$$

15. $$
\begin{aligned}
a_2 - a_1 &= d \\
\tfrac{1}{2} - \tfrac{1}{3} &= \tfrac{1}{6} = d \\
a_n &= a_1 + (n-1)d \\
a_{11} &= \tfrac{1}{3} + (11-1)\tfrac{1}{6} \\
&= 2 \\
S_n &= \tfrac{n}{2}(a_1 + a_n) \\
S_{11} &= \tfrac{11}{2}(\tfrac{1}{3} + 2) \\
&= \tfrac{77}{6}
\end{aligned}
$$

17. $$
\begin{aligned}
a_n &= a_1 + (n-1)d \\
a_{10} &= a_1 + 9d \\
a_3 &= a_1 + 2d
\end{aligned}
$$

Eliminating d between these two statements by addition, we have

$$
\begin{aligned}
2a_{10} &= 2a_1 &+ &&18d \\
-9a_3 &= -9a_1 &- &&18d \\
\hline
2a_{10} & &-9a_3 &= &-7a_1
\end{aligned}
$$

$$
\begin{aligned}
a_1 &= \frac{2a_{10}-9a_2}{-7} = \frac{2(55)-9(13)}{-7} \\
&= 1
\end{aligned}
$$

19. $$
\begin{aligned}
a_n &= a_n + (n-1)d \\
d &= a_2 - a_1 \\
&= (3 \cdot 2 + 3) - (3 \cdot 1 + 3) = 3 \\
a_{51} &= a_1 + (51-1)d \\
&= (3 \cdot 1 + 3) + 50 \cdot 3 \\
&= 156 \\
S_n &= \tfrac{n}{2}(a_1 + a_n) \\
S_{51} &= \tfrac{51}{2}(3 \cdot 1 + 3 + 156) \\
&= \tfrac{51}{2} \cdot 162 \\
&= 4,131
\end{aligned}
$$

21. $$
\begin{aligned}
g(t) &= 5 - t \\
g(1) &= 5 - 1 = 4 \\
g(51) &= 5 - 51 = -46
\end{aligned}
$$

$$
g(1) + g(2) + g(3) + \cdots + g(51) = S_{51}
$$

$$
\begin{aligned}
S_n &= \tfrac{n}{2}(a_1 + a_n) \\
S_{51} &= \tfrac{51}{2}(g(1) + g(51)) \\
&= \tfrac{51}{2}[4 + (-46)] \\
&= -1,071
\end{aligned}
$$

23.   First, find n:

$$a_n = a_1 + (n-1)d$$
$$134 = 22 + (n-1)2$$
$$n = 57$$

Now, find S_{57}

$$S_n = \frac{n}{2}(a_1 + a_n)$$
$$S_{57} = \frac{57}{2}(22 + 134)$$
$$= 4,446$$

25.   To prove: $1 + 3 + 5 + \cdots + (2n-1) = n^2$

The sequence $1, 3, 5, \cdots$ is an arithmetic

sequence, with $d = 2$. We are to find S_n.

But, $S_n = \frac{n}{2}(a_1 + a_n)$
$$= \frac{n}{2}[1 + (2n-1)]$$
$$= \frac{n}{2} \cdot 2n$$

So, $S_n = n^2$

27.   Note that $a_n - a_{n-1} = 3$.

Hence, this is an arithmetic

sequence, with $d = 3$. We are

to find a_n.

But, $a_n = a_1 + (n-1)d$

So, $a_n = -3 + (n-1)3$

or $3n - 6$

29.   Prove: $a_n = a_1 + (n-1)d, n \in N$

Write: $P_n : a_n = a_1 + (n-1)d, S = \{n \in N \mid P_n \text{ is true }\}$

To show $S = N$, we show:

Part 1: $1 \in S$

$$P_1 : a_1 = a_1 + (1-1)d$$
$$a_1 = a_1 \text{ is true}$$

Part 2: S is inductive

$$P_k : a_k = a_1 + (k-1)d$$
$$P_{k+1} : a_{k+1} = a_1 + kd$$

We start with P_k :

$$a_k = a_1 + (k-1)d$$

Adding d to both sides:

$$a_k + d = a_1 + (k-1)d + d$$
$$a_{k+1} = a_1 + d[k-1+1]$$
$$a_{k+1} = a_1 + kd$$

Thus, $P_k \Rightarrow P_{k+1}, k \in S \Rightarrow k+1 \in S$ and S is inductive.

Conclusion: $S = N$

31.   Assume x, y, z are consecutive terms of an arithmetic progression.

Then, $y - x = d, z - y = d$

To show $a = x^2 + xy + y^2, b = z^2 + xz + x^2, c = y^2 + yz + z^2$, are

consecutive terms of an arithmetic progression, we need only show that

$$b - a = c - b.$$
$$b - a = z^2 + xz + x^2 - (x^2 + xy + y^2)$$
$$= z^2 + xz - xy - y^2$$
$$= z^2 + x(z - y) - y^2$$
$$= z^2 - y^2 + x(z - y)$$
$$= (z - y)(z + y + x)$$
$$= d(x + y + z)$$
$$c - b = y^2 + yz + z^2 - (z^2 + xz + x^2)$$
$$= y^2 + yz - xz - x^2$$
$$= y^2 - x^2 + yz - xz$$
$$= y^2 - x^2 + z(y - x)$$
$$= (y - x)[y + x + z]$$
$$= d(x + y + z)$$

Hence, a, b, c, that is, $x^2 + xy + y^2, z^2 + xz + x^2, y^2 + yz + z^2$ are consecutive terms of an arithmetic progression.

33. Given a, b, c, d, e, f is an arithmetic progression, let D be the common difference. Then, assuming $D \neq 0$:

$$b = a + D, c = a + 2D, d = a + 3D, e = a + 4D, f = a + 5D$$

We know from Cramer's Rule that there will be a unique solution if

$$\begin{vmatrix} a & b \\ d & e \end{vmatrix} \neq 0$$

$$\begin{vmatrix} a & b \\ d & e \end{vmatrix} = \begin{vmatrix} a & a + D \\ a + 3D & a + 4D \end{vmatrix} \begin{aligned} &= a(a + 4D) - (a + 3D)(a + D) \\ &= a^2 + 4aD - (a^2 + 4aD + 3D^2) \\ &= -3D^2 \neq 0 \end{aligned}$$

We can now calculate x and y, since

$$x = \frac{\begin{vmatrix} c & b \\ f & e \end{vmatrix}}{-3D^2} = \frac{ce - fb}{-3D^2} = \frac{(a+2D)(a+4D) - (a+5D)(a+D)}{-3D^2}$$

$$= \frac{a^2 + 6aD + 8D^2 - (a^2 + 6aD + 5D^2)}{-3D^2}$$

$$= \frac{3D^2}{-3D^2}$$

$$= -1$$

$$y = \frac{\begin{vmatrix} a & c \\ d & f \end{vmatrix}}{-3D^2} = \frac{af - dc}{-3D^2} = \frac{a(a+5D) - (a+3D)(a+2D)}{-3D^2}$$

$$= \frac{a^2 + 5aD - (a^2 + 5aD + 6D^2)}{-3D^2}$$

$$= \frac{-6D^2}{-3D^2}$$

$$= 2$$

35. With each firm, the salaries form an arithmetic sequence. We are asked for the sum of fifteen terms, or S_{15}, given a_1 and d.

$$S_n = \frac{n}{2}[2a_1 + (n - 1)d]$$
$$S_{15} = \frac{15}{2}[2a_1 + 14d] = 15(a_1 + 7d)$$

Firm $A: a_1 = 25,000 \quad d = 1,200 \quad S_{15} = 15(25,000 + 7 \cdot 1,200) = \$501,000$

College Algebra

Firm $B : a_1 = 28,000 \quad d = 800 \qquad S_{15} = 15(28,000 + 7 \cdot 800) = \$504,000$

37. We have an arithmetic sequence, $16, 48, 80, \cdots$, with $a_1 = 16, d = 32$.

A. This requires a_{11} :
$$
\begin{aligned}
a_n &= a_1 + (n-1)d \\
a_{11} &= 16 + (11-1)32 \\
&= 336 \text{ feet}
\end{aligned}
$$

B. This requires s_{11} :
$$
\begin{aligned}
s_n &= \tfrac{n}{2}(a_1 + a_n) \\
s_{11} &= \tfrac{11}{2}(16 + 336) \\
&= 1,936 \text{ feet}
\end{aligned}
$$

C. This requires s_t :
$$
\begin{aligned}
s_n &= \tfrac{n}{2}[2a_1 + (n-1)d] \\
s_t &= \tfrac{t}{2}[2a_1 + (t-1)32] \\
&= \tfrac{t}{2}[32 + (t-1)32] \\
&= \tfrac{t}{2} \cdot 32t \\
&= 16t^2 \text{ feet}
\end{aligned}
$$

39.

   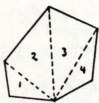

From the figure, and ordinary induction, it should be clear that the interior angles of an $n + 2$-sided polygon, $n = 1, 2, 3, \cdots$, are those of n triangles that is, $180°n$. Thus, the sequence of interior angles $\{a_n\}$,

$a_n - a_{n-1} = 180°n - 180°(n-1) = 180° = d.$

A detailed proof by mathematical induction is omitted.

For a 21-sided polygon, we use $a_n = a_1 + (n-1)d$ with $a_1 = 180°$, $d = 180°$, and $n + 2 = 21$, hence, $n = 19$.

$$
\begin{aligned}
a_{19} &= 180° + (19-1)180° \\
&= 3,420°
\end{aligned}
$$

Exercise 9-4

Key Ideas and Formulas

A sequence $a_1, a_2, a_3, \cdots, a_n$ is called a geometric sequence (or geometric progression) if there exists a constant r, called the common ratio, such that

$$\frac{a_n}{a_{n-1}} = r, \text{ that is } a_n = ra_{n-1}$$

Then $a_n = a_1 r^{n-1}$ for every $n > 1$

Sum of a geometric series: Let $S_n = \sum_{k=1}^n a_k$, then

$$S_n = \frac{a_1 - a_1 r^n}{1 - r} = \frac{a_1 - ra_n}{1-r} r \neq 1$$

Sum of an infinite geometric series

$S_\infty = \frac{a_1}{1-r} \mid r \mid < 1 only$. If $r > 1$ or $r < -1$ the infinite geometric series has no sum.

1.　A. $\frac{-4}{2} = -2, \frac{8}{-4} = 2$　Since these are equal, (A) is a geometric sequence with $r = -2, a_4 = ra_3 = (-2)8 = -16, a_5 = ra_4 = (-2)(-16) = 32$.

B. $\frac{6.5}{7} = \frac{13}{14}, \frac{6}{6.5} = \frac{12}{13}$.　Since these are different, (B) cannot be the first three terms of a geometric sequence.

C. $\frac{-16}{-11} = \frac{16}{11}, \frac{-21}{-16} = \frac{21}{16}$　Since these are different, (C) cannot be the first three terms of a geometric sequence.

D. $\frac{1}{6} \div \frac{1}{2} = \frac{1}{3}, \frac{1}{18} \div \frac{1}{6} = \frac{1}{3}$　Since these are equal, (D) is a geometric sequence with $r = \frac{1}{3}$. $a_4 = ra_3 = \frac{1}{3} \cdot \frac{1}{18} = \frac{1}{54}, a_5 = ra_4 = \frac{1}{3} \cdot \frac{1}{54} = \frac{1}{162}$

3.　$a_2 = a_1 r = (-6)(-\frac{1}{2}) = 3$　7.　$S_n = \frac{a_1 - ra_n}{1-r}$

$a_3 = a_2 r = 3(-\frac{1}{2}) = -\frac{3}{2}$　　$S_7 = \frac{3 - 3(2,187)}{1-3}$

$a_4 = a_3 r = (-\frac{3}{2})(-\frac{1}{2}) = \frac{3}{4}$　　$= 3,279$

5.　$a_n = a_1 r^{n-1}$

$a_{10} = 81(\frac{1}{3})^{10-1}$

$= \frac{1}{243}$

9. $a_n = a_1 r^{n-1}$

$1 = 100 r^{6-1}$

$\frac{1}{100} = r^5$

$r = \sqrt[5]{0.01} = 10^{-2/5}$

$r = 0.398$

11. $S_n = \frac{a_1 - a_1 r^n}{1-r}$

$S_{10} = \frac{5 - 5(-2)^{10}}{1 - (-2)}$

$= \frac{-5,115}{3}$

$= -1,705$

13. First, find r:

$a_n = a_1 r^{n-1}$

$\frac{8}{3} = 9 r^{4-1}$

$\frac{8}{27} = r^3$

$r = \frac{2}{3}$

$a_2 = r a_1 = \frac{2}{3}(9) = 6$

$a_3 = r a_2 = \frac{2}{3}(6) = 4$

15. $S_n = \frac{a_1 - a_1 r^n}{1-r}$

First, note $a_1 = (-3)^{1-1} = 1$

$r = \frac{a_2}{a_1} = \frac{(-3)^{2-1}}{(-3)^{1-1}} = -3$

$S_7 = \frac{1 - 1(-3)^7}{1 - (-3)}$

$= \frac{2,188}{4}$

$= 547$

17. $g(1) + g(2) + \cdots + g(10)$ is a geometric series, with

$g(1) = a_1 = \left(\frac{1}{2}\right)^1 = \frac{1}{2}$

$r = \frac{g(2)}{g(1)} = \frac{\left(\frac{1}{2}\right)^2}{\left(\frac{1}{2}\right)^1} = \frac{1}{2}$

$S_n = \frac{a_1 - a_1 r^n}{1-r}$

$S_{10} = \frac{\frac{1}{2} - \frac{1}{2}\left(\frac{1}{2}\right)^{10}}{1 - \frac{1}{2}}$

$= \frac{1 - \left(\frac{1}{2}\right)^{10}}{2^{-1}}$

$= \frac{1,023}{1,024}$

19. $\frac{a_2}{a_1} = \frac{a_3}{a_2} = r$ for $a_1 + a_2 + a_3$ to be a geometric series. Hence,

$\frac{x}{-2} = \frac{-6}{x}$

$x^2 = 12$

$x = 2\sqrt{3}$,

since x is specified positive.

23. $\frac{a_2}{a_1} = \frac{4}{2} = 2 = r \geq 1$

Therefore, this infinite geometric series has no sum.

25. $\frac{a_2}{a_1} = \left(-\frac{1}{2}\right) \div 2 = -\frac{1}{4} = r. \ |r| < 1.$

Therefore, this infinite geometric series has a sum.

$S_\infty = \frac{a_1}{1-r}$

$= \frac{2}{1 - \left(-\frac{1}{4}\right)}$

$= \frac{8}{5}$

27. $0.\bar{7} = 0.777\cdots = 0.7 + 0.07 + 0.007 + 0.0007 + \cdots$

This is an infinite geometric series with $a_1 = 0.7$ and $r = 0.1$.

Thus,

$S_\infty = \frac{a_1}{1-r} = \frac{0.7}{1 - 0.1} = \frac{0.7}{0.9} = \frac{7}{9}$

29. $0.\overline{54}\cdots = 0.54 + 0.0054 + 0.000054 + \cdots$

This is an infinte geometric series with $a_1 = 0.54$ and $r = 0.01.$. Thus

$S_\infty = \frac{a_1}{1-r} = \frac{0.54}{1 - 0.01} = \frac{0.054}{0.99} = \frac{6}{11}$

31. $3.\overline{216} = 3.216216216\cdots = 3 + 0.216 + 0.000216 + 0.000000216 + \cdot.$

Therefore, we note: $0.216 + 0.000216 + 0.000000216 + \cdots$ is an infinite geometric series with $a_1 = 0.216$ and $r = 0.001$. Thus,

$3.\overline{216} = 3 + S_\infty = 3 + \frac{a_1}{1-r} = 3 + \frac{0.216}{1 - 0.001} = 3 + \frac{0.216}{0.999} = 3\frac{8}{37}$ or $\frac{119}{37}$

33. This is a geometric sequence with ratio $\frac{a_n}{a_{n-1}} = -3$.

Hence, $a_n = a_1 r^{n-1}$

$= (-2)(-3)^{n-1}$

35. Prove: $a_n = a_1 r^{n-1}$, given $a_n = r a_{n-1}$ for every $n > 1$

Write: $P_n : a_n = a_1 r^{n-1}, n > 1, S = \{n \in N \mid P_n \text{ is true}\}$

To show $S = N$, we show:

Part 1: $1 \in S$

$$P_1 : a_1 = a_1 r^{1-1}$$
$$= a_1 r^0$$
$$= a_1 \quad \text{is true}$$

Part 2: S is inductive

$$P_k : a_k = a_1 r^{k-1}$$
$$P_{k+1} : a_{k+1} = a_1 r^k$$

We start with P_k :

$$a_k = a_1 r^{k-1}$$

Multiply both sides by r :

$$a_k r = a_1 r^{k-1} r$$
$$a_k r = a_1 r^k$$

But, we are given that $a_{k+1} = r a_k = a_k r$.

Hence, $a_{k+1} = a_1 r^k$

Thus, $P_k \Rightarrow P_{k+1}, k \in S \Rightarrow k+1 \in S$ and S is inductive.

Conclusion: $S = N, a_n = a_1 r^{n-1}, n \in N$

37. After one year, $P(1+r)$ is present. Hence, the geometric sequence has $a_1 = P(1+r)$. The ratio is given as $(1+r)$, hence

$$a_n = a_1 (1+r)^{n-1}$$
$$= P(1+r)(1+r)^{n-1}$$
$$= P(1+r)^n \quad \text{is the amount present after } n \text{ years.}$$

The time taken for P to double is represented by n years. We set $A = 2P, r = 0.06$, then solve

$$2P = P(1+0.06)^n$$
$$2 = (1.06)^n$$
$$\log 2 = n \log 1.06$$
$$n = \frac{\log 2}{\log 1.06}$$
$$\approx 12 \text{years}$$

39. We are asked for the sum of an infinite geometric series.

$a_1 =$ number of revolutions in the first minute $= 300$

$r = \frac{2}{3} \cdot \mid r \mid < 1$, so the series has a sum,

$S_\infty = \frac{a_1}{1-r} = \frac{300}{1-\frac{2}{3}} = 900$ revolutions.

41. We are asked for the sum of an infinite geometric series.

$$a = \$800,000$$
$$r = 0.8 \cdot \mid r \mid \le 1, \text{ so the series has a sum,}$$
$$S_\infty = \frac{a_1}{1-r}$$
$$= \frac{\$800,000}{1-0.8}$$
$$= \frac{\$800,000}{0.2}$$
$$= \$4,000,000$$

Exercise 9-5

Key Ideas and Formulas

	Arithmetic Sequence $a_n = a_{n-1} + d$	Geometric Sequence $a_n = a_{n-1}r$	
Definition:	$a_n = a_{n-1} + d$	$a_n = a_{n-1}r$	
nth term:	$a_n = a_1 + (n-1)d$	$a_n = a_1 r^{n-1}$	
Sum of Series:	$S_n = \frac{n}{2}[2a_1 + (n-1)d]$ $= \frac{n}{2}(a_1 + a_n)$	$S_n = \frac{a_1 - a_1 r^n}{1-r} r \neq 1$ $= \frac{a_1 - ra_n}{1-r} r \neq 1$	
		$S_\infty = \frac{a_1}{1-r} \mid r \mid < 1$	Sum of an infinite geometric series

1.   This involves an arithmetic sequence. Let d = increase in earnings each year.

$$a_1 = 7,000$$
$$a_{11} = 14,000$$
$$a_n = a_1 + (n-1)d$$
$$14,000 = 7,000 + (11-1)d$$
$$d = \$700$$

The amount of money received over the 11 years is

$$S_{11} = a_1 + a_2 + \cdots + a_{11}$$
$$S_n = \frac{n}{2}(a_1 + a_n)$$
$$S_{11} = \frac{11}{2}(7,000 + 14,000) = \$115,500$$

3.   This involves a geometric sequence. Let $a_1 = 15$ = pressure at sea level, $r = \frac{1}{10}$ = factor of decrease. We require a_4 = pressure after four 10-mile increases in altitude.

$$a_4 = 15(\tfrac{1}{10})^{4-1}$$
$$= 0.015 \quad \text{pounds per square inch.}$$

5.   This involves a geometric sequence. Let $a_n = 2,000$ calories. There are five stages: $n = 5$. We require a_1 on the assumption that $r = 20\% = \frac{1}{5}$.

$$a_n = a_1 r^{n-1}$$
$$2000 = a_1(\tfrac{1}{5})^{5-1}$$
$$a_1 = 2,000 \cdot 5^4$$
$$= 1,250,000 \quad \text{calories}$$

7.   This involves a geometric sequence. Let $a_1 = 2A_0$ = number present after 1 half-hour period. In t hours, $2t$ half-hours will have elapsed, hence, $n = 2t; r = 2$, since the number of bacteria doubles in each period.

$$a_n = a_1 r^{n-1}$$
$$a_{2t} = 2A_0 2^{2t-1}$$
$$= A_0 2^1 2^{2t-1}$$
$$= A_0 2^{2t}$$

9.   If b_n is the brightness of an $n-th$ magnitude star, we find r for the geometric progression, $b_1, b_2, b_3, \cdots$, given $b_1 = 100b_6$.

$$b_n = b_1 r^{n-1}$$
$$b_6 = b_1 r^{6-1}$$
$$b_6 = b_1 r^5$$
$$b_6 = 100 b_6 r^5$$
$$\frac{1}{100} = r^5$$
$$r = \sqrt[5]{0.01} = 10^{-0.4} = 0.398$$

11. This involves an infinite geometric series. a_1 = perimeter of first
triangle = 1, $r = \frac{1}{2}, |\, r\, | < 1$, so the series has a sum.
$$S_\infty = \frac{a_1}{1-r} = \frac{1}{1-\frac{1}{2}} = 2$$

13. This invovles a geometric sequence. a_1 = amount of money on first square;
$a_1 = \$0.01, r = 2; a_n$ = amount of money on n-th square.
$$a_n = a_1 r^{n-1}$$
$$a_{64} = 0.01 \cdot 2^{64-1}$$
$$= 9.22 \times 10^{16} \quad \text{dollars}$$

The amount of money on the whole board $= a_1 + a_2 + a_3 + \cdots + a_{64} = S_{64}$.
$$S_n = \frac{a_1 - r a_n}{1-r}$$
$$S_{64} = \frac{.01 - 2(9.223 \times 10^{16})}{1-2}$$
$$= 1.845 \times 10^{17} \quad \text{dollars}$$

Exercise 9-6

Key Ideas and Formulas

$$\binom{n}{r} = \frac{n!}{r!(n-r)!} = \frac{n(n-1)(n-2)\cdots(n-r+1)}{r(r-1)\cdots2\cdot1}$$

For $n \in N$, $n! = n(n-1)\cdots2\cdot1$
$$1! = 0! = 1$$
$$n! = n\cdot(n-1)!$$

Binomial Formula
$$(a+b)^n = \sum_{k=0}^{n}\binom{n}{k}a^{n-k}b^k \quad n \in N, n \geq 1$$

$$(a+b)^0 = 1$$
$$(a+b)^1 = a+b$$
$$(a+b)^2 = a^2 + 2ab + b^2$$
$$(a+b)^3 = a^3 + 3a^2b + 3ab^2 + b^3$$
$$(a+b)^4 = a^4 + 4a^3b + 6a^2b^2 + 4ab^3 + b^4$$
$$(a+b)^5 = a^5 + 5a^4b + 10a^3b^2 + 10a^2b^3 + 5ab^4 + b^5$$

---

1. $6! = 6\cdot5\cdot4\cdot3\cdot2\cdot1 = 720$

3. $\frac{20!}{19!} = \frac{20\cdot19!}{19!} = 20$

5. $\frac{10!}{7!} = \frac{10\cdot9!}{7!} = \frac{10\cdot9\cdot8!}{7!} = \frac{10\cdot9\cdot8\cdot7!}{7!} = 10\cdot9\cdot8 = 720$

7. $\frac{6!}{4!2!} = \frac{6\cdot5\cdot4\cdot3\cdot2\cdot1}{4\cdot3\cdot2\cdot1\cdot2\cdot1} = \frac{6\cdot5}{2\cdot1} = 15$

9. $\frac{9!}{0!(9-0)!} = \frac{9!}{1(9!)} = 1$

11. $\frac{8!}{2!(8-2)!} = \frac{8!}{2!6!} = \frac{8\cdot7\cdot6\cdot5\cdot4\cdot3\cdot2\cdot1}{2\cdot1\cdot6\cdot5\cdot4\cdot3\cdot2\cdot1}$
$$= \frac{8\cdot7}{2\cdot1} = 28$$

13.   Since $n! = n(n-1)!$, $\frac{n!}{(n-1)!} = n$. Hence, $9 = \frac{9!}{(9-1)!} = \frac{9!}{8!}$.

15. $6\cdot7\cdot8 = \frac{1\cdot2\cdot3\cdot4\cdot5\cdot6\cdot7\cdot8}{1\cdot2\cdot3\cdot4\cdot5} = \frac{8!}{5!}$

17. $\binom{9}{5} = \frac{9!}{5!(9-5)!} = \frac{9!}{5!4!} = \frac{9\cdot8\cdot7\cdot6\cdot5!}{5!4\cdot3\cdot2\cdot1} = \frac{9\cdot8\cdot7\cdot6}{4\cdot3\cdot2\cdot1} = 126$

19. $\binom{6}{5} = \frac{6!}{5!(6-5)!} = \frac{6!}{5!1!} = \frac{6\cdot5!}{5!\cdot1} = 6$

21. $\binom{9}{9} = \frac{9!}{9!(9-9)!} = \frac{9!}{9!0!} = \frac{1}{0!} = \frac{1}{1} = 1$

23. $\binom{17}{13} = \frac{17!}{13!(17-13)!} = \frac{17!}{13!4!} = \frac{17\cdot16\cdot15\cdot14\cdot13!}{13!4\cdot3\cdot2\cdot1} = \frac{17\cdot16\cdot15\cdot14}{4\cdot3\cdot2\cdot1} = 2,380$

25. $(m+n)^3 = \sum_{k=0}^{3}\binom{3}{k}m^{3-k}n^k$
$$= \binom{3}{0}m^3 + \binom{3}{1}m^2n + \binom{3}{2}mn^2 + \binom{3}{3}n^3$$
$$= m^3 + 3m^2 + 3mn^2 + n^3$$

27. $(2x-3y)^3 = [2x+(-3y)]^3 = \sum_{k=0}^{3}\binom{3}{k}(2x)^{3-k}(-3y)^k$
$$= \binom{3}{0}(2x)^3 + \binom{3}{1}(2x)^2(-3y)^1 + \binom{3}{2}(2x)(-3y)^2 + \binom{3}{3}(-3y)^3$$
$$= 8x^3 + 3(4x^2)(-3y) + 3(2x)(9y^2) + (-27y^3)$$
$$= 8x^3 - 36x^2y + 54xy^2 - 27y^3$$

29. $(x-2)^4 = [x+(-2)]^4 = \sum_{k=0}^{4}\binom{4}{k}x^{4-k}(-2)^k$
$= \binom{4}{0}x^4 + \binom{4}{1}x^3(-2)^1 + \binom{4}{2}x^2(-2)^2 + \binom{4}{3}x(-2)^3 + \binom{4}{4}(-2)^4$
$= x^4 - 8x^3 + 24x^2 - 32x + 16$

31. $(m+3n)^4 = \sum_{k=0}^{4}m^{4-k}(3n)^k$
$= \binom{4}{0}m^4 + \binom{4}{1}m^3(3n)^1 + \binom{4}{2}m^2(3n)^2 + \binom{4}{3}m(3n)^3 + \binom{4}{4}(3n)^4$
$= m^4 + 12m^3 + 54m^2n^2 + 108mn^3 + 81n^4$

33. $(2x-y)^5 = [2x+(-y)]^5 = \sum_{k=0}^{5}\binom{5}{k}(2x)^{5-k}(-y)^k$
$= \binom{5}{0}(2x)^5 + \binom{5}{1}(2x)^4(-y)^1 + \binom{5}{2}(2x)^3(-y)^2$
$\quad + \binom{5}{3}(2x)^2(-y)^3 + \binom{5}{4}(2x)(-y)^4 + \binom{5}{5}(-y)^5$
$= 32x^5 - 80x^4y + 80x^3y^2 - 40x^2y^3 + 10xy^4 - y^5$

35. $(m+2n)^6 = \sum_{k=0}^{6}m^{6-k}(2n)^k$
$= \binom{6}{0}m^6 + \binom{6}{1}m^5(2n)^1 + \binom{6}{2}m^4(2n)^2 + \binom{6}{3}m^3(2n)^3 + \binom{6}{4}m^2(2n)^4$
$\quad + \binom{6}{5}m(2n)^5 + \binom{6}{6}(2n)^6$
$= m^6 + 12m^5n + 60m^4n^2 + 160m^3n^3 + 240m^2n^4 + 192mn^5 + 64n^6$

37. In the expansion of $(a+b)^n$, the exponent of b in the r-th term is $r-1$ and the exponent of a is $n-(r-1)$. Here, $r=7, n=15$.
Seventh term $= \binom{15}{6}u^9v^6$
$= \frac{15!}{9!6!}u^6v^6$
$= \frac{15\cdot14\cdot13\cdot12\cdot11\cdot10}{6\cdot5\cdot4\cdot3\cdot2\cdot1}u^9v^6$
$= 5,005u^9v^6$

39. In the expansion of $(a+b)^n$, the exponent of b in the r-th term is $r-1$ and the exponent of a is $n-(r-1)$. Here, $r=11, n=12$.
Eleventh term $= \binom{12}{10}(2m)^2n^{10}$
$= \frac{12!}{10!2!}4m^2n^{10}$
$= \frac{12\cdot11}{2\cdot1}4m^2n^{10}$
$= 264m^2n^{10}$

41. In the expansion of $(a+b)^n$, the exponent of b in the r-th term is $r-1$ and the exponent of a is $n-(r-1)$. Here, $r=7, n=12$.
Seventh term $= \binom{12}{6}(\frac{w}{2})^6(-2)^6$
$= \frac{12!}{6!6!}\frac{w^6}{2^6}2^6$
$= \frac{12\cdot11\cdot10\cdot9\cdot8\cdot7\cdot6!}{6\cdot5\cdot4\cdot3\cdot2\cdot1\cdot6!}w^6$
$= 924w^6$

43. In the expansion of $(a+b)^n$, the exponent of b in the r-th term is $r-1$ and the exponent of a is $n-(r-1)$. Here, $r=6, n=8$.

$$\text{Sixth term} = \binom{8}{5}(3x)^3(-2y)^5$$

$$= \frac{8!}{5!3!}(3x)^3(-2y)^5$$

$$= \frac{8\cdot7\cdot6\cdot5!}{5!\cdot3\cdot2\cdot1}(27x^3)(-32y^5)$$

$$= -(56)(27)(32)x^3y^5$$

$$= -48,384x^3y^5$$

45. $(1.01)^{10} = (1+0.01)^{10} = \sum_{k=0}^{10}\binom{10}{k}1^{10-k}(0.01)^k, 1^{10-k} = 1$

$$= \sum_{k=0}^{10}\binom{10}{k}(0.01)^k$$

$$= \binom{10}{0} + \binom{10}{1}(0.01)^1 + \binom{10}{2}(0.01)^2 + \binom{10}{3}(0.01)^3$$

$$+\binom{10}{4}(0.01)^4 + \binom{10}{5}(0.01)^5 + \binom{10}{6}(0.01)^6$$

$$+\binom{10}{7}(0.01)^7 + \binom{10}{8}(0.01)^8 + \binom{10}{9}(0.01)^9$$

$$+\binom{10}{10}(0.01)^{10}$$

$$= 1 + 0.1 + 0.0045 + 0.00012 + (0.0000021 + \text{other terms with}$$
$$\text{no effect in fourth decimal place})$$

$$= 1.1046$$

47. $\binom{n}{r} = \frac{n!}{r!(n-r)!} = \frac{n!}{(n-r)!r!} = \frac{n!}{(n-r)![n-(n-r)]!} = \binom{n}{n-r}$

49. $\binom{k}{r-1} + \binom{k}{r} = \frac{k!}{(r-1)!(k-r+1)!} + \frac{k!}{r!(k-r)!}$

$$= \frac{rk!+(k-r+1)k!}{r!(k-r+1)!} = \frac{(r+k-r+1)k!}{r!(k-r+1)!}$$

$$= \frac{(k+1)k!}{r!(k-r+1)!} = \frac{(k+1)!}{r!(k-r+1)!}$$

$$= \binom{k+1}{r}$$

51. $\binom{k}{k} = \frac{k!}{k!(k-k)!} = 1 = \frac{(k+1)!}{(k+1)![(k+1)-(k+1)]!} = \binom{k+1}{k+1}$

53. $2^n = (1+1)^n = \sum_{k=0}^{n}\binom{n}{k}1^{n-k}1^k, 1^{n-k} = 1^k = 1$

$$= \sum_{k=0}^{n}\binom{n}{k}$$

$$= \binom{n}{0} + \binom{n}{1} + \binom{n}{2} + \cdots + \binom{n}{n}$$

Exercise 9-7

CHAPTER REVIEW

1. A. Since $\frac{-8}{16} = \frac{4}{-8} = -\frac{1}{2}$, this could start a geometric sequence.

B. Since $7 - 5 = 9 - 7 = 2$, this could start an arithmetic sequence.

C. Since $-5 - (-8) = -2 - (-5) = 3$, this could start an arithmetic sequence.

D. Since $\frac{3}{2} \neq \frac{5}{3}$ and $3 - 2 \neq 5 - 3$, this could start neither an arithmetic nor a geometric sequence.

E. Since $\frac{2}{-1} = \frac{-4}{2} = -2$, this could start a geometric sequence.

$(9 - 1, 9 - 3, 9 - 4)$

2. $a_n = 2n + 3$

A.
$$\begin{aligned}
a_1 &= 2 \cdot 1 + 3 = 5 \\
a_2 &= 2 \cdot 2 + 3 = 7 \\
a_3 &= 2 \cdot 3 + 3 = 9 \\
a_4 &= 2 \cdot 4 + 3 = 11
\end{aligned}$$

B. This is an arithmetic sequence with $d = 2$. Hence,
$$\begin{aligned}
a_n &= a_1 + (n - 1)d \\
a_{10} &= 5 + (10 - 1)d \\
&= 23
\end{aligned}$$

C.
$$\begin{aligned}
s_n &= \frac{n}{2}(a_1 + a_n) \\
s_{10} &= \frac{10}{2}(5 + 23) \\
&= 140
\end{aligned}$$

$(9 - 1, 9 - 3)$

3. $a_n = 32(\frac{1}{2})^n$

A.
$$\begin{aligned}
a_1 &= 32(\tfrac{1}{2})^1 = 16 \\
a_2 &= 32(\tfrac{1}{2})^2 = 8 \\
a_3 &= 32(\tfrac{1}{2})^3 = 4 \\
a_4 &= 32(\tfrac{1}{2})^4 = 2
\end{aligned}$$

B. This is a geometric sequence with $r = \frac{1}{2}$. Hence,
$$\begin{aligned}
a_n &= a_1 r^{n-1} \\
a_{10} &= 16(\tfrac{1}{2})^{10-1} \\
&= \frac{1}{32}
\end{aligned}$$

C.
$$\begin{aligned}
S_n &= \frac{a_1 - r a_n}{1 - r} \\
S_{10} &= \frac{16 - \frac{1}{2}(\frac{1}{32})}{-\frac{1}{2}} = \frac{16 - \frac{1}{64}}{\frac{1}{2}} = 31\tfrac{31}{32}
\end{aligned}$$

$(9 - 1, 9 - 4)$

4. $a_1 = -8, a_n = a_{n-1} + 3, n \geq 2$

A.
$$\begin{aligned}
a_1 &= -8 \\
a_2 &= a_1 + 3 = -8 + 3 = -5 \\
a_3 &= a_2 + 3 = -5 + 3 = -2 \\
a_4 &= a_3 + 3 = -2 + 3 = 1
\end{aligned}$$

B. This is an arithmetic sequence with $d = 3$. Hence,
$$\begin{aligned}
a_n &= a_1 + (n - 1)d \\
a_{10} &= -8 + (10 - 1)3 \\
&= 19
\end{aligned}$$

C.
$$\begin{aligned}
S_n &= \frac{n}{2}(a_1 + a_n) \\
S_{10} &= \frac{10}{2}(-8 + 19) \\
&= 55
\end{aligned}$$

$(9 - 1, 9 - 3)$

5. $a_1 = -1, a_n = (-2)a_{n-1}, n \geq 2$

A.
$$\begin{aligned}
a_1 &= -1 \\
a_2 &= (-2)a_1 = (-2)(-1) = 2 \\
a_3 &= (-2)a_2 = (-2)2 = -4 \\
a_4 &= (-2)a_3 = (-2)(-4) = 8
\end{aligned}$$

B. This is a geometric sequence with $r = -2$. Hence
$$\begin{aligned}
a_n &= a_1 r^{n-1} \\
a_{10} &= (-1)(-2)^{10-1} \\
&= 512
\end{aligned}$$

C.
$$\begin{aligned}
S_n &= \frac{a_1 - r a_n}{1 - r} \\
S_{10} &= \frac{-1 - (-2)(512)}{1 - (-2)} \\
&= 341
\end{aligned}$$

$(9 - 1, 9 - 4)$

6.   This is a geometric sequence with $a_1 = 16$ and $r = \frac{1}{2} < 1$, so the sum exists:

$$\begin{aligned} S_\infty &= \frac{a_1}{1-r} \\ &= \frac{16}{1-\frac{1}{2}} \\ &= 32 \quad (9-4) \end{aligned}$$

7. $6! = 6 \cdot 5 \cdot 4 \cdot 3 \cdot 2 \cdot 1 = 720$

8. $\frac{22!}{19!} = \frac{22 \cdot 21 \cdot 20 \cdot 19!}{19!} = 9,240$ $(9-6)$

9. $\frac{7!}{2!(7-2)!} = \frac{7!}{2!5!} = \frac{7 \cdot 6 \cdot 5!}{2 \cdot 1 \cdot 5!} = 21$ $(9-6)$

10.
$$\begin{aligned} P_1 &: 5 = 1^2 + 4 \cdot 1 = 5 \\ P_2 &: 5 + 7 = 2^2 + 4 \cdot 2 \\ &\quad\; 12 = 12 \\[4pt] P_3 &: 5 + 7 + 9 = 3^2 + 4 \cdot 3 \\ &\qquad\;\; 21 = 21 \end{aligned}$$
$(9-2)$

11.
$$\begin{aligned} P_1 &: 2 = 2^{1+1} - 2 = 4 - 2 = 2 \\ P_2 &: 2 + 4 = 2^{2+1} - 2 \\ &\quad\;\; 6 = 6 \\[4pt] P_3 &: 2 + 4 + 8 = 2^{3+1} - 2 \\ &\qquad\quad 14 = 14 \quad (9-2) \end{aligned}$$

12. $P_1 : 49^1$ is divisible by 6

$48 = 6 \cdot 8$ true

$P_2 : 49^2 - 1$ is divisible by 6

$2,400 = 6 \cdot 400$ true

$P_3 : 49^3 - 1$ is divisible by 6

$117,648 = 19,608 \cdot 6$ true $(9-2)$

13. $P_k : 5 + 7 + 9 + \cdots + (2k + 3) = k^2 + 4k$

$P_{k+1} : 5 + 7 + 9 + \cdots + (2k + 3) + (2k + 5)$

$= (k+1)^2 + 4(k+1) \quad (9-2)$

14. $P_k : 2 + 4 + 8 + \cdots + 2^k = 2^{k+1} - 2$

$P_{k+1} : 2 + 4 + 8 + \cdots + 2^k + 2^{k+1} = 2^{k+2} - 2 \quad (9-2)$

15. $P_k : 49^k - 1$ is divisible by 6.

$P_{k+1} : 49^{k+1} - 1$ is divisible by 6 $(9-2)$.

16.
$$\begin{aligned} S_{10} &= (2 \cdot 1 - 8) + (2 \cdot 2 - 8) + (2 \cdot 3 - 8) + (2 \cdot 4 - 8) + (2 \cdot 5 - 8) \\ &\quad + (2 \cdot 6 - 8) + (2 \cdot 7 - 8) + (2 \cdot 8 - 8) + (2 \cdot 9 - 8) + (2 \cdot 10 - 8) \\ &= (-6) + (-4) + (-2) + 0 + 2 + 4 + 6 + 8 + 10 + 12 \\ &= 30 \quad (9-3) \end{aligned}$$

17.
$$\begin{aligned} S_7 &= \frac{16}{2^1} + \frac{16}{2^2} + \frac{16}{2^3} + \frac{16}{2^4} + \frac{16}{2^5} + \frac{16}{2^6} + \frac{16}{2^7} \\ &= 8 + 4 + 2 + 1 + \frac{1}{2} + \frac{1}{4} + \frac{1}{8} \\ &= 15\frac{7}{8} \quad (9-4) \end{aligned}$$

18.   This is an infinite geometric sequence with $a_1 = 27$.

$$r = \frac{18}{27} = -\frac{2}{3} \quad \left|-\frac{2}{3}\right| < 1, \text{ hence the sum exists}$$

$$\begin{aligned} S_\infty &= \frac{a_1}{1-r} \\ &= \frac{27}{1-(-\frac{2}{3})} \\ &= \frac{81}{5} \quad (9-4) \end{aligned}$$

19. $S_n = \sum_{k=1}^{\infty} \frac{(-1)^{k+1}}{3k}$

This geometric sequence has $a_1 = \frac{1}{3}$.

$r = (-\frac{1}{9}) \div \frac{1}{3} = -\frac{1}{3}$ $\left|-\frac{1}{3}\right| < 1$,

hence the sum exists.

$$\begin{aligned} S_\infty &= \frac{a_1}{1-r} \\ &= \frac{\frac{1}{3}}{1-(-\frac{1}{3})} \\ &= \frac{1}{4} \end{aligned}$$

(9 − 4)

20.   First, find d :

$$\begin{aligned} a_n &= a_1 + (n-1)d \\ 31 &= 13 + (7-1)d \\ 31 &= 13 + 6d \\ d &= 3 \end{aligned}$$

Hence; $\begin{aligned} a_5 &= 13 + (5-1)3 \\ &= 25 \quad (1-3) \end{aligned}$

21. $0.\overline{72} = 0.72 + 0.0072 + 0.000072 + \cdots$

This is an infinite geometric sequence

with $a_1 = 0.72$ and $r = 0.01$.

$$\begin{aligned} 0.\overline{72} = S_\infty &= \frac{a_1}{1-r} \\ &= \frac{0.72}{1-0.01} \\ &= \frac{0.72}{0.99} \\ &= \frac{8}{11} \end{aligned}$$

(9 − 4)

22. $\begin{aligned} \frac{20!}{18!(20-18)!} &= \frac{20!}{18!2!} \\ &= \frac{20 \cdot 19 \cdot 18!}{18!2 \cdot 1} \\ &= 190 \end{aligned}$

(9 − 6)

23. $\begin{aligned} \binom{16}{12} &= \frac{16!}{12!(16-12)!} \\ &= \frac{16!}{12!4!} \\ &= \frac{16 \cdot 15 \cdot 14 \cdot 13 \cdot 12!}{12! \cdot 4 \cdot 3 \cdot 2 \cdot 1} \\ &= 1,820 \end{aligned}$

(9 − 6)

24. $\binom{11}{11} = \frac{11!}{11!(11-11)!} = \frac{11!}{11!0!} = 1$

(9 − 6)

25. $\begin{aligned} (x-y)^5 = [x+(-y)]^5 &= \sum_{k=0}^{5} \binom{5}{k} x^{5-k}(-y)^k \\ &= \binom{5}{0}x^5 + \binom{5}{1}x^4(-y)^1 + \binom{5}{2}x^3(-y)^2 \\ &\quad + \binom{5}{3}x^2(-y)^3 + \binom{5}{4}x(-y)^4 + \binom{5}{5}(-y)^5 \\ &= x^5 - 5x^4y + 10x^3y^2 - 10x^2y^3 + 5xy^4 - y^5 \end{aligned}$

(9 − 6)

26.   In the expansion of $(a+b)^n$, the exponent of b in the r-th term is $r-1$ and the exponent of a is $n-(r-1)$. Here, $r = 10, n = 12$.

$$\begin{aligned} \text{Tenth term} &= \binom{12}{9}(2x)^3(-y)^9 \\ &= \frac{12!}{9!3!}(8x^3)(-y^9) \\ &= \frac{12 \cdot 11 \cdot 10 \cdot 9!}{9! \cdot 3 \cdot 2 \cdot 1}(-8x^3y^9) \\ &= -1760x^3y^9 \end{aligned}$$

(9 − 6)

27.   Prove: $P_n : 5 + 7 + 9 + \cdots + (2n+3) = n^2 + 4n$. Write $S = \{n \in N \mid P_n \text{ is true}\}$

To show $S = N$, we show:

Part 1: $1 \in S$

$$P_1 : 5 = 1^2 + 4 \cdot 1 \text{ is true (Problem 10)}$$

Part 2: S is inductive

$$P_k : 5 + 7 + 9 + \cdots + (2k + 3) = k^2 + 4k$$

$$P_{k+1} : 5 + 7 + 9 + \cdots + (2k + 3) + (2k + 5) = (k + 1)^2 + 4(k + 1)$$

We start with P_k :

$$5 + 7 + 9 + \cdots + (2k + 3) = k^2 + 4k$$

Adding $2k + 5$ to both sides:

$$
\begin{aligned}
5 + 7 + 9 + \cdots + (2k + 3) + (2k + 5) &= k^2 + 4k + 2k + 5 \\
&= k^2 + 6k + 5 \\
&= k^2 + 2k + 1 + 4k + 4 \\
&= (k + 1)^2 + 4(k + 1)
\end{aligned}
$$

Thus, $P_k \Rightarrow P_{k+1}, k \in S \Rightarrow k + 1 \in S$, and S is inductive

Conclusion: $S = N$.

$(9 - 2)$

28.   Prove: $P_n : 2 + 4 + 8 + \cdots + 2^n = 2^{n+1} - 2$. Write $S = \{n \in N \mid P_n \text{ is true}\}$

To show $S = N$, we show:

Part 1: $1 \in S$

$$P_1 : 2 = 2^{1+1} - 2 \text{ is true (Problem 11)}$$

Part 2: S is inductive.

$$P_k : 2 + 4 + 8 + \cdots + 2^k = 2^{k+1} - 2$$

$$P_{k+1} : 2 + 4 + 8 + \cdots + 2^k + 2^{k+1} = 2^{k+2} - 2$$

We start with P_k :

$$2 + 4 + 8 + \cdots + 2^k = 2^{k+1} - 2$$

Adding 2^{k+1} to both sides:

$$
\begin{aligned}
2 + 4 + 8 + \cdots + 2^k + 2^{k+1} &= 2^{k+1} - 2 + 2^{k+1} \\
&= 2^{k+1} + 2^{k+1} - 2 \\
&= 2 \cdot 2^{k+1} - 2 \\
&= 2^{k+2} - 2
\end{aligned}
$$

Thus, $P_k \Rightarrow P_{k+1}, k \in S \Rightarrow k + 1 \in S$, and S is inductive.

Conclusion: $S = N$.

$(9 - 2)$

29.   Prove: $P_n : 49^n - 1$ is divisible by 6.

Write: $P_n : 49^n - 1 = 6r$ for some $r \in N$. $S = \{n \in N \mid P_n \text{ is true }\}$

To show $S = N$, we show:

Part 1: $1 \in S$

$\qquad P_1 : 49^1 - 1 = 6r$ is true (Problem 12)

Part 2: S is inductive.

$\qquad P_k : 49^k - 1 = 6r$ for some $r \in N$

$\qquad P_{k+1} : 49^{k+1} - 1 = 6s$ for some $s \in N$

We start with P_k :

$\qquad 49^k - 1 = 6r, r \in N$

Now, $\quad 49^{k+1} - 1 \;=\; 49^{k+1} - 49^k + 49^k - 1$

$\qquad\qquad\qquad\quad\; = \; 49^k(49 - 1) + 49^k - 1$

$\qquad\qquad\qquad\quad\; = \; 49^k \cdot 8 \cdot 6 + 6r$

$\qquad\qquad\qquad\quad\; = \; 6(49^k \cdot 8 + r)$

$\qquad\qquad\qquad\quad\; = \; 6s \text{ with } s = 49^k \cdot 8 + r$

Thus, $P_k \Rightarrow P_{k+1}, k \in S \Rightarrow k + 1 \in S$, and S is inductive.

Conclusion: $S = N$.

(9 − 2)

30. An arithmetic sequence is involved, with $a_1 = \frac{9}{2}, d = \frac{3g}{2} - \frac{g}{2} = g$.

Distance fallen during the twenty-fifth second $= a_{25}$.

$\quad a_n \;=\; a_1 + (n - 1)d$

$\quad a_{25} \;=\; \frac{g}{2} + (25 - 1)g$

$\qquad\; = \; \frac{49g}{2} \text{ feet}$

Total distance fallen after twenty-five seconds $= a_1 + a_2 + a_3 + \cdots + a_{25} = S_{25}$

$\quad S_n \;=\; \frac{n}{2}(a_1 + a_n)$

$\quad S_{25} \;=\; \frac{25}{2}\left(\frac{g}{2} + \frac{49g}{2}\right)$

$\qquad\;\; = \; \frac{625g}{2} \text{ feet} \quad (9 - 3)$

31. $(x + i)^6 \;=\; \sum_{k=0}^{6}\binom{6}{k}x^{6-k}i^k$

$\qquad\qquad\; = \; \binom{6}{0}x^6 + \binom{6}{1}x^5 i^1 + \binom{6}{2}x^4 i^2 + \binom{6}{3}x^3 i^3 + \binom{6}{4}x^2 i^4 + \binom{6}{5}x i^5 + \binom{6}{6}i^6$

$\qquad\qquad\; = \; x^6 + 6ix^5 - 15x^4 - 20ix^3 + 15x^2 + 6ix - 1$

(9 − 6)

32. Prove $\sum_{k=1}^{n} k^3 = \left(\sum_{k=1}^{n} k\right)^2$

Write: $P_n : \sum_{k=1}^{n} k^3 = \left(\sum_{k=1}^{n} k\right)^2, S = \{n \in N \mid P_n \text{ is true}\}$

To show $S = n$, we write:

Part 1: $1 \in S$

$\qquad \sum_{k=1}^{1} k^3 = 1^3 = 1^2 = \left(\sum_{k=1}^{1} k\right)^2$

Part 2: S is inductive.

$P_j : \sum_{k=1}^{j} k^3 = (\sum_{k=1}^{j} k)^2$

$P_{j+1} : \sum_{k=1}^{j+1} k^3 = (\sum_{k=1}^{j+1} k)^2$

We start with P_j :

$\sum_{k=1}^{j} k^3 = (\sum_{k=1}^{j} k)^2$

Adding $(j+1)^3$ to both sides:

$\sum_{k=1}^{j} k^3 + (j+1)^3 = (\sum_{k=1}^{j} k)^2 + (j+1)^3$

$$
\begin{aligned}
\sum_{k=1}^{j+1} k^3 &= (1 + 2 + 3 + \cdots + j)^2 + (j+1)^3 \\
&= [\tfrac{j(j+1)}{2}]^2 + (j+1)^3 \text{ using Matched Problem 5, Section } 9-2 \\
&= (j+1)^2 \tfrac{j^2}{4} + (j+1)^2(j+1) \\
&= (j+1)^2[\tfrac{j^2}{4} + j + 1] \\
&= (j+1)^2[\tfrac{j^2+4j+4}{4}] \\
&= (j+1)^2 \tfrac{(j+2)^2}{2^2} \\
&= [\tfrac{(j+1)(j+2)}{2}]^2 \\
&= [1 + 2 + 3 + \cdots + (j+1)]^2 \text{ using Matched Problem 5, Section } 9-2 \\
&= (\sum_{k=1}^{j+1} k)^2
\end{aligned}
$$

Thus, $P_k \Rightarrow P_{k+1}, k \in S \Rightarrow k+1 \in S$, and S is inductive.

Conclusion: $S = N$.

$(9-2)$

33. Prove: $x^{2n} - y^{2n} = (x-y)Q_n(x,y)$, where $Q_n(x,y)$ denotes some polynomial in x and y

Write: $P_n : x^{2n} - y^{2n} = (x-y)Q_n(x,y), S = \{n \in N \mid P_n \text{ is true}\}$

To show $S = N$, we show:

Part 1: $1 \in S$

$$
\begin{aligned}
x^{2\cdot 1} - y^{2\cdot 1} = x^2 - y^2 &= (x-y)(x+y) \\
&= (x-y)Q_1(x,y)
\end{aligned}
$$

P_1 is true.

Part 2: S is inductive.

$P_k : x^{2k} - y^{2k} = (x-y)Q_k(x,y)$

$P_{k+1} : x^{2k+2} - y^{2k+2} = (x-y)Q_{k+1}(x,y)$

We start with P_k :

$x^{2k} - y^{2k} = (x-y)Q_k(x,y)$

$$
\begin{aligned}
\text{Now,} \ x^{2k+2} - y^{2k+2} &= x^{2k+2} - x^{2k}y^2 + x^{2k}y^2 - y^{2k+2} \\
&= x^{2k}(x^2 - y^2) + y^2(x^{2k} - y^{2k}) \\
&= (x-y)x^{2k}(x+y) + y^2(x-y)Q_k(x,y) \text{ by } P_k \\
&= (x-y)[x^{2k}(x+y) + y^2 Q_k(x,y)] \\
&= (x-y)Q_{k+1}(x,y)
\end{aligned}
$$

Thus, $P_k \Rightarrow P_{k+1}, k \in S \Rightarrow k+1 \in S$, and S is inductive.

Conclusion: $S = N$.　$(9-2)$

34.　Let m be an arbitrary integer. We will prove

$\frac{a^n}{a^m} = a^{n-m}, n > m$ by induction on n

Write: $P_n : \frac{a^n}{a^m} = a^{n-m}, n > m,$　$S_m = \{n \in N \mid P_n \text{ is true}\}$

To show $S = \{n \in N \mid n > m\}$, we show:

Part 1: $m + 1 \in S_m$

$$P_{m+1} : \begin{aligned} \frac{a^{m+1}}{a^m} &= a^{m+1-m} \\ \frac{a^m a}{a^m} &= a^1 \quad \text{by the redefinition of } a^n \\ a &= a \end{aligned}$$

Part 2: S is inductive.

$$P_k : \frac{a^k}{a^m} = a^{k-m}$$
$$P_{k+1} : \frac{a^{k+1}}{a^m} = a^{k-m+1}$$

We start with P_k :

$$\frac{a^k}{a^m} = a^{k-m}$$

Multiplying both sides by a :

$$\frac{a^k}{a^m} a = a^{k-m} a$$
$$\frac{a^k a}{a^m} = a^{k-m+1}$$
$$\frac{a^{k+1}}{a^m} = a^{k-m+1}$$

Thus, $P_k \Rightarrow P_{k+1}, k \in S_m \Rightarrow k + 1 \in S_m$.

Thus, $S_m = \{n \in N \mid n > m\}$. This is true for arbitrary m.

Thus, $\frac{a^n}{a^m} = a^{n-m}, n > m$ is true for any $m, n \in N$.

$(9-2)$

35.　Write: $P_n : a_n = b_n, S = \{n \in N \mid P_n \text{ is true}\}$

To show $S = N$, we show:

Part 1: $1 \in S$

$$a_1 = -3 = -5 + 2 = -5 + 2 = -5 + 2 \cdot 1 = b_1$$

Part 2: S is inductive.

$$P_k : a_k = b_k$$
$$P_{k+1} : a_{k+1} = b_{k+1}$$

We start with P_k :

$$a_k = b_k$$

Now, $\begin{aligned} a_{k+1} = a_k + 2 &= b_k + 2 \\ &= -5 + 2k + 2 \\ &= -5 + 2(k+1) \\ &= b_{k+1} \end{aligned}$

Thus, $P_k \Rightarrow P_{k+1}, k \in S \Rightarrow k + 1 \in S$, and S is inductive.

Conclusion: $a_n = b_n$ for all $n \in N$

$\{a_n\} = \{b_n\}$

$(9-2)$

36.　Write: $P_n : (1!)1 + (2!)2 + (3!)3 + \cdots + (n!)n = (n+1)! - 1$,

$S = \{n \in N \mid P_n \text{ is true}\}$

To show $S = N$, we show:

Part 1: $1 \in S$

$\quad\quad P_1 : (1!)1 = 1 = 2 - 1 = 2! - 1$ is true

Part 2: S is inductive.

$\quad\quad P_k : (1!)1 + (2!)2 + (3!)3 + \cdots + (k!)k = (k+1)! - 1$

$\quad\quad P_{k+1} : (1!)1 + (2!)2 + (3!)3 + \cdots + (k!)k + (k+1)!(k+1) = (k+2)! - 1$

We start with P_k :

$\quad (1!)1 + (2!)2 + (3!)3 + \cdots + (k!)k = (k+1)! - 1$

Adding $(k+1)!(k+1)$ to both sides:

$\quad (1!)1 + (2!)2 + (3!)3 + \cdots + (k!)k \quad + \quad (k+1)!(k+1)$

$$
\begin{aligned}
&= (k+1)! - 1 + (k+1)!(k+1) \\
&= (k+1)!(1 + k + 1) - 1 \\
&= (k+2)(k+1)! - 1 \\
&= (k+2)! - 1
\end{aligned}
$$

Thus, $P_k \Rightarrow P_{k+1}, k \in S \Rightarrow k + 1 \in S$, and S is inductive.

Conclusion: $S = N \quad\quad (9-2)$.

CHAPTER 10

Exercise 10-1

Key Ideas and Formulas

The graphs of a second degree equation in two variables $Ax^2 + Bxy + Cy^2 + Dx + Ey + F = 0$ (for different values of the coefficients: A, B, C not all zero) are plane curves called conic sections. They include the circle, parabola, ellipse, and hyperbola. A parabola is the set of all points in a plane equidistant from a fixed point F (the focus) and a fixed line L (the directrix) in the plane. A line through the focus perpendicular to the plane is called the axis and the point on the axis halfway between the focus and directrix is called the vertex.

Standard Equations of a Parabola

1. $y^2 = 4ax$
   Vertex: $(0,0)$
   Focus: $(a, 0)$
   Directrix: $x = -a$

   $a < 0$ (opens left) $a > 0$ (opens right)

   Symmetric with respect to the x axis.

2. $x^2 = 4ay$
   Vertex: $(0,0)$
   Focus: $(0, a)$
   Directrix: $y = -a$

   $a < 0$ (opens down) $a > 0$ (opens up)

   Symmetric with respect to the y axis.

---

1. To graph $y^2 = 4x$, assign x values that make the right side a perfect square (x must be nonnegative for y to be real) and solve for y. Since the coefficient of x is positive, a must be positive, and the parabola opens right.

x	0	1	4
y	0	$\pm$	± 4

To find the focus and directrix, solve
$$4a = 4$$
$$a = 1$$
Focus: $(1,0)$ Directrix: $x = -1$.

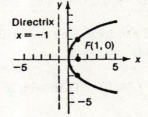

3. To graph $x^2 = 8y$, assign y values that make the right side a perfect square (y must be

nonnegative for x to be real) and solve for x. Since the coefficient of y is positive, a must be positive, and the parabola opens up.

x	0	± 4	± 2
y	0	2	$\frac{1}{2}$

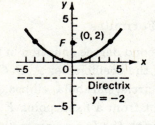

To find the focus and directrix, solve

$4a = 8$

$a = 2$

Focus: $(0,2)$ Directrix: $y = -2$

5. To graph $y^2 = -12x$, assign x values that make the right side a perfect square (x must be zero or negative for y to be real) and solve for y. Since the coefficient of x is negative, a must be negative, and the parabola opens left.

x	0	-3	$-\frac{1}{3}$
y	0	± 6	± 2

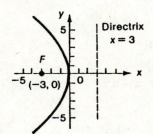

To find the focus and directrix,

Solve $4a = -12$

$a = -3$

Focus: $(-3,0)$ Directrix: $x = -(-3) = 3$

7. To graph $x^2 = -4y$, assign y values that make the right side a perfect square (y must be zero or negative for x to be real) and solve for x. Since the coefficient of y is negative, a must be negative, and the parabola opens down.

x	0	± 2	± 4
y	0	-1	-4

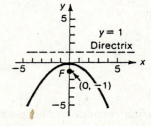

To find the focus and directrix,

Solve $4a = -4$

$a = -1$

Focus: $(0,-1)$ Directrix: $y = -(-1) = 1$

9. To graph $y^2 = -20x$, we may proceed as in problem 5. Alternatively, after noting that x must be zero or negative   for y to be real, we may pick convenient values for x and solve for y using a calculator. Since the coefficient of x is negative, a must be negative, and the parabola opens left.

x	0		-1	-2
y	0	$\pm\sqrt{20} \approx \pm 4.5$		± 6.3

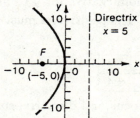

To find the focus and directrix,

Solve $4a = -20$

$a = -5$

Focus: $(-5,0)$ Directrix: $x = -(-5) = 5$

11.   To graph $x^2 = 10y$, we may proceed as in Problem 3. Alternatively, after noting that y must be nonnegative for x to be real, we may pick convenient values for y and solve for x using a calculator. Since the coefficient of y is positive, a must be positive, and the parabola opens up.

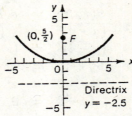

x	0	$\pm\sqrt{10} \approx \pm 3.2$	$\pm\sqrt{2} \approx \pm 4.5$
y	0	1	2

To find the focus and directrix,
Solve $4a = 10$
$a = 2.5$
Focus: $(0, 2.5)$ Directrix: $y = -2.5$

13.   Comparing $y^2 = 39x$ with $y^2 = 4ax$, the standard equation of a parabola symmetric with respect to the x axis, we have
$4a = 39$ Focus on x axis
$a = 9.75$
Focus: $(9.75, 0)$

15.   Comparing $x^2 = -105y$ with $x^2 = 4ay$, the standard equation of a parabola symmetric with respect to the y axis, we have
$4a = -105$ Focus on y axis
$a = -26.25$
Focus: $(0, -26.25)$

17.   Comparing $y^2 = -77x$ with $y^2 = 4ax$, the standard equation of a parabola symmetric with respect to the x axis, we have
$4a = -77$ Focus on x axis
$a = -19.25$
Focus: $(-19.25, 0)$

19.   The parabola is opening up and has an equation of the form $x^2 = 4ay$.
Since $(4, 2)$ is on the graph, we have:
$$4 = 4ay$$
$$(4)^2 = 4a(2)$$
$$16 = 8a$$
$$2 = a$$
Thus, the equation of the parabola is
$$x^2 = 4(2)y$$
$$x^2 = 8y$$

21.   The parabola is opening left and has an equation of the form $y^2 = 4ax$. Since $(-3, 6)$ is on the graph, we have:
$$y^2 = 4ax$$
$$(6)^2 = 4a(-3)$$
$$-3 = a$$
Thus, the equation of the parabola is
$$y^2 = 4(-3)x$$
$$y^2 = -12x$$

23.   The parabola is opening down and has an equation of the form $x^2 = 4ay$. Since $(-6, -9)$ is on the graph, we have:
$$x^2 = 4ay$$
$$(-6)^2 = 4a(-9)$$
$$36 = -36a$$
$$-1 = a$$
Thus, the equation of the parabola is
$$x^2 = 4(-1)y$$
$$x^2 = -4y$$

25. From the figure, we see that the point $P(x, y)$ is a point on the parabola if and only if

$$
\begin{aligned}
d_1 &= d_2 \\
d(P, N) &= d(P, F) \\
\sqrt{(x - x)^2 + (y - 4)^2} &= \sqrt{(x - 2)^2 + (y - 2)^2} \\
(y - 4)^2 &= (x - 2)^2 + (y - 2)^2 \\
y^2 - 8y + 16 &= x^2 - 4x + 4 + 4y^2 - 4y + 4 \\
0 &= x^2 - 4x + 4y - 8 \\
x^2 - 4x + 4y - 8 &= 0
\end{aligned}
$$

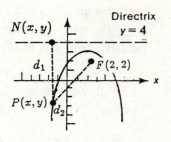

Directrix
$y = 4$

$N(x, y)$

d_1

$F(2, 2)$

$P(x, y)$

d_2

27. From the figure, we see that the coordinates of P must be $(-100, -50)$. The parabola is opening down with axis the y axis, hence it has an equation of the form $x^2 = 4ay$. Since $(-100, -50)$ is on the graph, we have

$$
\begin{aligned}
(-100)^2 &= 4a(-50) \\
10,000 &= -200a \\
a &= -50
\end{aligned}
$$

Thus, the equation of the parabola is

$$
\begin{aligned}
x^2 &= 4(-50)y \\
x^2 &= -200y
\end{aligned}
$$

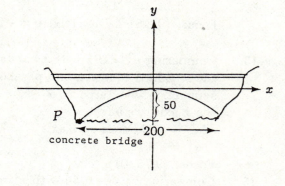

y

50

P

200

concrete bridge

29. A. From the figure, we see that the parabola is opening up with axis the y-axis, hence it has an equation of the form $x^2 = 4ay$. Since the focus is at $(0, a) = (0, 100), a = 100$ and the equation of the parabola is $x^2 = 400y$ or $y = 0.0025x^2$

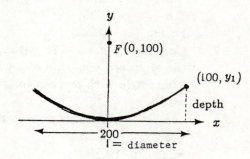

y

$F(0, 100)$

$(100, y_1)$

depth

x

200

$| = $ diameter

B. Since the depth represents the y coordinate y, of a point on the parabola with $x = 100$, we have

$$
\begin{aligned}
y_1 &= 0.0025(100)^2 \\
\text{depth} &= 25 \text{ feet}
\end{aligned}
$$

Exercise 10-2

Key Ideas and Formulas

An ellipse is the set of all points P in a plane such that the sum of the distances of P from two fixed points (the foci) in the plane is constant.

The line through the foci intersects the ellipse at two points called vertices; the line segment connecting the vertices is called the major axis; its perpendicular bisector is called the minor axis. The axes intersect at a point called the center of the ellipse.

Standard Equations of an Ellipse - Center at (0,0)

1. $\frac{x^2}{a^2} + \frac{y^2}{b^2} = 1 \quad a > b > 0$

   x intercepts: $\pm a$ (vertices)
   y intercepts: $\pm b$
   Foci: $F'(-c,0), F(c,0)$
   $c^2 = a^2 - b^2$
   Major axis length $= 2a$
   Minor axis length $= 2b$

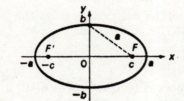

2. $\frac{x^2}{b^2} + \frac{y^2}{a^2} = 1 \quad a > b > 0$
   x intercepts: $\pm b$
   y intercepts: $\pm a$ (vertices)
   Foci: $F'(0,-c), F(0,c)$
   $c^2 = a^2 - b^2$
   Major axis length $= 2a$
   Minor axis length $= 2b$

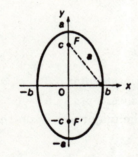

---

1. When $y = 0, \frac{x^2}{25} = 1$. x intercepts: ± 5
   When $x = 0, \frac{y^2}{4} = 1$. y intercepts: ± 2
   Thus $a = 5, b = 2$, and the major axis is on the x axis.
   Foci: $c^2 = a^2 - b^2$
   $\quad\quad c^2 = 25 - 4$
   $\quad\quad c^2 = 21$
   $\quad\quad c = \sqrt{21}$ Foci : $F'(-\sqrt{21},0), F(\sqrt{21},0)$
   Major axis length $= 2(5) = 10$ Minor axis length $= 2(2) = 4$
   Common Error:
   The relationship $c^2 = a^2 + b^2$ does not apply to to a, b, c as defined for ellipses.

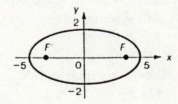

3. When $y = 0, \frac{x^2}{4} = 1$. x intercepts: ± 2
   When $x = 0, \frac{y^2}{25} = 1$. y intercepts: ± 5

Thus $a = 5, b = 2$, and the major axis is on the y axis.

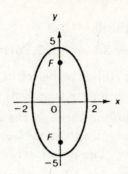

Foci: $c^2 = a^2 - b^2$
$c^2 = 25 - 4$
$c^2 = 21$
$c = \sqrt{21}$ Foci: $F'(0, -\sqrt{21}), F(0, \sqrt{21})$

Major axis length $= 2(5) = 10$ Minor axis length $= 2(2) = 4$

5.   First, write the equation in standard form by dividing both sides by 9.

$x^2 + 9y^2 = 9$
$\frac{x^2}{9} + \frac{y^2}{1} = 1$

Locate the intercepts.

When $y = 0, \frac{x^2}{9} = 1$. x intercepts: ± 3

When $x = 0, \frac{y^2}{1} = 1$. y intercepts: ± 1

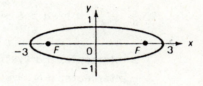

Thus $a = 3, b = 1$, and the major axis is on the x axis.

Foci: $c^2 = a^2 - b^2$
$c^2 = 9 - 1$
$c^2 = 8$
$c = \sqrt{8}$ Foci: $F'(-\sqrt{8}, 0), F(\sqrt{8}, 0)$

Major axis length $= 2(3) = 6$ Minor axis length $= 2(1) = 2$

7.   First, write the equation in standard form by dividing both sides by 225.

$25x^2 + 9y^2 = 225$
$\frac{x^2}{9} + \frac{y^2}{225} = 1$

Locate the intercepts.

When $y = 0, \frac{x^2}{9} = 1$. x intercepts: ± 3

When $x = 0, \frac{y^2}{25} = 1$. y intercepts: ± 5

Thus $a = 5, b = 3$, and the major axis is on the y axis.

Foci: $c^2 = a^2 - b^2$
$c^2 = 25 - 9$
$c^2 = 16$
$c = 4$ Foci: $F'(0, -4), F(0, 4)$

Major axis length $= 2(5) = 10$ Minor axis length $= 2(3) = 6$

9.   First, write the equation in standard form by dividing both sides by 12.

$2x^2 + y^2 = 12$
$\frac{x^2}{6} + \frac{y^2}{12} = 1$

Locate the intercepts.

When $y = 0, \frac{x^2}{6} = 1$. x intercepts: $\pm\sqrt{6}$

When $x = 0, \frac{y^2}{12} = 1$. y intercepts: $\pm\sqrt{12}$

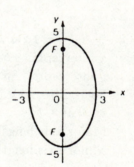

Thus $a = \sqrt{12}, b = \sqrt{6}$, and the major axis is on the y axis.

Foci: $c^2 = a^2 - b^2$
$c^2 = 12 - 6$
$c^2 = 6$
$c = \sqrt{6}$ Foci: $F'(0, -\sqrt{6}), F(0, \sqrt{6})$

Major axis length $= 2\sqrt{12} \approx 6.93$ Minor axis length $= 2\sqrt{6} \approx 4.90$

11. First, write the equation in standard form by dividing
    both sides by 28.

    $$4x^2 + 7y^2 = 28$$
    $$\frac{x^2}{7} + \frac{y^2}{4} = 1$$

    Locate the intercepts.

    When $y = 0$, $\frac{x^2}{7} = 1$. x intercepts: $\pm\sqrt{7}$

    When $x = 0$, $\frac{y^2}{4} = 1$. y intercepts: ± 2

    Thus $a = \sqrt{7}, b = 2$, and the major axis is on the x axis.

    Foci: $c^2 = a^2 - b^2$

    $\qquad c^2 = 7 - 4$

    $\qquad c^2 = 3$

    $\qquad c = \sqrt{3}$ Foci : $F'(-\sqrt{3}, 0), F(\sqrt{3}, 0)$

    Major axis length $= 2\sqrt{7} \approx 5.29$ Minor axis length $= 2(2) = 4$

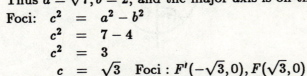

13. Make a rough sketch of the ellipse and
    compute x and y intercepts.

    $\frac{x^2}{a^2} + \frac{y^2}{b^2} = 1$

    $a = \frac{8}{2} = 4, b = \frac{6}{2} = 3$

    $\frac{x^2}{16} + \frac{y^2}{9} = 1$

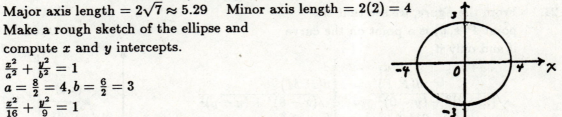

15. Make a rough sketch of the ellipse
    and compute x and y intercepts.

    $\frac{x^2}{b^2} + \frac{y^2}{a^2} = 1$

    $a = \frac{22}{2} = 11, b = \frac{16}{2} = 8$

    $\frac{x^2}{64} + \frac{y^2}{121} = 1$

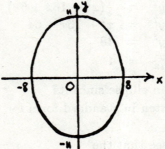

17. Make a rough sketch of the ellipse
    locate focus and x intercepts,
    then determine y intercepts using the
    special triangle relationship

    $\frac{x^2}{a^2} + \frac{y^2}{b^2} = 1$

    $a = \frac{16}{2} = 8$

    $b^2 = 8^2 - 6^2 = 64 - 36 = 28$

    $b = \sqrt{28}$

    $\frac{x^2}{64} + \frac{y^2}{28} = 1$

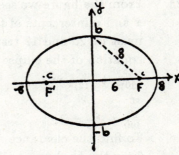

19. **Make a rough sketch of the ellipse locate focus and x intercepts, then determine y intercepts using the special triangle relationship**

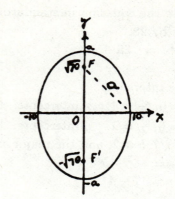

$$\frac{x^2}{b^2} + \frac{y^2}{a^2} = 1$$
$$b = \frac{20}{2} = 10$$
$$a^2 = 10^2 + (\sqrt{70})^2 = 100 + 70 = 170$$
$$a = \sqrt{170}$$
$$\frac{x^2}{100} + \frac{y^2}{170} = 1$$

21. **From the figure, we see that the point $P(x, y)$ is a point on the curve if and only if**

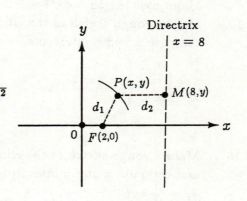

$$
\begin{aligned}
d_1 &= \tfrac{1}{2}d_2 \\
d(P, F) &= \tfrac{1}{2}d(P, M) \\
\sqrt{(x-2)^2 + (y-0)^2} &= \tfrac{1}{2}\sqrt{(x-8)^2 + (y-y)^2} \\
(x-2)^2 + y^2 &= \tfrac{1}{4}(x-8)^2 \\
x^2 - 4x + 4 + y^2 &= \tfrac{1}{4}(x^2 - 16x + 64) \\
4x^2 - 16x + 16 + 4y^2 &= x^2 - 16x + 64 \\
3x^2 + 4y^2 &= 48 \\
\frac{x^2}{16} + \frac{y^2}{12} &= 1
\end{aligned}
$$

The curve must be an ellipse since its equation can be written in standard form for an ellipse.

23. **From the figure we see that the x and y intercepts of the ellipse must be 20 and 12 respectively. Hence the equation of the ellipse must be**

$$\frac{x^2}{(20)^2} + \frac{y^2}{(12)^2} = 1$$

or

$$\frac{x^2}{400} + \frac{y^2}{144} = 1.$$

To find the clearance above the water 5 feet from the bank, we need the y coordinate y_1 of the point P whose x coordinate is $a - 5 = 15$. Since P is on the ellipse, we have

$$
\begin{aligned}
\frac{15^2}{400} + \frac{y_1^2}{144} &= 1 \\
\frac{225}{400} + \frac{y_1^2}{144} &= 1 \\
0.5625 + \frac{y_1^2}{144} &= 1 \\
\frac{y_1^2}{144} &= 0.4375 \\
y_1^2 &= 144(0.4375) \\
y_1^2 &= 63 \\
y_1 &\approx \pm 7.94
\end{aligned}
$$

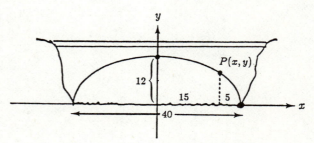

Therefore, the clearance is 7.94 feet, approximately.

25. From the figure we see that the x
intercept of the ellipse must be 24.0. Hence the
equation of the ellipse must have the form
$$\frac{x^2}{(24.0)^2} + \frac{y^2}{b^2} = 1$$
Since the point $(23.0, 1.14)$ is on the ellipse,
its coordinates must satisfy the equation of the
ellipse. Hence

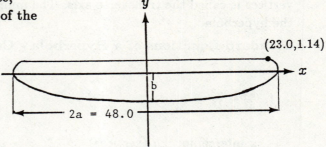

$$\frac{(23.0)^2}{(24.0)^2} + \frac{(1.14)^2}{b^2} = 1$$
$$\frac{(1.14)^2}{b^2} = 1 - \frac{(23.0)^2}{(24.0)^2}$$
$$\frac{(1.14)^2}{b^2} = \frac{47}{576}$$
$$b^2 = \frac{576(1.14)^2}{47}$$
$$b^2 = 15.9$$

Thus, the equation of the ellipse must be
$$\frac{x^2}{576} + \frac{y^2}{15.9} = 1$$
B. From the figure, we can see that the width
of the wing must equal
$$1.14 + b = 1.14 + \sqrt{15.9} = 5.13 \text{ feet.}$$

Exercise 10-3

Key Ideas and Formulas

A hyperbola is the set of all points P in a plane such that the absolute value of the difference of the distances of P to two-fixed points (the foci) is a positive constant. The line through the foci intersects the hyperbola at two points called vertices; the line segment connecting the vertices is called the transverse axis. The midpoint of the transverse axis is called the center of the hyperbola.

Standard Equations of a Hyperbola - Center at (0,0)

1. $\frac{x^2}{a^2} - \frac{y^2}{b^2} = 1$

x intercepts: $\pm a$ (vertices)
y intercepts: none
Foci: $F'(-c,0), F(c,0)$
$c^2 = a^2 + b^2$
Transverse axis length $= 2a$
Conjugate axis length $= 2b$

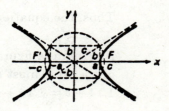

Asymptotes: $y = \pm \frac{b}{a}x$

2. $\frac{y^2}{a^2} - \frac{x^2}{b^2} = 1$
x intercepts: none
y intercepts: $\pm a$ (vertices)
Foci: $F'(0,-c), F(0,c)$
$c^2 = a^2 + b^2$
Transverse axis length $= 2a$
Conjugate axis length $= 2b$

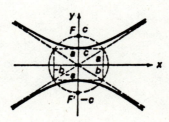

Asymptotes: $y = \pm \frac{a}{b}x$

[Note: Both graphs are symmetric with respect to the x axis, y axis, and origin.]

---

1.   When $y = 0, \frac{x^2}{9} = 1$. x intercepts: ± 3 $a = 3$
When $x = 0, -\frac{y^2}{4} = 1$. There are no y intercepts, but $b = 2$.
Sketch the asymptotes using the asymptote rectangle,
then sketch in the hyperbola.
Foci: $c^2 = 3^2 - 2^2$
         $c^2 = 13$
         $c^2 = \sqrt{13}$
$F'(-\sqrt{13}, 0), F(\sqrt{13}, 0)$

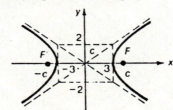

Transverse axis length $= 2(3) = 6$

Conjugate axis length $= 2(2) = 4$

3.   When $y = 0, -\frac{x^2}{9} = 1$. There are no x intercepts, but $b = 3$.

When $x = 0, \frac{y^2}{4} = 1$. y intercepts: ± 2 $a = 2$

Sketch the asymptotes using the asymptote rectangle, then

sketch in the hyperbola.

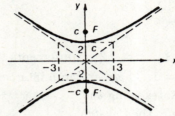

Foci: $c^2 \;=\; 2^2 + 3^2$

         $c^2 \;=\; 13$

         $c \;=\; \sqrt{13}$

$F'(0, -\sqrt{13}), F(0, \sqrt{13})$

Transverse axis length $= 2(2) = 4$

Conjugate axis length $= 2(3) = 6$

5.   First, write the equation in standard form by dividing

both sides by 16.

$$4x^2 - y^2 \;=\; 16$$
$$\frac{x^2}{4} - \frac{y^2}{16} \;=\; 1$$

Locate intercepts:

When $y = 0, x = \pm 2$. x intercepts: ± 2 $a = 2$

When $x = 0, -\frac{y^2}{16} = 1$. There are no y intercepts, but $b = 4$.

Sketch the asymptotes using the asymptote rectangle

then sketch in the hyperbola.

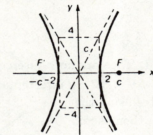

Foci: $c^2 \;=\; 2^2 + 4^2$

         $c^2 \;=\; 20$

         $c \;=\; \sqrt{20}$

$F'(-\sqrt{20}, 0), F(\sqrt{20}, 0)$

Transverse axis length $= 2(2) = 4$

Conjugate axis length $= 2(4) = 8$

7.   First, write the equation in

standard form by dividing both

sides by 144.

$$9y^2 - 16x^2 \;=\; 144$$
$$\frac{y^2}{16} - \frac{x^2}{9} \;=\; 1$$

Locate intercepts:

When $y = 0, -\frac{x^2}{9} = 1$.

There are no x intercepts, but $b = 3$.

When $x = 0, \frac{y^2}{16} = 1, y = \pm 4$.

y intercepts: ± 4 $a = 4$

Sketch the asymptotes using the

asymptote rectangle, then sketch

in the hyperbola.

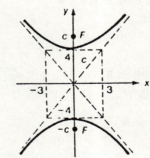

Foci: $c^2 \;=\; 4^2 + 3^2$

         $c^2 \;=\; 25$

         $c \;=\; 5$

$F'(0, -5), F(0, 5)$

Transverse axis length $= 2(4) = 8$

Conjugate axis length $= 2(3) = 6$

9.   First write the equation in standard form by dividing both sides by 12.

$$3x^2 - 2y^2 = 12$$

$$\frac{x^2}{4} - \frac{y^2}{6} = 1$$

Locate intercepts:

When $y = 0$, $\frac{x^2}{4} = 1$. $x = \pm 2$ x intercepts ± 2 $a = 2$

When $x = 0$, $-\frac{y^2}{6} = 1$, There are no y intercepts, but $b = \sqrt{6}$.

Sketch the asymptotes using the
asymptote rectangle, then sketch
in the hyperbola.

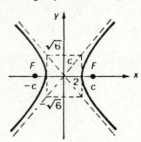

Foci: $c^2 = 2^2 + (\sqrt{6})^2$

$$c^2 = 10$$

$$c = \sqrt{10}$$

$F'(-\sqrt{10}, 0), F(\sqrt{10}, 0)$

Transverse axis length $= 2(2) = 4$

Conjugate axis length $= 2\sqrt{6} \approx 4.90$

11.   First, write the equation in standard
form by dividing both sides by 28.

$$7y^2 - 4x^2 = 28$$

$$\frac{y^2}{4} - \frac{x^2}{7} = 1$$

Locate intercepts:

When $y = 0$, $-\frac{x^2}{7} = 1$. Thre are no x intercepts, but $b = \sqrt{7}$.

When $x = 0$, $\frac{y^2}{4} = 1$. $y = \pm 2$ y intercepts: ± 2 $a = 2$

Sketch the asymptotes using the
asymptote rectangle, then sketch
in the hyperbola.

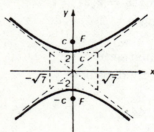

Foci: $c^2 = 2^2 + (\sqrt{7})^2$

$$c^2 = 11$$

$$c = \sqrt{11}$$

$F'(0, -\sqrt{11}), F(0, \sqrt{11})$

Transverse axis length $= 2(2) = 4$

Conjugate axis length $= 2\sqrt{7} \approx 5.29$

13.   Since the transverse axis is on
the x axis, start with

$$\frac{x^2}{a^2} - \frac{y^2}{b^2} = 1$$

and find a and b

$a = \frac{14}{2} = 7$ and $b = \frac{10}{2} = 5$

Thus, the equation is

$$\frac{x^2}{49} - \frac{y^2}{25} = 1$$

15. Since the transverse axis is on the y axis, start with

$$\frac{y^2}{a^2} - \frac{x^2}{b^2} = 1$$

and find a and b

$$a = \frac{24}{2} = 12 \text{ and } b = \frac{18}{2} = 9$$

Thus, the equation is

$$\frac{y^2}{144} - \frac{x^2}{81} = 1$$

17. Since the transverse axis is on the x axis, start with

$$\frac{x^2}{a^2} - \frac{y^2}{b^2} = 1$$

and find a and b

$$a = \frac{18}{2} = 9$$

To find b, sketch the asymptote rectangle, label known parts, and use the Pythagorean Theorem.

$$b^2 = 11^2 - 9^2$$
$$b^2 = 40$$
$$b = \sqrt{40}$$

Thus, the equation is

$$\frac{x^2}{81} - \frac{y^2}{40} = 1$$

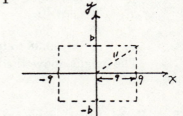

19. Since the conjugate axis is on the x axis, start with

$$\frac{y^2}{a^2} - \frac{x^2}{b^2} = 1$$

and find a and b

$$b = \frac{14}{2} = 7$$

To find a, sketch the asymptote rectangle, label known parts, and use the Pythagorean Theorem.

$$a^2 = (\sqrt{200})^2 - 7^2$$
$$a^2 = 151$$
$$a = \sqrt{151}$$

Thus, the equation is

$$\frac{y^2}{151} - \frac{x^2}{49} = 1$$

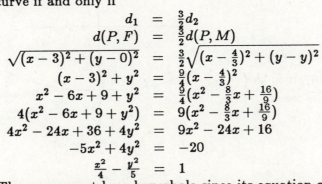

21. From the figure, we see that the point $P(x,y)$ is a point on the curve if and only if

$$
\begin{aligned}
d_1 &= \tfrac{3}{2} d_2 \\
d(P,F) &= \tfrac{3}{2} d(P,M) \\
\sqrt{(x-3)^2 + (y-0)^2} &= \tfrac{3}{2}\sqrt{(x-\tfrac{4}{3})^2 + (y-y)^2} \\
(x-3)^2 + y^2 &= \tfrac{9}{4}(x-\tfrac{4}{3})^2 \\
x^2 - 6x + 9 + y^2 &= \tfrac{9}{4}(x^2 - \tfrac{8}{3}x + \tfrac{16}{9}) \\
4(x^2 - 6x + 9 + y^2) &= 9(x^2 - \tfrac{8}{3}x + \tfrac{16}{9}) \\
4x^2 - 24x + 36 + 4y^2 &= 9x^2 - 24x + 16 \\
-5x^2 + 4y^2 &= -20 \\
\frac{x^2}{4} - \frac{y^2}{5} &= 1
\end{aligned}
$$

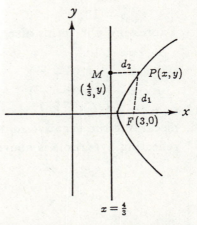

The curve must be a hyperbola since its equation can be written in standard form for a hyperbola.

23. From the figure we see that the transverse axis of the hyperbola must be on the y axis and $a = 4$. Hence the equation of the hyperbola must have the form

$$\frac{y^2}{4^2} - \frac{x^2}{b^2} = 1.$$

To find b, we note that the point $(8,12)$ is on the hyperbola, hence its coordinates satisfy the equation. Substituting, we have

$$\frac{12^2}{4^2} - \frac{8^2}{b^2} = 1$$
$$9 - \frac{64}{b^2} = 1$$
$$-\frac{64}{b^2} = -8$$
$$-64 = -8b^2$$
$$b^2 = 8$$

The equation required is

$$\frac{y^2}{16} - \frac{x^2}{8} = 1$$

Using this equation, we can compute y when $x = 6$ to answer the question asked (see figure).

$$\frac{y^2}{16} - \frac{6^2}{8} = 1$$
$$\frac{y^2}{16} - \frac{36}{8} = 1$$
$$y^2 - 72 = 16$$
$$y^2 = 88$$
$$y = 9.38 \text{ to two decimal places}$$

$$
\begin{aligned}
\text{The height above the vertex} &= y - \text{height of vertex} \\
&= 9.38 - 4 \\
&= 5.38 \text{ feet}
\end{aligned}
$$

25.   From the figure we can see:

$$FF' = 2c = 120 - 20 = 100$$

Thus $c = 50$

$$FV = c - a = 120 - 110 = 10$$

Thus, $a = c - 10 = 50 - 10 = 40$

Since
$$c^2 = a^2 + b^2$$
$$50^2 = 40^2 + b^2$$
$$b = 30$$

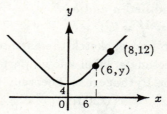

Hyperbola part of dome

Thus the equation of the hyperbola, in standard form, is

$$\frac{y^2}{40^2} - \frac{x^2}{30^2} = 1$$

Expressing y in terms of x, we have

$$\frac{y^2}{40^2} = 1 + \frac{x^2}{30^2}$$
$$\frac{y^2}{40^2} = \frac{1}{30^2}(30^2 + x^2)$$
$$y^2 = \frac{40^2}{30^2}(30^2 + x^2)$$
$$y = \frac{4}{3}\sqrt{x^2 + 30^2}$$

discarding the negative solution, since the reflecting hyperbola is above the x-axis.

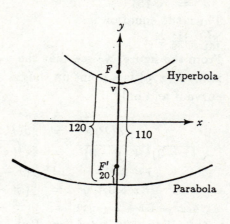

Exercise 10-4

Key Ideas and Formulas

A translation of coordinates occurs when the new coordinate axes have the same direction as and are parallel to the old coordinate axes.

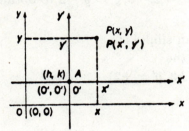

If the coordinates of the origin of the translated system are (h, k) relative to the old system, then (x, y) and (x', y') are related by:

1. $x = x' + h \quad y = y' + k$	*Common Error:*
2. $x' = x - h \quad y' = y - k$	The signs are very easily confused here, especially if h or k is negative in a problem.

---

1. A. $x' = x - 3, y' = y - 5, (h, k) = (3, 5)$ B. $x'^2 + y'^2 = 81$ C. Circle

3. A. $x' = x + 7, y' = y - 4, (h, k) = (-7, 4)$ B. $\frac{x'^2}{9} + \frac{y'^2}{16} = 1$ C. Ellipse

5. A. $x' = x - 4, y' = y + 9, (h, k) = (4, -9)$ B. $y'^2 = 16x'^2$ C. Parabola

7. A. $x' = x + 8, y' = y + 3, (h, k) = (-8, -3)$ B. $\frac{x'^2}{12} + \frac{y'^2}{8} = 1$ C. Ellipse

9. The translation required is $x' = x - 3, y' = y + 2$. The equation then becomes
$16x'^2 - 9y'^2 = 144$
We divide both sides by 144 to convert to standard form:
A. $\frac{x'^2}{9} - \frac{y'^2}{16} = 1$. This is the equation of a hyperbola.
B. $x' = x - 3, y' = y + 2, (h, k) = (3, -2)$

11. The translation required is $x' = x + 5, y' = y + 7$. The equation then becomes
$6x'^2 + 5y'^2 = 30$
We divide both sides by 30 to convert to standard form:
A. $\frac{x'^2}{30} + \frac{y'^2}{6} = 1$. This is the equation of an ellipse.
B. $x' = x + 5, y' = y + 7, (h, k) = (-5, -7)$

13. The translation required is $x' = x + 6, y' = y - 4$. The equation then becomes
$x'^2 + 24y' = 0$
Rewriting this in standard form, we have
A. $x'^2 = -24y'$. This is the equation of a parabola.
B. $x' = x + 6, y' = y - 4, (h, k) = (-6, 4)$

15.
$$4x^2 + 9y^2 - 16x - 36y + 16 = 0$$
$$4x^2 - 16x + 9y^2 - 36y = -16$$
$$4(x^2 - 4x + ?) + 9(y^2 - 4y + ?) = -16$$
$$4(x^2 - 4x + 4) + 9(y^2 - 4y + 4) = -16 + 16 + 36$$
$$4(x - 2)^2 + 9(y - 2)^2 = 36$$
$$\frac{(x-2)^2}{9} + \frac{(y-2)^2}{4} = 1$$

We use the translation $x' = x - 2, y' = y - 2$ to obtain
$$\frac{x'^2}{9} + \frac{y'^2}{4} = 1$$
This is the equation of an ellipse.
Sketch the intercepts in the
translated system $(x' = \pm 3, y' = \pm 2)$,
then sketch the graph.

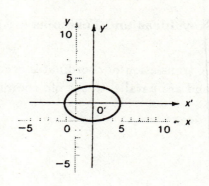

$(h, k) = (2, 2)$ is the new origin $0'$

17.
$$x^2 + 8x + 8y = 0$$
$$x^2 + 8x = -8y$$
$$x^2 + 8x + 16 = -8y + 16$$
$$(x + 4)^2 = 4(-2)(y - 2)$$

We use the translation $x' = x + 4, y' = y - 2$ to obtain
$$x'^2 = -8y'$$
This is the equation of a
parabola; since $a = -2$ and
the axis of symmetry is the y'
axis, it opens down.

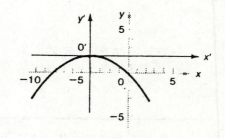

x'	0	± 4
y'	0	-2

$(h, k) = (-4, 2)$ is the new origin $0'$.

19.
$$x^2 + y^2 + 12x + 10y + 45 = 0$$
$$x^2 + 12x + y^2 + 10y = -45$$
$$x^2 + 12x + 36 + y^2 + 10y + 25 = -45 + 36 + 25$$
$$(x + 6)^2 + (y + 5)^2 = 16$$

We use the translation $x' = x + 6, y' = y + 5$ to obtain
$$x'^2 + y'^2 = 16$$
This is a circle of radius 4.
$(h, k) = (-6, -5)$ is the new origin
$0'$, and the center of the circle.

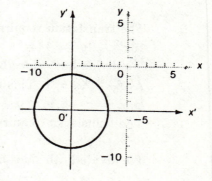

21.
$$-9x^2 + 16y^2 - 72x - 96y - 144 = 0$$
$$-9x^2 - 72x + 16y^2 - 96y = 144$$
$$-9(x^2 + 8x + ?) + 16(y^2 - 6y + ?) = 144$$
$$-9(x^2 + 8x + 16) + 16(y^2 - 6y + 9) = 144 - 144 + 144$$
$$-9(x + 4)^2 + 16(y - 3)^2 = 144$$
$$\frac{(y-3)^2}{9} - \frac{(x+4)^2}{16} = 1$$

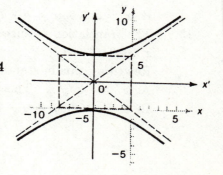

We use the translation $x' = x + 4, y' = y - 3$ to obtain

$$\frac{y'^2}{9} - \frac{x'^2}{16} = 1$$

This is the equation of a hyperbola with transverse axis on the y' axis.

Sketch the asymptote rectangle $(a = 3, b = 4)$, asymptotes, and vertices in the translated system, then sketch in the hyperbola. $(h, k) = (-4, 3)$ is the new origin $0'$.

23. First find the coordinates of the foci in the translated system.

$$\begin{aligned}
c'^2 &= 3^2 - 2^2 = 5 \\
c' &= \sqrt{5} \\
-c' &= -\sqrt{5}
\end{aligned}$$

Thus the coordinates in the translated system are

$F'(-\sqrt{5}, 0)$ and $F(\sqrt{5}, 0)$

Now use

$$\begin{aligned}
x &= x' + h = x' + 2 \\
y &= y' + k = y' + 2
\end{aligned}$$

to obtain

$F'(-\sqrt{5} + 2, 2)$ and $F(\sqrt{5} + 2, 2)$

as the coordinates of the foci in the original system.

25. First find the coordinates of the focus in the translated system. Since $a = -2$, and the parabola opens down, the coordinates are $(0, -2)$. Now use

$$\begin{aligned}
x &= x' + h = x' - 4 = 0 - 4 \\
y &= y' + k = y' + 2 = -2 + 2
\end{aligned}$$

to obtain $(-4, 0)$ as the coordinates of the focus in the original system.

27. First find the coordinates of the foci in the translated system.

$$\begin{aligned}
c'^2 &= 3^2 + 4^2 = 25 \\
c' &= 5 \\
-c' &= -5
\end{aligned}$$

Thus the coordinates in the translated system are:

$F'(0, -5)$ and $F(0, 5)$

Now use:

$$\begin{aligned}
x &= x' + h = x' - 4 \\
y &= y' + k = y' + 3
\end{aligned}$$

to obtain $F'(0 - 4, -5 + 3) = F'(-4, -2)$ and $F(0 - 4, 5 + 3) = F(-4, 8)$ as the coordinates of the foci in the original system.

Exercise 10-5

CHAPTER REVIEW

1.   First, write the equation in standard form by dividing both sides by 225.

$$9x^2 + 25y^2 = 225$$
$$\frac{x^2}{25} + \frac{y^2}{9} = 1$$

In this form the equation is identifiable as that of an ellipse. Locate the intercepts.

When $y = 0, \frac{x^2}{25} = 1$. x intercepts : ± 5

When $x = 0, \frac{y^2}{9} = 1$. y intercepts : ± 3

Thus, $a = 5, b = 3$, and the major axis is on the x axis.

Common Error:
The relationship $c^2 = a^2 + b^2$ applies to a, b, c as defined for hyperbolas but not for ellipses.

Foci: $c^2 = a^2 - b^2$
$$c^2 = 25 - 9$$
$$c^2 = 16$$
$$c = 4 \quad \text{Foci}: F'(-4, 0), F(4, 0)$$

Major axis length $= 2(5) = 10$

Minor axis length $= 2(3) = 6 \qquad (10-2)$

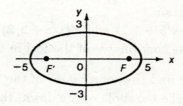

2. $x^2 = -12y$ is the equation of a parabola. To graph, assign y values that make the right side a perfect square (y must be zero or negative for x to be real) and solve for x. Since the coefficient of x is negative, a must be negative, and the parabola opens down.

x	0	± 6	± 2
y	0	-3	$-\frac{1}{3}$

To find the focus and directrix, solve

$$4a = -12$$
$$a = -3$$

Focus: $(0, -3)$. Directrix: $y = -(-3) = 3$

$$(10-1)$$

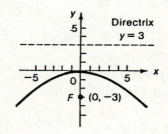

3. First, write the equation in standard form by dividing both sides by 225.
$$25y^2 - 9x^2 = 225$$
$$\frac{y^2}{9} - \frac{x^2}{25} = 1$$
In this form the equation is identifiable as that of a hyperbola.

When $y = 0, -\frac{x^2}{25} = 1$. There are no x intercepts, but $b = 5$.

When $x = 0, \frac{y^2}{9} = 1$. y intercepts, ± 3

Sketch the asymptotes using the asymptote rectangle, then sketch in the hyperbola.

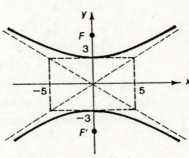

Foci: $c^2 = 3^2 + 5^2$
$$c^2 = 34$$
$$c = \sqrt{34} \quad \text{Foci}: F'(0, -\sqrt{34}), F(0, \sqrt{34})$$
Transverse axis length $= 2(3) = 6$
Conjugate axis length $= 2(5) = 10$
$$(10 - 3)$$

4. The translation required is $x' = x - 4, y' = y + 2$. The equation then becomes
$$4y'^2 - 25x'^2 = 100$$
We divide both sides by 100 to convert to standard form.
A. $\frac{y'^2}{25} - \frac{x'^2}{4} = 1$
This is the equation of a hyperbola.
B. $x' = x - 4, y' = y + 2,$
$(h, k) = (4, -2) \quad (10 - 4)$

5. The translation required is $x' = x + 5, y' = y + 4.$
The equation then becomes
A. $x'^2 = -12y'$
This is the equation of a parabola.
B. $x' = x + 5, y' = y + 4,$
$(h, k) = (-5, -4) \quad (10 - 4)$

6. The translation required is $x' = x - 6, y' = y - 4$. The equation then becomes
$$16x'^2 + 9y'^2 = 144$$
We divide both sides by 144 to convert to standard form:
A. $\frac{x'^2}{9} + \frac{y'^2}{16} = 1$
This is the equation of an ellipse.
B. $x' = x - 6, y' = y - 4, (h, k) = (6, 4) \quad (10 - 4)$

7. The parabola is opening either left or right and has an equation of the form $y^2 = 4ax$. Since $(-4, -2)$ is on the graph, we have:
$$(-2)^2 = 4a(-4)$$
$$4 = -16a$$
$$-\frac{1}{4} = a$$
Thus, the equation of the parabola is
$$y^2 = 4(-\tfrac{1}{4})x$$
$$y^2 = -x \quad (10 - 1)$$

8. Make a rough sketch of the ellipse, locate focus and x
    intercepts, then determine y intercepts using the special
    triangle relationship.

    $$\frac{x^2}{b^2} + \frac{y^2}{a^2} = 1$$
    $$b = \frac{6}{2} = 3$$
    $$a^2 = 4^2 + 3^2 = 16 + 9 = 25$$
    $$a = 5$$
    $$\frac{x^2}{9} + \frac{y^2}{25} = 1$$
    $(10 - 2)$

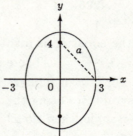

9. Start with $\frac{y^2}{a^2} - \frac{x^2}{b^2} = 1$ and find a and b.
    $b = \frac{8}{2} = 4$

    To find a, sketch the asymptote rectangle, label known parts, and use
    the Pythagorean Theorem.

    $$a^2 = 5^2 - 4^2$$
    $$a^2 = 9$$
    $$a = 3$$

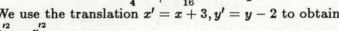

    Thus, the equation is

    $$\frac{y^2}{9} - \frac{x^2}{16} = 1 \quad (10 - 2)$$

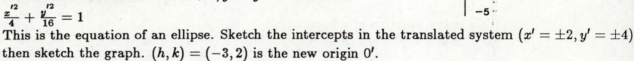

10.
    $$16x^2 + 4y^2 + 96x - 16y + 96 = 0$$
    $$16x^2 + 96x + 4y^2 - 16y = -96$$
    $$16(x^2 + 6x + ?) + 4(y^2 - 4y + ?) = -96$$
    $$16(x^2 + 6x + 9) + 4(y^2 - 4y + 4) = -96 + 144 + 16$$
    $$16(x + 3)^2 + 4(y - 2)^2 = 64$$
    $$\frac{(x+3)^2}{4} + \frac{(y-2)^2}{16} = 1$$

    We use the translation $x' = x + 3, y' = y - 2$ to obtain

    $$\frac{x'^2}{4} + \frac{y'^2}{16} = 1$$

    This is the equation of an ellipse. Sketch the intercepts in the translated system $(x' = \pm 2, y' = \pm 4)$
    then sketch the graph. $(h, k) = (-3, 2)$ is the new origin $0'$.
    $(10 - 4)$

11.
    $$x^2 - 4x - 8y - 20 = 0$$
    $$x^2 - 4x = 8y + 20$$
    $$x^2 - 4x + 4 = 8y + 24$$
    $$(x - 2)^2 = 4(2)(y + 3)$$

    We use the translation
    $x' = x - 2, y' = y + 3$ to obtain
    $x'^2 = 8y'$

    This is the equation of a parabola; since $a = 2$ and the axis of symmetry is the y'
    axis, it opens up.

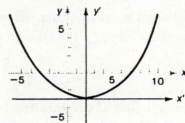

x'	0	± 4
y'	0	2

    $(h, k) = (2, -3)$ is the new origin $0'$. $(10 - 4)$

12.

$$4x^2 - 9y^2 + 24x - 36y + 36 \ = \ 0$$
$$4x^2 + 24x - 9y^2 - 36y \ = \ 36$$
$$4(x^2 + 6x + ?) - 9(y^2 + 4y + ?) \ = \ 36$$
$$4(x^2 + 6x + 9) - 9(y^2 + 4y + 4) \ = \ 36 + 36 - 36$$
$$4(x + 3)^2 - 9(y + 2)^2 \ = \ 36$$
$$\frac{(x+3)^2}{9} - \frac{(y-2)^2}{4} \ = \ 1$$

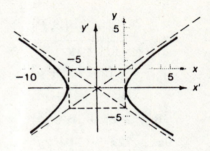

We use the translation $x' = x + 3, y' = y + 2$

to obtain

$$\frac{x'^2}{9} - \frac{y'^2}{4} = 1$$

This is the equation of a hyperbola with transverse axis on the x' axis. Sketch the asymptote rectangle $(a = 3, b = 2)$, asymptotes, and vertices in the translated system, then sketch in the hyperbola. $(h, k) = (-3, -2)$ is the new origin $0'$.
$(10 - 4)$

13.   From the figure we see that the
point $P(x, y)$ is a point on the curve
if and only if:

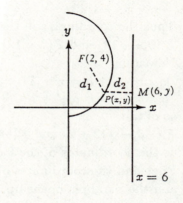

$$d_1 \ = \ d_2$$
$$d(F, P) \ = \ d(M, P)$$
$$\sqrt{(x - 2)^2 + (y - 4)^2} \ = \ \sqrt{(x - 6)^2 + (y - y)^2}$$
$$(x - 2)^2 + (y - 4)^2 \ = \ (x - 6)^2$$
$$x^2 - 4x + 4 + (y - 4)^2 \ = \ x^2 - 12x + 36$$
$$(y - 4)^2 \ = \ -8x + 32$$
$$(y - 4)^2 \ = \ -8(x - 4) \text{ or}$$
$$y^2 - 8y + 16 \ = \ -8x + 32$$
$$y^2 - 8y + 8x - 16 \ = \ 0 \quad (10 - 1)$$

14.   From the figure we see that the
point $P(x, y)$ is a point on the curve
if and only if:

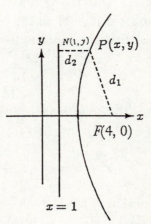

$$d_1 \ = \ 2d_2$$
$$d(F, P) \ = \ 2d(N, P)$$

$$\sqrt{(x - 4)^2 + (y - 0)^2} \ = \ 2\sqrt{(x - 1)^2 + (y - y)^2}$$
$$(x - 4)^2 + y^2 \ = \ 4(x - 1)^2$$
$$x^2 - 8x + 16 + y^2 \ = \ 4(x^2 - 2x + 1)$$
$$x^2 - 8x + 16 + y^2 \ = \ 4x^2 - 8x + 4$$
$$-3x^2 + y^2 \ = \ -12$$
$$\frac{x^2}{4} - \frac{y^2}{12} = 1$$

This is the equation of a hyperbola. $(10 - 3)$

15.   From the figure, we see that the point $P(x, y)$ is a point on the curve if and only if

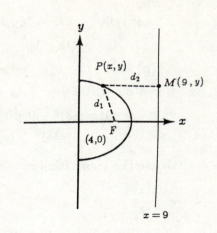

$$d_1 = \tfrac{2}{3}d_2$$
$$d(F,P) = \tfrac{2}{3}d(M,P)$$
$$\sqrt{(x-4)^2+(y-0)^2} = \tfrac{2}{3}\sqrt{(x-9)^2+(y-y)^2}$$
$$(x-4)^2+y^2 = \tfrac{4}{9}(x-9)^2$$
$$9[(x-4)^2+y^2] = 4(x-9)^2$$
$$9(x^2-8x+16+y^2) = 4(x^2-18x+81)$$
$$9x^2-72x+144+9y^2 = 4x^2-72x+324$$
$$5x^2+9y^2 = 180$$
$$\tfrac{x^2}{36}+\tfrac{y^2}{20} = 1$$

This is the equation of an ellipse.   (10 − 2)

16.   First find the coordinates of the foci in the translated system.
$$c'^2 = 4^2 - 2^2 = 12$$
$$c' = \sqrt{12}$$
$$-c' = -\sqrt{12}$$
Thus the coordinates in the translated system are
$F'(0,-\sqrt{12})$ and $F(0,\sqrt{12})$
Now use
$$x = x'+h = x'-3$$
$$y = y'+k = y'+2$$
to obtain
$F'(-3,-\sqrt{12}+2)$ and $F(-3,\sqrt{12}+2)$
as the coordinates of the foci in the original system.   (10 − 4)

17.   First find the coordinates of the focus in the translated system. Since $a = 2$
and the parabola opens up, they are (0,2). Now use
$$x = x'+h = x'+2 = 0+2 = 2$$
$$y = y'+k = y'-3 = 2-3 = -1$$
to obtain $(2,-1)$ as the coordinates of the focus in the original system.   (10 − 4)

18.   First find the coordinates of the foci in the original system.
$$c'^2 = 3^2 + 2^2 = 13$$
$$c' = \sqrt{13}$$
$$-c' = -\sqrt{13}$$
Thus the coordinates in the translated system are
$F'(-\sqrt{13},0)$ and $F(\sqrt{13},0)$
Now use
$$x = x'+h = x'-3$$
$$y = y'+k = y'-2$$
to obtain $F'(-\sqrt{13}-3,-2)$ and $F(\sqrt{13}-3,-2)$ as the coordinates of the foci in the
original system.   (10-4)

19.   From the figure, we see that the parabola opens up, hence its
equation must be of the form $x^2 - 4ay$. Since (4,1) is on
the graph, we have

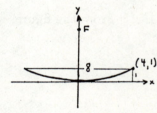

$$4^2 = 4a \cdot 1$$
$$16 = 4a$$
$$a = 4$$

Thus a, the distance of the focus from the vertex, is 4 feet.   (10-1)

20.   From the figure, we see that the x-intercepts must be at $(-5,0)$ and $(5,0)$, the foci at $(-4,0)$ and $(4,0)$. Hence $a = 5$ and $c = 4$. We can determine the y intercepts using the special triangle relationship

$$5^2 = 4^2 + b^2$$
$$25 = 16 + b^2$$
$$b = 3$$

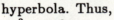

Hence, the equation of the ellipse is
$$\frac{x^2}{5^2} + \frac{y^2}{3^2} = 1 \quad (10-2)$$

21.   From the figure, we can see:

$d + a = y_1 =$ the y coordinate of the point on the hyperbola with x coordinate 15. From the equation of the hyperbola $\frac{y^2}{40^2} - \frac{x^2}{30^2} = 1$ we have $a = 40$, hence $d + 40 = y_1$, or $d = y_1 - 40$.

Since the point $(15, y_1)$ is on the hyperbola, its coordinates must satisfy the equation of the hyperbola. Thus,

$$\frac{y_1^2}{40^2} - \frac{15^2}{30^2} = 1$$
$$\frac{y_1^2}{40^2} = 1 + \frac{15^2}{30^2}$$
$$\frac{y_1^2}{40^2} = \frac{5}{4}$$
$$y_1^2 = 2000$$
$$y_1 = 44.72$$
$$\text{depth} = y_1 - 40 = 4.72 \text{ feet.} \quad (10-3)$$

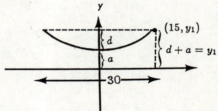

CHAPTER 11

Exercise 11-1

Key Ideas and Formulas

Multiplication Principle (Fundamental Counting Principle)

1. If two operations O_1 and O_2 are performed in order, with N_1 possible outcomes for the first operation and N_2 possible outcomes for the second operation, then there are

$$N_1 \cdot N_2$$

possible combined operations of the first operation followed by the second.

2. In general, if n operations $O_1, O_2, \cdots, O_n$ are performed in order, with possible number of outcomes $N_1, N_2, \cdots, N_n$, respectively, then there are

$$N_1 \cdot N_2 \cdot \cdots \cdot N_n$$

possible combined outcomes of the operations performed in the given order.

Number of Permutations of n objects: $P_{n,n} = n!$

Number of Permutations of n objects, taken r at a time:
$$P_{n,r} = n(n-1)\cdots(n-r+1) = \frac{n!}{(n-r)!}$$
$$0 \leq r \leq n \quad (0! = 1)$$

Number of Combinations of n objects, taken r at a time:
$$C_{n,r} = \binom{n}{r} = \frac{P_{n,r}}{r!} = \frac{n!}{r!(n-r)!} \quad 0 \leq r \leq n$$

In a permutation, the order of the objects counts. In combinations, the order of the objects does not count.

---

1. $\frac{11!}{8!} = \frac{11 \cdot 10 \cdot 9 \cdot 8!}{8!} = 990$

3. $\frac{5!}{2!3!} = \frac{5 \cdot 4 \cdot 3!}{2 \cdot 1 \; 3!} = 10$

5. $\frac{7!}{4!(7-4)!} = \frac{7!}{4!3!} = \frac{7 \cdot 6 \cdot 5 \cdot 4!}{4!3 \cdot 2 \cdot 1} = 35$

7. $\frac{7!}{7!(7-7)!} = \frac{7!}{7!0!} = \frac{7!}{7!(1)} = 1$

9. $P_{5,3} = \frac{5!}{(5-3)!} = \frac{5!}{2!} = \frac{5 \cdot 4 \cdot 3 \cdot 2!}{2!} = 60$

11. $P_{52,4} = \frac{52!}{(52-4)!} = \frac{52!}{48!} = \frac{52 \cdot 51 \cdot 50 \cdot 49 \cdot 48!}{48!} = 6,497,400$

13. $C_{5,3} = \frac{5!}{3 \; (5-3)!} = \frac{5!}{3!2!} = 10.$ (problem 3)

15. $C_{52,4} = \frac{52!}{4!(52-4!)} = \frac{P_{52,4}}{4!} = \frac{6,497,400}{4 \cdot 3 \cdot 2 \cdot 1} = 270,725$

17.

O_1 :	Selecting the color	N_1 :	5 ways
O_2 :	Selecting the transmission	N_2 :	3 ways
O_3 :	Selecting the interior	N_3 :	4 ways
O_4 :	Selecting the engine	N_4 :	2 ways

Applying the multiplication principle, there are $5 \cdot 3 \cdot 4 \cdot 2 = 120$ variations of the car.

19. Order is important here. We use permutations, selecting, in order, three horses
     out of ten:
     $P_{10,3} = 10 \cdot 9 \cdot 8 = 720$ different finishes

21. For the subcommittee, order is not important We use combinations, selecting
     three persons out of seven:
     $C_{7,3} = \frac{7!}{3!(7-3)!} = \frac{7!}{3!4!} = \frac{7 \cdot 6 \cdot 5 \cdot 4!}{3 \cdot 2 \cdot 1 \cdot 4!} = 35$ subcommittees
     In choosing a president, a vice-president, and a secretary, we can use permutations,
     or apply the multiplication principle.

     | O_1 : | Selecting the president | N_1 : | 7 ways |
     | O_2 : | Selecting the vice – president | N_2 : | 6 ways (the president is not considered) |
     | O_3 : | Selecting the secretary | N_3 : | 5 ways (the president is not considered) |

     Thus, there are $N_1 \cdot N_2 \cdot N_3 = 7 \cdot 6 \cdot 5 (= P_{7,3}) = 210$ ways

23. For each game, we are selecting two teams out of ten to be opponents. Since the
     order of the opponents does not matter (this has nothing to do with the
     order in which the games might be played, which is not under discussion here), we
     use combinations.
     $C_{10,2} = \frac{10!}{2!(10-2)!} = \frac{10!}{2!8!} = \frac{10 \cdot 9 \cdot 8!}{2 \cdot 1 \cdot 8!} = 45$ games

25.

No letter can be repeated			Allowing letters to repeat	
O_1:	Selecting first letter	N_1: 6 ways	N_1:	6 ways
O_2:	Selecting second letter	N_2: 5 ways	N_2:	6 ways
O_3:	Selecting third letter	N_3: 4 ways	N_3:	6 ways
O_4:	Selecting fourth letter	N_4: 3 ways	N_4:	6 ways

$P_{6,4} = 6 \cdot 5 \cdot 4 \cdot 3 = 360$ $6 \cdot 6 \cdot 6 \cdot 6 = 1,296$
possible code words   possible code words

27.

No digit can be repeated			Allowing digits to repeat	
O_1:	Selecting first digit	N_1 : 10 ways	N_1:	10 ways
O_2:	Selecting second digit	N_2 : 9 ways	N_2:	10 ways
$\vdots$		$\vdots$	$\vdots$	
O_5:	Selecting fifth digit	N_5 : 6 ways	N_5:	10 ways

$P_{10,5} = 10 \cdot 9 \cdot 8 \cdot 7 \cdot 6 \cdot = 30,240$ $10 \cdot 10 \cdot 10 \cdot 10 \cdot 10 = 100,000$
lock combinations   lock combinations.

29. We are selecting five cards out of the 13 hearts in the deck. The order is not important,
     so we use combinations.
     $C_{13,5} = \frac{13!}{5!(13-5)!} = \frac{13!}{5!8!} = \frac{13 \cdot 12 \cdot 11 \cdot 10 \cdot 9 \cdot 8!}{5 \cdot 4 \cdot 3 \cdot 2 \cdot 1 \cdot 8!} = 1,287$

31.   Repeats allowed   No repeats allowed

	Repeats allowed	No repeats allowed
O_1: Selecting first letter	N_1: 26 ways	N_1: 26 ways
O_2: Selecting second letter	N_2: 26 ways	N_1: 25 ways
O_3: Selecting third letter	N_3: 26 ways	N_1: 24 ways
O_4: Selecting first digit	N_4: 10 ways	N_4: 10 ways
O_5: Selecting second digit	N_5: 10 ways	N_4: 9 ways
O_6: Selecting third digit	N_6: 10 ways	N_6: 8 ways

$26 \cdot 26 \cdot 26 \cdot 10 \cdot 10 \cdot 10 = 17,576,000$ $26 \cdot 25 \cdot 24 \cdot 10 \cdot 9 \cdot 8 = 11,232,000$

license plates   license plates.

33. O_1: Choosing 5 spades out of 13 possible (order is not important)

     N_1: $C_{13,5}$

     O_2: Choosing 2 hearts out of 13 possible (order is not important)

     N_2: $C_{13,2}$

     Using the multiplication principle, we have:

$$\begin{aligned} \text{Number of hands} = C_{13,5} \cdot C_{13.2} &= \tfrac{13!}{5!(13-5)!} \cdot \tfrac{13!}{2!(13-2)!} \\ &= 1,287 \cdot 78 \\ &= 100,386 \end{aligned}$$

35. O_1: Choosing 3 appetizers out of 8 possible (order is not important)

     N_1: $C_{8,3}$

     O_2: Choosing 4 main courses out of 10 possible (order is not important)

     N_2: $C_{10,4}$

     O_3 : Choosing 2 desserts out of 7 possible (order is not important)

     N_3: $C_{7,2}$

     Using the multiplication principle, we have:

$$\begin{aligned} \text{Number of banquets} = C_{8,3} \cdot C_{10,4} \cdot C_{7,2} &= \tfrac{8!}{3!(8-3)!} \cdot \tfrac{10!}{4!(10-4)!} \cdot \tfrac{7!}{2!(7-2)!} \\ &= 56 \cdot 210 \cdot 21 \\ &= 246,960 \end{aligned}$$

37. O_1: Choosing a left glove out of 12 possible

     N_1: 12

     O_2: Choosing a right glove out of all the right gloves that do *not* match the left glove already chosen

     N_2: 11

     Using the multiplication principle, we have:

     Number of ways to mismatch gloves $= 12 \cdot 11 = 132$.

39.   A. We are choosing 2 points out of the 8 to join by a chord. Order is not important.

     $C_{8,2} = \frac{8!}{2!(8-2)!} = 28$ chords

     B. No three of the points can be collinear, since no line intersects a circle in more than two points. Thus, we can select any three of the 8 to use as vertices of the triangle. Order is not important.

     $C_{8,3} = \frac{8!}{3!(8-3)!} 56$ triangles

     C. We can select any four of the eight points to use as vertices of a quadrilateral. Order is not important.

$C_{8,4} = \frac{8!}{4!(8-4)!} = 70$ quadrilaterals

41.   To seat two people, we can seat one person, then the second person.

O_1: Seat the first person in any chair

N_1: 5 ways

O_2: Seat the second person in any remaining chair

N_2: 4 ways

Thus, applying the multiplication principle, we can seat two persons in $N_1 \cdot N_2 = 5.4 = 20$ ways.

We can continue this reasoning for a third person.

O_3: Seat the third person in any of the three remaining chairs

N_3: 3 ways

Thus we can seat 3 persons in $5 \cdot 4 \cdot 3 = 60$ ways.

For a fourth person:

O_4: Seat the fourth person in any of the two remaining chairs

N_4: 2 ways

Thus we can seat 4 persons in $5 \cdot 4 \cdot 3 \cdot 2 = 120$ ways

For a fifth person

O_5: Seat the fifth person. There will be only one chair remaining.

N_5: 1 ways

Thus we can seat 5 persons in $5 \cdot 4 \cdot 3 \cdot 2 \cdot 1 = 120$ ways

43.   A. Order is important, so we use permutations, selecting, in order, 5 persons out of 8:

$P_{8,5} = 8 \cdot 7 \cdot 6 \cdot 5 \cdot 4 = 6,720$ teams.

B. Order is not important, so we use combinations, selecting 5 persons out of 8

$C_{8,5} = \frac{8!}{5!(8-5)!} = \frac{8!}{5!3!} = \frac{8 \cdot 7 \cdot 6 \cdot 5}{5! \cdot 3 \cdot 2 \cdot 1} = 56$ teams

C. O_1: Selecting either Mike or Ken out of { Mike, Ken }

N_1: $C_{2,1}$

O_2: Selecting the 4 remaining players out of the

6 possibilities that do not include either Mike or Ken.

N_2: $C_{6,4}$

Using the multiplication principle, we have

$$
\begin{aligned}
N_1 \cdot N_2 = C_{2,1} \cdot C_{6,4} &= \frac{2!}{1!(2-1)!} \cdot \frac{6!}{4!(6-4)!} \\
&= \frac{2!}{1!1!} \cdot \frac{6!}{4!2!} \\
&= 2 \cdot 15 \\
&= 30 \text{ teams}
\end{aligned}
$$

Exercise 11-2

Key Ideas and Formulas

A simple event is the outcome of an experiment.

A sample space S for an experiment is the set of simple events possible. An event E is any subspace of S. If E is simple, it has one element.

If E has more than one element, it is a compound event. We say an event E occurs if any of the simple events in E occurs.

Given an sample space $S = \{e_1, e_2, \cdots, e_n\}$, to each simple event e_i we assign a real number $P(e_i)$ called the probability of the event e_i. An acceptable probability assignment is any assignment of the numbers e_i that satisfies the following two conditions:

1. $0 \le P(e_i) \le 1$

2. $P(e_1) + P(e_2) + \cdots + P(e_n) = 1$

Given an acceptable probability assignment for the simple events in a sample space S, we define the probability of an arbitrary event E, denoted by $P(E)$, as follows.

1. If E is the empty set, then $P(E) = 0$

2. If E is a simple event, then $P(E)$ has already been assigned.

3. If E is a compound event, then $P(E)$ is the sum of the probabilities of all the simple events in E.

4. If $E = S$, then $P(E) = P(S) = 1$

If we can assume each simple event in sample space S is as likely to occur as any other, than the probability of an arbitrary event E in S is given by

$$P(E) = \frac{\text{Number of elements in E}}{\text{Number of elements in S}} = \frac{n(E)}{n(S)}$$

---

1.   The occurrence of E is certain.

3.   Let $S = \{BB, BG, GB, GG\}. n(S) = 4$

Here $E = \{BG, BG\}$ the event that the two children have opposite sexes. $n(E) = 2$.

Then $P(E) = \frac{n(E)}{n(S)} = \frac{2}{4} = \frac{1}{2}$

5.   A. Must be rejected. It violates condition 1; no probability can be negative.

        B. Must be rejected. It violates condition 2, since $P(R) + P(G) + P(Y) + P(B) \neq 1$.

        C. Is an acceptable probability assignment.

7.   This is a compound event made up of the simple events R and Y.

        $P(E) = P(R) + P(Y) = .26 + .30 = .56$

9.   Let $S = \{BBB, BBG, BGB, BGG, GBB, GBG, GGB, GGG\}. n(S) = 8$

        The event E is the simple event $BBG. n(E) = 1$. Then $P(E) = \frac{n(E)}{n(S)} = \frac{1}{8}$.

11.   The sample space S is the set of all possible sequences of three digits with no repeats.

        $n(S) = P_{10,3} = 10 \cdot 9 \cdot 8 = 720$.

        The event is the simple event of one particular sequence.

        $n(E) = 1$. Then $P(E) = \frac{n(E)}{n(S)} = \frac{1}{720} \approx .0014$.

13.   The sample space S is the set of all possible 5-card hands, chosen from a deck of 52 cards,

        $n(S) = C_{52,5}$ The event E is the set of all possible 5-card hands that are all black,

        thus, are chosen from the 26 black cards. $n(E) = C_{26,5}$.

        $P(E) = \frac{n(E)}{n(S)} = \frac{C_{26,5}}{C_{52,5}} = \frac{26!}{5!(26-5)!} \div \frac{52!}{5!(52-5)!} \approx .025$

15.   The sample space S is the set of all possible 5-card hands, chosen from a deck of 52 cards.

        $n(S) = C_{52,5}$. The event E is the set of all possible 5-card hands that are all

        face cards, thus, are chosen from the 4 face cards in each of 4 suits, or

        16 face cards. $n(E) = C_{16,5}$

        $P(E) = \frac{n(E)}{n(S)} = \frac{C_{16,5}}{C_{52,5}} = \frac{16!}{5!(16-5)!} \div \frac{52!}{5!(52-5)!} \approx .0017$

17.   The sample space S is the set of all possible four digit numbers less than 5,000

        formed from the digits 1,3,5,7,9.

        $n(S) \quad = \quad N_1 \cdot N_2 \cdot N_3 \cdot N_4$

        where

        $N_1 \quad = \quad$ 2 (digits 1 or 3 only for the first digit)

        $N_2 \quad = \quad N_3 = N_4 = 5$ (digits 1, 3, 5, 7, 9)

        $n(S) \quad = \quad 2 \cdot 5 \cdot 5 \cdot 5 = 250$

        The event E is the set of those numbers in S which are divisible by 5.

        These must end in 5.

        $n(E) \quad = \quad N_1 \cdot N_2 \cdot N_3 \cdot N_4'$

        where

        $N_4' \quad = \quad$ 1 (digit 5 only)

        $n(E) \quad = \quad 2 \cdot 5 \cdot 5 \cdot 1 = 50$

        $P(E) = \frac{n(E)}{n(S)} = \frac{50}{250} = .2$

19.   The same space S is the set of all arrangements of the five notes in the five envelopes.

        $n(S) = P_{5,5} = 5! = 120$ The event E is the one correct arrangement. $n(E) = 1$

        $P(E) = \frac{n(E)}{n(S)} = \frac{1}{120} \approx .008$

        In problems 21-35, using Figure 1 of the text, we have that $S =$ the set of all ordered

        pairs of possible dots showing,

        $n(S) = 36$.

21.   The sum of the dots can be 2 in exactly one way. $E = \{(1,1)\}$ $n(E) = 1$

$P(E) = \frac{n(E)}{n(S)} = \frac{1}{36}$

23.   The sum of the dots can be 6 in 5 ways:

$E = \{(5,1),(4,2),(3,3),(2,4),(1,5)\}$

$n(E) = 5$

$p(E) = \frac{n(E)}{n(S)} = \frac{5}{36}$

25.   The sum of the dots can be less than 5 in 6 ways.

(upper left hand corner of figure) $n(E) = 6$.

$P(E) = \frac{n(E)}{n(S)} = \frac{6}{36} = \frac{1}{6}$

27.   The sum of the dots can be 7 in 6 ways and 11 in 2 ways. Thus

$n(E) = 36 - (6+2) = 28$

$P(E) = \frac{n(E)}{n(S)} = \frac{28}{36} = \frac{7}{9}$

29.   The sum of the dots cannot be 1. Thus $p(E) = 0$.

31.   The sum of the dots is divisible by 3 if the sum is 3,6,9, or 12. 3 can occur in 2 ways, 6 in 5 ways, 9 in 4 ways, and 12 in 1 way.

Thus $n(E) = 2 + 5 + 4 + 1 = 12$.

$P(E) = \frac{n(E)}{n(S)} = \frac{12}{36} = \frac{1}{3}$

33.   The sum of the dots can be 7 in 6 ways and 11 in 2 ways.

Thus $n(E) = 6 + 2 = 8$.

$P(E) = \frac{n(E)}{n(S)} = \frac{8}{36} = \frac{2}{9}$

35.   The sum of the dots will *not* be divisible by 2 or 3 if the sum is 5,7, or 11. 5 can occur in 4 ways, 7 in 6 ways, and 11 in 2 ways. Otherwise the sum will be divisible by 2 or 3.

Thus, $n(E) = 36 - (4 + 6 + 2) = 36 - 12 = 24$.

$P(E) = \frac{n(E)}{n(S)} = \frac{24}{36} = \frac{2}{3}$

37.   The sample space for this experiment $= \{HTH, HHH, THH, TTH\}$ assuming that the third coin is the one with a head on both sides.

Thus $n(S) = 4$. E is the event $\{TTH\}$. Thus $n(E) = 1$

$P(E) = \frac{n(E)}{n(S)} = \frac{1}{4}$

39.   See problem 37. $n(S) = 4$. E is the event $\{HHH\}$. Thus $n(E) = 1$

$P(E) = \frac{n(E)}{n(S)} = \frac{1}{4}$

41.   See problem 37. $n(S) = 4$ E is the event $\{HTH, HHH, THH\}$.

Thus $n(E) = 3$

$P(E) = \frac{n(E)}{n(S)} = \frac{3}{4}$

43.   Since it is equally likely that each die shows a   1,a   2 or a   3, we can draw the following figure to represent S, the set of all possibilities:

First   Die

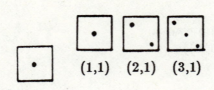

		(1,1)	(2,1)	(3,1)
Second Die		(1,2)	(2,2)	(3,2)
		(1,3)	(2,3)	(3,3)

$n(S) = 9$

The sum of the dots can be 2 in exactly one way. $n(E) = 1$

$P(E) = \frac{n(E)}{n(S)} = \frac{1}{9}$

45. See figure, problem 43, $n(S) = 9$. The sum of the dots can be 4 in 3 ways:

$E = \{(1,3),(2,2),(3,1)\}$

$n(E) = 3$

$P(E) = \frac{n(E)}{n(S)} = \frac{3}{9} = \frac{1}{3}$

47. See figure, problem 43. $n(S) = 9$. The sum of the dots can be 6 in exactly one way:

$E = \{(3,3)\}$. $n(E) = 1$

$P(E) = \frac{n(E)}{n(S)} = \frac{1}{9}$

49. See figure, problem 43. $n(S) = 9$. The sum of the dots will be odd if the sum
is 3 or 5.

$E = \{(1,2),(2,1)(2,3),(3,2)\}$ $n(E) = 4$

$P(E) = \frac{n(E)}{n(S)} = \frac{4}{9}$

51. The sample space S is the set of all possible 5-card hands, chosen from a deck of
52 cards. $n(S) = C_{52,5} = 2,598,960$. The event E is the set of all possible 5-card hands that
are all jacks through aces, thus, are chosen from the 16 cards
(4 jacks, 4 queens, 4 kings, 4 aces) of this type.

$n(E) = C_{16,5} = 4,368$.

$P(E) = \frac{n(E)}{n(S)} = \frac{C_{16,5}}{C_{52,5}} = \frac{4,368}{2,598,960} \approx .00168$

53. The sample space S is the set of all possible 5-card hands chosen from a deck of 52 cards.
$n(S) = C_{52,5}$. The event E is the set of all possible 5-card hands of the type $AAAAx$
where x is any of the 52-4=48 cards that are not aces. $n(E) = 48$

$P(E) = \frac{n(E)}{n(S)} = \frac{48}{C_{52,5}} = \frac{48}{2,598,960} \approx .000\ 0185$

55. The sample space S is the set of all possible 5-card hands, chosen from a deck of 52 cards.
$n(S) = C_{52,5}$. The event E is the set consisting of the cards ace, king, queen, jack, ten,
in each of the 4 suits, abbreviated:

{club AKQJ10, diamond AKQJ10, heart AKQJ10, spade AKQJ10}. Thus $n(E) = 4$.

$P(E) = \frac{n(E)}{n(S)} = \frac{4}{C_{52,5}} = \frac{4}{2,598,960} \approx .000\ 0015$

57. The sample space S is the set of all possible 5-card hands, chosen from a deck of 52 cards.
$n(S) = C_{52,5}$ The event E is the set of all possible 5-card hands chosen as follows:
O_1 : Choose 2 aces out of the set of 4 aces
N_1 : $C_{4,2}$
O_2 : Choose 3 queens out of the set of 4 queens
N_2 : $C_{4,3}$
Thus, applying the multiplication principle, $n(E) = N_1 N_2 = C_{4,2} C_{4,3}$
$$P(E) = \frac{n(E)}{n(S)} = \frac{C_{4,2} \cdot C_{4,3}}{C_{52,5}} = \frac{6 \cdot 4}{2,598,960} = .000\ 009$$

Exercise 11-3

Key Ideas and Formulas

If an experiment is conducted n times and an event occurs with frequency $f(E)$, then the ratio $\frac{f(E)}{n}$ is called the relative frequency of the occurrence of event E in n trials. We define the empirical probability of E, denoted by $P(E)$, by the number (if it exists) that $\frac{f(E)}{n}$ gets closer and closer to as n gets larger and larger.

Then $P(E)$ is approximately equal to $\frac{f(E)}{n}$ for any n, with the approximation getting better as n gets larger.

---

1. $P(E) \approx \frac{f(E)}{n} = \frac{25}{250} = .1$

3. $P(E) \approx \frac{f(E)}{n} = \frac{189}{420} = .45$

5. $P \text{ (point down)} \approx \frac{389}{1,000} = .389 \quad P \text{ (point up)} \approx \frac{611}{1,000} = .611$

7. A. $P(\text{ two girls }) \approx \frac{2,351}{10,000} = .2351$
       $P(\text{ one girl }) \approx \frac{5,435}{10,000} = .5435$
       $P(\text{ no girls }) \approx \frac{2,214}{10,000} = .2214$

    B. A sample space of equally likely events is:
    $S = \{BB, BG, GB, GG\}$

    Let $E_1 = 2 \text{ girls} = \{GG\}$
         $E_2 = 1 \text{ girl} = \{BG, GB\}$
         $E_3 = 0 \text{ girls} = \{BB\}$

    then
    $P(E_1) = \frac{n(E_1)}{n(S)} = \frac{1}{4} = .25 \quad P(E_2) = \frac{n(E_2)}{n(S)} = \frac{2}{4} = .50 \quad P(E_3) = \frac{n(E_3)}{n(S)} = \frac{1}{4} = .25$

9. A. $P \text{ (three heads)} \approx \frac{132}{1,000} = .132$
       $P \text{ (two heads)} \approx \frac{368}{1,000} = .368$
       $P \text{ (one head)} \approx \frac{380}{1,000} = .380$
       $P \text{ (no heads)} \approx \frac{120}{1,000} = .12$

    B. A sample space of equally likely events is:
    $S = \{HHH, HHT, HTH, HTT, THH, THT, TTH, TTT\}$

    Let $E_1 = 3 \text{ heads} = \{HHH\}$
         $E_2 = 2 \text{ heads} = \{HHT, HTH, THH\}$
         $E_3 = 1 \text{ head} = \{HHT, THT, TTH\}$
         $E_4 = 0 \text{ heads} = \{TTT\}$

    then

$$P(E_1) = \frac{n(E_1)}{n(S)} = \frac{1}{8} = .125$$

$$P(E_2) = \frac{n(E_2)}{n(S)} = \frac{3}{8} = .375$$

$$P(E_3) = \frac{n(E_3)}{n(S)} = \frac{3}{8} = .375$$

$$P(E_4) = \frac{n(E_4)}{n(S)} = \frac{1}{8} = .125$$

C. 　　　　Expected frequency $= P(E) \cdot$ number of trials

Expected frequency of 3 heads $= P(E_1) \cdot 1,000 = .125(1,000) = 125$

Expected frequency of 2 heads $= P(E_2) \cdot 1,000 = .375(1,000) = 375$

Expected frequency of 1 head $= P(E_3) \cdot 1,000 = .375(1,000) = 375$

Expected frequency of 0 heads $= P(E_4) \cdot 1,000 = .125(1,000) = 125$

11.　A sample space of equally likely events is:

$$S = \begin{array}{l} \{HHHH, HHHT, HHTH, HHTT, HTHH, HTHT, HTTH, HTTT, \\ THHH, THHT, THTH, THTT, TTHH, TTHT, TTTH, TTTT\} \end{array}$$

Let

$E_1 = $ 4 heads $= \{HHHH\}$

$E_2 = $ 3 heads $= \{HHHT, HHTH, HTHH, THHH\}$

$E_3 = $ 2 heads $= \{HHTT, HTHT, HTTH, THHT, THTH, TTHH\}$

$E_4 = $ 1 head $= \{HTTT, THTT, TTHT, TTTH\}$

$E_5 = $ 0 heads $= \{TTTT\}$

then Expected frequency $= P(E) \cdot$ number of trials $= \frac{n(E)}{n(S)} \cdot 80$

Expected frequency of 4 heads $= \frac{n(E_1)}{n(S)} \cdot 80 = \frac{1}{16} \cdot 80 = 5$

Expected frequency of 3 heads $= \frac{n(E_2)}{n(S)} \cdot 80 = \frac{4}{16} \cdot 80 = 20$

Expected frequency of 2 heads $= \frac{n(E_3)}{n(S)} \cdot 80 = \frac{6}{16} \cdot 80 = 30$

Expected frequency of 1 head $= \frac{n(E_4)}{n(S)} \cdot 80 = \frac{4}{16} \cdot 80 = 20$

Expected frequency of 0 heads $= \frac{n(E_5)}{n(S)} \cdot 80 = \frac{1}{16} \cdot 80 = 5$

13.　A. P(this event) $= \frac{15}{1,000} = .015$

B. P(this event) $= \frac{130+80+12}{1,000} = .222$

C. P(this event) $= \frac{30+32+25+10+21+12+1}{1,000} = .169$

D. P(this event) $= 1 - P$(owning zero television sets) $= 1 - \frac{2+10+30}{1,000} = .958$

Exercise 11-4

CHAPTER REVIEW

1. A. The outcomes can be displayed in a tree diagram as follows:

(1,H)
(1,T)
(2,H)
(2,T)
(3,H)
(3,T)
(4,H)
(4,T)
(5,H)
(5,T)
(6,H)
(6,T)

B. O_1 : Rolling the die
N_1 : 6 outcomes
O_2 : Flipping the coin
N_2 : 2 outcomes

Applying the multiplication principle, there are $6 \cdot 2 = 12$ combined outcomes $(11-1)$

2. $C_{6,2} = \frac{6!}{2!(6-2)!} = \frac{6!}{2!4!} = \frac{6 \cdot 5 \cdot 4!}{2 \cdot 1 \cdot 4!} = 15$ $P_{6,2} = 6 \cdot 5 = 30$ $(11-1)$

3. O_1 : Seating the first person N_1 : 6 ways
O_2 : Seating the second person N_2 : 5 ways
O_3 : Seating the third person N_3 : 4 ways
O_4 : Seating the fourth person N_4 : 3 ways
O_5 : Seating the fifth person N_5 : 2 ways
O_6 : Seating the sixth person N_6 : 1 way

Applying the multiplication principle, there are $6 \cdot 5 \cdot 4 \cdot 3 \cdot 2 \cdot 1 = 720$ arrangements.
$(11-1)$

4. Order is important here. We use permutations to determine the number of arrangements of 6 objects. $P_{6,6} = 6! = 720$
$(11-1)$

5. The sample space S is the set of all possible 5-card hands, chosen from a deck of 52 cards. $n(S) = C_{52,5}$

The event E is the set of all possible 5-card hands that are all clubs, thus, are chosen from the 13 clubs. $n(E) = C_{13,5}$

$$p(E) = \frac{n(E)}{n(S)} = \frac{C_{13,5}}{C_{52,5}} = \frac{13!}{5!(13-5)!} \div \frac{52!}{5!(52-5)!} \approx .0005$$

$(11-2)$

6.   The sample space S is the set of all possible arrangements of two persons, chosen from a set of 15 persons. Order is important here, so we use permutations. $n(S) = P_{15,2}$.

The event E is one of these arrangements. $n(E) = 1$.

$$p(E) = \frac{n(E)}{n(S)} = \frac{1}{P_{15,2}} = \frac{1}{15 \cdot 14} \approx .005 \quad (11-2)$$

7.   In the first case, order is important. The sample space is the set of all possible arrangements of three letters, drawn from a set of 10 letters. $n(S) = P_{10,3}$.

The event E is one of these arrangements. $n(E) = 1$.

$$p(E) = \frac{n(E)}{n(S)} = \frac{1}{P_{10,3}} = \frac{1}{10 \cdot 9 \cdot 8} \approx .0014$$

In the second case, order is not important. The sample space is the set of all possible three-card hands, chosen from a deck of ten cards. $n(S) = C_{10,3}$

The event E is one of these hands. $n(E) = 1$.

$$p(E) = \frac{n(E)}{n(S)} = \frac{1}{C_{10,3}} = 1 \div \frac{10!}{3!(10-3)!} \approx .0083$$

$(11-2)$

8. $P(E) \approx \frac{f(E)}{n} = \frac{50}{1,000} = .05$

$(11-3)$

9.   The probability of an event cannot be negative, but $P(e_2)$ is given as negative. The sum of the probabilities of the simple events must be 1, but it is given as 2.5. The probability of an event cannot be greater than 1, but $P(e_4)$ is given as 2.

$(11-2)$

10.   We can select any three of the six points to use as vertices of the triangle. Order is not important.

$C_{6,3} = \frac{6!}{3!(6-3)!} = 20$ triangles

11.

	Case 1	Case 2	Case 3
O_1: select the first letter	N_1: 8 ways	8 ways	8 ways
O_2: select the first letter	N_2: 7 ways	8 ways	7 ways
O_3: select the first letter	N_3: 6 ways (exclude first and second letter)	8 ways	7 ways (exclude second letter.)
	$8 \cdot 7 \cdot 6 = 336$ words	$8 \cdot 8 \cdot 8 = 512$ words	$8 \cdot 7 \cdot 7 = 392$ words

$(11-1)$

12.   A. Order is important here. We are selecting 3 digits out of 6 possible.

$P_{6,3} = 6 \cdot 5 \cdot 4 = 120$ lock combinations

B. Order is not important here. We are selecting 2 players out of 5.

$C_{5,2} = \frac{5!}{2!(5-2)!} = 10$ games

$(11-1)$

13.   A. $P(2 \text{ heads}) \approx \frac{210}{1,000} = .21$

$P(1 \text{ head}) \approx \frac{480}{1,000} = .48$

$P(0 \text{ heads}) \approx \frac{310}{1,000} = .31$

B. A sample space of equally likely events is:

$S = \{HH, HT, TH, TT\}$

$\text{Let} E_1 = 2 \text{ heads} = \{HH\}$

$E_2 = 1 \text{ head} = \{HT, TH\}$

$E_3 = 0 \text{ heads} = \{TT\}$

then $P(E_1) = \frac{n(E_1)}{n(S)} = \frac{1}{4} = .25$

$P(E_2) = \frac{n(E_2)}{n(S)} = \frac{2}{4} = .5$

$P(E_3) = \frac{n(E_3)}{n(S)} = \frac{1}{4} = .25$

C. Expected frequency $= P(E) \cdot$ number of trials

Expected frequency of 2 heads $= P(E_1) \cdot 1,000 = .25(1,000) = 250$

Expected frequency of 1 head $= P(E_2) \cdot 1,000 = .5(1,000) = 500$

Expected frequency of 0 heads $= P(E_3) \cdot = .25(1,000) = 250.$

$(11-3)$

14.   The sample space S is the set of all possible 5-card hands chosen from a deck of 52 cards. $n(S) = C_{52,5}.$

A. The event E is the set of all possible 4-card hands that are all diamonds, thus, are chosen from the 13 diamonds. $n(E) = C_{13,5}$

$p(E) = \frac{n(E)}{n(S)} = \frac{C_{13,5}}{C_{52,5}}$

B. The event E is the set of all possible 5-card hands chosen as follows:

O_1 : Choose 3 diamonds out of the 13 in the deck. N_1: $C_{13,3}$

O_2 : Choose 2 spades out of the 13 in the deck. N_2 : $C_{13,2}$

Applying the multiplication principle, $n(E) = N_1 \cdot N_2 = C_{13,3} \cdot C_{13 \cdot 2}$

$p(E) = \frac{n(E)}{n(S)} = \frac{C_{13,3} \cdot C_{13,2}}{C_{52,5}}$ $(11-2)$

15.   The sample space S is the set of all possible 4-person choices out of the 10 people.

$n(S) = C_{10,4}$

The event E is the set of all of those choices that include 2 particular people, thus

O_1:   Choose the 2 married people $N_1 : 1$

O_2:   Choose 2 more people out of the 8 remaining $N_2 : C_{8,2}$

Applying the multiplication principle, $n(E) = N_1 \cdot N_2 = 1 \cdot C_{8,2} = C_{8,2}$

$p(E) = \frac{n(E)}{n(S)} = \frac{C_{8,2}}{C_{10,4}} = \frac{8!}{2!(8-2)!} \div \frac{10!}{4!(10-4)!} = 28 \div 210 = \frac{2}{15}$

$(11-2)$

16.   The sample space S is the set of all possible results of spinning the device twice:

$\{ (1,1), (1,2), (1,3), (2,1), (2,2), (2,3), (3,1), (3,2), (3,3)\}.$ $n(S) = 9.$

A. The event E is the set of all possible results in which both spins are the same.

$E = \{(1,1), (2,2), (3,3)\}. n(E) = 3$

$p(E) = \frac{n(E)}{n(S)} = \frac{3}{9} = \frac{1}{3}$

B. The event E is the set of those spin sequences that add up to 5:

$E = \{(2,3), (3,2)\}. \quad n(E) = 2$

$p(E) = \frac{n(E)}{n(S)} = \frac{2}{9}$

$(11-2)$

17.   In the first case, the order matters. We have five successive events, each of

which can happen two ways (girl or boy). Applying the multiplication principle, there are $2 \cdot 2 \cdot 2 \cdot 2 \cdot 2 = 32$ possible families. In the second case, the possible families can be listed as {0 girls, 1 girl, 2 girls, 3 girls, 4 girls, 5 girls}, thus, there are 6 possibilities. (11 − 1)

18. Since there are several ways in which at least one woman can be selected, it is simplest to note: probability of selecting of at least one woman = 1− probability of selecting 0 women. To calculate the probability of selecting 0 women, that is, three men, we note:

The sample space S is the set of all 3-person subsets, chosen from the ten people.

$n(S) = C_{10,3}$

The event E is the set of all 3-man subsets, chosen from the 7 men.

$n(E) = C_{7,3}$

Then $p(E) = \frac{C_{7,3}}{C_{10,3}}$

Hence p (at least one woman) $= 1 - p$ (0 women) $= 1 - p(E) = 1 - \frac{C_{7,3}}{C_{10,3}} = 1 - \frac{7}{24} = \frac{17}{24}$

(11 − 2)

19. To seat two people, we can seat one person, then the second person.

O_1: Seat the first person in any chair.

N_1: 4 ways

O_2: Seat the second person in any remaining chair.

N_2: 3 ways

Thus, applying the multiplication principle, there are $4 \cdot 3 = 12$ ways to seat two persons.

20. A. $p(E) \approx \frac{f(E)}{n} = \frac{350}{1,000} = .350$

B. A sample space of equally likely events is:

$S = \{HHH, HHT, HTH, HTT, THH, THT, TTH, TTT\}$

Let $E = 2$ heads $= \{HHT, HTH, THH\}$

then $P(E) = \frac{n(E)}{n(S)} = \frac{3}{8} = .375$

C. Expected frequency $= P(E) \cdot$ number of trials $= .375(1,000) = 375$

(11 − 1)

21. A route plan can be regarded as a series of choices of stores, thus an arrangement of the 5 stores. Since the order matters, we use permutations:

$P_{5,5} = 5! = 120$ route plans. (11 − 1)

22. A. $P($ this event$) = \frac{40}{1,000} = .04$

B. $P($this event$) = \frac{100+60}{1,000} = .16$

C. $P($this event$) = 1 - P($ not$12 - 18$ and buys 0 or 1 cassette annually$)$

$= 1 - \frac{60+70+70+110+100+50}{1,000} = 1 - \frac{460}{1,000} = .54$

(11 − 3)

23. P (shipment returned) $= 1 - P$ (no substandard part found).

The sample space S is the set of all possible 4-part subsets of the 12 part set.

$n(S) = C_{12,4}$

The event, no substandard part found, is the set of all possible 4-part subsets of the 10 acceptable parts. $n(E) = C_{10,4}$.

Then P (shipment returned) $= 1 - \frac{n(E)}{n(S)} = 1 - \frac{C_{10,4}}{C_{12,4}} = 1 - \frac{14}{33} \approx .576.$ (11 − 2)

APPENDICES
APPENDIX B
Key Ideas and Formulas

y varies directly as x means $y = kx$ $k \neq 0$

y varies inversely as x means $y = \frac{k}{x}$ $k \neq 0$

w varies jointly as x and y means $w = kxy$ $k \neq 0$

To solve a typical variation problem, there are two methods.

Method 1:

Write the equation of variation. Find k using the first set of values. Answer the question using k and the second set of values.

Method 2:

Write the equation of variation and solve for k. Then apply subscripts to the variables and equate the results. Substitute all known values and solve for the unknown.

(Thus, if $y = kx$, $k = \frac{y}{x}$, and $\frac{y_1}{x_1} = \frac{y_2}{x_2}$.)

---

1. $F = kv^2$

3. $f = k\sqrt{T}$

5. $y = \frac{k}{\sqrt{x}}$

7. $t = \frac{k}{T}$

9. $R = kSTV$

11. $V = khr^2$

13.
$$u = k\sqrt{v}$$
$$\frac{u}{\sqrt{v}} = k$$
$$\frac{u_1}{\sqrt{v_1}} = \frac{u_2}{\sqrt{v_2}}$$
$$\frac{2}{\sqrt{2}} = \frac{u_2}{\sqrt{8}}$$
$$\frac{2\sqrt{8}}{\sqrt{2}} = u_2$$
$$4 = u_2$$
$$u_2 = 4$$

15.
$$L = \frac{k}{\sqrt{M}}$$
$$L\sqrt{M} = k$$
$$L_1\sqrt{M_1} = L_2\sqrt{M_2}$$
$$9\sqrt{9} = L_2\sqrt{3}$$
$$27 = L_2\sqrt{3}$$
$$\frac{27}{\sqrt{3}} = L_2$$
$$L_2 = 9\sqrt{3}$$

17. $V = k\frac{ab}{c^3}$

19. $L = k\frac{wh^2}{\ell}$

21.
$$Q = k\frac{mn^2}{p}$$
$$\frac{PQ}{mn^2} = k$$
$$\frac{P_1Q_1}{m_1n_1^2} = \frac{P_2Q_2}{m_2n_2^2}$$
$$\frac{12(-4)}{6(2)^2} = \frac{6Q_2}{4(3)^2}$$
$$-2 = \frac{Q_2}{6}$$
$$Q_2 = -12$$

23.
$$w = \frac{k}{d^2}$$
$$wd^2 = k$$
$$w_1d_1^2 = w_2d_2^2$$
$$d_1 = 4000 \text{ on the surface}$$
$$d_2 = 4000 + 400 = 4400$$
$$100(4000)^2 = w_2(4400)^2$$
$$\frac{100(4000)^2}{(4400)^2} = w_2$$
$$w_2 = 83 \text{ pounds}$$

25.
$$I = k\frac{E}{R}$$
$$\frac{RI}{E} = k$$
$$\frac{R_1I_1}{E_1} = \frac{R_2I_2}{E_2}$$
$$\frac{5(22)}{110} = \frac{11I_2}{220}$$
$$1 = \frac{I_2}{20}$$
$$I_2 = 20 \text{ amperes}$$

27. $P = kv^3$. If v is replaced by $2v$, $P = k(2v)^3 = 8kv^3$, so P is multiplied by 8. The new horsepower must be 8 times the old.

29. $f = \frac{k\sqrt{T}}{L}$. If T is replaced by $4T$ and L by $2L$, then $f = \frac{k\sqrt{4T}}{2L} = \frac{KT}{L}$ will be unchanged. No effect.

31. $t^2 = kd^3$.

33.
$$t = k\frac{r}{v}$$
$$\frac{tv}{r} = k$$
$$\frac{t_1 v_1}{r_1} = \frac{t_2 v_2}{r_2}$$
$$\frac{(1.42)(18,000)}{4050} = \frac{t_2(18,500)}{4300}$$
$$\frac{(1.42)(18,000)(4,300)}{(4050)(18,500)} = t_2$$
$$t_2 = 1.47 \text{ hours}$$

35.
$$d = kh$$
$$\frac{d}{h} = k$$
$$\frac{d_1}{h_1} = \frac{d_2}{h_2}$$
$$\frac{4}{500} = \frac{d_2}{2500}$$
$$\frac{2500(4)}{500} = d_2$$
$$d_2 = 20 \text{days}$$

37. $L = kv^2$. If v is replaced by $2v$, then $L = k(2v)^2 = 4kv^2$. The length will be quadrupled.

39.
$$P = kAV^2$$
$$\frac{P}{Av^2} = k$$
$$\frac{P_1}{A_1 v_1^2} = \frac{P_2}{A_2 v_2^2}$$
$$\frac{120}{(100)(20)^2} = \frac{P_2}{(200)(30)^2}$$
$$\frac{120}{40,000} = \frac{P_2}{180,000}$$
$$P_2 = 540 \text{ pounds}$$

41. A. $\Delta S = kS$

B.
$$k = \frac{\Delta S}{S}$$
$$\frac{\Delta S_1}{S_1} = \frac{\Delta S_2}{S_2}$$
$$\frac{1}{50} = \frac{\Delta S_2}{500}$$
$$\Delta S_2 = 10 \text{ ounces}$$

C.
$$\frac{1}{60} = \frac{\Delta S_2}{480}$$
$$\Delta S_2 = 8 \text{ candlepower}$$

43. Let f = frequency, ℓ = length
$$f = \frac{k}{\ell}$$
$$f\ell = k$$
$$f_1 \ell_1 = f_2 \ell_2$$
$$(16)(32) = f_2(16)$$
$$f_2 = 32 \text{ times per second}$$

45. $N = k\frac{F}{d}$

47. $v = \frac{k}{\sqrt{w}}$

Let
$$w_1 = \text{weight of oxygen molecule}$$
$$\text{Then } \tfrac{1}{16}w_1 = \text{weight of hydrogen molecule}$$
$$v\sqrt{w} = k$$
$$v_1\sqrt{w_1} = v_2\sqrt{w_2}$$
$$0.3\sqrt{w_1} = v_2\sqrt{\tfrac{1}{16}w_1}$$
$$v_2 = \frac{0.3\sqrt{w_1}}{\sqrt{\tfrac{1}{16}w_1}}$$
$$v_2 = \frac{0.3\sqrt{w_1}}{\tfrac{1}{4}\sqrt{w_1}}$$
$$v_2 = \frac{0.3}{1/4} = (0.3)4$$
$$v_2 = 1.2 \text{ miles per second}$$

49.
$$A = kWt$$
$$\frac{A}{Wt} = k$$
$$\frac{A_1}{W_1 t_1} = \frac{A_2}{W_2 t_2}$$

In this problem $A_1 = A_2$
$$\frac{A_1}{(10)(8)} = \frac{A_1}{4t}$$
$$\frac{A_1}{80} = \frac{A_1}{4t}$$
$$80 = 4t$$
$$20 = t$$
$$20 \text{ days}$$

51. $v = kr^2$. If r is replaced by $2r$,
$$v = k(2r)^3 = 8kv^3,$$
so v is increased by a factor of 8.

APPENDIX C

Key Ideas and Formulas

A function f is a rational function if $f(x) = \frac{n(x)}{d(x)}$ $d(x) \neq 0$ where $n(x)$ and $d(x)$ are polynomials. The domain of f is the set of all real numbers such that $d(x) \neq 0$.

If $f(x) = \frac{n(x)}{d(x)}$ and $d(a) = 0$, then f is discontinuous at $x = a$ and the graph of f has a hole or break at $x = a$. If $n(c) = 0$ and $d(c) \neq 0$, then $x = c$ is an x intercept for the graph of f.

Asymptotes: The line $x = a$ is a vertical asymptote for the graph of $y = f(x)$ if $f(x)$ either increases or decreases without bound as x approaches a from the right or from the left. $f(x) \to \infty$ or $f(x) \to -\infty$ as $x \to a^+$ or $x \to a^-$.

If $f(x) = \frac{n(x)}{d(x)}, d(a) = 0$ and $n(a) \neq, 0$ then the line $x = a$ is a vertical asymptote.

The line $y = b$ is a horizontal asymptote for the graph of $y = f(x)$ if $f(x)$ approaches b as x increases without bound or as x decreases without bound. $f(x) \to b$ as $x \to \infty$ or $x \to -\infty$.

If $f(x) = \frac{a_m x^m + \cdots + a_1 x + a_0}{b_n x^n + \cdots + b_1 x + b_0} a_m, b_n \neq 0$, then

1. for $m < n$ the x axis a horizontal asymptote
2. for $m = n$ the line $y = \frac{a_m}{b_n}$ is a horizontal asymptote
3. for $m > n$ there are no horizontal asymptotes

If m, the degree of $n(x)$, is one more than n, the degree of $d(x)$, then $f(x)$ can be written in the form $f(x) = mx + b + \frac{r(x)}{d(x)}$ where the degree of $r(x)$ is less than the degree of $d(x)$. Then the line $y = mx + b$ is an oblique asymptote for the graph of f.

$$[f(x) - (mx + b)] \to 0 \text{ as } x \to -\infty \text{ or } x \to \infty$$

To graph a rational function, follow the steps
1. Intercepts
2. Vertical asymptotes
3. Sign chart
4. Horizontal asymptotes
5. Symmetry
6. Complete the sketch. For details, see text.

---

1. $\frac{2x-4}{x+1}$ *Domain* : $d(x) = x + 1$ zero : $x = -1$ domain : $(-\infty, -1) \cup (-1, \infty)$
    x *intercepts* : $n(x) = 2x - 4$ zero : $x = 2$ x intercept : 2

3. $\frac{x^2-1}{x^2-16}$ *Domain* : $d(x) = x^2 - 16$ zeros : $x^2 - 16 = 0$
    $$x^2 = 16$$
    $$x = \pm 4$$
    domain:$(-\infty, -4) \cup (-4, 4) \cup (4, \infty)$
    x *intercepts* : $n(x) = x^2 - 1$ zeros : $x^2 - 1 = 0$
    $$x^2 = 1$$
    $$x = \pm 1$$
    x intercepts:$-1, 1$

5. $\frac{x^2-x-6}{x^2-x-12}$ Domain : $d(x) = x^2 - x - 12$ zeros : $x^2 - x - 12 = 0$
$$(x+3)(x-4) = 0$$
$$x = -3, 4$$

domain : $(-\infty, -3) \cup (-3, 4) \cup (4, \infty)$

x intercepts : $n(x) = x^2 - x - 6$ zeros : $x^2 - x - 6 = 0$
$$(x+2)(x-3) = 0$$
$$x = -2, 3$$

x intercepts: $-2, 3$

7. $\frac{x}{x^2+4}$ Domain : $d(x) = x^2 + 4$ no real zeros

Domain: R

x intercepts : $n(x) = x$ zero : $x = 0$

x intercept: 0

9. $\frac{2x}{x-4}$ vertical asymptotes : $d(x) = x - 4$ zero : $x = 4$

vertical asymptote $x = 4$

Common Error.

$x = 0$ is not a vertical asymptote.

horizontal asymptotes: Since $n(x)$ and $d(x)$ have the same degree,
the line $y = 2$ is a horizontal asymptote.

11. $\frac{2x^2+3x}{3x^2-48}$ vertical asymptotes : $d(x) = 3x^2 - 48$ Zeros : $3x^2 - 48 = 0$
$$3(x+4)(x-4) = 0$$
$$x = -4, 4$$

vertical asymptotes: $x = -4, x = 4$

horizontal asymptotes: Since $n(x)$ and $d(x)$ have the same degree, the line
$y = \frac{2}{3}$ is a horizontal asymptote.

13. $\frac{2x}{x^4+1}$ vertical asymptotes: $d(x) = x^4 + 1$ No real zeros:

No vertical asymptotes

horizontal asymptotes: Since the degree of $n(x)$ is less than the degree of
$d(x)$, the x - axis is a horizontal asymptote

horizontal asymptote: $y = 0$

15. $\frac{6x^4}{3x^2-2x-5}$ vertical asymptotes : $d(x) = 3x^2 - 2x - 5$ zeros :
$$3x^2 - 2x - 5 = 0$$
$$(3x-5)(x+1) = 0$$
$$x = \frac{5}{3}, -1$$

vertical asymptotes: $x = -1, x = \frac{5}{3}$

horizontal asymptotes: Since the degree of $n(x)$ is greater than the degree of $d(x)$,
there are no horizontal asymptotes.

17. $f(x) = \frac{1}{x-4} = \frac{n(x)}{d(x)}$

Intercepts. There are no real zeros of $n(x) = 1$. No x intercept

$f(0) = -\frac{1}{4}$ $y = -\frac{1}{4}$ y intercept

Vertical asymptotes $d(x) = x - 4$ zeros: 4 $x = 4$

Sign Chart.

Test numbers	3	5
Value of f	-1	1
Sign of f	$-$	$+$

Since $x = 4$ is a vertical asymptote and $f(x) < 0$ for $x < 4$,

    $f(x) \to -\infty$ as $x \to 4^-$

Since $x = 4$ is a vertical asymptote and $f(x) > 0$ for $x < 4$,

    $f(x) \to \infty$ as $x \to 4^+$

Horizontal asymptotes. Since the degree of $n(x)$ is less than the degree of $d(x)$, the x axis is a horizontal asymptote.

Symmetry. $f(-x) = \frac{1}{-x-4}$. Since $f(-x) \neq f(x)$ and $f(-x) \neq -f(x)$, the graph is not symmetric with respect to the y axis or the origin

Complete the sketch.

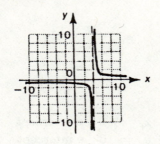

19. $p(x) = \frac{-1}{x-4} = \frac{n(x)}{d(x)}$

Intercepts. There are no real zeros of $n(x) = -1$. No x intercept

    $p(0) = \frac{1}{4}$ $y = \frac{1}{4}$ y intercept

Vertical asymptotes: $d(x) = x - 4$ zeros: 4 $x = 4$

Sign Chart.

Test numbers	3	5
Value of p	1	-1
Sign of p	$+$	$-$

Since $x = 4$ is a vertical asymptote and $p(x) > 0$ for $x < 4$

$p(x) \to \infty$ as $x \to 4^-$

Since $x = 4$ is a vertical asymptote and $p(x) < 0$ for $x > 4$,

$p(x) \to -\infty$ as $x \to 4^+$

Horizontal asymptotes. Since the degree of $n(x)$ is less than the degree of $d(x)$,

the x axis is a horizontal asymptote.

Symmetry. $p(-x) = \frac{-1}{-x-4} = \frac{1}{x+4}$. Since $p(-x) \neq p(x)$ and $p(-x) \neq -p(x)$, the graph is not symmetric with respect to the y axis or the origin.

Complete the sketch.

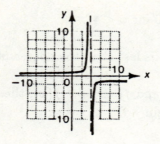

21. $f(x) = \frac{x}{x+1} = \frac{n(x)}{d(x)}$

   Intercepts. Real zeros of $n(x) = x$ $x = 0$ x intercept

                               $f(0) = 0$ $y = 0$ y intercept

The graph crosses the coordinate axes only at the origin.

   Vertical asymptotes. $d(x) = x + 1$ zeros: -1 $x = -1$

   Sign Chart.

Test numbers	-2	$-\frac{1}{2}$	1
Value of f	2	-1	$\frac{1}{2}$
Sign of f	$+$	$-$	$+$

$$+ \;+ \quad \underset{-1}{\circ} \quad - \;- \quad \underset{0}{\bullet} \quad + \;+ \qquad x$$

Since $x = -1$ is a vertical asymptote and $f(x) > 0$ for $x < -1$,

$f(x) \to \infty$ as $x \to -1^-$

Since $x = -1$ is a vertical asymptote and $f(x) < 0$ for $-1 < x < 0$

$f(x) \to -\infty$ as $x \to -1^+$

Horizontal asymptotes. Since $n(x)$ and $d(x)$ have the same degree,

the line $y = 1$ is a horizontal asymptote.

Symmetry. $f(x) = \frac{-x}{-x+1} = \frac{x}{x-1}$. Since $f(-x) \neq f(x)$ and $f(-x) \neq -f(x)$,

the graph is not symmetric with respect to the y axis or the origin.

Complete the sketch.

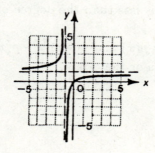

23. $q(x) = \frac{2x-1}{x} = \frac{n(x)}{d(x)}$

    Intercepts. Real zeros of $n(x) = 2x - 1$ $x = \frac{1}{2}$ x intercept

    $q(0)$ is not defined   no y intercept

    Vertical asymptotes $d(x) = x$ zeros: 0 $x = 0$

    Sign Chart.

Test numbers	-1	$\frac{1}{4}$	1
Value of q	3	-2	1
Sign of q	$+$	$-$	$+$

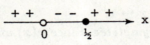

Since $x = 0$ is a vertical asymptote and $q(x) > 0$ for $x < 0$

$q(x) \to \infty$ as $x \to 0^-$

Since $x = 0$ is a vertical asymptote and $q(x) < 0$ for $0 < x < \frac{1}{2}$

$q(x) \to -\infty$ as $x \to 0^+$

Horizontal asymptotes. Since $n(x)$ and $d(x)$ have the same degree, the line $y = 2$ is a horizontal asymptote.

Symmetry. $q(-x) = \frac{-2x-1}{-x} = \frac{2x+1}{x}$

Since $q(-x) \neq q(x)$ and $q(-x) \neq -q(x)$ the graph is not symmetric with respect to the y axis or the origin.

Complete the sketch.

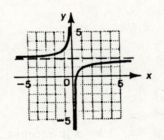

25. $h(x) = \frac{x}{2x-2} = \frac{n(x)}{d(x)}$

    Intercepts. Real zeros of $n(x)$ $=$ x x $=$ 0 x intercept

                        $h(0)$ $= 0$ y $=$ 0 y intercept

    The graph crosses the coordinate axes only at the origin.

    Vertical asymptotes. $d(x) = 2x - 2$ zeros: 1 $x = 1$

Sign Chart.

Test numbers	-1	$\frac{1}{2}$	2
Value of h	$\frac{1}{4}$	$-\frac{1}{2}$	1
Sign of h	$+$	$-$	$+$

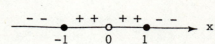

Since $x = 1$ is a vertical asymptote and $h(x) < 0$ for $0 < x < 1$,
$h(x) \to -\infty$ as $x \to 1^{-}$
Since $x = 1$ is a vertical asymptote and $h(x) > 0$ for $x > 1$
$h(x) \to \infty$ as $x \to 1^{+}$
Horizontal asymptotes. Since $n(x)$ and $d(x)$ have the same degree, the line
$y = \frac{1}{2}$ is a horizontal asymptote.
Symmetry. $h(-x) = \frac{-x}{-2x-2} = \frac{x}{2x+2}$.
Since $h(-x) \neq h(x)$ and $h(-x) \neq -h(x)$, the graph is not symmetric
with respect to the y axis or the origin.
Complete the sketch.

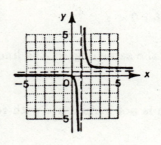

27. $g(x) = \frac{1-x^2}{x^2} = \frac{n(x)}{d(x)}$

Intercepts. Real zeros of $n(x) = 1 - x^2$ $1 - x^2 = 0$
$$x^2 = 1$$
$$x = \pm 1 \quad x \text{ intercepts}$$

$g(0)$ is not defined   no y intercepts

Vertical asymptotes. $d(x) = x^2$ zeros: 0 $x = 0$

Sign Chart.

Test numbers	-2	$-\frac{1}{2}$	$\frac{1}{2}$	2
Value of g	$-\frac{3}{4}$	3	3	$-\frac{3}{4}$
Sign of g	$-$	$+$	$+$	$-$

Since $x = 0$ is a vertical asymptote and $g(x) > 0$ for $-1 < x < 0$ and $0 < x < 1$,
$g(x) \to \infty$ as $x \to 0^{+}$ and as $x \to 0^{-}$
Horizontal asymptotes. Since $n(x)$ and $d(x)$ have the same degree, the line

$y = -1$ is a horizontal asymptote

Symmetry. $g(-x) = \frac{1-(-x)^2}{(-x)^2} = \frac{1-x^2}{x^2} = g(x)$. The graph has y axis is symmetry.

Complete the sketch.

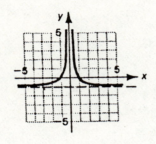

29. $f(x) = \frac{9}{x^2-9} = \frac{n(x)}{d(x)}$

Intercepts. There are no real zeros of $n(x) = 9$. No x intercept

$\quad\quad\quad f(0) = -1 \quad\quad y$ intercept

Vertical asymptotes. $d(x) = x^2 - 9 \quad\quad$ zeros : $x^2 - 9 = 0$

$$x^2 = 9$$
$$x = \pm 3$$

Sign Chart.

Test numbers	-4	0	4
Value of f	$\frac{9}{7}$	-1	$\frac{9}{7}$
Sign of f	$+$	$-$	$+$

Since $x = -3$ is a vertical asymptote and $f(x) > 0$ for $x < -3$ and $f(x) < 0$ for $-3 < x < 3$,
$f(x) \to \infty$ as $x \to -3^-$ and $f(x) \to -\infty$ as $x \to -3^+$
Since $x = 3$ is a vertical asymptote and $f(x) < 0$ for $-3 < x < 3$ and $f(x) > 0$ for $x > 3$,
$f(x) \to -\infty$ as $x \to 3^-$ and $f(x) \to -\infty$ as $x \to 3^-$
Horizontal asymptotes. Since the degree of $n(x)$ is less than the degree of $d(x)$,
the x axis is a horizontal asymptote.
Symmetry. $f(-x) = \frac{9}{(-x)^2-9} = \frac{9}{x^2-9} = f(x)$.
The graph has y axis symmetry.

Complete the sketch.

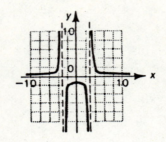

31. $f(x) = \frac{x}{x^2-1} = \frac{n(x)}{d(x)}$

Intercepts. Real zeros of $n(x) = x$ $x = 0$ x intercept

$\qquad\qquad\qquad f(0) = 0$ $\qquad\qquad\qquad y = 0$ y intercept

The graph crosses the coordinate axes only at the origin.

Vertical asymptotes. $d(x) = x^2 - 1$ zeros $x^2 - 1 = 0$

$$x^2 = 1$$
$$x = \pm 1$$

Sign Chart.

Test numbers	-2	$-\frac{1}{2}$	$\frac{1}{2}$	2
Value of f	$-\frac{2}{3}$	$\frac{2}{3}$	$-\frac{2}{3}$	$\frac{2}{3}$
Sign of f	$-$	$+$	$-$	$+$

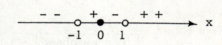

Since $x = -1$ is a vertical asymptote and $f(x) < 0$ for $x < -1$ and $f(x) > 0$ for $x > -1$
$f(x) \to -\infty$ as $x \to -1^-$ and $f(x) \to \infty$ as $x \to -1^+$
Since $x = 1$ is a vertical asymptote and $f(x) < 0$ for
$0 < x < 1$ and $f(x) > 0$ for $x > 1$
$f(x) \to -\infty$ as $x \to -1^-$ and $f(x) \to \infty$ as $x \to 1^+$
Horizontal asymptotes. Since the degree of $n(x)$ is less than the degree of $d(x)$,
the x axis is a horizontal asymptote.
Symmetry. $f(-x) = \frac{-x}{(-x)^2-1} = \frac{-x}{x^2-1} = -f(x)$.
The graph has origin symmetry.

Complete the sketch.

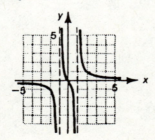

33. $g(x) = \frac{2}{x^2+1} = \frac{n(x)}{d(x)}$

    Intercepts.   There are no real zeros of $n(x) \;=\; 2.$ No x intercept
$$g(0) \;=\; 2 \quad y \text{ intercept}$$

Vertical asymptotes. There are no real zeros of $d(x) = x^2 + 1$

       No vertical asymptotes

Sign behavior: $g(x)$ is always positive.

Horizontal asymptotes. Since the degree of $n(x)$ is less than the degree of $d(x)$, the x axis is a horizontal asymptote.

Symmetry. $g(-x) = \frac{2}{(-x)^2+1} = \frac{2}{x^2+1} = g(x)$

The graph has y axis symmetry

Complete the sketch

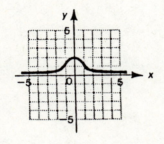

35. $h(x) = \frac{2x^2}{x^2+1} = \frac{n(x)}{d(x)}$

    Intercepts.

    Real zeros of $n(x) = 2x^2$ $x \;=\; 0$ x intercept

  $h(0) = 0$ $y \;=\; 0$ y intercept

The graph crosses the coordinate axes only at the origin.

Vertical asymptotes. There are no real zeros of $d(x) = x^2 + 1$.

No vertical asymptotes.

Sign Chart.

Test numbers	-1	1
Value of h	1	1
Sign of h	$+$	$+$

$$\xrightarrow[\qquad\;\;0\qquad\qquad]{+\;+\qquad\qquad+\;+} \; x$$

Horizontal asymptotes. Since $n(x)$ and $d(x)$ have the same degree, the line $y = 2$ is a horizontal asymptote.

Symmetry. $h(-x) = \frac{2(-x)^2}{(-x)^2+1} = \frac{2x^2}{x^2+1} = h(x)$.

The graph has y axis symmetry.

Complete the sketch.

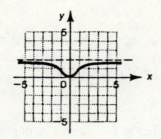

37. $f(x) = \frac{2x^2}{x-1} = \frac{n(x)}{d(x)}$

Vertical asymptotes: Real zeroes of $d(x) = x - 1$ $x = 1$

Horizontal asymptote: Since the degree of $n(x)$ is greater then the degree of $d(x)$, there is no horizontal asymptote.

Oblique asymptote:

$$
\begin{array}{r}
2x \quad + \quad 2 \\
x - 1 \overline{\big)\ 2x^2 } \\
2x^2 \ - \ 2x \\
\overline{ 2x } \\
2x \ \ -2 \\
\overline{ 2 }
\end{array}
$$

Thus $f(x) = 2x + 2 + \frac{2}{x-1}$

Hence, the line $y = 2x + 2$ is an oblique asymptote.

39. $h(x) = \frac{x^2+1}{x+1} = \frac{n(x)}{d(x)}$

    Vertical asymptotes :   Real zeros of $d(x) = x^3 + 1$

$$
\begin{aligned}
x^3 + 1 &= 0 \\
(x+1)(x^2 - x + 1) &= 0 \\
x + 1 &= 0 \quad x^2 - x + 1 = 0 \\
x &= -1 \quad \text{no real solutions}
\end{aligned}
$$

                       $x = -1$

Horizontal asymptote: Since the degree of $n(x)$ is less than the degree of $d(x)$, the line $y = 0$ (the x axis) is a horizontal asymptote.

41. $p(x) = \frac{x^3}{x^2+1} = \frac{n(x)}{d(x)}$

Vertical asymptotes: There are no real zeros of $d(x) = x^2 + 1$.

No vertical asymptotes.

Horizontal asymptote: Since the degree of $n(x)$ is greater than the degree of $d(x)$ there is no horizontal asymptote.

Oblique asymptote:

$$
\begin{array}{r}
x \\
x^2 + 1 \overline{\big)\ x^3 } \\
x^3 \ + \ x \\
\overline{ - \ x}
\end{array}
$$

Thus $p(x) = x + \frac{-x}{x^2+1}$

Hence, the line $y = x$ is an oblique asymptote.

43. $r(x) = \frac{2x^2 - 3x + 5}{x} = \frac{n(x)}{d(x)}$

Vertical asymptotes: Real zeros of $d(x) = x$ $x = 0$

Horizontal asymptote: Since the degree of $n(x)$ is greater than the degree of $d(x)$,
there is no horizontal asymptote.

Oblique asymptote: $\begin{aligned} \frac{2x^2 - 3x + 5}{x} &= \frac{2x^2}{x} - \frac{3x}{x} + \frac{5}{x} \\ &= 2x - 3 + \frac{5}{x} \end{aligned}$

Thus $r(x) = 2x - 3 + \frac{5}{x}$

Hence, the line $y = 2x - 3$ is an oblique asymptote.

45. $f(x) = \frac{x^2 + 1}{x} = \frac{n(x)}{d(x)}$

Intercepts.   There are no real zeros of $n(x) = x^2 + 1$.   No x intercept
            $f(0)$ is not defined   No y intercept

Vertical Asymptotes: Real zeros of $d(x) = x$. $x = 0$

Sign Chart.

Test numbers	-1	1
Value of f	-2	2
Sign of f	$-$	$+$

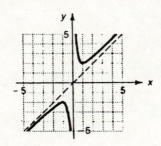

Since $x = 0$ is a vertical asymptote and $f(x) < 0$ for $x < 0$ and
$f(x) > 0$ for $x > 0$,
$f(x) \to -\infty$ as $x \to 0^-$ and $f(x) \to \infty$ as $x \to 0^+$

Horizontal asymptote: Since the degree of $n(x)$ is greater than the
degree of $d(x)$, there is no horizontal asymptote.

Oblique asymptote: $f(x) = \frac{x^2 + 1}{x} = \frac{x^2}{x} + \frac{1}{x} = x + \frac{1}{x}$

Hence, the line $y = x$ is an oblique asymptote.

Symmetry: $f(-x) = \frac{(-x)^2 + 1}{-x} = \frac{x^2 + 1}{-x} = -f(x)$

The graph has origin symmetry.

Complete the sketch.

47. $k(x) = \frac{x^2 - 4x + 3}{2x - 4} = \frac{n(x)}{d(x)}$

Intercepts.   Real zeros of $n(x) = x^2 - 4x + 3$ $x^2 - 4x + 3 = 0$

$$(x - 1)(x - 3) = 0$$

$$x = 1, 3 \quad x \text{ intercepts}$$

$k(0) = -\frac{3}{4}$ y intercept

Vertical asymptotes.   Real zeros of $d(x) = 2x - 4$ $2x - 4 = 0$

$$2x = 4$$

$$x = 2$$

Sign Chart.

Test numbers	0	$2\frac{1}{2}$	$3\frac{1}{2}$	4
Value of f	$-\frac{3}{4}$	$\frac{3}{4}$	$-\frac{3}{4}$	$\frac{3}{4}$
Sign of f	$-$	$+$	$-$	$+$

Since $x = 2$ is a vertical asymptote and $k(x) > 0$ for $1 < x < 2$ and $k(x) < 0$ for $2 < x < 3$,

$k(x) \to \infty$ as $x \to 2^-$ and $k(x) \to -\infty$ as $x \to 2^+$

Horizontal asymptote. Since the degree of $n(x)$ is greater than the degree of $d(x)$, there is no horizontal asymptote.

Oblique asymptote:

$$
\begin{array}{r}
\frac{1}{2}x \quad - \quad 1 \\
2x - 4\overline{\smash{)}\ x^2 \quad - \quad 4x \quad + \quad 3} \\
\underline{x^2 \quad - \quad 2x } \\
-2x \quad + \quad 3 \\
\underline{-2x \quad + \quad 4} \\
- \quad 1
\end{array}
$$

Thus, $k(x) = \frac{1}{2}x - 1 + \frac{-1}{2x-4}$.

Hence, the line $y = \frac{1}{2}x - 1$ is an oblique asymptote.

Symmetry $k(-x) = \frac{(-x)^2 - 4(-x) + 3}{2(-x) - 4} = \frac{x^2 + 4x + 3}{-2x - 4}$.

Since $k(-x) \neq k(x)$ and $k(-x) \neq -k(x)$,

the graph is not symmetric with

respect to the y axis or the origin.

Complete the sketch.

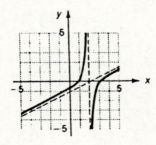

49. $F(x) = \frac{8-x^3}{4x^2} = \frac{n(x)}{d(x)}$

 Intercepts. Real zeros of $n(x) = 8 - x^3$

$$8 - x^3 = 0$$
$$(2 - x)(4 + 2x + x^2) = 0$$
$$2 - x = 0 \quad 4 + 2x + x^2 = 0$$
$$x = 2 \quad \text{No real zeros}$$

$x = 2$ x intercept

$F(0)$ is not defined.   No y intercept.

Vertical asymptotes. Real zeros of $d(x) = 4x^2$. $x = 0$

Sign Chart.

Test numbers	-1	1	3
Value of F	$\frac{9}{4}$	$\frac{7}{4}$	$-\frac{19}{36}$
Sign of F	$+$	$+$	$-$

Since $x = 0$ is a vertical asymptote and $F(x) > 0$ for $x < 0$ and $0 < x < 2$,

$F(x) \to \infty$ as $x \to 0^-$ and $F(x) \to \infty$ as $x \to 0^+$

Horizontal asymptote. Since the degree of $n(x)$ is greater than the degree

of $d(x)$, there is no horizontal asymptote.

Oblique asymptote: $F(x) = \frac{8-x^3}{4x^2} = \frac{8}{4x^2} - \frac{x^3}{4x^2} = -\frac{1}{4}x + \frac{2}{x^2}$

Hence, the line $y = -\frac{1}{4}x$ is an oblique asymptote.

Symmetry. $F(-x) = \frac{8-(-x)^3}{4(-x)^2} = \frac{8+x^3}{4x^2}$.

Since $F(-x) \neq F(x)$ and $F(-x) \neq -F(x)$,

the graph is not symmetric with respect to the y axis or the origin.

Complete the sketch.

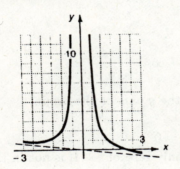

51. $f(x) = \frac{x^2-4}{x-2}$. $f(x)$ is not defined if $x - 2 = 0$, that is, $x = 2$

   Domain :$(-\infty, 2) \cup (2, \infty)$

$f(x) = \frac{(x-2)(x+2)}{(x-2)}$

$f(x) = x + 2$

The graph is a straight

line with slope 1 and
y intercept 2, except
that the point $(2,4)$
is not on the graph.

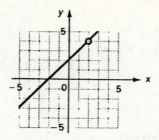

53. $r(x) = \frac{x+2}{x^2-4}$ $f(x)$ is not defined if
$$x^2 - 4 = 0, \text{ that is,}$$
$$x^2 = 4$$
$$x = \pm 2$$
Domain: $(-\infty, -2) \cup (-2, 2) \cup (2, \infty)$
$$r(x) = \frac{x+2}{(x+2)(x-2)}$$
$$r(x) = \frac{1}{x-2}$$
The graph is the same
as the graph of
the function $\frac{1}{x-2}$, except
that the point $(-2, -\frac{1}{4})$
is not on the graph.
Intercepts: $y = -\frac{1}{2}$ No x intercept.
Vertical asymptote: $x = 2$
Sign Chart:

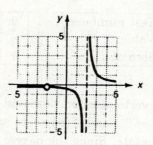

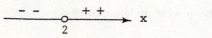

As $x \to 2^-, r(x) \to -\infty$. As $x \to 2^+, r(x) \to \infty$
Horizontal asymptote: $y = 0$
Symmetry: No symmetry with respect to the y axis or the origin

55. $N(t) = \frac{50t}{t+4}$ $t \geq 0$
Intercepts: Real zeros of $50t : t = 0$ $N(0) = 0$
Vertical asymptotes: None, since -4, the only zero of $t + 4$, is not in the domain of N.
Sign behavior: $N(t)$ is always positive.
Horizontal asymptote: $N = 50$ As $t \to \infty, N \to 50$
Symmetry: No obvious symmetry.

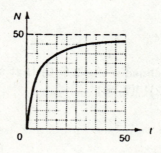

57. $N(t) = \frac{5t+30}{t}$ $t \geq 1$

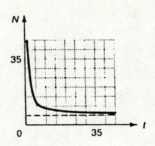

Intercepts: Real zeros of $5t + 30, t \geq 1$.
None, since -6, the only zero of
$5t + 30$, is not in the domain of N.
Vertical asymptotes: None, since 0,
the only zero of t, is not in the domain of N.
Sign behavior: $N(t)$ is always positive.
Horizontal asymptote: $N = 5$. As $t \to \infty, N \to 5$.
Symmetry: No obvious symmetry.

59.   A. $\bar{C}(n) = \frac{C(n)}{n} = \frac{2,500+175n+25n^2}{n} = 25n + 175 + \frac{2,500}{n}$
     B. The minimum value of the function $\bar{C}n$ is $\bar{C}(\sqrt{\frac{c}{a}})$, where
     $a = 25$ and $c = 2,500$
     min $\bar{C}(n) = \bar{C}(\sqrt{\frac{2,500}{25}}) = \bar{C}(\sqrt{100}) = \bar{C}(10)$
     This minimum occurs when $n = 10$, after 10 years.
     C. Intercepts: Real zeros of $2,500 + 175n + 25n^2$.
     None. No n intercepts
     0 is not in the domain of n, so there are no $\bar{C}$ intercepts.

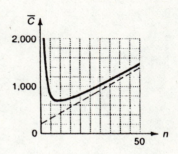

     Vertical asymptotes: Real zeros of n.
     The line $n = 0$ is a vertical asymptote.
     Sign behavior: $\bar{C}$ is always positive since $n \geq 0$.
     Horizontal asymptote: None, since the degree of
     $C(n)$ is greater than the degree of n.
     Oblique asymptote; The line $\bar{C} = 25n + 175$
     is an oblique asymptote.
     Symmetry: No obvious symmetry.

61.   A. Since Area $=$ length $\times$ width, length $= \frac{\text{Area}}{\text{width}} = \frac{225}{x}$
     Then total length of fence $=$ 2x width $+$ 2 x length
     $L(x) = 2x + \frac{450}{x} = \frac{2x^2+450}{x}$
     B. x can be any positive number, thus, domain $= (0, \infty)$
     C. The minimum value of the function $L(x)$ is $L(\sqrt{\frac{c}{a}})$ where
     $a = 2$ and $c = 450$.
     $min L(x) = L(\sqrt{\frac{450}{2}}) = L(\sqrt{225}) = L(15)$

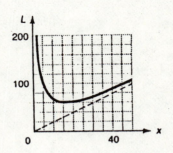

     This minimum occurs when $x = 15$.
     Width $= 15$ feet Length $= \frac{225}{15} = 15$ feet.
     D. Intercepts: Real zeros of $2x^2 + 450$.
     None, hence, no x intercepts,
     0 is not in the domain of L, so there are no L intercepts.
     Vertical asymptotes: Real zeros of x.
     The line $x = 0$ is a vertical asymptote.
     Sign behavior: $L(x)$ is always positive since $x > 0$.
     Horizontal asymptote: None, since the degree of
     $2x^2 + 450$ is greater than the degree of x.
     Oblique asymptote: The line $L = 2x$ is an oblique asymptote.
     Symmetry: No obvious symmetry.

APPENDIX D

Key Ideas and Formulas

Two polynomials are equal to each other if and only if the coefficients of like-degree terms are equal.

For a polynomial with real coefficients, there always exists a complete factoring involving only linear and/or quadratic factors with real coefficients where the linear and quadratic factors are prime relative to the real numbers.

Any proper fraction $P(x)/D(x)$ reduced to lowest terms can be decomposed into a sum of partial fractions as follows:

If $D(x)$ has:	then the decomposition of $P(x)/D(x)$ has a term of form:
1. a non-repeating linear factor of the form $ax + b$	$\frac{A}{ax+b}$ A a constant
2. a κ-repeating linear factor of the form $(ax + b)^\kappa$	$\frac{A_1}{ax+b} + \frac{A_2}{(ax+b)^2} + \cdots + \frac{A_\kappa}{(ax+b)^\kappa}$ $A_1, A_2 \cdots, A_\kappa$ constant
3. a non-repeating quadratic factor of the form $ax^2 + bx + c$, prime relative to the real numbers	$\frac{Ax+B}{ax^2+bx+c}$ A and B constants
4. a κ- repeating quadratic factor of the form $(ax^2 + bx + c)^\kappa$, where $ax^2 + bx + c$ is prime relative to the real numbers	$\frac{A_1x+B_1}{ax^2+bx+c} + \frac{A_2x+B_2}{(ax^2+bx+c)^2} + \cdots + \frac{A_\kappa x+B_\kappa}{(ax^2+bx+c)^\kappa}$ $A_1,\cdots,A_\kappa$ and $B_1,\cdots,B_\kappa$ constants

Common Errors:

1.   Confusing $(ax + b)^2$ and $ax^2 + bx + c$. The first is a repeated linear factor, and leads to a partial fraction of form

$$\frac{A_1}{ax+b} + \frac{A_2}{(ax+b)^2}$$

The second is a quadratic factor and leads to a partial fraction of form

$$\frac{Ax+B}{ax^2+bx+c}$$

2.   Neglecting the first-degree terms in the numerator in cases 3 and 4.

Always consider $\frac{Ax+B}{ax^2+bx+c}$. Never write $\frac{A}{ax^2+bx+c}$.

---

1. $\frac{7x-14}{(x-4)(x+3)} = \frac{A}{x-4} + \frac{B}{x+3} = \frac{A(x+3)+B(x-4)}{(x-4)(x+3)}$

Thus, for all x

$$7x - 14 = A(x + 3) + B(x - 4)$$

If $x = -3$, then
$$-35 = -7B$$
$$B = 5$$

If $x = 4$, then

$14 \;=\; 7A$

$A \;=\; 2$

$A = 2, B = 5$

3. $\dfrac{17x-1}{(2x-3)(3x-1)} = \dfrac{A}{2x-3} + \dfrac{B}{3x-1} = \dfrac{A(3x-1)+B(2x-3)}{(2x-3)(3x-1)}$

Thus, for all x

$17x - 1 = A(3x - 1) + B(2x - 3)$

If $x = \frac{1}{3}$, then

$17(\frac{1}{3}) - 1 \;=\; B[2(\frac{1}{3}) - 3]$

$\frac{14}{3} \;=\; -\frac{7}{3}B$

$B \;=\; -2$

If $x = \frac{3}{2}$, then

$17(\frac{3}{2}) - 1 \;=\; A[3(\frac{3}{2}) - 1]$

$\frac{49}{2} \;=\; \frac{7}{2}A$

$A \;=\; 7$

$A = 7, B = -2$

5. $\dfrac{3x^2+7x+1}{x(x+1)^2} = \dfrac{A}{x} + \dfrac{B}{x+1} + \dfrac{C}{(x+1)^2} = \dfrac{A(x+1)^2+Bx(x+1)+Cx}{x(x+1)^2}$

Thus, for all x

$3x^2 + 7x + 1 = A(x + 1)^2 + Bx(x + 1) + Cx$

If $x = -1$, then

$-3 \;=\; -C$

$C \;=\; 3$

If $x = 0$, then

$1 = A$

If $x = 1$, then using $A = 1$ and $C = 3$, we have

$11 \;=\; 4 + 2B + 3$

$B \;=\; 2$

$A = 1, B = 2, C = 3$

7. $\dfrac{3x^2+x}{(x-2)(x^2+3)} = \dfrac{A}{x-2} + \dfrac{Bx+C}{x^2+3} = \dfrac{A(x^2+3)+(Bx+C)(x-2)}{(x-2)(x^2+3)}$

Thus, for all x

$3x^2 + x = A(x^2 + 3) + (Bx + C)(x - 2)$

If $x = 2$, then

$14 \;=\; 7A$

$A \;=\; 2$

If $x = 0$, then using $A = 2$, we have

$0 \;=\; 6 - 2C$

$C \;=\; 3$

If $x = 1$, then using $A = 2$, and $C = 3$, we have

$4 \;=\; 8 + (B + 3)(-1)$

$-4 \;=\; (-1)(B + 3)$

$B \;=\; 1$

$A = 2, B = 1, C = 3$

9. $\dfrac{2x^2+4x-1}{(x^2+x+1)^2} = \dfrac{Ax+B}{x^2+x+1} + \dfrac{Cx+D}{(x^2+x+1)^2} = \dfrac{(Ax+B)(x^2+x+1)+Cx+D}{(x^2+x+1)^2}$

Thus, for all x

$2x^2 + 4x - 1 = (Ax + B)(x^2 + x + 1) + Cx + D$

Multiplying out the right side, we have

$$\begin{aligned} 2x^2 + 4x - 1 &= Ax^3 + Ax^2 + Ax + Bx^2 + Bx + B + Cx + D \\ &= Ax^3 + (A + B)x^2 + (A + B + C)x + B + D \end{aligned}$$

Equating coefficients of like terms we have

$$\begin{aligned} 0 &= A \\ 2 &= A + B \\ 4 &= A + B + C \\ -1 &= B + D \end{aligned}$$

Hence $A = 0, B = 2, C = 2, D = 3$

11.   Since $x^2 - 2x - 8 = (x + 2)(x - 4)$, we write

$\dfrac{-x+22}{x^2-2x-8} = \dfrac{A}{x+2} + \dfrac{B}{x-4} = \dfrac{A(x-4)+B(x+2)}{(x+2)(x-4)}$

Thus for all x

$-x + 22 = A(x - 4) + B(x + 2)$

If $x = 4$

$$\begin{aligned} 18 &= 6B \\ B &= 3 \end{aligned}$$

If $x = -2$

$$\begin{aligned} 24 &= -6A \\ A &= -4 \end{aligned}$$

So $\dfrac{-x+22}{x^2-2x-8} = \dfrac{-4}{x+2} + \dfrac{3}{x-4}$

13.   Since $6x^2 - x - 12 = (3x + 4)(2x - 3)$, we write

$\dfrac{3x-13}{6x^2-x-12} = \dfrac{A}{3x+4} + \dfrac{B}{2x-3} = \dfrac{A(2x-3)+B(3x+4)}{(3x+4)(2x-3)}$

Thus, for all x

$3x - 13 = A(2x - 3) + B(3x + 4)$

If $x = \frac{3}{2}$

$$\begin{aligned} 3(\tfrac{3}{2}) - 13 &= B[3(\tfrac{3}{2}) + 4] \\ -\tfrac{17}{2} &= \tfrac{17}{2}B \\ B &= -1 \end{aligned}$$

If $x = -\frac{4}{3}$

$$\begin{aligned} 3(-\tfrac{4}{3}) - 13 &= A[2(-\tfrac{4}{3}) - 3] \\ -17 &= -\tfrac{17}{3}A \\ A &= 3 \end{aligned}$$

So $\dfrac{3x-13}{6x^2-x-12} = \dfrac{3}{3x+4} - \dfrac{1}{2x-3}$

15.   Since $x^3 - 6x^2 + 9x = x(x^2 - 6x + 9) = x(x - 3)^2$, we write

$\dfrac{x^2-12x+18}{x^3-6x^2+9x} = \dfrac{A}{x} + \dfrac{B}{x-3} + \dfrac{C}{(x-3)^2} = \dfrac{A(x-3)^2+Bx(x-3)+Cx}{x(x-3)^2}$

Thus for all x

$$x^2 - 12x + 18 = A(x-3)^2 + Bx(x-3) + Cx$$

If $x = 3$

$$-9 = 3c$$
$$C = -3$$

If $x = 0$

$$18 = 9A$$
$$A = 2$$

If $x = 2$

$$-2 = A - 2B + 2C$$
$$-2 = 2 - 2B - 6 \text{ using } A = 2 \text{ and } C = 3$$
$$2 = -2B$$
$$B = -1$$

So $\frac{x^2 - 12x + 18}{x^3 - 6x^2 + 9x} = \frac{2}{x} - \frac{1}{x-3} - \frac{3}{(x-3)^2}$

17.   Since $x^3 + 2x^2 + 3x = x(x^2 + 2x + 3)$, we write

$$\frac{5x^2 + 3x + 6}{x^3 + 2x^2 + 3x} = \frac{A}{x} + \frac{Bx + C}{x^2 + 2x + 3} = \frac{A(x^2 + 2x + 3) + (Bx + C)x}{x(x^2 + 2x + 3)}$$

Common Error:

Writing $\frac{B}{x^2 + 2x + 3}$. Since $x^2 + 2x + 3$ is quadratic, the numerator must be first degree.

Thus for all x

$$5x^2 + 3x + 6 = A(x^2 + 2x + 3) + (Bx + C)x$$
$$= (A + B)x^2 + (2A + C)x + 3A$$

Equating coefficients of like terms, we have

$$5 = A + B$$
$$3 = 2A + C$$
$$3A = 6$$

Hence $A = 2, C = -1, B = 3$

So $\frac{5x^2 + 3x + 6}{x^3 + 2x^2 + 3x} = \frac{2}{x} + \frac{3x - 1}{x^2 + 2x + 3}$

19.   Since $x^4 + 4x^2 + 4 = (x^2 + 2)^2$, we write

$$\frac{2x^3 + 7x + 5}{x^4 + 4x^2 + 4} = \frac{Ax + B}{x^2 + 2} + \frac{Cx + D}{(x^2 + 2)^2} = \frac{(Ax + B)(x^2 + 2) + Cx + D}{(x^2 + 2)^2}$$

Thus for all x

$$2x^3 + 7x + 5 = (Ax + B)(x^2 + 2) + Cx + D$$
$$= Ax^3 + Bx^2 + (2A + C)x + 2B + D$$

Equating coefficients of like terms, we have

$$2 = A$$
$$0 = B$$
$$7 = 2A + C$$
$$5 = 2B + D$$

Hence $A = 2, B = 0, C = 3, D = 5$

$$\frac{2x^3 + 7x + 5}{x^4 + 4x^2 + 4} = \frac{2x}{x^2 + 2} + \frac{3x + 5}{(x^2 + 2)^2}$$

21.   First we divide to obtain a polynomial plus a proper fraction.

$$
\begin{array}{r}
x \quad - \quad 2 \\
x^2 - 5x + 6\, \overline{)\, x^3 \;-\; 7x^2 \;+\; 17x \;-\; 17} \\
x^3 \;-\; 5x^2 \;+\; 6x \\
\hline
-\,2x^2 \;+\; 11x \;-\; 17 \\
-\,2x^2 \;+\; 10x \;-\; 12 \\
\hline
x \;-\; 5
\end{array}
$$

So, $\dfrac{x^3 - 7x^2 + 17x - 17}{x^2 - 5x + 6} = x - 2 + \dfrac{x-5}{x^2 - 5x + 6}$

To decompose the proper fraction, we note $x^2 - 5x + 6 = (x-2)(x-3)$ and we write:

$\dfrac{x-5}{x^2-5x+6} = \dfrac{A}{x-2} + \dfrac{B}{x-3} = \dfrac{A(x-3) + B(x-2)}{(x-2)(x-3)}$

Thus for all x

$x - 5 = A(x-3) + B(x-2)$

If $x = 3$

$-2 = B$

If $x = 2$

$-3 = -A$

$A = 3$

So $\dfrac{x^3 - 7x^2 + 17x - 17}{x^2 - 5x + 6} = x - 2 + \dfrac{3}{x-2} - \dfrac{2}{x-3}$

23. First, we must factor $x^3 - 6x - 9$. Possible rational zeros of this polynomial are $\pm 1, \pm 3, \pm 9$; Descartes' rule of signs gives the following possible combinations of zeros:

+	−	I
1	2	0
1	0	2

Forming a synthetic division table, we see

	1	0	−6	−9
1	1	1	−5	−14
3	1	3	3	0

Thus $x^3 - 6x - 9 = (x-3)(x^2 + 3x + 3)$

$x^2 + 3x + 3$ cannot be factored further in the real numbers.

So $\dfrac{4x^2 + 5x - 9}{x^3 - 6x - 9} = \dfrac{A}{x-3} + \dfrac{Bx+C}{x^2+3x+3} = \dfrac{A(x^2+3x+3) + (Bx+C)(x-3)}{(x-3)(x^2+3x+3)}$

Thus for all x

$$
\begin{aligned}
4x^2 + 5x - 9 &= A(x^2 + 3x + 3) + (Bx + C)(x - 3) \\
&= Ax^2 + 3Ax + 3A + Bx^2 - 3Bx + Cx - 3C \\
&= (A + B)x^2 + (3A - 3B + C)x + 3A - 3C
\end{aligned}
$$

Before equating coefficients of like terms, we note that if $x = 3$

$$4(3)^2 + 5(3) - 9 = A(3^2 + 3 \cdot 3 + 3)$$
$$\text{So } 42 = 21A$$
$$A = 2$$
$$\text{Since } 4 = A + B$$
$$5 = 3A - 3B + C$$
$$-9 = 3A - 3C$$

we have $A = 2, B = 2, C = 5$

So $\frac{4x^2 + 5x - 9}{x^3 - 6x - 9} = \frac{2}{x-3} + \frac{2x+5}{x^2+3x+3}$

25. First we must factor $x^3 + 2x^2 - 15x - 36$. The possible rational zeros are $\pm 1, \pm 2, \pm 3, \pm 4, \pm 6, \pm 9, \pm 12, \pm 18, \pm 36$. Descartes' rule of signs gives the following possible combinations of zeros:

+	−	I
1	2	0
1	0	2

We form a synthetic division table:

	1	2	−15	−36
1	1	3	−12	−48
2	1	4	−7	−50
3	1	5	0	−36
4	1	6	9	0

Hence $x^3 + 2x^2 - 15x - 36 = (x-4)(x^2 + 6x + 9) = (x-4)(x+3)^2$.

So
$$\frac{x^2 + 16x + 18}{x^3 + 2x^2 - 15x - 36} = \frac{A}{x-4} + \frac{B}{x+3} + \frac{C}{(x+3)^2}$$
$$= \frac{A(x+3)^2 + B(x-4)(x+3) + C(x-4)}{(x-4)(x+3)^2}$$

Thus for all x
$$x^2 + 16x + 18 = A(x+3)^2 + B(x-4)(x+3) + C(x-4)$$
If $x = -3$
$$-21 = -7C$$
$$C = 3$$
If $x = 4$
$$98 = 49A$$
$$A = 2$$
If $x = 5$
$$123 = 64A + 8B + C$$
$$= 128 + 8B + 3 \text{ using } A = 2 \text{ and } C = 3$$
$$-8 = 8B$$
$$B = -1$$

So $\frac{x^2 + 16x + 18}{x^3 + 2x^2 - 15x - 36} = \frac{2}{x-4} - \frac{1}{x+3} + \frac{3}{(x+3)^2}$

27. First we must factor $x^4 - 5x^3 + 9x^2 - 8x + 4$. The possible rational zeros are

$\pm 1, \pm 2, \pm 4$. Descartes' rule of signs gives the following possible combinations of zeros:

+	−	I
4	0	0
2	0	2
0	0	4

We form a synthetic division table:

	1	−5	9	−8	4
1	1	−4	5	−3	1
2	1	−3	3	−2	0

We examine $x^3 - 3x^2 + 3x - 2$; the only remaining possibility is 2.

	1	−3	3	−2
2	1	−1	1	0

Thus $x^4 - 5x^3 + 9x^2 - 8x + 4 = (x-2)^2(x^2 - x + 1)$. $x^2 - x + 1$ cannot be factored further in the real numbers, so

$$\frac{-x^2+x-7}{x^4-5x^3+9x^2-8x+4} = \frac{A}{x-2} + \frac{B}{(x-2)^2} + \frac{Cx+D}{x^2-x+1}$$

$$= \frac{A(x-2)(x^2-x+1)+B(x^2-x+1)+(Cx+D)(x-2)^2}{(x-2)^2(x^2-x+1)}$$

Thus for all x

$$-x^2 + x - 7 = A(x-2)(x^2 - x + 1) + B(x^2 - x + 1) + (Cx + D)(x - 2)^2$$

If $x = 2$

$$-9 = 3B$$
$$B = -3$$

$$-x^2 + x - 7 = A(x^3 - 3x^2 + 3x - 2) + B(x^2 - x + 1) + (Cx + D)(x^2 - 4x + 4)$$
$$= (A+C)x^3 + (-3A + B - 4C + D)x^2 + (3A - B + 4C - 4D)x - 2A + B + 4D$$

We have already $B = -3$, so equating coefficients of like terms,

$$0 = A + C$$
$$-1 = -3A - 3 - 4C + D$$
$$1 = 3A + 3 + 4C - 4D$$
$$-7 = -2A - 3 + 4D$$

Since $C = -A$, we can write

$$-1 = -3A - 3 + 4A + D$$
$$1 = 3A + 3 - 4A - 4D$$
$$-7 = -2A - 3 + 4D$$

So $D = 0$ (adding the first equations), $A = 2, C = -2$.

Hence $\dfrac{-x^2+x-7}{x^4-5x^3+9x^2-8x+4} = \dfrac{2}{x-2} - \dfrac{3}{(x-2)^2} - \dfrac{2x}{x^2-x+1}$

29.   First we divide to obtain a polynomial plus a proper fraction.

$$x + 2$$

$$4x^4 + 4x^3 - 5x^2 + 5x - 2 \;\big|\; \begin{array}{rrrrrrr} 4x^5 & + 12x^4 & - x^3 & + 7x^2 & - 4x & + 2 \\ 4x^5 & + 4x^4 & - 5x^3 & + 5x^2 & - 2x & \\ \hline & 8x^4 & + 4x^3 & + 2x^2 & - 2x & + 2 \\ & 8x^4 & + 8x^3 & - 10x^2 & + 10x & - 4 \\ \hline & & - 4x^3 & + 12x^2 & - 12x & + 6 \end{array}$$

Now we must decompose $\frac{-4x^3 + 12x^2 - 12x + 6}{4x^4 + 4x^3 - 5x^2 + 5x - 2}$, starting by factoring $4x^4 + 4x^3 - 5x^2 + 5x - 2$.

Possible rational zeros are $\pm 1, \pm 2, \pm\frac{1}{2}, \pm\frac{1}{4}$.

Descartes' rule of signs gives the following possible combinations of zeros:

+	−	I
3	1	0
1	1	2

We form a synthetic division table:

	4	4	−5	5	−2	
1	4	8	3	8	6	1 is an upper bound, eliminating 2
−1	4	0	−5	0	−2	
−2	4	−4	3	−1	0	−2 is a zero

We investigate $4x^3 - 4x^2 + 3x - 1$; the only remaining rational zeros are $\pm\frac{1}{2}$, $\pm\frac{1}{4}$.
Descartes' rule of signs gives the following possible combinations of zeros:

+	−	I
3	0	0
1	0	2

The lack of negative zeros eliminates $-\frac{1}{2}$ and $-\frac{1}{4}$. We form a synthetic division table:

	4	−4	3	−1
$\frac{1}{2}$	4	−2	2	0

So $\begin{aligned} 4x^4 + 4x^3 - 5x^2 + 5x - 2 &= (x + 2)(x - \tfrac{1}{2})(4x^2 - 2x + 2) \\ &= (x + 2)(x - \tfrac{1}{2})2(2x^2 - x + 1) \\ &= (x + 2)(2x - 1)(2x^2 - x + 1) \end{aligned}$

$2x^2 - x + 1$ cannot be factored further in the real numbers, so we write:

$$\frac{-4x^3+12x^2-12x+6}{4x^4+4x^3-5x^2+5x-2} = \frac{A}{x+2} + \frac{B}{2x-1} + \frac{Cx+D}{2x^2-x+1}$$

$$= \frac{A(2x-1)(2x^2-x+1)+B(x+2)(2x^2-x+1)+(Cx+D)(x+2)(2x-1)}{(x+2)(2x-1)(2x^2-x+1)}$$

Thus for all x

$$4x^3 + 12x^2 - 12x + 6 = A(2x-1)(2x^2-x+1) + B(x+2)(2x^2-x+1) + (Cx+D)(x+2)(2x-1)$$

If $x = -2$

$$-4(-2)^3 + 12(-2)^2 - 12(-2) + 6 = A(-5)[2(-2)^2 - (-2) + 1]$$
$$32 + 48 + 24 + 6 = = -55A$$
$$110 = -55A$$
$$A = -2$$

If $x = \frac{1}{2}$

$$-4(\tfrac{1}{2})^3 + 12(\tfrac{1}{2})^2 - 12(\tfrac{1}{2}) + 6 = B(\tfrac{5}{2})[2(\tfrac{1}{2})^2 - \tfrac{1}{2} + 1]$$

$$-\tfrac{1}{2} + 3 - 6 + 6 = B(\tfrac{5}{2})(1)$$

$$B = 1$$

If $x = 0$

$$6 = A(-1)(1) + B(2)(1) + D(2)(-1)$$
$$6 = -A + 2B - 2D$$
$$6 = 2 + 2 - 2D$$
$$D = -1$$

If $x = 1$

$$-4 + 12 - 12 + 6 = A(1)(2) + B(3)(2) + (C+D)(3)(1)$$
$$2 = 2A + 6B + 3C + 3D$$
$$= -4 + 6 + 3C - 3$$
$$C = 1$$

So $\dfrac{4x^5+12x^4-x^3+7x^2-4x+2}{4x^4+4x^3-5x^2+5x-2} = x + 2 - \dfrac{2}{x+2} + \dfrac{1}{2x-1} + \dfrac{x-1}{2x^2-x+1}$